Vector Optimization

Series Editor:

Johannes Jahn
University of Erlangen-Nürnberg
Department of Mathematics
Martensstr. 3
91058 Erlangen
Germany
jahn@am.uni-erlangen.de

Vector Optimization

The series in Vector Optimization contains publications in various fields of optimization with vector-valued objective functions, such as multiobjective optimization, multi criteria decision making, set optimization, vector-valued game theory and border areas to financial mathematics, biosystems, semidefinite programming and multiobjective control theory. Studies of continuous, discrete, combinatorial and stochastic multiobjective models in interesting fields of operations research are also included. The series covers mathematical theory, methods and applications in economics and engineering. These publications being written in English are primarily monographs and multiple author works containing current advances in these fields.

Radu Ioan Boţ · Sorin-Mihai Grad · Gert Wanka

Duality in Vector Optimization

Springer

Dr. Radu Ioan Boţ
Faculty of Mathematics
Chemnitz University of Technology
Reichenhainer Str. 39
09126 Chemnitz
Germany
radu.bot@mathematik.tu-chemnitz.de

Dr. Sorin-Mihai Grad
Faculty of Mathematics
Chemnitz University of Technology
Reichenhainer Str. 39
09126 Chemnitz
Germany
sorin-mihai.grad@mathematik.tu-chemnitz.de

Professor Dr. Gert Wanka
Faculty of Mathematics
Chemnitz University of Technology
Reichenhainer Str. 39
09126 Chemnitz
Germany
gert.wanka@mathematik.tu-chemnitz.de

ISSN 1867-8971 e-ISSN 1867-898X
ISBN 978-3-642-26936-3 e-ISBN 978-3-642-02886-1
DOI 10.1007/978-3-642-02886-1
Springer Dordrecht Heidelberg London New York

Cover design: WMXDesign GmbH

Printed on acid-free paper

Springer is part of Springer Science+Business Media (www.springer.com)

Radu Ioan Boţ dedicates this book to Cassandra and Nina
Sorin-Mihai Grad dedicates this book to Carmen Lucia
Gert Wanka dedicates this book to Johanna

Preface

The continuous and increasing interest concerning vector optimization perceptible in the research community, where contributions dealing with the theory of duality abound lately, constitutes the main motivation that led to writing this book. Decisive was also the research experience of the authors in this field, materialized in a number of works published within the last decade. The need for a book on duality in vector optimization comes from the fact that despite the large amount of papers in journals and proceedings volumes, no book mainly concentrated on this topic was available so far in the scientific landscape. There is a considerable presence of books, not all recent releases, on vector optimization in the literature. We mention here the ones due to Chen, Huang and Yang (cf. [49]), Ehrgott and Gandibleux (cf. [65]), Eichfelder (cf. [66]), Goh and Yang (cf. [77]), Göpfert and Nehse (cf. [80]), Göpfert, Riahi, Tammer and Zălinescu (cf. [81]), Jahn (cf. [104]), Kaliszewski (cf. [108]), Luc (cf. [125]), Miettinen (cf. [130]), Mishra, Wang and Lai (cf. [131, 132]) and Sawaragi, Nakayama and Tanino (cf. [163]), where vector duality is at most tangentially treated. We hope that from our efforts will benefit not only researchers interested in vector optimization, but also graduate and undergraduate students.

The framework we consider is taken as general as possible, namely we work in (locally convex) topological vector spaces, going to the usual finite dimensional setting when this brings additional insights or relevant connections to the existing literature. We tried to add a certain order in the not always correct or rigorous results one can meet in the different segments of the vast literature addressed here. The investigations we perform in the book are always accompanied by the well-developed apparatus of conjugate duality for scalar convex optimization problems. Actually, a whole chapter is dedicated to classical results, but also to new achievements in this field. An additional motivation for this, as well as for displaying a consistent preliminary chapter on convex analysis and vector optimization, was our intention to keep the book as self-contained as possible. Four chapters remained for the vector duality itself, two of them directly extending the conjugate duality from the scalar case, another one focusing on the Wolfe and Mond-Weir duality concepts, while the last one deals with the broader class of set-valued optimization problems.

S.-M. Grad and G. Wanka are grateful to R. I. Boţ for the improvements he brought during the correction process. The authors want to express their sincere thanks to Ernö Robert Csetnek for reading a preliminary version of this book and for providing useful comments and suggestions that enhanced its quality. Thanks are also due to André Heinrich, Ioan Bogdan Hodrea and Catrin Schönyan for typewriting parts of the manuscript. We would like to thank our families for their unconditioned support and patience during writing this book. Without this background the authors would not have found the time and energy to bring this work to an end.

For updates and errata we refer the reader to

`http://www.tu-chemnitz.de/mathematik/approximation/dvo`

Chemnitz, Germany,
April 2009

Radu Ioan Boţ
Sorin-Mihai Grad
Gert Wanka

Contents

List of symbols and notations

Sets and elements

A_i	the i-th row of the matrix $A \in \mathbb{R}^{n \times m}$, $i = 1, \dots, n$
e^i	the i-th unit vector of $\mathbb{R}^n$, $i = 1, \dots, n$
e	the vector $(1, \dots, 1)^T$
Δ_{X^m}	the set $\{(x, \dots, x) \in X^m : x \in X\}$
$\mathrm{Pr}_X(U)$	the projection of the set $U \subseteq X \times Y$ on X
$l(K)$	linearity space of the cone K
$N(U, x)$	normal cone to the set U at x
$T(U, x)$	Bouligand tangent cone to the set U at x
$\mathrm{lin}(U)$	linear hull of the set U
$\mathrm{aff}(U)$	affine hull of the set U
$\mathrm{co}(U)$	convex hull of the set U
$\mathrm{cone}(U)$	conical hull of the set U
$\mathrm{coneco}(U)$	convex conical hull of the set U
$\mathrm{core}(U)$	algebraic interior of the set U
$\mathrm{icr}(U)$	intrinsic core of the set U
$\mathrm{int}(U)$	interior of the set U
$\mathrm{cl}(U)$	closure of the set U
$\overline{\mathrm{co}}(U)$	closed convex hull of the set U
$\mathrm{qri}(U)$	quasi relative interior of the set U
$\mathrm{qi}(U)$	quasi interior of the set U
$\mathrm{sqri}(U)$	strong quasi relative interior of the set U
$\mathrm{ri}(U)$	relative interior of the set U
X^*	topological dual space of X
$w(X, X^*)$	weak topology on X induced by X^*
$w(X^*, X)$	weak* topology on X^* induced by X
K^*	topological dual cone K^* of the cone K
K^{*0}	quasi interior of the dual cone of K
$\widehat{K}$	the cone $\mathrm{core}(K) \cup \{0\}$, where K is a convex cone

Functions and operators

id_X	identity function on X
$\mathcal{L}(X,Y)$	the set of linear continuous mappings from X to Y
$\mathcal{L}_+(X,Y)$	the set of positive mappings from X to Y
$\langle x^*, x\rangle$	the value of $x^* \in X^*$ at $x \in X$
A^*	adjoint mapping of $A \in \mathcal{L}(X,Y)$
$f_1 \square \ldots \square f_m$	infimal convolution of the functions f_i, $i = 1, \ldots, m$
δ_U	indicator function of the set U
σ_U	support function of the set U
δ_U^V	vector indicator function of the set U
γ_U	gauge of the set U
$\mathrm{dom}\, f$	(effective) domain of the (vector) function f
$\mathrm{epi}\, f$	epigraph of the function f
$\mathrm{epi}_K\, h$	K-epigraph of the vector function h
(v^*h)	the function $\langle v^*, h\rangle$, where h is a vector function and $v^* \in K^*$
Tf	infimal function of the function f through $T \in \mathcal{L}(X,Y)$
$h(\cdot)$	the scalar infimal value function
$\mathrm{co}\, f$	convex hull of the function f
$\bar{f}$	lower semicontinuous hull of the function f
$\overline{\mathrm{co}} f$	lower semicontinuous convex hull of the function f
f^*	conjugate function of the function f
f_S^*	conjugate function of the function f with respect to the set S
$\partial f(x)$	subdifferential of the function f at $x \in X$
$\nabla f(x)$	gradient of the function f at $x \in X$
F^*	conjugate map of the set-valued map F
$\mathrm{dom}\, F$	domain of the set-valued map F
$\mathrm{gph}\, F$	graph of the set-valued map F
$\mathrm{epi}_K\, F$	K-epigraph of the set-valued map F
$\partial F(x; v)$	subdifferential of the set-valued map F at $(x, v) \in \mathrm{gph}\, F$
$\partial F(x)$	subdifferential of the set-valued map F at $x \in X$
$H(\cdot)$	the set-valued minimal or infimal value map
F_k^*	k-conjugate map of the set-valued map F
$\partial_k F(x; v)$	k-subdifferential of the set-valued map F at $(x, v) \in \mathrm{gph} F$
$\partial_k F(x)$	k-subdifferential of the set-valued map F at $x \in X$

Partial orderings

$\leqq_K$	the partial ordering induced by the convex cone K
$x \leq_K y$	$x \leqq_K y$ and $x \neq y$
$x <_K y$	$y - x \in \mathrm{core}(K)$ (or $y - x \in \mathrm{int}(K)$), where K is a convex cone with $\mathrm{core}(K) \neq \emptyset$ ($\mathrm{int}(K) \neq \emptyset$)

$+\infty_K$	a greatest element with respect to the ordering cone K attached to a space
$-\infty_K$	a smallest element with respect to the ordering cone K attached to a space
$\overline{V}$	the space V to which the elements $\pm\infty_K$ are added
$A(M)$	the set of elements above the set M
$B(M)$	the set of elements below the set M

Minimality notions (with respect to the cone $K \subseteq V$)

$\operatorname{Min}(M, K)$	the set of minimal elements of the set M
$\operatorname{Max}(M, K)$	the set of maximal elements of the set M
$\operatorname{WMin}(M, K)$	the set of weakly minimal elements of the set M
$\operatorname{WMax}(M, K)$	the set of weakly maximal elements of the set M
$\operatorname{PMin}(M, K)$	generic notation for sets of properly minimal elements of the set M
$\operatorname{Min} M$	abbreviation for the minimal set of the set $M \subseteq \overline{V}$
$\operatorname{Max} M$	abbreviation for the maximal set of the set $M \subseteq \overline{V}$
$\operatorname{WMin} M$	abbreviation for the weak minimum of the set $M \subseteq \overline{V}$
$\operatorname{WMax} M$	abbreviation for the weak maximum of the set $M \subseteq \overline{V}$
$\operatorname{WInf} M$	abbreviation for the weak infimum of the set $M \subseteq \overline{V}$
$\operatorname{WSup} M$	abbreviation for the weak supremum of the set $M \subseteq \overline{V}$

Generic notations

$(P^{\cdots})$	primal optimization problem
$(PV^{\cdots})$	primal vector optimization problem
$(PSV^{\cdots})$	primal set-valued optimization problem
$v(P^{\cdots})$	the optimal objective value of the problem $(P^{\cdots})$
$(D^{\cdots})$	dual optimization problem
$(DV^{\cdots})$	dual vector optimization problem
$(DSV^{\cdots})$	dual set-valued optimization problem
$\mathcal{A}^{\cdots}$	feasible set of a primal vector problem
$\mathcal{B}^{\cdots}$	feasible set of a dual vector problem
$h^{\cdots}$	objective function of a dual vector problem
$\Phi^{\cdots}$	(vector) perturbation function or set-valued perturbation map
$(RC^{\cdots})$	regularity condition
$\gamma^{\cdots}$	gap function

1

Introduction

The conception of this book had a twofold motivation. The lack of a book or monograph intensively dedicated to duality in vector optimization, different to the scalar case where the theory is widely treated and well-founded, and, on the other hand, the continuously increasing number of publications dealing with this topic from different points of view. With this monograph we provide an overview of the major duality concepts in vector optimization, concomitantly emphasizing the achievements we brought to this field during the last decade. Working in a general framework, we encompass the majority of the contributions to this topic in the literature. We mainly work in (locally convex) topological vector spaces, resorting to the usual finite dimensional setting especially when we relate to situations met in the literature. Additionally, we followed the aim of bringing a certain order in the diversity of results not always having the necessary rigor. Nevertheless, we did not go through all the branches and ramifications of the main classes of vector duality results, leaving the general setting only for pointing out the ones with major impact on the development of the field. Thus the list of references is far from being complete, containing mainly some representative works connected to this area, only a few vis-á-vis the large number of publications touching this topic, though.

Given a vector (minimum) optimization problem, by *vector duality* we understand attaching dual vector (maximum) optimization problems to it and investigating the existence of weak, strong and, sometimes, converse duality. When the values attained by the objective function of the dual problem over its feasible set do not surpass the ones of the primal objective function, we say that we have *weak duality*. Starting from a solution to the primal problem, when a solution to the dual problem is discovered, such that the two objective functions coincide, we are in the situation called *strong duality*. *Converse duality* means that the existence of a solution to the dual problem allows to prove that the primal problem has a solution such that both of the objective values coincide. A variety of types of solutions can be considered to a vector optimization problem, each of them giving rise to different vector duals to the

R.I. Boţ et al., *Duality in Vector Optimization*, Vector Optimization,
DOI 10.1007/978-3-642-02886-1_1,

primal. When the vector problem is specialized to the scalar case, the vector dual turns out to be a corresponding known scalar dual.

Three major directions in vector duality are brought into attention in this book. The first one has its roots in the conjugate duality for scalar optimization problems. It is characterized by the fact that in the structure of the vector dual problems the formulation of a conjugate dual problem to the scalarized problem one can attach to the primal vector optimization problem can always be recognized. Studying vector duality is strongly based on the well-developed duality in scalar optimization. In this context are included the classical duality concepts due to Jahn (cf. [101, 104]) for *Lagrange duality*, Breckner and Kolumbán (cf. [42, 43]) for *Fenchel duality*, as well as the one due to Nakayama (cf. [142, 144]) which is based on *geometric duality*. The celebrated *linear vector duals* from the pioneering works due to Gale, Kuhn and Tucker (cf. [70]), Kornbluth (cf. [118]), Schönefeld (cf. [167]), Rödder (cf. [161]) and Isermann (cf. [96, 97]) belong here, too. The second vector duality concept considered here gravitates around *Wolfe* (cf. [166, 202]) and *Mond-Weir duality* (cf. [138, 195, 197]). The formulation of the vector duals is again based on scalar duality, but this time the optimality conditions for the scalarized primal-dual pair appear explicitly. A characteristic of this vector duality principle is the possibility to employ different types of generalized convexity concepts for the functions involved. This direction currently enjoys a blossoming development, nevertheless we restrict ourselves to the classical setting, working under hypotheses of *convexity*, *pseudoconvexity* or *quasiconvexity*, respectively. *Invexity* assumptions are considered here, too, but we do not go beyond, as the techniques used when working with its many and sometimes quite artificial and too complicated generalizations are the same. The third direction concerns *set-valued optimization problems*, which actually are extensions of the vector optimization ones. The duality considered here meets the philosophy from the scalar case, too, being based on the notion of *vector conjugacy*. By employing two different minimality notions, corresponding set-valued conjugate theories are developed, in each of them set-valued duality being introduced and investigated. The first one is based on works due to Tanino and Sawaragi (cf. [180]) and Sawaragi, Nakayama and Tanino (cf. [163]), while the second one has its roots in contributions due to Tanino (cf. [177, 178]), Kawasaki (cf. [114]) and Song (cf. [168–170]).

Besides this introductive one, the book contains six chapters, whose descriptions are given in the following.

Chapter 2. Basic notions and results in convex analysis, as well as minimality concepts for sets are introduced here. The first section deals with preliminaries on *convex sets*, for which algebraical as well as topological properties are displayed. The concepts of *partial ordering* and *cone* are intensively investigated. Different generalized *interiority* notions for convex sets are introduced, as they play important roles in formulating regularity conditions. Some basic *separation theorems* are recalled, since their usage is propagated through the whole book. A section on *convex functions* follows, where basic algebraic and topological properties in Hausdorff locally convex spaces are presented. For *vector functions* notions which extend the scalar convexity and lower semicontinuity are discussed. *Conjugate functions* constitute the core of the third section. Their basic properties are presented and the proof of the classical *Fenchel-Moreau theorem* is given. *Subdifferentiability* of convex functions is considered, too, and its connections to the conjugate functions are outlined. In the fourth section of this chapter we introduce several classes of *minimality notions* for a subset of a vector space partially ordered by a convex cone. We mention here the classical *Pareto minimality*, the *weak minimality*, as well as the *proper minimality notions* in the sense of *Geoffrion*, *Hurwicz*, *Borwein*, *Benson*, *Henig and Lampe* and *linear scalarization*, respectively. The relations between them are stressed, and sufficient conditions which guarantee their coincidence are provided. The situations when one can characterize these minimality notions via linear scalarization are taken into discussion. The last section concerns the formulation of a *general vector minimization problem* and different *efficiency notions* for it, in connection to the minimality concepts treated previously.

Chapter 3. The aim of this chapter is to describe the *conjugate duality theory* for *scalar optimization problems*, a cornerstone for the later vector duality investigations. We begin with a section in which we describe the *general perturbation approach* for constructing a *dual problem*, employed to two different classes of problems, namely the unconstrained one having a composition with a linear continuous mapping as objective function and the one with geometric and cone constraints. For the first one we consider the classical *Fenchel dual* problem, while for the latter we deal with three different dual problems, the *Lagrange dual*, the *Fenchel dual* and the *Fenchel-Lagrange dual*. The next section is dedicated to formulating regularity conditions for achieving *strong duality*, namely the situation when the optimal objective values of the primal and dual problem coincide and the latter has an optimal solution. First we deal with the general optimization problem, for which two kinds of such conditions are given, namely *interiority type* ones and *closedness type* ones. The latter arose mainly from the research carried out by the authors in the last years. Each of these regularity conditions is particularized to the two classes of problems mentioned above. *Necessary and sufficient optimality conditions* expressed via conjugate functions, as well as subdifferentials, and *saddle point assertions* are delivered in the third section, first for the general scalar optimization problem, afterwards for its particular instances considered throughout this chapter. The next section is dedicated to duality for the *composed convex optimization problem*, to which we assign two different conjugate dual problems. We also provide in each case corresponding regularity conditions, strong duality statements, optimality conditions and saddle point assertions. For all these classes of scalar problems, including the general one, we give in the last section of this chapter *stable strong duality* results and corresponding *subdifferential formulae*. By stable strong duality we understand the situation when the strong duality is not violated when adding any linear continuous functional to the objective function of the primal problem.

Chapter 4. This is the first chapter where we deal with duality for vector optimization problems, considered in a very general framework where the image space is an arbitrary partially ordered Hausdorff locally convex space. Throughout the whole chapter we parallelly deal with vector duals constructed by means of *linear scalarization* with respect to properly and weakly efficient solutions, respectively. We begin with *Fenchel type vector duality* for the optimization problem having as objective function the sum of a vector function with the composition of another vector function with a linear continuous mapping. For the primal-dual pair considered here we give *weak, strong and converse duality* assertions. We also put the dual in relation to the classical one in [42, 43]. In the second section we consider the problem with geometric and cone constraints and the duality developed here extends the geometric approach from [142, 144], new results with respect to it being achieved. In analogy to the scalar case, in the third section we introduce a *vector duality scheme* based on a general perturbation approach, particularizing it to the vector minimization problem with geometric and cone constraints. To the latter we assign different vector dual problems and we investigate the relations between them. Among these we rediscover the celebrated vector dual problem from [101, 104]. In the general scheme we provide, the vector geometric dual from the previous section is also incorporated. The fourth section deals with vector duality via a *general scalarization*. The vector duals are constructed by using different scalarization functions, where the investigations from the previous chapter made for composed convex problems play an important role. Besides *linear scalarization*, as special cases we consider here the *maximum(-linear) scalarization*, the *set scalarization* and the *(semi)norm scalarization*. We close the chapter by dealing with a *linear vector optimization problem* introduced in general spaces, by means of the duality schemes developed in the previous sections. We discuss the situation when the sets of maximal elements attached to all these duals become equal.

Chapter 5. The investigations on vector duality from the previous chapter are completed here by considering new Fenchel type and Fenchel-Lagrange type vector duals for the case when the image space of the objective functions is *finite dimensional*, culminating in a review on *linear vector duality*. The initial section introduces two new *Fenchel type vector duals* to the problem of minimizing the sum of a vector function with a composition of another vector function with a linear continuous mapping, one with respect to properly efficient solution, the other concerning weakly efficient solutions. Comparisons of the new duals with the ones considered in the previous chapter to the same primal are also given. In the second section we extend the family of *Fenchel-Lagrange vector dual problems* to the problem of minimizing a vector function with respect to both geometric and cone constraints introduced in [24, 36, 184] concerning properly efficient solutions for the situation when the functions involved are defined on Hausdorff locally convex spaces, giving moreover corresponding Fenchel-Lagrange vector duals with respect to weakly efficient solutions. We also show that in the case when the cone constraints are linear equalities each of the two families of vector duals we introduced consists of a single dual. We compare the vector duals given in the previous section with the ones from the previous chapter and with some new vector duals we introduce here for the same primal in the third section, stressing that under certain regularity conditions the sets of maximal, respectively weakly maximal, solutions of these vector duals coincide. In the fourth section we investigate what happens to the vector duals introduced in the previous two sections when the primal problem has a linear vector objective function and geometric and linear constraints. In the last section we consider all the spaces to be finite dimensional, treating the so-called *classical linear vector duality*. The vector duals considered for vector problems with constraints concerning efficient solutions in this chapter and in the previous one are written in this particular framework, being compared among them and also with the linear vector duals considered in [84, 96]. A *general scheme* regarding their *sets of maximal elements* is also provided. Moreover, we recall the *classical linear vector duals* in [70, 118]. When working with weakly efficient elements we compare the vector duals introduced so far, giving again a general scheme regarding their *weakly maximal elements.*

Chapter 6. An overview on the most important aspects regarding Wolfe duality and Mond-Weir duality is brought in this chapter. The first section is dedicated to *scalar Wolfe* and *Mond-Weir duality.* We start by working with convex optimization problems in Hausdorff locally convex spaces, then when particularizing the framework to finite dimensional spaces, we rediscover the dual in [166]. Taking the functions involved to be moreover *differentiable*, the classical duals in [138, 202] are recalled. Replacing in the differentiable case the *convexity* hypotheses by *pseudoconvexity*, *quasiconvexity* and *invexity* assumptions for the functions involved, respectively, *weak* and *strong duality statements* are also proven. In the second section we deal with vector Wolfe duality and Mond-Weir duality. We cover first the nondifferentiable case, when convexity plays a key role, then we take the functions involved to be differentiable, replacing afterwards the convexity assumptions by the mentioned generalized convexity hypotheses on the functions involved. We work parallelly with vector duals concerning both *properly* and *weakly efficient solutions*, delivering weak and strong duality statements. In this way we cover the most important results published so far in the field, see [62, 191, 192, 197], correcting moreover some inaccuracies intermingled in the literature. Two kinds of special Wolfe duality type and Mond-Weir duality type investigations are presented in the third section. Following papers like [14, 140, 196] we present other Wolfe and Mond-Weir type dual problems for which strong duality holds *without* assuming any *regularity condition*, both in the *nondifferentiable* and *differentiable case.* We treat similarly the vector case, starting from [63, 198]. *Symmetric Wolfe type duality* and *Mond-Weir type duality* can be found in this book, too, and here the primal problems have a special formulation, the functions involved being taken twice differentiable. We begin with the scalar case, via [136, 138], turning then to the vector case where we recall results based on [115, 171]. *Wolfe* and *Mond-Weir fractional duality* are taken into consideration in the fourth section, following the two main directions in the literature, introduced in [100, 164] and [13], respectively, in the scalar case, and [193, 194] in the vector case. In the last section, turning again to convex functions defined on general spaces, we generalize the Wolfe and Mond-Weir duality concepts, introducing for the first time in the literature a *perturbation approach* in connection to them. Thus one can treat via generalized Wolfe duality and Mond-Weir duality both classes of optimization problems considered in the scalar case. The duals obtained via the Lagrange perturbation turn out to be, in the finite dimensional case, the classical Wolfe and Mond-Weir duals from the literature, respectively. We also deliver a *Wolfe type* and a *Mond-Weir type duality scheme* for general vector optimization problems.

Chapter 7. The last chapter of the book entails investigations on set-valued optimization problems, by involving the so-called *vector conjugacy*, considered here with respect to two different minimality notions. In the first section we begin by introducing the *conjugate*, *biconjugate* and *subdifferential* of a *set-valued map*, following the approach from [163, 180]. The minimality notions are extended to sets in topological vector spaces to which infinite elements are attached. Some basic properties of these notions are discussed by outlining a certain analogy to their scalar correspondents. A *perturbation approach* for introducing a set-valued dual to a general set-valued optimization problem is employed and sufficient conditions for *stability* and *strong duality* are provided. By means of the same minimality notions a second approach, based on the so-called *vector k-conjugacy*, is considered, while similar issues are addressed. In the second section, by dealing in parallel with the two vector conjugacy approaches mentioned above, *Lagrange*, *Fenchel* and *Fenchel-Lagrange set-valued duals* are introduced for the set-valued optimization problem with constraints. For all these duals corresponding stability criteria, strong duality statements and necessary and sufficient optimality conditions are delivered. The set-valued problem having a composition with a linear continuous mapping as objective map and its Fenchel set-valued dual are the object of similar investigations in the next section. An application to constructing *set-valued gap maps* for *vector variational inequalities* closes the section. A further set-valued duality scheme, this time based on weak minimality in the same sense as in [168–170, 178] is presented in the next section. Its advantages opposite to the approach from the previous sections as concerns the conjugate calculus are outlined. Also here, a general set-valued dual via the perturbation approach is introduced to a general set-valued optimization problem, stability and strong duality statements being provided. As a particular instance of the general duality scheme, we deal first with the set-valued optimization problem with constraints to which we attach the same three types of set-valued duals as in the second section. Regularity conditions for strong duality and necessary and sufficient optimality conditions are stressed.

From the same point of view, investigations on the set-valued optimization problem having a composition with a linear continuous mapping as objective map, regarding its Fenchel dual, are performed. Their implementation in the construction of *set-valued gap maps* for *set-valued equilibrium problems* closes this last section.

2

Preliminaries on convex analysis and vector optimization

In this chapter we introduce some basic notions and results in convex analysis and vector optimization in order to make the book as self-contained as possible. The reader is supposed to have basic notions of functional analysis.

2.1 Convex sets

This section is dedicated mainly to the presentation of convex sets and their properties. With some exceptions the results we present in this section are given without proofs, as these can be found in the books and monographs on this topic mentioned in the bibliographical notes at the end of the chapter. All around this book we denote by $\mathbb{R}^n$ the *n-dimensional real vector space*, while by $\mathbb{R}^n_+ = \{x = (x_1, \ldots, x_n)^T \in \mathbb{R}^n : x_i \geq 0 \text{ for all } i = 1, \ldots, n\}$ we denote its *nonnegative orthant*. By $\mathbb{N} = \{1, 2, ...\}$ we denote the *set of natural numbers*, while $\emptyset$ is the *empty set*. All the vectors are considered as column vectors. An upper index ^{T} transposes a column vector to a row one and vice versa. By $\mathbb{R}^{m \times n}$ we denote the space of the *$m \times n$ matrices with real entries*. When we have a matrix $A \in \mathbb{R}^{m \times n}$, by A_i, $i = 1, \ldots, m$, we denote its rows and, naturally, by A^T its *transpose*. By $e^i \in \mathbb{R}^n$ we denote the *i-th unit vector of* $\mathbb{R}^n$, while by $e := \sum_{i=1}^n e^i \in \mathbb{R}^n$ we understand the vector having all entries equal to 1. If a function f takes everywhere the value $a \in \overline{\mathbb{R}}$ we write $f \equiv a$.

2.1.1 Algebraic properties of convex sets

Let X be a real nontrivial vector space. A *linear subspace* of X is a nonempty subset of it which is invariant with respect to the addition and the scalar multiplication on X. Note that an intersection of linear subspaces is itself a linear subspace. The *algebraic dual* space of X is defined as the set of all linear functionals on X and it is denoted by $X^\#$. Given any linear functional $x^\# \in X^\#$, we denote its value at $x \in X$ by $\langle x^\#, x \rangle$.

R.I. Boţ et al., *Duality in Vector Optimization*, Vector Optimization,
DOI 10.1007/978-3-642-02886-1_2,

For $x^\# \in X^\# \backslash \{0\}$ and $\lambda \in \mathbb{R}$ the set $\mathcal{H} := \{x \in X : \langle x^\#, x\rangle = \lambda\}$ is called *hyperplane*. The sets $\{x \in X : \langle x^\#, x\rangle \leq \lambda\}$ and $\{x \in X : \langle x^\#, x\rangle \geq \lambda\}$ are the *closed halfspaces* determined by the hyperplane $\mathcal{H}$, while $\{x \in X : \langle x^\#, x\rangle < \lambda\}$ and $\{x \in X : \langle x^\#, x\rangle > \lambda\}$ are the *open halfspaces* determined by $\mathcal{H}$. In order to simplify the presentation, the origins of all spaces will be denoted by 0, since the space where this notation is used always arises from the context. By Δ_{X^m} we denote the set $\{(x, \ldots, x) \in X^m : x \in X\}$, which is a linear subspace of $X^m := X \times \ldots \times X = \{(x_1, \ldots, x_m) : x_i \in X, i = 1, \ldots, m\}$.

If U and V are two subsets of X, their *Minkowski sum* is defined as $U + V := \{u + v : u \in U, \ v \in V\}$. For $U \subseteq X$ we define also $x + U = U + x := U + \{x\}$ when $x \in X$, $\lambda U := \{\lambda u : u \in U\}$ when $\lambda \in \mathbb{R}$ and $\Lambda U := \cup_{\lambda \in \Lambda} \lambda U$ when $\Lambda \subseteq \mathbb{R}$. According to these definitions one has that $U + \emptyset = \emptyset + U = \emptyset$ and $\lambda\emptyset = \emptyset$ whenever $U \subseteq X$ and $\lambda \in \mathbb{R}$. Moreover, if $U \subseteq V \subseteq X$ and $U \neq V$ we write $U \subsetneqq V$.

Some important classes of subsets of a real vector space X follow. Let be $U \subseteq X$. If $[-1,1]U \subseteq U$, then U is said to be a *balanced* set. When $U = -U$ we say that U is *symmetric*, while U is called *absorbing* if for all $x \in X$ there is some $\lambda > 0$ such that one has $x \in \lambda U$.

Affine and convex sets. Before introducing the notions of affine and convex sets, some necessary prerequisites follow. Taking some $x_i \in X$ and $\lambda_i \in \mathbb{R}$, $i = 1, \ldots, n$, the sum $\sum_{i=1}^n \lambda_i x_i$ is said to be a *linear combination* of the vectors $\{x_i : i = 1, \ldots, n\}$. The vectors $x_i \in X$, $i = 1, \ldots, n$, are called *linearly independent* if from $\sum_{i=1}^n \lambda_i x_i = 0$ follows $\lambda_i = 0$ for all $i = 1, \ldots, n$. The *linear hull* of a set $U \subseteq X$,

$$\operatorname{lin}(U) := \left\{ \sum_{i=1}^n \lambda_i x_i : \ n \in \mathbb{N}, x_i \in U, \lambda_i \in \mathbb{R}, \ i = 1, \ldots, n \right\},$$

is the intersection of all linear subspaces containing U, being the smallest linear subspace having U as a subset.

The set $U \subseteq X$ is called *affine* if $\lambda x + (1 - \lambda)y \in U$ whenever $\lambda \in \mathbb{R}$. The intersection of arbitrarily many affine sets is affine, too. The smallest affine set containing U or, equivalently, the intersection of all affine sets having U as a subset is the *affine hull* of U,

$$\operatorname{aff}(U) := \left\{ \sum_{i=1}^n \lambda_i x_i : \ n \in \mathbb{N}, x_i \in U, \lambda_i \in \mathbb{R}, \ i = 1, \ldots, n, \sum_{i=1}^n \lambda_i = 1 \right\}.$$

A set $U \subseteq X$ is called *convex* if

$$\{\lambda x + (1 - \lambda)y : \lambda \in [0, 1]\} \subseteq U \text{ for all } x, y \in U.$$

Obviously, $\emptyset$ and the whole space X are convex sets, as well as the hyperplanes, linear subspaces, affine sets and any set containing a single element. An example of a convex set in $\mathbb{R}^n$ is the *standard* $(n-1)$*-simplex* which is the set $\Delta_n := \{x = (x_1, \ldots, x_n)^T \in \mathbb{R}^n_+ : \sum_{i=1}^n x_i = 1\}$. Given $x_i \in X$, $i = 1, \ldots, n$, and $(\lambda_1, \ldots, \lambda_n)^T \in \Delta_n$, the sum $\sum_{i=1}^n \lambda_i x_i$ is said to be a *convex combination* of the elements x_i, $i = 1, \ldots, n$. The intersection of arbitrarily many convex sets is convex, while in general the union of convex sets is not convex. Note also that when U and V are convex subsets of X, for all $\alpha, \beta \in \mathbb{R}$ the set $\alpha U + \beta V$ is convex, too.

If X_i, $i = 1, \ldots, m$, are nontrivial real vector spaces, then $U_i \subseteq X_i$, $i = 1, \ldots, m$, are convex sets if and only if $\prod_{i=1}^m U_i$ is a convex set in $\prod_{i=1}^m X_i$. When X and Y are nontrivial real vector spaces and $U \subseteq X \times Y$, the *projection of U on X* is the set $\Pr_X(U) := \{x \in X : \exists y \in Y \text{ such that } (x, y) \in U\}$. If U is convex then $\Pr_X(U)$ is convex, too.

When U is a subset of the real vector space X, the intersection of all convex sets containing U is the *convex hull* of U,

$$\operatorname{co}(U) := \left\{ \sum_{i=1}^n \lambda_i x_i : \ n \in \mathbb{N}, x_i \in U, \ i = 1, \ldots, n, (\lambda_1, \ldots, \lambda_n)^T \in \Delta_n \right\},$$

which is the smallest convex set with U as a subset. If U and V are subsets of X, for all $\alpha, \beta \in \mathbb{R}$ one gets $\operatorname{co}(\alpha U + \beta V) = \alpha \operatorname{co}(U) + \beta \operatorname{co}(V)$.

A special case of convex sets are the *polyhedral sets* which are finite intersections of closed halfspaces. If U and V are polyhedral sets, then for all $\lambda, \mu \in \mathbb{R}$ the set $\lambda U + \mu V$ is polyhedral, too.

Consider another nontrivial real vector space Y and let $T : X \to Y$ be a given mapping. The *image* of a set $U \subseteq X$ through T is the set $T(U) := \{T(u) : u \in U\}$, while the *counter image* of a set $W \subseteq Y$ through T is $T^{-1}(W) := \{x \in X : T(x) \in W\}$. The mapping A is called *linear* if $A(x+y) = Ax + Ay$ and $A(\lambda x) = \lambda Ax$ for all $x, y \in X$ and all $\lambda \in \mathbb{R}$ or, equivalently, if

$$A(\alpha x + \beta y) = \alpha Ax + \beta Ay \ \forall x, y \in X \ \forall \alpha, \beta \in \mathbb{R}.$$

If $A : X \to Y$ is a linear mapping and $U \subseteq X$ is a linear subspace, then $A(U)$ is a linear subspace, too. On the other hand, if $W \subseteq Y$ is a linear subspace, then $A^{-1}(W)$ is a linear subspace, too. A special linear mapping is the *identity function* on X, $\operatorname{id}_X : X \to X$ defined by $\operatorname{id}_X(x) = x$ for all $x \in X$.

The mapping $T : X \to Y$ is said to be *affine* if

$$T(\lambda x + (1 - \lambda)y) = \lambda T(x) + (1 - \lambda)T(y) \ \forall x, y \in X \ \forall \lambda \in \mathbb{R}.$$

If $T : X \to Y$ is an affine mapping and the set $U \subseteq X$ is affine (or convex), then $T(U)$ is affine (or convex), too. Moreover, if $W \subseteq Y$ is affine (or convex), then $T^{-1}(W)$ is affine (or convex), too.

Cones. A nonempty set $K \subseteq X$ which satisfies the condition $\lambda K \subseteq K$ for all $\lambda \geq 0$ is said to be a *cone*. Throughout this book we assume, as follows by the definition, that the considered cones always contain the origin. The intersection of a family of cones is a cone, too.

A *convex cone* is a cone which is a convex set. One can prove that a cone K is convex if and only if $K + K \subseteq K$. If K is a convex cone, then its *linearity space* $l(K) = K \cap (-K)$ is a linear subspace. A cone K is said to be *pointed* if $l(K) = \{0\}$. The cones $K = \{0\}$ and $K = X$ are called *trivial cones*. Typical examples of nontrivial cones which occur in optimization are, when $X = \mathbb{R}^n$, the nonnegative orthant $\mathbb{R}^n_+$ and the *lexicographic cone*

$$\mathbb{R}^n_{lex} := \{0\} \cup \{x \in \mathbb{R}^n : x_1 > 0\} \cup$$
$$\{x \in \mathbb{R}^n : \exists k \in \{2, ..., n\} \text{ such that } x_i = 0 \ \forall i \in \{1, \ldots, k-1\} \text{ and } x_k > 0\},$$

while for $X = \mathbb{R}^{n \times n}$ the *cone of symmetric positive semidefinite matrices* $\mathcal{S}^n_+ := \{A \in \mathbb{R}^{n \times n} : A = A^T, \langle x, Ax \rangle \geq 0 \ \forall x \in \mathbb{R}^n\}$. Note that in $\mathbb{R}$ one can find only four cones: $\{0\}$, $\mathbb{R}_+$, $-\mathbb{R}_+$ and $\mathbb{R}$.

The *conical hull* of a set $U \subseteq X$, denoted by $\operatorname{cone}(U)$, is the intersection of all the cones which contain U, being the smallest cone in X that contains U. One can show that $\operatorname{cone}(U) = \cup_{\lambda \geq 0} \lambda U$. When U is convex, then $\operatorname{lin}(U - x) = \operatorname{cone}(U - x)$ and, consequently, $\operatorname{aff}(U) = x + \operatorname{cone}(U - U)$, whenever $x \in U$.

The *convex conical hull* of a set $U \subseteq X$,

$$\operatorname{coneco}(U) := \left\{ \sum_{i=1}^{n} \lambda_i x_i : \ n \in \mathbb{N}, x_i \in U, \lambda_i \geq 0, \ i = 1, \ldots, n \right\},$$

is the intersection of all the convex cones that contain U, being the smallest convex cone having U as a subset. One has $\operatorname{coneco}(U) = \operatorname{cone}(\operatorname{co}(U)) = \operatorname{co}(\operatorname{cone}(U))$. Due to the *Minkowski-Weyl* theorem, a set $U \subseteq \mathbb{R}^n$ is polyhedral if and only if there are two finite sets $V, W \subseteq \mathbb{R}^n$ such that $U = \operatorname{co}(V) + \operatorname{coneco}(W)$.

If K is a nontrivial convex cone, then $U \subseteq K$ is called a *base of the cone* K if each $x \in K \setminus \{0\}$ has an unique representation of the form $x = \lambda u$ for some $\lambda > 0$ and $u \in U$. Each nontrivial convex cone with a base in a nontrivial real vector space is pointed.

If $K \subseteq X$ is a given cone, its *algebraic dual cone* is $K^\# := \{x^\# \in X^\# : \langle x^\#, x \rangle \geq 0 \text{ for all } x \in K\}$. The set $K^\#$ is a convex cone. If C and K are cones in X, one has $(C + K)^\# = C^\# \cap K^\# = (C \cup K)^\#$ and $C^\# + K^\# \subseteq (C \cap K)^\#$. If the two cones satisfy $C \subseteq K$, then $C^\# \supseteq K^\#$.

Given a set $U \subseteq X$ and $x \in U$ we consider the *normal cone to* U *at* x,

$$N(U, x) = \{x^\# \in X^\# : \langle x^\#, y - x \rangle \leq 0 \ \forall y \in U\},$$

which is a convex cone.

Partial orderings. Very important, not only in convex analysis, is to consider certain orderings on the spaces one works with. Let the nonempty set $R \subseteq X \times X$ be a so-called *binary relation* on X. The elements $x, y \in X$ are said in this case to be *in relation* R if $(x, y) \in R$ and we write also xRy. A *binary relation* R is said to be a *partial ordering* on the vector space X if it satisfies the following axioms

(i) *reflexivity*: for all $x \in X$ it holds xRx;

(ii) *transitivity*: for all $x, y, z \in X$ from xRy and yRz follows xRz;
(iii) *compatibility with the linear structure*:
- for all $x, y, z, w \in X$ from xRy and zRw follows $(x+z)R(y+w)$;
- for all $x, y \in X$ and $\lambda \in \mathbb{R}_+$ from xRy follows $(\lambda x)R(\lambda y)$.

In such a situation it is common to use the symbol "$\leq$" and to write $x \leq y$ for xRy. The partial ordering "$\leq$" is called *antisymmetric* if for $x, y \in X$ fulfilling $x \leq y$ and $y \leq x$ there is $x = y$. A real vector space equipped with a partial ordering is called a *partially ordered vector space*.

If there is a partial ordering "$\leq$" on X, then the set $\{x \in X : 0 \leq x\}$ is a convex cone. If the partial ordering "$\leq$" is moreover antisymmetric, this cone is also pointed. Vice versa, having a convex cone $K \subseteq X$, it induces on X a partial ordering relation "$\leqq_K$" defined as follows

$$\leqq_K := \big\{(x, y) \in X \times X : y - x \in K\big\}.$$

If K is pointed, then "$\leqq_K$" is antisymmetric. To write $x \leqq_K y$, also the notation $y \geqq_K x$ is used, while $x \nleqq_K y$ means $y - x \notin K$. We denote also $x \leq_K y$ if $x \leqq_K y$ and $x \neq y$, while $x \nleq_K y$ is used when $x \leq_K y$ is not fulfilled. A convex cone which induces a partial ordering on X is called *ordering cone*. For the natural partial ordering on $\mathbb{R}^n$, which is introduced by $\mathbb{R}^n_+$, we use "$\leqq$" instead of "$\leqq_{\mathbb{R}^n_+}$" and also "$\leq$" for "$\leq_{\mathbb{R}^n_+}$".

By $\overline{\mathbb{R}}$ we denote the *extended real space* which consists of $\mathbb{R} \cup \{\pm\infty\}$. The operations on $\overline{\mathbb{R}}$ are the usual ones on $\mathbb{R}$ to which we add the following natural ones: $\lambda+(+\infty) = (+\infty)+\lambda := +\infty\ \forall \lambda \in (-\infty, +\infty]$, $\lambda+(-\infty) = (-\infty)+\lambda := -\infty\ \forall \lambda \in [-\infty, +\infty)$, $\lambda \cdot (+\infty) := +\infty\ \forall \lambda \in (0, +\infty]$, $\lambda \cdot (+\infty) := -\infty\ \forall \lambda \in [-\infty, 0)$, $\lambda \cdot (-\infty) := -\infty\ \forall \lambda \in (0, +\infty]$ and $\lambda \cdot (-\infty) := +\infty\ \forall \lambda \in [-\infty, 0)$. We also assume by convention that

$$(+\infty) + (-\infty) = (-\infty) + (+\infty) := +\infty, 0(+\infty) := +\infty \text{ and } 0(-\infty) := 0.$$

In analogy to the extended real space we attach to X a *greatest* and a *smallest* element with respect to "$\leqq_K$", denoted by $+\infty_K$ and $-\infty_K$, respectively, which do not belong to X and let $\overline{X} := X \cup \{\pm\infty_K\}$. Then for $x \in \overline{X}$ it holds $-\infty_K \leqq_K x \leqq_K +\infty_K$. Similarly, we assume that $-\infty_K \leq_K x \leq_K +\infty_K$ for any $x \in X$. On $\overline{X}$ we consider the following operations, in analogy to the ones stated above for the extended real space: $x + (+\infty_K) = (+\infty_K) + x := +\infty_K$ $\forall x \in X \cup \{+\infty_K\}$, $x + (-\infty_K) = (-\infty_K) + x := -\infty_K\ \forall x \in X \cup \{-\infty_K\}$, $\lambda \cdot (+\infty_K) := +\infty_K\ \forall \lambda \in (0, +\infty]$, $\lambda \cdot (+\infty_K) := -\infty_K\ \forall \lambda \in [-\infty, 0)$, $\lambda \cdot (-\infty_K) := -\infty_K\ \forall \lambda \in (0, +\infty]$ and $\lambda \cdot (-\infty_K) := +\infty_K\ \forall \lambda \in [-\infty, 0)$. We consider also the following conventions

$$\begin{aligned}(+\infty_K) + (-\infty_K) = (-\infty_K) + (+\infty_K) := +\infty_K,\\ 0(+\infty_K) := +\infty_K \text{ and } 0(-\infty_K) := 0.\end{aligned} \tag{2.1}$$

Moreover, if $x^\# \in K^\#$ we let $\langle x^\#, +\infty_K \rangle := +\infty$.

Algebraic interiority notions. Even without assuming a topological structure on X, different algebraic interiority notions can be considered for its subsets, as follows. The *algebraic interior*, also called *core*, of a set $U \subseteq X$ is

$$\text{core}(U) := \{x \in X : \text{ for every } y \in X \exists \delta > 0 \text{ such that } x + \lambda y \in U \forall \lambda \in [0, \delta]\}.$$

It is clear that $\text{core}(U) \subseteq U$. The algebraic interior with respect to the affine hull of U is called the *intrinsic core* of U, being the set

$$\text{icr}(U) := \{x \in X : \text{ for every } y \in \text{aff}(U) \exists \delta > 0 \text{ such that } x + \lambda y \in U \forall \lambda \in [0, \delta]\}.$$

There is $\text{core}(U) \subseteq \text{icr}(U)$. If $x \in U$ and U is convex, then $x \in \text{core}(U)$ if and only if $\text{cone}(U - x) = X$ and, on the other hand, $x \in \text{icr}(U)$ if and only if $\text{cone}(U - x)$ is a linear subspace, or, equivalently, $\text{cone}(U - x) = \text{cone}(U - U)$.

Taking two subsets U and V of X we have $U + \text{core}(V) \subseteq \text{core}(U + V)$, with equality if $V = \text{core}(V)$. The equality holds also in case U and V are convex and $\text{core}(V) \neq \emptyset$, as proved in [176]. Note also that a set $U \subseteq X$ is absorbing if and only if $0 \in \text{core}(U)$. If K is a cone in X with $\text{core}(K) \neq \emptyset$, then $K - K = X$ and, consequently, $K^{\#}$ is pointed. When K is a convex cone then $\text{core}(K) \cup \{0\}$ is a convex cone, too, and $\text{core}(K) = \text{core}(K) + K$. If K is a convex cone with nonempty algebraic interior, then one has $\text{core}(K) = \{x \in X : \langle x^{\#}, x \rangle > 0 \ \forall x^{\#} \in K^{\#} \setminus \{0\}\}$.

2.1.2 Topological properties of convex sets

Further we consider X being a real *topological vector space*, i.e. a real vector space endowed with a topology $\mathcal{T}$ which renders continuous the following functions

$$(x, y) \mapsto x + y, \ x, y \in X \ \text{ and } (\lambda, x) \mapsto \lambda x, \ x \in X, \lambda \in \mathbb{R}.$$

Throughout the book, if we speak about (topological) vector spaces, we always mean real nontrivial (topological) vector spaces, this means not equal to $\{0\}$. Moreover we agree to omit further the word "real" in such contexts. A topological space for which any two different elements have disjoint neighborhoods is said to be *Hausdorff*. A topological vector space X is said to be *metrizable* if it can be endowed with a metric which is compatible with its topology. Every metrizable vector space is Hausdorff.

For a set $U \subseteq X$ we denote by $\text{int}(U)$ the *interior* of U and by $\text{cl}(U)$ its *closure*. Then $\text{bd}(U) = \text{cl}(U) \setminus \text{int}(U)$ is called the *boundary* of U.

If Y is a topological vector space and $T : X \to Y$ is a linear mapping, then there is $T(\text{cl}(U)) \subseteq \text{cl}(T(U))$ for every $U \subseteq X$. If U is a convex subset of X, $x \in \text{int}(U)$ and $y \in \text{cl}(U)$, then $\{\lambda x + (1 - \lambda) y : \lambda \in (0, 1]\} \subseteq \text{int}(U)$. For $U \subseteq X$ there is $\text{int}(U) \subseteq \text{core}(U)$. If U is convex and one of the following conditions is fulfilled: $\text{int}(U) \neq \emptyset$; X is a Banach space and U is closed; X is finite dimensional, then $\text{int}(U) = \text{core}(U)$. If U is convex and $\text{int}(U) \neq \emptyset$, then it holds $\text{int}(U) = \text{int}(\text{cl}(U))$ and $\text{cl}(\text{int}(U)) = \text{cl}(U)$. The interior and the closure of a convex set in a topological vector space are convex, too. If U is a subset of X, then the intersection of all closed convex sets containing U is the *closed convex hull* of U, denoted by $\overline{\text{co}}(U)$, and it is the smallest closed convex set containing U. It is also the closure of the convex hull of U.

When $K \subseteq X$ is a convex cone with $\text{core}(K) \neq \emptyset$ we denote $\widehat{K} := \text{core}(K) \cup \{0\}$ and, for $x, y \in X$ which satisfy $y - x \in \text{core}(K)$ we write $x <_K y$. When $\text{int}(K) \neq \emptyset$, $x <_K y$ means $y - x \in \text{int}(K)$. Concerning the elements $+\infty_K$ and $-\infty_K$ introduced in the previous subsection we assume that for all $x \in X$ one has $-\infty_K <_K x <_K +\infty_K$.

In $\mathbb{R}^n$ we work with the *Euclidean topology* induced by the *Euclidean norm*. The *open ball* centered in $x \in \mathbb{R}^n$ and with radius $\varepsilon > 0$ is denoted by $B(x, \varepsilon)$, while the *closed ball* centered in $x \in \mathbb{R}^n$ and with radius $\varepsilon > 0$ is denoted by $\overline{B}(x, \varepsilon)$.

Dual spaces. The set of all linear continuous mappings defined on X and taking values in the topological vector space Y is denoted by $\mathcal{L}(X, Y)$.

The topological vector space $\mathcal{L}(X, \mathbb{R})$ is said to be the *topological dual space* of X, being denoted by X^*. Further, we refer with "dual" to topological duals, not to algebraical ones, unless otherwise specified. Analogously to vector spaces, by $\langle x^*, x \rangle$ we denote the value taken at $x \in X$ by the linear continuous functional $x^* \in X^*$. The hyperplane $\mathcal{H} := \{x \in X : \langle x^{\#}, x \rangle = \lambda\}$ with $x^{\#} \in X^{\#}$ and $\lambda \in \mathbb{R}$ is closed if and only if $x^{\#}$ is continuous.

For a mapping $A \in \mathcal{L}(X, Y)$ we consider its *adjoint mapping* $A^* \in \mathcal{L}(Y^*, X^*)$ defined by $\langle A^* y^*, x \rangle := \langle y^*, Ax \rangle$ for all $x \in X$ and $y^* \in Y^*$. When $X = \mathbb{R}^n$ and $Y = \mathbb{R}^m$, A can be identified with an $m \times n$ matrix and A^* coincides with A^T.

For every $x^* \in X^*$ let the seminorm $p_{x^*} : X \to \mathbb{R}$ defined by $p_{x^*}(x) := |\langle x^*, x \rangle|$. The coarsest topology on X which makes all the seminorms p_{x^*}, for $x^* \in X^*$, continuous is called the *weak topology* on X induced by X^*, being denoted $w(X, X^*)$. Every *weakly closed* set in X, i.e. closed in the weak topology, is closed also in the original topology on X, while the reverse assertion does not always hold.

Considering for all $x \in X$ the seminorms $p_x : X^* \to \mathbb{R}$, $p_x(x^*) = |\langle x^*, x \rangle|$, one defines analogously a topology on X^*, called the *weak* topology*, denoted $w(X^*, X)$. When one works with X^* endowed with the topology $w(X^*, X)$, the *bidual space* X^{**} of X, defined as the topological dual of X^*, can be identified with X.

Locally convex spaces. By a *local base* $\mathcal{B}$ of the topological vector space X endowed with the topology $\mathcal{T}$ we understand a collection of neighborhoods of zero from $\mathcal{T}$ such that every neighborhood of zero contains an element of $\mathcal{B}$. Then a set belongs to $\mathcal{T}$ if and only if it can be written as a union of translates of members of $\mathcal{B}$. A topological vector space is called *locally convex* if it has a local base whose members are convex sets. In a Hausdorff locally convex space the weakly closed convex sets are identical with the closed convex sets.

A locally convex space is called *Fréchet* if it is complete and metrizable by a metric which is invariant to translations.

If X is a Hausdorff locally convex space, to a nonempty subset U of it one can introduce the *Bouligand tangent cone* at $x \in \mathrm{cl}(U)$, which is

$$T(U,x) := \Big\{ y \in X : \exists (x_l)_{l\geq 1} \in U \text{ and } (\lambda_l)_{l\geq 1} > 0 \text{ such that } \lim_{l\to+\infty} x_l = x \text{ and } \lim_{l\to+\infty} \lambda_l (x_l - x) = y \Big\}.$$

Whenever $U \neq \emptyset$ and $x \in \mathrm{cl}(U)$, $T(U,x)$ is a cone and $T(U,x) \subseteq \mathrm{cl}(\mathrm{cone}(U - x))$. For a convex set $U \subseteq X$ there is $\mathrm{cone}(U - x) \subseteq T(U,x)$, which yields in this case that $\mathrm{cl}(T(U,x)) = \mathrm{cl}(\mathrm{cone}(U - x))$. If X is metrizable, then $T(U,x)$ is closed and, thus, if U is convex one has $T(U,x) = \mathrm{cl}(\mathrm{cone}(U - x))$ for all $x \in \mathrm{cl}(U)$.

Topological dual cones. Analogously to the algebraic dual cone used when working in vector spaces, one can consider a dual cone in topological vector spaces, too. When K is a cone in X, its *topological dual cone*, further called simply *dual cone*, is

$$K^* := \big\{ x^* \in X^* : \langle x^*, x\rangle \geq 0 \text{ for all } x \in K \big\}.$$

The cone K^* is always convex and weak* closed. If K is a convex cone with nonempty interior, then there is $\mathrm{int}(K) = \big\{ x \in X : \langle x^*, x\rangle > 0 \ \forall x^* \in K^* \backslash \{0\} \big\}$. If C and K are convex closed cones in X, then $(C \cap K)^* = \mathrm{cl}_{w(X^*,X)}(C^* + K^*)$ and the closure can be removed, for instance, when $C \cap \mathrm{int}(K) \neq \emptyset$.

The *bidual cone* of a cone $K \subseteq X$ is

$$K^{**} := \{ x \in X : \langle x^*, x\rangle \geq 0 \text{ for all } x^* \in K^* \}.$$

Note that $K^{**} = \overline{\mathrm{co}}(K)$. When X^* is endowed with the weak* topology then K^{**} is nothing but the dual cone of K^*.

Topological interiority notions. Let, unless otherwise specified, X be a Hausdorff locally convex space and X^* its topological dual space endowed with the weak* topology. Besides the already introduced interiority notions, which are defined only by algebraical means, we deal in this book also with topological notions of generalized interiors for a set.

The *quasi relative interior* of $U \subseteq X$ is

$$\mathrm{qri}(U) := \big\{ x \in U : \mathrm{cl}(\mathrm{cone}(U - x)) \text{ is a linear subspace} \big\}.$$

If U is convex, then $x \in \mathrm{qri}(U)$ if and only if $x \in U$ and $N(U,x)$ is a linear subspace of X^* (cf. [21]). The *quasi interior* of a set $U \subseteq X$ is the set

$$\mathrm{qi}(U) := \big\{ x \in U : \mathrm{cl}(\mathrm{cone}(U - x)) = X \big\}.$$

Note that $\mathrm{qi}(U)$ is a subset of $\mathrm{qri}(U)$. When U is convex one has $x \in \mathrm{qi}(U)$ if and only if $x \in U$ and $N(U,x) = \{0\}$ (cf. [26,27]) and also that if $\mathrm{qi}(U) \neq \emptyset$ then $\mathrm{qi}(U) = \mathrm{qri}(U)$. The next result provides a characterization for the quasi interior of the dual cone of a convex closed cone.

Proposition 2.1.1. *If $K \subseteq X$ is a convex closed cone, then*

$$\text{qi}(K^*) = \{x^* \in K^* : \langle x^*, x\rangle > 0 \ \text{ for all } x \in K\backslash\{0\}\}. \tag{2.2}$$

Proof. Assume first that there is some $x^* \in \text{qi}(K^*)$ not belonging to set in the right-hand side of (2.2). Then there is some $x \in K\backslash\{0\}$ such that $\langle x^*, x\rangle = 0$. As $\langle y^*, x\rangle \geq 0$ for all $y^* \in K^*$, we obtain $\langle y^* - x^*, -x\rangle \leq 0$ for all $y^* \in K^*$, i.e. $-x \in N(K^*, x^*) = \{0\}$. As this cannot take place because $x \neq 0$, it follows that $\text{qi}(K^*) \subseteq \{x^* \in K^* : \langle x^*, x\rangle > 0 \, \forall x \in K\backslash\{0\}\}$. Assume now the existence of some $x^* \in K^* \backslash \text{qi}(K^*)$ which fulfills $\langle x^*, x\rangle > 0$ whenever $x \in K\backslash\{0\}$. Then there is some $y \in X\backslash\{0\}$ such that $\langle y^* - x^*, y\rangle \leq 0$ for all $y^* \in K^*$. This yields $\langle y^*, y\rangle \leq \langle x^*, y\rangle$ for all $y^* \in K^*$. Taking into consideration that K^* is a cone, this implies $\langle y^*, y\rangle \leq 0$ whenever $y^* \in K^*$, i.e. $y \in -K^{**}$. As K is convex and closed we get $y \in -K\backslash\{0\}$, thus $\langle x^*, y\rangle < 0$, which if false. Consequently, (2.2) holds. □

Whenever $K \subseteq X$ is a convex cone, even if not necessarily closed, the above proposition motivates the use of the name *quasi interior of the dual cone* of K for the set

$$K^{*0} := \{x^* \in K^* : \langle x^*, x\rangle > 0 \ \text{ for all } x \in K\backslash\{0\}\}.$$

In case X is a separable normed space and K is a pointed convex closed cone, the Krein-Rutman theorem guarantees the nonemptiness of K^{*0} (see [104, Theorem 3.38]). Considering $X = l^2$ and $K = l^2_+$, it can be noted that $(l^2_+)^{*0}$ is nonempty, different to $\text{int}((l^2_+)^*) = \text{int}(l^2_+)$ which is an empty set. If $K^{*0} \neq \emptyset$ then K is pointed. If K is closed and $\text{int}_{w(X^*,X)}(K^*) \neq \emptyset$, then $\text{int}_{w(X^*,X)}(K^*) = K^{*0}$.

The *strong quasi relative interior* of a set $U \subseteq X$ is

$$\text{sqri}(U) := \{x \in U : \text{cone}(U - x) \text{ is a closed linear subspace}\}.$$

It is known that $\text{core}(U) \subseteq \text{sqri}(U) \subseteq \text{icr}(U)$. If U is convex, then $u \in \text{sqri}(U)$ if and only if $u \in \text{icr}(U)$ and $\text{aff}(U - u)$ is a closed linear subspace. Assuming additionally that $X = \mathbb{R}^n$ and $U \subseteq \mathbb{R}^n$, there is $\text{qi}(U) = \text{int}(U)$ and $\text{icr}(U) = \text{sqri}(U) = \text{qri}(U) = \text{ri}(U)$, where

$$\text{ri}(U) := \{x \in \text{aff}(U) : \exists \varepsilon > 0 \text{ such that } B(x, \varepsilon) \cap \text{aff}(U) \subseteq U\}$$

is the *relative interior* of the set U.

Separation theorems. Separation statements are very important in convex analysis and optimization, being crucial in the proofs of some of the basic results. In the following we present the ones which we need later in this book. We begin with a classical result in topological vector spaces followed by its version for real vector spaces and a consequence.

Theorem 2.1.2. *(Eidelheit) Let U and V be nonempty convex subsets of the topological vector space X with $\operatorname{int}(U) \neq \emptyset$. Then $\operatorname{int}(U) \cap V = \emptyset$ if and only if there are some $x^* \in X^* \backslash \{0\}$ and $\lambda \in \mathbb{R}$ such that*

$$\sup_{x \in U} \langle x^*, x \rangle \leq \lambda \leq \inf_{x \in V} \langle x^*, x \rangle$$

and $\langle x^*, x \rangle < \lambda$ for all $x \in \operatorname{int}(U)$.

Theorem 2.1.3. *Let U and V be nonempty convex subsets of a vector space X with $\operatorname{core}(U) \neq \emptyset$. Then $\operatorname{core}(U) \cap V = \emptyset$ if and only if there are some $x^\# \in X^\# \backslash \{0\}$ and $\lambda \in \mathbb{R}$ such that*

$$\sup_{x \in U} \langle x^\#, x \rangle \leq \lambda \leq \inf_{x \in V} \langle x^\#, x \rangle$$

and $\langle x^\#, x \rangle < \lambda$ for all $x \in \operatorname{core}(U)$.

Corollary 2.1.4. *Let U and V be nonempty convex subsets of the topological vector space X such that $\operatorname{int}(U - V) \neq \emptyset$. Then $0 \notin \operatorname{int}(U - V)$ if and only if there exists an $x^* \in X^* \backslash \{0\}$ such that*

$$\sup_{x \in U} \langle x^*, x \rangle \leq \inf_{x \in V} \langle x^*, x \rangle.$$

When working in locally convex spaces one has the following separation result.

Theorem 2.1.5. *(Tuckey) Let U and V be nonempty convex subsets of the locally convex space X, one compact and the other closed. Then $U \cap V = \emptyset$ if and only if there exists an $x^* \in X^* \backslash \{0\}$ such that*

$$\sup_{x \in U} \langle x^*, x \rangle < \inf_{x \in V} \langle x^*, x \rangle.$$

Corollary 2.1.6. *Let U and V be nonempty convex subsets of the locally convex space X. Then $0 \notin \operatorname{cl}(U - V)$ if and only if there exists an $x^* \in X^* \backslash \{0\}$ such that*

$$\sup_{x \in U} \langle x^*, x \rangle < \inf_{x \in V} \langle x^*, x \rangle.$$

In finite dimensional spaces, i.e. when $X = \mathbb{R}^n$, we also have the following separation statement involving relative interiors.

Theorem 2.1.7. *Let U and V be nonempty convex sets in $\mathbb{R}^n$. Then $\operatorname{ri}(U) \cap \operatorname{ri}(V) = \emptyset$ if and only if there exists an $x^* \in X^* \backslash \{0\}$ such that*

$$\sup_{x \in U} \langle x^*, x \rangle \leq \inf_{x \in V} \langle x^*, x \rangle$$

and

$$\inf_{x \in U} \langle x^*, x \rangle < \sup_{x \in V} \langle x^*, x \rangle.$$

2.2 Convex functions

In this section X and Y are considered, unless otherwise specified, Hausdorff locally convex spaces and X^* and Y^* their topological dual spaces, respectively. We list some well-known basic results on convex functions, but, as in the previous section, without proofs concerning the most of them. They can be found in different textbooks and monographs devoted to convex analysis, functional analysis, optimization theory, etc. (cf. [67,90,104,157,207]).

2.2.1 Algebraic properties of convex functions

We begin with some basic definitions and results.

Definition 2.2.1. *A function $f : X \to \overline{\mathbb{R}}$ is called convex if for all $x, y \in X$ and all $\lambda \in [0,1]$ one has*

$$f(\lambda x + (1-\lambda)y) \leq \lambda f(x) + (1-\lambda)f(y). \tag{2.3}$$

A function $f : X \to \overline{\mathbb{R}}$ is said to be concave if $(-f)$ is convex.

Remark 2.2.1. Given a convex set $U \subseteq X$ we say that a function $f : U \to \mathbb{R}$ is *convex on U* if (2.3) holds for all $x, y \in U$ and every $\lambda \in [0,1]$. The function f is said to be *concave on U* if $(-f)$ is convex on U. The *extension* of the function f to the whole space is the function

$$\tilde{f} : X \to \overline{\mathbb{R}},\ \tilde{f}(x) := \begin{cases} f(x), & \text{if } x \in U, \\ +\infty, & \text{otherwise.} \end{cases}$$

It is a simple verification to prove that $\tilde{f}$ is convex if and only if U is a convex set and f is convex on U. Thus the theory built for functions defined on the whole space X and having values in $\overline{\mathbb{R}}$ can be employed for real-valued functions defined on subsets of X, too.

In case $X = \mathbb{R}$ the following convexity criterion can be useful.

Remark 2.2.2. Consider $(a,b) \subseteq \mathbb{R}$ and the twice differentiable function $f : (a,b) \to \mathbb{R}$. Then f is convex (concave) on (a,b) if and only if $f''(x) \geq (\leq) 0$ for all $x \in (a,b)$.

Definition 2.2.2. *A function $f : X \to \overline{\mathbb{R}}$ is called strictly convex if for all $x, y \in X$ with $x \neq y$ and all $\lambda \in (0,1)$ one has (2.3) fulfilled as a strict inequality. A function $f : X \to \overline{\mathbb{R}}$ is called strictly concave if $(-f)$ is strictly convex.*

Example 2.2.1. (a) The *indicator function*

$$\delta_U : X \to \overline{\mathbb{R}},\ \delta_U(x) := \begin{cases} 0, & \text{if } x \in U, \\ +\infty, & \text{otherwise,} \end{cases}$$

of a set $U \subseteq X$ is convex if and only if U is convex.

(*b*) Let A be a $n \times n$ positive semidefinite matrix with real entries. Then the function $f : \mathbb{R}^n \to \mathbb{R}$, $f(x) = x^T Ax$, is convex. If A is positive definite, then f is strictly convex.

(*c*) If $\|\cdot\|$ denotes a norm on a vector space X, then $x \mapsto \|x\|$ is a convex function.

One can easily prove that a function $f : X \to \overline{\mathbb{R}}$ is convex if and only if for any $n \in \mathbb{N}$, $x_i \in X$ and $\lambda_i \in \mathbb{R}_+$, $i = 1, \ldots, n$, such that $\sum_{i=1}^n \lambda_i = 1$, *Jensen's inequality* is satisfied, namely

$$f\left(\sum_{i=1}^n \lambda_i x_i\right) \leq \sum_{i=1}^n \lambda_i f(x_i).$$

For a function $f : X \to \overline{\mathbb{R}}$ we consider the *(effective) domain* $\operatorname{dom} f := \{x \in X : f(x) < +\infty\}$ and the *epigraph* $\operatorname{epi} f := \{(x, r) \in X \times \mathbb{R} : f(x) \leq r\}$. The *strict epigraph* of f is $\operatorname{epi}_s f := \{(x, r) \in X \times \mathbb{R} : f(x) < r\}$. A function $f : X \to \overline{\mathbb{R}}$ is called *proper* if $f(x) > -\infty$ for all $x \in X$ and $\operatorname{dom} f \neq \emptyset$. Otherwise f is said to be *improper*.

A characterization of the convexity of a function through the convexity of its epigraph is given in the next result.

Proposition 2.2.1. *Let the function $f : X \to \overline{\mathbb{R}}$. The following assertions are equivalent:*

(i) f *is convex;*
(ii) $\operatorname{epi} f$ *is convex;*
(iii) $\operatorname{epi}_s f$ *is convex.*

Remark 2.2.3. For $f : X \to \overline{\mathbb{R}}$ we have $\Pr_X(\operatorname{epi} f) = \operatorname{dom} f$. Thus, if f is convex, then its domain is a convex set.

If $f : X \to \overline{\mathbb{R}}$ and $\lambda \in \mathbb{R}$, we call $\{x \in X : f(x) \leq \lambda\}$ the *level set* of f at λ and $\{x \in X : f(x) < \lambda\}$ is said to be the *strict level set* of f at λ. If f is convex, then the level sets and the strict level sets of f at λ are convex, for all $\lambda \in \mathbb{R}$. The opposite assertion is not true in general.

Definition 2.2.3. *A function $f : X \to \overline{\mathbb{R}}$ is called*

(a) subadditive if for all $x, y \in X$ one has $f(x + y) \leq f(x) + f(y)$;
(b) positively homogenous if $f(0) = 0$ and for all $x \in X$ and all $\lambda > 0$ one has $f(\lambda x) = \lambda f(x)$;
(c) sublinear if it is subadditive and positively homogenous.

Example 2.2.2. Given a nonempty set $U \subseteq X$, its *support function* $\sigma_U : X^* \to \overline{\mathbb{R}}$ defined by $\sigma_U(x^*) := \sup\{\langle x^*, x\rangle : x \in U\}$ is sublinear.

Notice that a convex function $f : X \to \overline{\mathbb{R}}$ is sublinear if and only if it is also positively homogenous.

Let be given the convex functions $f, g : X \to \overline{\mathbb{R}}$. Then $f+g$ and λf, $\lambda \geq 0$, are convex. One should notice that, due to the way the operations on the extended real space are defined, there is $0f = \delta_{\operatorname{dom} f}$.

Proposition 2.2.2. *Given a family of functions $f_i : X \to \overline{\mathbb{R}}$, $i \in I$, where I is an arbitrary index set, one has $\operatorname{epi}(\sup_{i\in I} f_i) = \cap_{i\in I} \operatorname{epi} f_i$. Consequently, the pointwise supremum $f : X \to \overline{\mathbb{R}}$ of a family of convex functions $f_i : X \to \overline{\mathbb{R}}$, $i \in I$, defined by $f(x) = \sup_{i\in I} f_i(x)$ is a convex function, too.*

Consider the Hausdorff locally convex spaces X_i, $i = 1, \ldots, m$, and take $X = \prod_{i=1}^m X_i$. Given the convex functions $f_i : X_i \to \overline{\mathbb{R}}$, $i = 1, \ldots, m$, the function $f : X \to \overline{\mathbb{R}}$ defined by $f(x^1, \ldots, x^m) = \sum_{i=1}^m f_i(x^i)$ is convex, too. Obviously, $\operatorname{dom} f = \prod_{i=1}^m \operatorname{dom} f_i$.

Consider $U \subseteq X \times \mathbb{R}$ a given set. To U we associate the so-called *lower bound function* $\phi_U : X \to \overline{\mathbb{R}}$ defined as

$$\phi_U(x) := \inf\{t \in \mathbb{R} : (x,t) \in U\}.$$

For an arbitrary function $f : X \to \overline{\mathbb{R}}$ it holds $f(x) = \phi_{\operatorname{epi} f}(x)$ for all $x \in X$. If U is a convex set, then ϕ_U is a convex function. By means of the lower bound function we introduce in the following the *convex hull* of a function.

Definition 2.2.4. *Consider a function $f : X \to \overline{\mathbb{R}}$. The function $\operatorname{co} f : X \to \overline{\mathbb{R}}$, defined by*

$$\operatorname{co} f(x) := \phi_{\operatorname{co}(\operatorname{epi} f)}(x) = \inf\{t \in \mathbb{R} : (x,t) \in \operatorname{co}(\operatorname{epi} f)\},$$

is called the convex hull of f.

It is clear from the construction that the convex hull of a function $f : X \to \overline{\mathbb{R}}$ is convex and it is the greatest convex function less than or equal to f. Consequently,

$$\operatorname{co} f = \sup\{g : X \to \overline{\mathbb{R}} : g(x) \leq f(x) \text{ for all } x \in X \text{ and } g \text{ is convex}\}.$$

Thus, f is convex if and only if $f = \operatorname{co} f$. Regarding the convex hull of f we have also the following result.

Proposition 2.2.3. *Let the function $f : X \to \overline{\mathbb{R}}$ be given. Then the convex hull of its domain coincides with the domain of its convex hull, namely $\operatorname{co}(\operatorname{dom} f) = \operatorname{dom}(\operatorname{co} f)$. Moreover, there is $\operatorname{epi}_s(\operatorname{co} f) \subseteq \operatorname{co}(\operatorname{epi} f) = \operatorname{epi}(\operatorname{co} f)$.*

Next we consider some notions which extend the classical monotonicity to functions defined on partially ordered spaces.

Definition 2.2.5. *Let be the vector space V partially ordered by the convex cone K, a nonempty set $W \subseteq V$ and $g : V \to \overline{\mathbb{R}}$ a given function.*

(a) If $g(x) \leq g(y)$ for all $x, y \in W$ such that $x \leqq_K y$, the function g is called K-increasing on W.
(b) If $g(x) < g(y)$ for all $x, y \in W$ such that $x \leq_K y$, the function g is called strongly K-increasing on W.
(c) If g is K-increasing on W, $\text{core}(K) \neq \emptyset$ and for all $x, y \in W$ fulfilling $x <_K y$ follows $g(x) < g(y)$, the function g is called strictly K-increasing on W.
(d) When $W = V$ we call these classes of functions K-increasing, strongly K-increasing and strictly K-increasing, respectively.

Remark 2.2.4. When $X = \mathbb{R}$, the $\mathbb{R}_+$-increasing functions are actually the *increasing functions*, while the strongly and the strictly $\mathbb{R}_+$-increasing functions are actually the *strictly increasing functions.*

Remark 2.2.5. For a cone $K \subseteq V$ with $\text{core}(K) \neq \emptyset$, we defined $\widehat{K} := \text{core}(K) \cup \{0\}$. Then the definition of the strictly K-increasing functions on a set $W \subseteq V$ coincide with the strongly $\widehat{K}$-increasing functions on W. When $\text{int}(K) \neq \emptyset$ one has $\text{int}(K) = \text{core}(K)$, thus the core of the cone K can be replaced in Definition 2.2.5 by the interior of K.

Example 2.2.3. Consider a vector space V and a linear functional $v^\# \in V^\#$. If $v^\# \in K^\#$, then the definition of the algebraic dual cone secures that for all $v_1, v_2 \in V$ such that $v_1 \leqq_K v_2$ we have $\langle v^\#, v_2 - v_1 \rangle \geq 0$. Therefore $\langle v^\#, v_1 \rangle \leq \langle v^\#, v_2 \rangle$ and this means that the elements of $K^\#$ are actually K-increasing linear functions on the vector space V.

If $v^\# \in K^{\#0} := \{x^\# \in K^\# : \langle x^\#, x \rangle > 0 \text{ for all } x \in K \backslash \{0\}\}$, which can be seen as the analogous of K^{*0} in vector spaces, then for all $v_1, v_2 \in V$ such that $v_1 \leq_K v_2$ it holds $\langle v^\#, v_2 - v_1 \rangle > 0$. According to the previous definition this means that the elements of $K^{\#0}$ are strongly K-increasing linear functions on V.

On the other hand, if $\text{core}(K) \neq \emptyset$, then, according to the representation $\text{core}(K) = \{v \in V : \langle v^\#, v \rangle > 0 \ \forall v^\# \in K^\# \backslash \{0\}\}$, every $v^\# \in K^\# \backslash \{0\}$ is strictly K-increasing on V.

There are notions given for functions with extended real values that can be formulated also for functions mapping from X into vector spaces. Let V be Hausdorff locally convex space partially ordered by the convex cone K and $\overline{V} = V \cup \{\pm\infty_K\}$.

The *domain* of a vector function $h : X \to \overline{V}$ is the set $\text{dom}\, h := \{x \in X : h(x) \neq +\infty_K\}$. When $h(x) \neq -\infty_K$ for all $x \in X$ and $\text{dom}\, h \neq \emptyset$ we call h *proper*. The *K-epigraph* of a vector function $h : X \to \overline{V}$ is the set $\text{epi}_K h := \{(x, v) \in X \times V : h(x) \leqq_K v\}$.

Definition 2.2.6. *A vector function $h : X \to \overline{V}$ is said to be K-convex if $\text{epi}_K h$ is a convex set.*

One can easily prove that a function $h : X \to V \cup \{+\infty_K\}$ is K-convex if and only if

$$h(\lambda x + (1-\lambda)y) \leqq_K \lambda h(x) + (1-\lambda)h(y) \ \forall x, y \in X \ \forall \lambda \in [0,1].$$

For a convex set $U \subseteq X$ we say that the function $h : U \to V$ is *K-convex on U* if $h(\lambda x + (1-\lambda)y) \leqq_K \lambda h(x) + (1-\lambda)h(y)$ for all $x, y \in U$ and all $\lambda \in [0,1]$. Considering the function

$$\tilde{h} : X \to \overline{V},\ \tilde{h}(x) := \begin{cases} h(x), & \text{if } x \in U, \\ +\infty_K, & \text{otherwise,} \end{cases}$$

note that $\tilde{h}$ is K-convex if and only if U is convex and h is K-convex on U.

Having a set $U \subseteq X$, its *vector indicator function* is

$$\delta_U^V : X \to \overline{V},\ \delta_U^V(x) := \begin{cases} 0, & \text{if } x \in U, \\ +\infty_K, & \text{otherwise.} \end{cases}$$

Then δ_U^V is K-convex if and only if U is convex.

Remark 2.2.6. Let $h : X \to \overline{V}$ be a given vector function. For $v^* \in K^*$ we shall use the notation $(v^*h) : X \to \overline{\mathbb{R}}$ for the function defined by $(v^*h)(x) := \langle v^*, h(x) \rangle$ and one can easily notice that $\operatorname{dom}(v^*h) = \operatorname{dom} h$.

The proof of the following result is straightforward.

Theorem 2.2.4. *Let be the convex and K-increasing function $f : V \cup \{+\infty_K\} \to \overline{\mathbb{R}}$ defined with the convention $f(+\infty_K) = +\infty$ and consider the proper K-convex function $h : X \to \overline{V}$. Then the function $f \circ h : X \to \overline{\mathbb{R}}$ is convex.*

Corollary 2.2.5. *Let be the convex function $f : V \to \overline{\mathbb{R}}$ and the affine mapping $T : X \to V$. Then the function $f \circ T : X \to \overline{\mathbb{R}}$ is convex.*

Proof. For $K = \{0\}$, the mapping T is K-convex and the result follows by Theorem 2.2.4. □

Another important function attached to a given function $\Phi : X \times Y \to \overline{\mathbb{R}}$ is the so-called *infimal value function* to it, defined as follows

$$h : Y \to \overline{\mathbb{R}},\ h(y) := \inf\{\Phi(x,y) : x \in X\}.$$

Theorem 2.2.6. *Given a convex function $\Phi : X \times Y \to \overline{\mathbb{R}}$, its infimal value function is convex, too.*

Proof. One can prove that $\operatorname{epi}_s h = \Pr_{Y \times \mathbb{R}}(\operatorname{epi}_s \Phi)$. As the projection preserves the convexity and $\operatorname{epi}_s \Phi$ is convex, it follows that $\operatorname{epi}_s h$ is convex, too. By Proposition 2.2.1, h is convex. □

Remark 2.2.7. It can also be proven that

$$\mathrm{Pr}_{Y\times\mathbb{R}}(\mathrm{epi}\,\Phi) \subseteq \mathrm{epi}\,h \subseteq \mathrm{cl}(\mathrm{Pr}_{Y\times\mathbb{R}}(\mathrm{epi}\,\Phi)).$$

As a special case of Theorem 2.2.6 we obtain the following result.

Theorem 2.2.7. *Let be the convex function $f : X \to \overline{\mathbb{R}}$ and $T \in \mathcal{L}(X, V)$. Then the infimal function of f through T,*

$$Tf : V \to \overline{\mathbb{R}},\ (Tf)(y) := \inf\{f(x) : Tx = y\}$$

is convex, too.

For an arbitrary function $f : X \to \overline{\mathbb{R}}$ and $T \in \mathcal{L}(X, V)$ it holds $\mathrm{dom}(Tf) = T(\mathrm{dom}\,f)$.

The following notion can also be introduced as a particular instance of the infimal function of a given function through a suitable linear continuous mapping, as can be seen below. Though, we introduce it directly because of its importance in convex analysis and optimization.

Definition 2.2.7. *The infimal convolution of the functions $f_i : X \to \overline{\mathbb{R}}$, $i = 1, \ldots, m$, is the function*

$$f_1\square\ldots\square f_m : X \to \overline{\mathbb{R}}, (f_1\square\ldots\square f_m)(x) := \inf\left\{\sum_{i=1}^m f_i(x^i) : x^i \in X, \sum_{i=1}^m x^i = x\right\}.$$

When for $x \in X$ the infimum within is attained we say that the infimal convolution is exact at x. When the infimal convolution is exact everywhere we call it simply exact.

For $f_i : X \to \overline{\mathbb{R}}$, $i = 1, \ldots, m$, given functions, $f : X^m \to \overline{\mathbb{R}}$ defined by $f(x^1, \ldots, x^m) = \sum_{i=1}^m f_i(x^i)$ and $A \in \mathcal{L}(X^m, X)$, $A(x^1, \ldots, x^m) = \sum_{i=1}^m x^i$ it holds $Af = f_1\square\ldots\square f_m$. Thus $\mathrm{dom}(f_1\square\ldots\square f_m) = \sum_{i=1}^m \mathrm{dom}\,f_i$. By Theorem 2.2.7 it follows that if $f_i : X \to \overline{\mathbb{R}}$, $i = 1, \ldots, m$, are convex, as stated in the following theorem, their infimal convolution is convex, too.

Theorem 2.2.8. *Given the convex functions $f_i : X \to \overline{\mathbb{R}}$, $i = 1, \ldots, m$, then their infimal convolution $f_1\square\ldots\square f_m : X \to \overline{\mathbb{R}}$ is convex, too.*

The notion we introduce next is a generalization of the K-convexity (see Definition 2.2.6).

Definition 2.2.8. *A vector function $h : X \to V \cup \{+\infty_K\}$ is called K-convexlike if for all $x, y \in X$ and all $\lambda \in [0, 1]$ there is some $z \in X$ such that $h(z) \leqq_K \lambda h(x) + (1 - \lambda)h(y)$.*

It is easy to see that $h : X \to V \cup \{+\infty_K\}$ is K-convexlike if and only if $h(\operatorname{dom} h) + K$ is a convex set.

For $U \subseteq X$ a given nonempty set we call $h : U \to V$ *K-convexlike on U* if for all $x, y \in U$ and all $\lambda \in [0,1]$ there is some $z \in U$ such that $h(z) \leqq_K \lambda h(x) + (1-\lambda)h(y)$. Note that h is K-convexlike on U if and only if $h(U)+K$ is a convex set.

Remark 2.2.8. Every K-convex function $h : X \to V \cup \{+\infty_K\}$ is K-convexlike, but not all K-convexlike functions are K-convex. Consider, for instance, $\mathbb{R}^2$ partially ordered by the cone $\mathbb{R}^2_+$. Take the function $h : \mathbb{R} \to \mathbb{R}^2 \cup \{+\infty_{\mathbb{R}^2_+}\}$ defined by $h(x) = (x, \sin x)$ if $x \in [-\pi, \pi]$ and $h(x) = +\infty_{\mathbb{R}^2_+}$ otherwise. It can be proven that h is $\mathbb{R}^2_+$-convexlike, but not $\mathbb{R}^2_+$-convex.

2.2.2 Topological properties of convex functions

In this section we deal with topological notions for functions, which alongside the convexity endow them with special properties.

Definition 2.2.9. *A function $f : X \to \overline{\mathbb{R}}$ is called lower semicontinuous at $\bar{x} \in X$ if* $\liminf_{x \to \bar{x}} f(x) \geq f(\bar{x})$*. A function f is said to be upper semicontinuous at $\bar{x}$ if $(-f)$ is lower semicontinuous at $\bar{x}$. When a function f is lower (upper) semicontinuous at all $x \in X$ we call it lower (upper) semicontinuous.*

Obviously, $f : X \to \overline{\mathbb{R}}$ is continuous at $\bar{x} \in X$ if and only if f is lower and upper semicontinuous at $\bar{x} \in X$.

In the following we give some equivalent characterizations of the lower semicontinuity of a function.

Theorem 2.2.9. *Let be the function $f : X \to \overline{\mathbb{R}}$. The following statements are equivalent:*

(i) f is lower semicontinuous;
(ii) $\operatorname{epi} f$ *is closed;*
(iii) the level set $\{x \in X : f(x) \leq \lambda\}$ is closed for all $\lambda \in \mathbb{R}$.

Example 2.2.4. Given a set $U \subseteq X$, its indicator function δ_U is lower semicontinuous if and only if U is closed, while the support function σ_U is always weak* lower semicontinuous.

Proposition 2.2.10. *The pointwise supremum of a family of lower semicontinuous functions $f_i : X \to \overline{\mathbb{R}}$, $i \in I$, where I is an arbitrary index set, $f : X \to \overline{\mathbb{R}}$ defined by $f(x) = \sup_{i \in I} f_i(x)$ is lower semicontinuous, too.*

Proposition 2.2.11. *If $f, g : X \to \overline{\mathbb{R}}$ are lower semicontinuous at $x \in X$ and $\lambda \in (0, +\infty)$, then $f + g$ and λf are lower semicontinuous at x, too.*

Via the lower bound function one can introduce the *lower semicontinuous hull* of a function as follows.

Definition 2.2.10. *Consider a function $f : X \to \overline{\mathbb{R}}$. The function $\bar{f} : X \to \overline{\mathbb{R}}$, defined by*

$$\bar{f}(x) := \phi_{\mathrm{cl}(\mathrm{epi}\, f)}(x) = \inf\{t : (x,t) \in \mathrm{cl}(\mathrm{epi}\, f)\}$$

is called the lower semicontinuous hull of f.

Example 2.2.5. If $f : \mathbb{R} \to \overline{\mathbb{R}}$, $f = \delta_{(-\infty,0)}$, then obviously $\bar{f} = \delta_{(-\infty,0]}$.

Theorem 2.2.12. *Let be the function $f : X \to \overline{\mathbb{R}}$. Then the following statements are true*

(a) $\mathrm{epi}\, \bar{f} = \mathrm{cl}(\mathrm{epi}\, f)$*;*
(b) $\mathrm{dom}\, f \subseteq \mathrm{dom}\, \bar{f} \subseteq \mathrm{cl}(\mathrm{dom}\, f)$*;*
(c) $\bar{f}(x) = \liminf_{y \to x} f(y)$ *for all* $x \in X$*.*

Remark 2.2.9. For a given function $f : X \to \overline{\mathbb{R}}$ $\bar{f}$ is the greatest lower semicontinuous function less than or equal to f. Consequently,

$$\bar{f} = \sup\{g : X \to \overline{\mathbb{R}} : g(x) \leq f(x)\ \forall x \in X \text{ and } g \text{ is lower semicontinuous}\}.$$

In the following we deal with functions that are both convex and lower semicontinuous and show some of the properties this class of functions is endowed with.

Theorem 2.2.13. *Let be $f : X \to \overline{\mathbb{R}}$ a convex function . Then f is lower semicontinuous if and only if it is weakly lower semicontinuous.*

Proposition 2.2.14. *If $f : X \to \overline{\mathbb{R}}$ is convex and lower semicontinuous, but not proper, then f cannot take finite values, i.e. f is everywhere equal to $+\infty$ or f takes the value $-\infty$ everywhere on its domain.*

Proposition 2.2.14 has as consequence the fact that if $f : X \to \overline{\mathbb{R}}$ is convex and lower semicontinuous and finite somewhere, then $f(x) > -\infty$ for all $x \in X$. By Proposition 2.2.1 and Theorem 2.2.12 follows that if $f : X \to \overline{\mathbb{R}}$ is convex then $\bar{f}$ is convex, too. Further, by Proposition 2.2.14 one has that if there is some $\bar{x} \in X$ such that $f(\bar{x}) = -\infty$, then $\bar{f}(x) = -\infty$ for all $x \in \mathrm{dom}\, \bar{f} \supseteq \mathrm{dom}\, f$.

We come now to a fundamental result linking a convex and lower semicontinuous function with the set of its affine minorants. A function $g : X \to \mathbb{R}$ is said to be *affine* if there are some $x^* \in X^*$ and $c \in \mathbb{R}$ such that $g(x) = \langle x^*, x\rangle + c$ for all $x \in X$. If $f : X \to \overline{\mathbb{R}}$ is a given function, then any affine function $g : X \to \mathbb{R}$ which fulfills $g(x) \leq f(x)$ for all $x \in X$ is said to be an *affine minorant* of f.

Theorem 2.2.15. *Let be the given function $f : X \to \overline{\mathbb{R}}$. Then f is convex and lower semicontinuous and takes nowhere the value $-\infty$ if and only if its set of affine minorants is nonempty and f is the pointwise supremum of this set.*

Proof. The sufficiency is obvious, as a pointwise supremum of a family of affine functions is convex, by Proposition 2.2.2, and lower semicontinuous, via Proposition 2.2.10, noting that a function having an affine minorant cannot take the value $-\infty$.

To verify the necessity we first prove that the set

$$M := \{(x^*, \alpha) \in X^* \times \mathbb{R} : \langle x^*, x\rangle + \alpha \leq f(x) \ \forall x \in X\}$$

is nonempty. If $f \equiv +\infty$ then for all $x^* \in X^*$ and $\alpha \in \mathbb{R}$ we have $\langle x^*, x\rangle + \alpha \leq f(x)$ for all $x \in X$, i.e. $(x^*, \alpha) \in M$.

Otherwise, there must be at least an element $y \in X$ such that $f(y) \in \mathbb{R}$. Then $\operatorname{epi} f \neq \emptyset$ and $(y, f(y) - 1) \notin \operatorname{epi} f$. As the hypotheses guarantee that $\operatorname{epi} f$ is convex and closed, applying Theorem 2.1.5 follows the existence of some $x^* \in X^*$ and $\alpha \in \mathbb{R}$, $(x^*, \alpha) \neq (0, 0)$, such that

$$\langle x^*, y\rangle + \alpha(f(y) - 1) < \langle x^*, x\rangle + \alpha r \ \forall (x, r) \in \operatorname{epi} f.$$

As $(y, f(y)) \in \operatorname{epi} f$, it follows $\alpha > 0$ and $(1/\alpha)\langle x^*, y - x\rangle + f(y) - 1 < r$ for all $(x, r) \in \operatorname{epi} f$. Taking into consideration that whenever $x \in \operatorname{dom} f$ there is $(x, f(x)) \in \operatorname{epi} f$, the last inequality yields $(1/\alpha)\langle x^*, y - x\rangle + f(y) - 1 < f(x)$ for all $x \in \operatorname{dom} f$, and one can easily note that this inequality is valid actually for all $x \in X$. Consequently, the function $x \mapsto \langle(-1/\alpha)x^*, x\rangle + (1/\alpha)\langle x^*, y\rangle + f(y) - 1$ is an affine minorant of f, thus $M \neq \emptyset$ in this case, too.

For all $x \in X$ one has

$$f(x) \geq \sup\{\langle x^*, x\rangle + \alpha : (x^*, \alpha) \in X^* \times \mathbb{R},\ \langle x^*, z\rangle + \alpha \leq f(z) \ \forall z \in X\}$$

and next we prove that this inequality is always fulfilled as equality. Assume that there are some $\bar{x} \in X$ and $\bar{r} \in \mathbb{R}$ such that

$$f(\bar{x}) > \bar{r} > \sup\{\langle x^*, \bar{x}\rangle + \alpha : (x^*, \alpha) \in X^* \times \mathbb{R},\ \langle x^*, z\rangle + \alpha \leq f(z) \ \forall z \in X\}. \tag{2.4}$$

Then $(\bar{x}, \bar{r}) \notin \operatorname{epi} f$. Applying again Theorem 2.1.5 we obtain some $\bar{x}^* \in X^*$ and $\bar{\alpha} \in \mathbb{R}$, $(\bar{x}^*, \bar{\alpha}) \neq (0, 0)$, and an $\varepsilon > 0$ such that

$$\langle \bar{x}^*, x\rangle + \bar{\alpha} r > \langle \bar{x}^*, \bar{x}\rangle + \bar{\alpha}\bar{r} + \varepsilon \ \forall (x, r) \in \operatorname{epi} f. \tag{2.5}$$

For $(z, s) \in \operatorname{epi} f$ we get $(z, s + t) \in \operatorname{epi} f$ for all $t \geq 0$, thus $\bar{\alpha} \geq 0$. Assume that $f(\bar{x}) \in \mathbb{R}$. Then we obtain $\bar{\alpha}(f(\bar{x}) - \bar{r}) > \varepsilon$, which yields $\bar{\alpha} > 0$. Thus for all $x \in \operatorname{dom} f$ one has $f(x) > (1/\bar{\alpha})\langle \bar{x}^*, \bar{x} - x\rangle + \bar{r} + (1/\bar{\alpha})\varepsilon > (1/\bar{\alpha})\langle \bar{x}^*, \bar{x} - x\rangle + \bar{r}$. As the function $x \mapsto \langle(-1/\bar{\alpha})\bar{x}^*, x\rangle + \langle(1/\bar{\alpha})\bar{x}^*, \bar{x}\rangle + \bar{r}$ is an affine minorant of f taking at $x = \bar{x}$ the value $\bar{r}$, we obtain a contradiction to (2.4). Consequently, $f(\bar{x}) = +\infty$. Assuming $\bar{\alpha} > 0$ we reach again a contradiction, thus $\bar{\alpha} = 0$. Consider then the function $x \mapsto -\langle \bar{x}^*, x - \bar{x}\rangle + \varepsilon$. By (2.5) one gets $-\langle \bar{x}^*, x - \bar{x}\rangle + \varepsilon \leq 0$ for all $x \in \operatorname{dom} f$. As $M \neq \emptyset$, there are some $y^* \in X^*$ and $\beta \in \mathbb{R}$ such that $\langle y^*, z\rangle + \beta \leq f(z)$ whenever $z \in X$. Denote $\gamma := (\bar{r} - \langle y^*, \bar{x}\rangle - \beta)/\varepsilon$. It is clear that $\gamma > 0$ and that the function

$x \mapsto \langle y^* - \gamma \bar{x}^*, x\rangle + \langle \gamma \bar{x}^*, \bar{x}\rangle + \beta + \gamma\varepsilon$ is affine. For all $x \in \operatorname{dom} f$ there is $\langle y^* - \gamma \bar{x}^*, x\rangle + \langle \gamma \bar{x}^*, \bar{x}\rangle + \beta + \gamma\varepsilon = \langle y^*, x\rangle + \beta + \gamma(\langle -\bar{x}^*, x - \bar{x}\rangle + \varepsilon) \leq \langle y^*, x\rangle + \beta \leq f(x)$, thus $x \mapsto \langle y^* - \gamma \bar{x}^*, x\rangle + \langle \gamma \bar{x}^*, \bar{x}\rangle + \beta + \gamma\varepsilon$ is an affine minorant of f and for $x = \bar{x}$ one gets $\langle y^* - \gamma \bar{x}^*, \bar{x}\rangle + \langle \gamma \bar{x}^*, \bar{x}\rangle + \beta + \gamma\varepsilon = \bar{r}$, which contradicts (2.4). Thus f is the pointwise supremum of the set of its affine minorants. □

The lower bound function can be also used to introduce the *lower semicontinuous convex hull* of a function.

Definition 2.2.11. *Consider a function $f : X \to \overline{\mathbb{R}}$. The function $\overline{\operatorname{co}} f : X \to \overline{\mathbb{R}}$, defined by*

$$\overline{\operatorname{co}} f(x) := \phi_{\overline{\operatorname{co}}(\operatorname{epi} f)}(x) = \inf\{t : (x,t) \in \overline{\operatorname{co}}(\operatorname{epi} f)\}$$

is called the lower semicontinuous convex hull of f.

Some properties this notion is endowed with follow.

Theorem 2.2.16. *Let $f : X \to \overline{\mathbb{R}}$ be a given function. Then the following statements are true*

(a) $\operatorname{epi}(\overline{\operatorname{co}} f) = \overline{\operatorname{co}}(\operatorname{epi} f)$;
(b) $\operatorname{dom}(\operatorname{co} f) = \operatorname{co}(\operatorname{dom} f) \subseteq \operatorname{dom}(\overline{\operatorname{co}} f) \subseteq \operatorname{cl}(\operatorname{dom}(\operatorname{co} f)) = \overline{\operatorname{co}}(\operatorname{dom} f)$.

Remark 2.2.10. It is clear from the construction that the lower semicontinuous convex hull of a function $f : X \to \overline{\mathbb{R}}$ is convex and lower semicontinuous and it is the greatest convex lower semicontinuous function less than or equal to f. Consequently,

$$\overline{\operatorname{co}} f = \sup\{g : X \to \overline{\mathbb{R}} : g(x) \leq f(x) \text{ for all } x \in X \text{ and } g \text{ is convex and lower semicontinuous}\}.$$

Now we turn our attention to continuity properties of convex functions.

Theorem 2.2.17. *Let be $f : X \to \overline{\mathbb{R}}$ a convex function. The following statements are equivalent:*

(i) there is a nonempty open subset of X on which f is bounded from above by a finite constant and is not everywhere equal to $-\infty$;
(ii) f is proper and continuous on the interior of its effective domain, which is nonempty.

As a consequence of Theorem 2.2.17 it follows that a convex function $f : X \to \overline{\mathbb{R}}$ is continuous on $\operatorname{int}(\operatorname{dom} f)$ if and only if $\operatorname{int}(\operatorname{epi} f) \neq \emptyset$. If we take $X = \mathbb{R}^n$, every proper and convex function $f : \mathbb{R}^n \to \overline{\mathbb{R}}$ is continuous on $\operatorname{ri}(\operatorname{dom} f)$. Consequently, every convex function $f : \mathbb{R}^n \to \mathbb{R}$ is continuous.

Besides associating new functions to a given function, the lower bound function can be used to define the notion of a *gauge* of a given set. For $U \subseteq X$, consider the set $\operatorname{cone}(U \times \{1\}) = \{(\lambda x, \lambda) : \lambda \in \mathbb{R}_+, x \in U\}$. If U is convex and closed, then $\operatorname{cone}(U \times \{1\})$ is convex and closed, too.

Definition 2.2.12. *Given a convex absorbing subset U of a vector space X, the gauge (or Minkowski function) associated to it is the function $\gamma_U : X \to \mathbb{R}$ defined by*

$$\gamma_U(x) := \phi_{\text{cone}(U\times\{1\})}(x) = \inf\{\lambda \geq 0 : x \in \lambda U\}.$$

In this situation, U is called the unit ball of the gauge γ_U.

Proposition 2.2.18. *(a) If X is a vector space and $U \subseteq X$ is absorbing and convex, then γ_U is sublinear and* $\text{core}(U) = \{x \in X : \gamma_U(x) < 1\}$. *If U is moreover symmetric, then γ_U is a seminorm.*
(b) If X is a topological vector space and $U \subseteq X$ is a convex neighborhood of 0, then γ_U is continuous, $\text{int}(U) = \{x \in X : \gamma_U(x) < 1\}$ *and* $\text{cl}(U) = \{x \in X : \gamma_U(x) \leq 1\}$.

There are several extensions of the notion of lower semicontinuity for vector functions based on the properties of the lower semicontinuous functions. We recall here three of them, which are mostly used in convex optimization. Like before, V is a Hausdorff locally convex space partially ordered by the convex cone K.

Definition 2.2.13. *A function $h : X \to V \cup \{+\infty_K\}$ is called*

(a) K-lower semicontinuous at $x \in X$ if for any neighborhood W of zero in V and for any $b \in V$ satisfying $b \leqq_K h(x)$, there exists a neighborhood U of x in X such that $h(U) \subseteq b + W + K \cup \{+\infty_K\}$;
(b) star K-lower semicontinuous at $x \in X$ if (k^*h) is lower semicontinuous at x for all $k^* \in K^*$.

Remark 2.2.11. The K-lower semicontinuity of a function $h : X \to V \cup \{+\infty_K\}$ was introduced by Penot ant Théra in [150], being later refined in [50]. For all $x \in \text{dom}\, h$ the definition of the K-lower semicontinuity of h at x amounts to asking for any neighborhood W of zero in V the existence of a neighborhood U of x in X such that $h(U) \subseteq h(x) + W + K \cup \{+\infty_K\}$. The notion of star K-lower semicontinuity was first considered in [106].

Definition 2.2.14. *A function $h : X \to V \cup \{+\infty_K\}$ is called*

(a) K-lower semicontinuous if it is K-lower semicontinuous at every $x \in X$;
(b) star K-lower semicontinuous if it is star K-lower semicontinuous at every $x \in X$;
(c) K-epi closed if $\text{epi}_K h$ *is closed.*

Proposition 2.2.19. *Let be the function $h : X \to V \cup \{+\infty_K\}$.*

(a) If h is K-lower semicontinuous at $x \in X$, then it is also star K-lower semicontinuous at x.
(b) If h is star K-lower semicontinuous, then it is also K-epi closed.

The following example shows that there are K-epi closed functions which are not star K-lower semicontinuous.

Example 2.2.6. Consider the function

$$h : \mathbb{R} \to \mathbb{R}^2 \cup \{+\infty_{\mathbb{R}^2_+}\}, \; h(x) = \begin{cases} (\frac{1}{x}, x), & \text{if } x > 0, \\ +\infty_{\mathbb{R}^2_+}, & \text{otherwise.} \end{cases}$$

It can be verified that h is $\mathbb{R}^2_+$-convex and $\mathbb{R}^2_+$-epi-closed, but not star $\mathbb{R}^2_+$-lower semicontinuous. For instance, for $k^* = (0,1)^T \in (\mathbb{R}^2_+)^* = \mathbb{R}^2_+$ one has

$$((0,1)^T h)(x) = \begin{cases} x, & \text{if } x > 0, \\ +\infty, & \text{otherwise,} \end{cases}$$

which is not lower semicontinuous.

Remark 2.2.12. When $V = \mathbb{R}$ and $K = \mathbb{R}_+$, the notions of K-lower semicontinuity, star K-lower semicontinuity and K-epi closedness collapse into the classical notion of lower semicontinuity.

2.3 Conjugate functions and subdifferentiability

Throughout this entire section we consider X to be a Hausdorff locally convex space with its topological dual space X^* endowed with the weak* topology.

2.3.1 Conjugate functions

Let $f : X \to \overline{\mathbb{R}}$ be a given function. In the following we deal with the notion of conjugate function of f, a basic one in the theory of convex analysis and very important for establishing a general duality theory for convex optimization problems (see chapter 3 for more on this topic).

Definition 2.3.1. *The function*

$$f^* : X^* \to \overline{\mathbb{R}}, \; f^*(x^*) := \sup_{x \in X} \{\langle x^*, x\rangle - f(x)\}$$

is said to be the (Fenchel) conjugate function of f.

Note that for all $x^* \in X^*$ it holds $f^*(x^*) = \sup_{x \in \operatorname{dom} f}\{\langle x^*, x\rangle - f(x)\}$.
Some of the investigations we make in this book will employ the *conjugate function of f with respect to the nonempty set $S \subseteq X$*, defined by

$$f^*_S : X^* \to \overline{\mathbb{R}}, f^*_S(x^*) := (f + \delta_S)^*(x^*) = \sup_{x \in S}\{\langle x^*, x\rangle - f(x)\}.$$

Lemma 2.3.1. *(a) If the function f is proper, then $f^*(x^*) > -\infty$ for all $x^* \in X^*$.*

(b) The function f^* *is proper if and only if* $\operatorname{dom} f \neq \emptyset$ *and* f *has an affine minorant.*

Proof. (*a*) If the function f is proper, then by definition there exists some $\bar{x} \in X$ such that $f(\bar{x}) \in \mathbb{R}$. Then for all $x^* \in X^*$ it holds $f^*(x^*) = \sup_{x \in X}\{\langle x^*, x\rangle - f(x)\} \geq \langle x^*, \bar{x}\rangle - f(\bar{x}) > -\infty$.

(*b*) Suppose first that the function f^* is proper. By definition there exists $\bar{x}^* \in X^*$ such that that $f^*(\bar{x}^*) \in \mathbb{R}$. Since $f^*(\bar{x}^*) = \sup_{x \in X}\{\langle \bar{x}^*, x\rangle - f(x)\} \geq \langle \bar{x}^*, x\rangle - f(x)$ for all $x \in X$, $x \mapsto \langle \bar{x}^*, x\rangle - f^*(\bar{x}^*)$ is an affine minorant of the function f. Assuming that $\operatorname{dom} f = \emptyset$ or, equivalently, $f \equiv +\infty$, one would have that $f^* \equiv -\infty$, which would contradict the assumption that f^* is proper. Thus $\operatorname{dom} f$ must be a nonempty set.

Assume now that $\operatorname{dom} f \neq \emptyset$ and that there exist $\bar{x}^* \in X^*$ and $c \in \mathbb{R}$ such that $f(x) \geq \langle \bar{x}^*, x\rangle + c$ for all $x \in X$. Obviously, the function f is proper and by (*a*) we get $f^* > -\infty$ on X^*. Moreover, $f^*(\bar{x}^*) = \sup_{x \in X}\{\langle \bar{x}^*, x\rangle - f(x)\} \leq -c$, whence f^* is proper. □

It is straightforward to verify that in case f is not proper one either has $f^* \equiv -\infty$ (if $f \equiv +\infty$) or $f^* \equiv +\infty$ (if there exists an $x \in X$ with $f(x) = -\infty$). In case $\operatorname{dom} f \neq \emptyset$ and f has an affine minorant $x \mapsto \langle x^*, x\rangle + c$, with $x^* \in X^*$ and $c \in \mathbb{R}$ it holds, as we have seen, $-c \geq f^*(x^*)$. Under these circumstances, $-f^*(x^*)$ represents the largest value $c \in \mathbb{R}$ for which $x \mapsto \langle x^*, x\rangle + c$ is an affine minorant of f.

Remark 2.3.1. It is a direct consequence of Definition 2.3.1 that f^* is the pointwise supremum of the family of affine functions $g_x : X^* \to \overline{\mathbb{R}}$, $g_x(x^*) = \langle x^*, x\rangle - f(x)$, $x \in \operatorname{dom} f$. Therefore f^* turns out to be a convex and lower semicontinuous function.

Next we collect some elementary properties of conjugate functions.

Proposition 2.3.2. *Let* f, g, $f_i : X \to \overline{\mathbb{R}}$, $i \in I$, *be given functions, where* I *is an arbitrary index set. Then the following statements hold*

(a) $f(x) + f^*(x^*) \geq \langle x^*, x\rangle \ \forall x \in X \ \forall x^* \in X^*$ *(Young-Fenchel inequality);*
(b) $\inf_{x \in X} f(x) = -f^*(0)$*;*
(c) $f \leq g$ *on* X *implies* $f^* \geq g^*$ *on* X^**;*
(d) $(\sup_{i \in I} f_i)^* \leq \inf_{i \in I} f_i^*$ *and* $(\inf_{i \in I} f_i)^* = \sup_{i \in I} f_i^*$*;*
(e) $(\lambda f)^*(x^*) = \lambda f^*((1/\lambda)x^*) \ \forall x^* \in X^* \ \forall \lambda > 0$*;*
(f) $(f + \beta)^* = f^* - \beta \ \forall \beta \in \mathbb{R}$*;*
(g) for $f_{x_0}(x) = f(x - x_0)$*, when* $x_0 \in X$*, there is* $(f_{x_0})^*(x^*) = f^*(x^*) + \langle x^*, x_0\rangle \ \forall x^* \in X^*$*;*
(h) for $f_{x_0^*}(x) = f(x) + \langle x_0^*, x\rangle$*, when* $x_0^* \in X^*$*, there is* $(f_{x_0^*})^*(x^*) = f^*(x^* - x_0^*) \ \forall x^* \in X^*$*;*
(i) for Y *a Hausdorff locally convex space and* $A : Y \to X$ *a linear continuous invertible mapping there is* $(f \circ A)^* = f^* \circ (A^{-1})^*$*;*
(j) $(f + g)^*(x^* + y^*) \leq f^*(x^*) + g^*(y^*) \ \forall x^*, y^* \in X^*$*;*

(k) $(\lambda f + (1-\lambda)g)^*(x^*) \leq \lambda f^*(x^*) + (1-\lambda)g^*(x^*)$ $\forall x^* \in X^*$ $\forall \lambda \in (0,1)$;

(l) *for* $f : X_1 \times \ldots \times X_m \to \overline{\mathbb{R}}$, $f(x_1, \ldots, x_m) = \sum_{i=1}^m f_i(x_i)$, *where* X_i *is a Hausdorff locally convex space and* $f_i : X_i \to \overline{\mathbb{R}}$, $i = 1, \ldots, m$, *there is* $f^*(x_1^*, \ldots, x_m^*) = \sum_{i=1}^m f_i^*(x_i^*)$ $\forall (x_1^*, \ldots, x_m^*) \in X_1^* \times \ldots \times X_m^*$.

Proof. The verification of the above assertions is an obvious consequence of Definition 2.3.1. Therefore we confine ourselves only to point out the proof of the statements *(d)* and *(i)*.

(d) For all $j \in I$ and $x^* \in X^*$ we have $(\sup_{i \in I} f_i)^*(x^*) = \sup_{x \in X}\{\langle x^*, x\rangle - \sup_{i \in I} f_i(x)\} \leq \sup_{x \in X}\{\langle x^*, x\rangle - f_j(x)\} = f_j^*(x^*)$. Taking the infimum over $j \in I$ at the right-hand side of this inequality yields the wanted result.

In the second part of the statement, for all $x^* \in X^*$ we have $(\inf_{i \in I} f_i)^*(x^*) = \sup_{x \in X}\{\langle x^*, x\rangle - \inf_{i \in I} f_i(x)\} = \sup_{i \in I}\sup_{x \in X}\{\langle x^*, x\rangle - f_i(x)\} = \sup_{i \in I} f_i^*(x^*)$.

(i) Let $x^* \in X^*$ be arbitrarily taken. It holds $(f \circ A)^*(x^*) = \sup_{x \in X}\{\langle x^*, x\rangle - f(Ax)\} = \sup_{y \in Y}\{\langle x^*, A^{-1}y\rangle - f(y)\} = \sup_{y \in Y}\{\langle (A^{-1})^* x^*, y\rangle - f(y)\} = f^*((A^{-1})^* x^*) = (f^* \circ (A^{-1})^*)(x^*)$ and the desired relation is proved. □

Remark 2.3.2. The convention $(+\infty)+(-\infty) = (-\infty)+(+\infty) = +\infty$ ensures that the Young-Fenchel inequality applies also to improper functions.

For a function defined on the dual space X^* one can introduce its conjugate function analogously. More precisely, if we consider $g : X^* \to \overline{\mathbb{R}}$, then

$$g^* : X \to \overline{\mathbb{R}}, \quad g^*(x) = \sup_{x^* \in X^*}\{\langle x^*, x\rangle - g(x^*)\}$$

is the conjugate function of g. In particular, to the function $f : X \to \overline{\mathbb{R}}$ we can attach the so-called *biconjugate* function of f, which is defined as the conjugate function of the conjugate f^*, i.e.

$$f^{**} : X \to \overline{\mathbb{R}}, \ f^{**}(x) := (f^*)^*(x) = \sup_{x^* \in X^*}\{\langle x^*, x\rangle - f^*(x^*)\}.$$

The next lemma is a direct consequence of the Young-Fenchel inequality.

Lemma 2.3.3. *For all* $x \in X$ *it holds* $f^{**}(x) \leq f(x)$.

Next the conjugates of some convex functions needed later are provided.

Example 2.3.1. Let $X = \mathbb{R}$ and $f : \mathbb{R} \to \overline{\mathbb{R}}$.

(a) If $f(x) = (1/2)x^2$, $x \in \mathbb{R}$, then $f^*(x^*) = (1/2)(x^*)^2$ for $x^* \in \mathbb{R}$.

(b) If $f(x) = \mathrm{e}^x$, $x \in \mathbb{R}$, then

$$f^*(x^*) = \begin{cases} x^*(\ln x^* - 1), & \text{if } x^* > 0, \\ 0, & \text{if } x^* = 0, \\ +\infty, & \text{if } x^* < 0. \end{cases}$$

(c) If

$$f(x)=\begin{cases} x(\ln x-1), & \text{if } x>0,\\ 0, & \text{if } x=0,\\ +\infty, & \text{if } x<0,\end{cases}$$

then $f^*(x^*)=\mathrm{e}^{x^*}$ for $x^*\in\mathbb{R}$.

Example 2.3.2. Let be $f:X\to\mathbb{R}$, $f(x)=\langle y^*,x\rangle+c$, with $y^*\in X^*$ and $c\in\mathbb{R}$. Then

$$f^*(x^*)=\delta_{\{y^*\}}(x^*)-c=\begin{cases}-c, & \text{if } x^*=y^*,\\ +\infty, & \text{otherwise.}\end{cases}$$

Example 2.3.3. Let be $U\subseteq X$. Then for all $x^*\in X^*$ there is

$$\delta_U^*(x^*)=\sup_{x\in X}\{\langle x^*,x\rangle-\delta_U(x)\}=\sup_{x\in U}\langle x^*,x\rangle=\sigma_U(x^*).$$

Example 2.3.4. (*a*) Given a convex absorbing subset U of X, the conjugate of its gauge $\gamma_U:X\to\mathbb{R}$ at some $x^*\in X^*$ is

$$\begin{aligned}(\gamma_U)^*(x^*)&=\sup_{x\in X}\left\{\langle x^*,x\rangle-\inf\{\lambda\geq 0: x\in\lambda U\}\right\}\\ &=\sup_{x\in X}\left\{\langle x^*,x\rangle+\sup_{\substack{\lambda\geq 0,\\ x\in\lambda U}}\{-\lambda\}\right\}=\sup_{\lambda\geq 0}\left\{-\lambda+\sup_{y\in U}\langle x^*,\lambda y\rangle\right\}\\ &=\sup_{\lambda\geq 0}\left\{\lambda\left(\sup_{y\in U}\langle x^*,y\rangle-1\right)\right\}=\begin{cases}0, & \text{if } \sigma_U(x^*)\leq 1,\\ +\infty, & \text{otherwise.}\end{cases}\end{aligned}$$

(*b*) Let $(X,\|\cdot\|)$ be a normed vector space and $(X^*,\|\cdot\|_*)$ its topological dual space. The conjugate of the norm function can be deduced from the one of the gauge corresponding to the set $U=\{x\in X:\|x\|\leq 1\}$, since $\gamma_U=\|\cdot\|$ and $\sigma_U=\|\cdot\|_*$. Therefore for $x^*\in X^*$ one has

$$(\|\cdot\|)^*(x^*)=\begin{cases}0, & \text{if } \|x^*\|_*\leq 1,\\ +\infty, & \text{otherwise.}\end{cases}$$

Example 2.3.5. Let X be a normed space and $f:X\to\mathbb{R}$, $f(x)=(1/p)\|x\|^p$, $1<p<\infty$. Then $f^*(x^*)=(1/q)\|x^*\|_*^q$ for $x^*\in X^*$, where $(1/p)+(1/q)=1$.

The results we prove next are necessary for deriving further statements, in particular concerning duality.

Proposition 2.3.4. *The following relations are always fulfilled*

(a) $f^*=(\bar{f})^*=(\overline{\text{co}}f)^*$ *on* X^**;*
(b) $f^{**}\leq\overline{\text{co}}f\leq\bar{f}\leq f$ *on* X*.*

Proof. (*a*) Taking a careful look at the way the functions $\overline{\text{co}}f$ and $\bar{f}$ are defined, it is not hard to see that on X the inequalities $\overline{\text{co}}f\leq\bar{f}\leq f$ are always fulfilled (see also Theorem 2.2.12(*a*)). Applying Proposition 2.3.2(*c*) we get $(\overline{\text{co}}f)^* \geq \bar{f}^* \geq f^*$ on X^*. In order to get the desired conclusion, we prove that $(\overline{\text{co}}f)^* \leq f^*$. Let $x^* \in X^*$ be arbitrarily taken. We treat further three cases.

If $f^*(x^*) = +\infty$ we get $(\overline{\text{co}}f)^*(x^*) = (\bar{f})^*(x^*) = f^*(x^*) = +\infty$.

If $f^*(x^*) = -\infty$, then $f^{**} \equiv +\infty$ and, by Lemma 2.3.3, one has $f \equiv +\infty$. This implies further $\bar{f} = \overline{\text{co}}f = f \equiv +\infty$ and, consequently, $(\overline{\text{co}}f)^*(x^*) = (\bar{f})^*(x^*) = f^*(x^*) = -\infty$.

It remains to consider the case $f^*(x^*) \in \mathbb{R}$. Consider the function $g : X \to \mathbb{R}$, $g(x) = \langle x^*, x\rangle - f^*(x^*)$. According to the Young-Fenchel inequality for all $x \in X$ we have $g(x) \leq f(x)$, which is equivalent to $\operatorname{epi} g \supseteq \operatorname{epi} f$. Since g is an affine function, $\operatorname{epi} g$ is a convex and closed set and it holds $\operatorname{epi} g \supseteq \overline{\text{co}}(\operatorname{epi} f)$. By Theorem 2.2.16($a$) we get $\operatorname{epi} g \supseteq \operatorname{epi}(\overline{\text{co}}f)$, and from here we deduce that $g(x) = \langle x^*, x\rangle - f^*(x^*) \leq \overline{\text{co}}f(x)$ for all $x \in X$. This implies $(\overline{\text{co}}f)^*(x^*) = \sup_{x \in X}\{\langle x^*, x\rangle - \overline{\text{co}}f(x)\} \leq f^*(x^*)$ and in this way the statement (a) has been verified.

(b) We only have to justify $f^{**} \leq \overline{\text{co}}f$ on X. As the equality $f^* = (\overline{\text{co}}f)^*$ is secured by (a), it holds $f^{**} = (\overline{\text{co}}f)^{**} \leq \overline{\text{co}}f$ on X, where the last inequality follows by Lemma 2.3.3. □

Because of the inequality $f^{**} \leq f$ on X, arises in a natural way the question when does the coincidence of f and f^{**} occur. The next statement gives an answer.

Theorem 2.3.5. *If $f : X \to \overline{\mathbb{R}}$ is proper, convex and lower semicontinuous, then f^* is proper and $f = f^{**}$.*

Proof. We prove first that f^* is proper. As f is proper, $\operatorname{dom} f \neq \emptyset$ and, by Theorem 2.2.15, f has an affine minorant. Thus Lemma 2.3.1(b) guarantees the properness of f^*. We prove next that $f = f^{**}$. For all $x \in X$ we have

$$f^{**}(x) = \sup_{x^* \in X^*}\{\langle x^*, x\rangle - f^*(x^*)\} = \sup_{\substack{x^* \in X^*, c \in \mathbb{R}, \\ f^*(x^*) \leq -c}}\{\langle x^*, x\rangle + c\} =$$

$$\sup\{\langle x^*, x\rangle + c : x^* \in X^*, c \in \mathbb{R}, \langle x^*, z\rangle + c \leq f(z)\ \forall z \in X\}$$

and this is equal, again by Theorem 2.2.15, to $f(x)$. □

The well-known *Fenchel-Moreau theorem* follows as a direct consequence of Theorem 2.3.5. Because of its fame and historical importance we cite it here as a separate statement.

Theorem 2.3.6. *(Fenchel-Moreau) Let $f : X \to \overline{\mathbb{R}}$ be a proper function. Then $f = f^{**}$ if and only if f is convex and lower semicontinuous.*

Corollary 2.3.7. *Let $f : X \to \overline{\mathbb{R}}$. If $\overline{\text{co}}f > -\infty$, then $f^{**} = \overline{\text{co}}f$.*

Proof. If $\overline{\text{co}}f$ is proper, then the conclusion follows by Proposition 2.3.4(a) and Theorem 2.3.5. If $\overline{\text{co}}f \equiv +\infty$, then $\overline{\text{co}}f = f^{**} \equiv +\infty$ and the result holds also in this case. □

Remark 2.3.3. It is an immediate conclusion of Theorem 2.3.5 that for a convex function $f : X \to \overline{\mathbb{R}}$ its conjugate function f^* is proper if and only if $\bar{f}$ is proper.

Remark 2.3.4. Until now we have attached to a function $f : X \to \overline{\mathbb{R}}$ the conjugate function f^* and the biconjugate function f^{**}. It is natural to ask if it makes sense to consider the conjugate of the latter, namely $f^{***} : X^* \to \overline{\mathbb{R}}$ defined by $f^{***} = (f^{**})^*$. Since we always have $f^* = f^{***}$, this is not the case. In order to prove this we treat two cases.

Let us assume first that the function f^* is proper. Since f^* is also convex and lower semicontinuous, Theorem 2.3.5 secures the equality $f^{***} = (f^*)^{**} = f^*$. Assume now that the function f^* is not proper. If $f^* \equiv +\infty$ then $f^{**} \equiv -\infty$ and this implies $f^{***} \equiv +\infty$. If there is an x^* such that $f^*(x^*) = -\infty$, then $f^{**} \equiv +\infty$, which yields $f^{***} \equiv -\infty$. Moreover, by Lemma 2.3.3 it is obvious that $f \equiv +\infty$ and so $f^* \equiv -\infty$.

In convex analysis it is very natural and often also very useful to reformulate results employing functions in the language of their epigraphs. This applies also to conjugacy properties and the corresponding operations.

Let Y be another Hausdorff locally convex space whose topological dual space Y^* is endowed with the weak* topology. For $f : X \to \overline{\mathbb{R}}$ a given function and $A \in \mathcal{L}(X, Y)$ we calculate in the following the conjugate of the infimal function of f through A and derive from it the formula for the conjugate of the infimal convolution of a finite family of functions $f_i : X \to \overline{\mathbb{R}}$, $i = 1, \ldots, m$.

Proposition 2.3.8. *(a) Let $f : X \to \overline{\mathbb{R}}$ be a given function and $A \in \mathcal{L}(X, Y)$. Then it holds $(Af)^* = f^* \circ A^*$.*

(b) Let $f_i : X \to \overline{\mathbb{R}}$, $i = 1, \ldots, m$, be given functions. Then $(f_1 \Box \ldots \Box f_m)^* = \sum_{i=1}^m f_i^*$.

Proof. (a) By definition there holds for any $y^* \in Y^*$

$$(Af)^*(y^*) = \sup_{y \in Y}\{\langle y^*, y\rangle - (Af)(y)\} = \sup_{y \in Y}\{\langle y^*, y\rangle - \inf_{x \in X, Ax = y} f(x)\} =$$

$$\sup_{x \in X}\{\langle y^*, Ax\rangle - f(x)\} = \sup_{x \in X}\{\langle A^* y^*, x\rangle - f(x)\} = (f^* \circ A^*)(y^*).$$

(b) Taking $f : X^m \to \overline{\mathbb{R}}$, $f(x^1, \ldots, x^m) = \sum_{i=1}^m f_i(x^i)$ and $A \in \mathcal{L}(X^m, X)$, $A(x^1, \ldots, x^m) = \sum_{i=1}^m x^i$, we have seen that $Af = f_1 \Box \ldots \Box f_m$. Applying the result from (a) we get $(f_1 \Box \ldots \Box f_m)^* = f^* \circ A^*$. The conclusion follows by using Proposition 2.3.2(*l*) and the fact that $A^* x^* = (x^*, \ldots, x^*)$ for all $x^* \in X^*$. $\Box$

Of course, it is also of interest, to give a formula for the conjugate of the sum of a finite number of functions. A first calculation shows that for $x^{i*} \in X^*$, $i = 1, \ldots, m$, there is

$$\left(\sum_{i=1}^m f_i\right)^*\left(\sum_{i=1}^m x^{i*}\right) = \sup_{x\in X}\left\{\sum_{i=1}^m \langle x^{i*}, x\rangle - \sum_{i=1}^m f_i(x)\right\}$$

$$\leq \sum_{i=1}^m \sup_{x\in X}\{\langle x^{i*}, x\rangle - f_i(x)\} = \sum_{i=1}^m f_i^*(x^{i*}).$$

Consequently, for $x^* \in X^*$,

$$\left(\sum_{i=1}^m f_i\right)^*(x^*) \leq \inf\left\{\sum_{i=1}^m f_i^*(x^{i*}) : \sum_{i=1}^m x^{i*} = x^*\right\} = (f_1^*\Box\ldots\Box f_m^*)(x^*). \tag{2.6}$$

In a natural way the question of the coincidence of both sides of (2.6) arises. We can give first an equivalent characterization of this situation (see [38]).

Proposition 2.3.9. *Let $f_i : X \to \overline{\mathbb{R}}$, $i = 1,\ldots,m$, be proper functions such that $\cap_{i=1}^m \operatorname{dom} f_i \neq \emptyset$. Then the following statements are equivalent:*

(i) $\operatorname{epi}\left(\sum_{i=1}^m f_i\right)^* = \sum_{i=1}^m \operatorname{epi} f_i^*$;
(ii) $\left(\sum_{i=1}^m f_i\right)^* = f_1^*\Box\ldots\Box f_m^*$ *and the infimal convolution is exact.*

Proof. *(i)* $\Rightarrow$ *(ii)* Let $x^* \in X^*$ be arbitrarily taken. Then $\left(\sum_{i=1}^m f_i\right)^*(x^*) > -\infty$. If $\left(\sum_{i=1}^m f_i\right)^*(x^*) = +\infty$ then *(ii)* is automatically fulfilled, thus we consider further that $\left(\sum_{i=1}^m f_i\right)^*(x^*) < +\infty$, i.e. $\left(x^*, \left(\sum_{i=1}^m f_i\right)^*(x^*)\right) \in \operatorname{epi}\left(\sum_{i=1}^m f_i\right)^*$. By *(i)* there exist $(x^{i*}, r_i) \in \operatorname{epi} f_i^*$, $i = 1,\ldots,m$, such that $x^* = \sum_{i=1}^m x^{i*}$ and $\left(\sum_{i=1}^m f_i\right)^*(x^*) = \sum_{i=1}^m r_i$. This implies $f_i^*(x^{i*}) \leq r_i$, $i = 1,\ldots,m$, followed by $\sum_{i=1}^m f_i^*(x^{i*}) \leq \left(\sum_{i=1}^m f_i\right)^*(x^*)$. Consequently, $(f_1^*\Box\ldots\Box f_m^*)(x^*) \leq \left(\sum_{i=1}^m f_i\right)^*(x^*)$, which, combined with (2.6), yields *(ii)*.

(ii) $\Rightarrow$ *(i)* Let the pairs $(x^{i*}, r_i) \in \operatorname{epi} f_i^*$, $i = 1,\ldots,m$, be given. Then $\left(\sum_{i=1}^m f_i\right)^*\left(\sum_{i=1}^m x^{i*}\right) \leq \sum_{i=1}^m f_i^*(x^{i*}) \leq \sum_{i=1}^m r_i$, i.e. $\left(\sum_{i=1}^m x^{i*}, \sum_{i=1}^m r_i\right) \in \operatorname{epi}\left(\sum_{i=1}^m f_i\right)^*$. Therefore $\operatorname{epi}\left(\sum_{i=1}^m f_i\right)^* \supseteq \sum_{i=1}^m \operatorname{epi} f_i^*$ and this inclusion is always valid. Taking now some arbitrary pair $(x^*, r) \in \operatorname{epi}\left(\sum_{i=1}^m f_i\right)^*$, we get $\left(\sum_{i=1}^m f_i\right)^*(x^*) \leq r$. By *(ii)* there exist some $x^{i*} \in X^*$, $i = 1,\ldots,m$, such that $\sum_{i=1}^m x^{i*} = x^*$ and $\sum_{i=1}^m f_i^*(x^{i*}) \leq r$. This yields the existence of some $r_i \in \mathbb{R}$, $i = 1,\ldots,m$, with $\sum_{i=1}^m r_i = r$, such that $f_i^*(x^{i*}) \leq r_i$ for all $i = 1,\ldots,m$. Then $(x^*, r^*) = \left(\sum_{i=1}^m x^{i*}, \sum_{i=1}^m r_i\right) \in \sum_{i=1}^m \operatorname{epi} f_i^*$ and the proof is complete. □

In order to give a sufficient condition for the equality in (2.6) note first that for any proper functions $f_i : X \to \overline{\mathbb{R}}$, $i = 1,\ldots,m$, fulfilling $\cap_{i=1}^m \operatorname{dom} f_i \neq \emptyset$, there is

$$\operatorname{cl}\left(\sum_{i=1}^m \operatorname{epi} f_i\right) \supseteq \operatorname{epi}(f_1\Box\ldots\Box f_m) \supseteq \sum_{i=1}^m \operatorname{epi} f_i,$$

which has as consequence that

$$\mathrm{cl}(\mathrm{epi}(f_1 \Box \ldots \Box f_m)) = \mathrm{epi}\, \overline{f_1 \Box \ldots \Box f_m} = \mathrm{cl}\left(\sum_{i=1}^m \mathrm{epi}\, f_i\right). \tag{2.7}$$

The following two results characterize the epigraph of the conjugate of the sum of finitely many functions.

Theorem 2.3.10. *Let be $f_i : X \to \overline{\mathbb{R}}$, $i = 1, \ldots, m$, proper, convex and lower semicontinuous functions fulfilling $\cap_{i=1}^m \mathrm{dom}\, f_i \neq \emptyset$. Then one has*

$$\left(\sum_{i=1}^m f_i\right)^* = \overline{f_1^* \Box \ldots \Box f_m^*},$$

and, consequently,

$$\mathrm{epi}\left(\sum_{i=1}^m f_i\right)^* = \mathrm{epi}\, \overline{f_1^* \Box \ldots \Box f_m^*} = \mathrm{cl}\left(\sum_{i=1}^m \mathrm{epi}\, f_i^*\right).$$

Proof. By Theorem 2.2.12(*b*) we get

$$\sum_{i=1}^m \mathrm{dom}\, f_i^* = \mathrm{dom}(f_1^* \Box \ldots \Box f_m^*) \subseteq \mathrm{dom}\, \overline{f_1^* \Box \ldots \Box f_m^*}.$$

Since Theorem 2.3.5 ensures that f_i^*, $i = 1, \ldots, m$, are proper functions, it holds $\sum_{i=1}^m \mathrm{dom}\, f_i^* \neq \emptyset$, consequently, $\mathrm{dom}\, \overline{f_1^* \Box \ldots \Box f_m^*} \neq \emptyset$. Assuming that there is some $x^* \in X^*$ such that $\overline{f_1^* \Box \ldots \Box f_m^*}(x^*) = -\infty$, we get $\sum_{i=1}^m f_i^{**} = \sum_{i=1}^m f_i \equiv +\infty$, which contradicts the hypothesis $\cap_{i=1}^m \mathrm{dom}\, f_i \neq \emptyset$. Therefore $\overline{f_1^* \Box \ldots \Box f_m^*}$ is a proper function. We can apply now Theorem 2.3.5, which yields $\overline{f_1^* \Box \ldots \Box f_m^*} = \left(\overline{f_1^* \Box \ldots \Box f_m^*}\right)^{**}$. Using Proposition 2.3.4(*a*) and Proposition 2.3.8(*b*), it follows $\overline{f_1^* \Box \ldots \Box f_m^*} = \left(f_1^* \Box \ldots \Box f_m^*\right)^{**} = \left(\sum_{i=1}^m f_i^{**}\right)^*$. Then the first formula follows via Theorem 2.3.5 and, together with (2.7), it yields the second one, too. □

Remark 2.3.5. For two proper, convex and lower semicontinuous functions $f, g : X \to \overline{\mathbb{R}}$ fulfilling $\mathrm{dom}\, f \cap \mathrm{dom}\, g \neq \emptyset$, Theorem 2.3.10 yields the classical *Moreau-Rockafellar formula*, namely

$$(f + g)^* = \overline{f^* \Box g^*}.$$

Turning to epigraphs, we get

$$\mathrm{epi}(f + g)^* = \mathrm{epi}\, \overline{f^* \Box g^*} = \mathrm{cl}(\mathrm{epi}\, f^* + \mathrm{epi}\, g^*).$$

Remark 2.3.6. As seen in Theorem 2.3.10 and Proposition 2.3.9, a sufficient condition to have equality in (2.6) for the proper functions $f_i : X \to \overline{\mathbb{R}}$, $i = 1, \ldots, m$, when $\cap_{i=1}^m \mathrm{dom}\, f_i \neq \emptyset$ and all these functions are convex and lower semicontinuous, is $\sum_{i=1}^m \mathrm{epi}\, f_i^*$ closed. For other sufficient conditions that guarantee the equality in (2.6) we refer to section 3.5.

We close this subsection by characterizing the conjugate of a K-increasing function, which will be useful in chapter 3 when dealing with composed convex optimization problems. Let V be a Hausdorff locally convex space, partially ordered by a convex cone $K \subseteq V$.

Proposition 2.3.11. *If $g : V \to \overline{\mathbb{R}}$ is a K-increasing function with* $\operatorname{dom} g \neq \emptyset$*, then $g^*(v^*) = +\infty$ for all $v^* \notin K^*$, i.e.* $\operatorname{dom} g^* \subseteq K^*$.

Proof. If $K = \{0\}$ the conclusion follows automatically. Assume that $K \neq \{0\}$ and take an arbitrary $v^* \notin K^*$. By definition there exists $\bar{v} \in K$ such that $\langle v^*, \bar{v} \rangle < 0$. Since for some arbitrary $\tilde{v} \in \operatorname{dom} g$ and for all $\alpha > 0$ we have $g(\tilde{v} - \alpha\bar{v}) \leq g(\tilde{v})$, it is straightforward to see that

$$g^*(v^*) = \sup_{v \in V}\{\langle v^*, v\rangle - g(v)\} \geq \sup_{\alpha>0}\{\langle v^*, \tilde{v} - \alpha\bar{v}\rangle - g(\tilde{v} - \alpha\bar{v})\}$$

$$\geq \sup_{\alpha>0}\{\langle v^*, \tilde{v} - \alpha\bar{v}\rangle - g(\tilde{v})\} = \langle v^*, \tilde{v}\rangle - g(\tilde{v}) + \sup_{\alpha>0}\{-\alpha\langle v^*, \bar{v}\rangle\} = +\infty,$$

and the proof is complete. □

2.3.2 Subdifferentiability

In nondifferentiable convex optimization the classical *(Gâteaux) differentiability* may be replaced by the so-called *subdifferentiability*. To have a differentiability notion is extremely beneficial in analysis and optimization not only from the theoretical, but also from the numerical point of view. It allows, for instance, to formulate functional equations to describe mathematical objects and models for practical problems or to give optimality conditions in different fields of mathematical programming, variational calculus, for optimal control problems etc.

In this book we consider in the most situations scalar and multiobjective programming problems which involve convex sets and convex functions, without making use of the classical differentiability which is included as a special case of the general setting.

Definition 2.3.2. *Let $f : X \to \overline{\mathbb{R}}$ be a given function and take an arbitrary $x \in X$ such that $f(x) \in \mathbb{R}$. The set*

$$\partial f(x) := \{x^* \in X^* : f(y) - f(x) \geq \langle x^*, y - x\rangle \ \forall y \in X\}$$

is said to be the (convex) subdifferential of f at x. Its elements are called subgradients of f at x. We say that the function f is subdifferentiable at x if $\partial f(x) \neq \emptyset$.

If $f(x) \notin \mathbb{R}$ we consider by convention $\partial f(x) := \emptyset$.

Example 2.3.6. For $U \subseteq X$ and $f = \delta_U : X \to \overline{\mathbb{R}}$, one can easily show that for all $x \in U$ there is $\partial\delta_U(x) = N(U, x)$.

If $f : X \to \overline{\mathbb{R}}$ is subdifferentiable at x with $f(x) \in \mathbb{R}$ and $x^* \in \partial f(x)$, then the function $h : X \to \overline{\mathbb{R}}$, $h(y) = \langle x^*, y\rangle + f(x) - \langle x^*, x\rangle$ is an affine minorant of f. Moreover, this affine minorant coincides at x with f. In the following statement we give a characterization of the elements $x^* \in \partial f(x)$ according to the fact that for x^* and x the Fenchel-Young inequality is fulfilled as equality.

Theorem 2.3.12. *Let the function $f : X \to \overline{\mathbb{R}}$ be given and $x \in X$. Then $x^* \in \partial f(x)$ if and only if $f(x) + f^*(x^*) = \langle x^*, x\rangle$.*

Proof. Let $x^* \in \partial f(x)$. Then $f(x) \in \mathbb{R}$ and $f^*(x^*) = \sup\{\langle x^*, y\rangle - f(y) : y \in X\} \leq \langle x^*, x\rangle - f(x)$. Since the opposite inequality is always true, $f(x) + f^*(x^*) = \langle x^*, x\rangle$ follows.

Vice versa, let $x \in X$ and $x^* \in X^*$ be such that $f(x) + f^*(x^*) = \langle x^*, x\rangle$. Then $f(x) \in \mathbb{R}$ and $\langle x^*, x\rangle - f(x) = f^*(x^*) = \sup_{y\in X}\{\langle x^*, y\rangle - f(y)\} \geq \langle x^*, y\rangle - f(y)$ for all $y \in X$, and hence $x^* \in \partial f(x)$. □

For $f : X \to \overline{\mathbb{R}}$ a given function one has that $x \in X$ with $f(x) \in \mathbb{R}$ is a solution of the optimization problem $\inf_{x\in X} f(x)$ if and only if $0 \in \partial f(x)$. In general one can express necessary and sufficient optimality conditions for optimization problems by means of subdifferentials as we shall see in section 3.3.

Proposition 2.3.13. *(a) For a given function $f : X \to \overline{\mathbb{R}}$, one has $\partial(f + \langle x^*, \cdot\rangle)(x) = \partial f(x) + x^*$ for all $x^* \in X^*$ and all $x \in X$.*
(b) For $f : X_1 \times \ldots \times X_m \to \overline{\mathbb{R}}$, $f(x^1, \ldots, x^m) = \sum_{i=1}^m f_i(x^i)$, where X_i is a Hausdorff locally convex space and $f_i : X_i \to \overline{\mathbb{R}}$, $i = 1, \ldots, m$, there is $\partial f(x^1, \ldots, x^m) = \prod_{i=1}^m \partial f_i(x^i)$ for all $(x^1, \ldots, x^m) \in X^1 \times \ldots \times X^m$.

Theorem 2.3.14. *Let $f : X \to \overline{\mathbb{R}}$ and $x \in X$. The subdifferential $\partial f(x)$ is a (possibly empty) convex and closed set in X^*.*

Proof. If $f(x) = \pm\infty$ there is nothing to prove. Let be $f(x) \in \mathbb{R}$. By the Young-Fenchel inequality and Theorem 2.3.12 it follows that $x^* \in \partial f(x)$ if and only if $f(x) + f^*(x^*) \leq \langle x^*, x\rangle$. Therefore one can rewrite the subdifferential of the function f at x as the level set of the convex and lower semicontinuous function $x^* \mapsto -\langle x^*, x\rangle + f^*(x^*)$ at $-f(x)$, i.e. $\partial f(x) = \{x^* \in X^* : -\langle x^*, x\rangle + f^*(x^*) \leq -f(x)\}$. This guarantees the convexity and, via Theorem 2.2.9, the closedness of $\partial f(x)$. □

The aim of the next theorem is to present some connections between the subdifferentials of the functions f, $\bar{f}$ and $\overline{\text{co}}f$.

Theorem 2.3.15. *Let be $f : X \to \overline{\mathbb{R}}$ and $x \in X$ be such that $\partial f(x) \neq \emptyset$. Then it holds*

(a) $\overline{\text{co}}f(x) = \bar{f}(x) = f(x)$ and the functions f, $\bar{f}$ and $\overline{\text{co}}f$ are proper and f is lower semicontinuous at x;
(b) $\partial(\overline{\text{co}}f)(x) = \partial\bar{f}(x) = \partial f(x)$;

(c) $f^{**} = \overline{\text{co}} f$.

Proof. (a) Let $x^* \in \partial f(x)$ be arbitrarily taken and consider the function $h : X \to \mathbb{R}$, $h(y) = \langle x^*, y\rangle + f(x) - \langle x^*, x\rangle$, which is an affine minorant of f. Note that $f(x) \in \mathbb{R}$. Since h is also convex and lower semicontinuous it holds $h \leq \overline{\text{co}} f \leq \bar{f} \leq f$. Taking into consideration that $f(x) = h(x)$ we deduce that $f(x) = h(x) \leq \overline{\text{co}} f(x) \leq \bar{f}(x) \leq f(x)$ and the desired equalities follow. This also implies that the function f is lower semicontinuous at x and the properness of f, $\bar{f}$ and $\overline{\text{co}} f$ follows easily.

(b) If $x^* \in \partial f(x)$ then, by definition, $f(y) \geq f(x) + \langle x^*, y - x\rangle$ for all $y \in X$. As $y \mapsto \langle x^*, y - x\rangle + f(x)$ is a convex and lower semicontinuous function which is everywhere less than or equal to f, using (a) we get $\overline{\text{co}} f(y) \geq \overline{\text{co}} f(x) + \langle x^*, y - x\rangle$ for all $y \in X$. Thus $x^* \in \partial(\overline{\text{co}} f)(x)$ and the inclusion $\partial f(x) \subseteq \partial(\overline{\text{co}} f)(x)$ follows. Assume now that $x^* \in \partial(\overline{\text{co}} f)(x)$. Because of (a), for all $y \in X$ we have $f(y) - f(x) \geq \overline{\text{co}} f(y) - \overline{\text{co}} f(x) \geq \langle x^*, y - x\rangle$, i.e. $x^* \in \partial f(x)$. Therefore $\partial(\overline{\text{co}} f)(x) \subseteq \partial f(x)$ and we actually have $\partial f(x) = \partial(\overline{\text{co}} f)(x)$. Following the same idea one can also prove that $\partial f(x) = \partial \bar{f}(x)$.

(c) The assertion follows from (a) and Corollary 2.3.7. □

Theorem 2.3.16. *Let be $f : X \to \overline{\mathbb{R}}$ and $x \in X$.*

(a) If $\partial f(x) \neq \emptyset$, then $f(x) = f^{**}(x)$.
(b) If $f(x) = f^{**}(x)$, then $\partial f(x) = \partial f^{**}(x)$.

Proof. (a) The statement follows directly from Theorem 2.3.15$(a), (c)$.

(b) If $f(x) = f^{**}(x) = \pm\infty$, then by convention we have $\partial f(x) = \partial f^{**}(x) = \emptyset$. Otherwise, Theorem 2.3.12 and Remark 2.3.4 allow to conclude that $x^* \in \partial f(x) \Leftrightarrow f^*(x^*) = -f(x) + \langle x^*, x\rangle \Leftrightarrow f^{***}(x^*) = -f^{**}(x) + \langle x^*, x\rangle \Leftrightarrow x^* \in \partial f^{**}(x)$. □

Our next aim is to point out that the calculation rules which are available for the classical differential can be applied in general only partially to the subdifferential. Using Definition 2.3.2 it is easy to prove that for a given function $f : X \to \overline{\mathbb{R}}$ and $x \in X$ it holds

$$\partial(\lambda f)(x) = \lambda \partial f(x) \text{ for all } \lambda > 0.$$

Coming now to the sum, for some given arbitrary proper functions $f_i : X \to \overline{\mathbb{R}}$, $i = 1, \ldots, m$, one can only prove in general that for $x \in X$ it holds

$$\sum_{i=1}^{m} \partial f_i(x) \subseteq \partial\left(\sum_{i=1}^{m} f_i\right)(x). \tag{2.8}$$

We refer the reader to section 3.5 for sufficient conditions which guarantee, when the functions f_i, $i = 1, \ldots, m$, are convex, equality in (2.8).

The next result displays some connections between the subdifferential of a given function f and the one of its conjugate.

Theorem 2.3.17. *Let be $f : X \to \overline{\mathbb{R}}$ and $x \in X$.*

(a) If $x^* \in \partial f(x)$, then $x \in \partial f^*(x^*)$.
(b) If $f(x) = f^{**}(x)$, then $x^* \in \partial f(x)$ if and only if $x \in \partial f^*(x^*)$.
(c) If f is proper, convex and lower semicontinuous, then $x^* \in \partial f(x)$ if and only if $x \in \partial f^*(x^*)$.

Proof. (*a*) Since $x^* \in \partial f(x)$, according to Theorem 2.3.12 we have $f(x) + f^*(x^*) = \langle x^*, x\rangle$. But $f^{**}(x) \leq f(x)$, by Lemma 2.3.3, and thus $f^{**}(x) + f^*(x^*) \leq \langle x^*, x\rangle$. As the reverse inequality is always fulfilled, using once more Theorem 2.3.12, we get $x \in \partial f^*(x^*)$.

(*b*) Because of (*a*) only the sufficiency must be proven. For any $x \in \partial f^*(x^*)$, again by Theorem 2.3.12, it holds $\langle x^*, x\rangle = f^*(x^*) + f^{**}(x) = f^*(x^*) + f(x)$ and therefore $x^* \in \partial f(x)$.

(*c*) Theorem 2.3.5 yields $f = f^{**}$ and the equivalence follows from (*b*). □

A classical assertion on the existence of a subgradient is given in the following statement (cf. [67]).

Theorem 2.3.18. *Let the convex function $f : X \to \overline{\mathbb{R}}$ be finite and continuous at some point $x \in X$. Then $\partial f(x) \neq \emptyset$, i.e. f is subdifferentiable at x.*

Theorem 2.3.18 follows easily as a consequence of the Fenchel duality statement Theorem 3.2.6, which we give in the next chapter. For further results concerning subdifferential calculus we refer to section 3.5 and the book [207].

We conclude this subsection by resuming the relations between the subdifferentiability and the rather classical notion of Gâteaux differentiability accompanied by some further properties of the Gâteaux differential.

Definition 2.3.3. *Let $f : X \to \overline{\mathbb{R}}$ be a proper function and $x \in \operatorname{dom} f$. If the limit*

$$\lim_{t \downarrow 0} \frac{f(x + ty) - f(x)}{t}$$

exists we call it the directional derivative of f at x in the direction $y \in X$ and we denote it by $f'(x; y)$. If there exists an $x^* \in X^*$ such that $f'(x; y) = \langle x^*, y\rangle$ for all $y \in X$, then f is said to be Gâteaux differentiable at x, x^* is called the Gâteaux differential of f at x and it is denoted by $\nabla f(x)$, i.e. $f'(x; y) = \langle \nabla f(x), y\rangle$ for all $y \in X$.

We need to note that if f is proper and convex and $x \in \operatorname{dom} f$ then $f'(x; y)$ exists for all $y \in X$ (cf. [207, Theorem 2.1.12]). If, additionally, f is continuous at $x \in \operatorname{dom} f$ then for all $y \in X$ there is $f'(x; y) = \max\{\langle x^*, y\rangle : x^* \in \partial f(x)\}$. For convex functions the Gâteaux differentiability and the uniqueness of the subgradient are closely related, as stated below.

Proposition 2.3.19. *Let $f : X \to \overline{\mathbb{R}}$ be a proper and convex function and $x \in \operatorname{dom} f$.*

(a) If $x \in \operatorname{core}(\operatorname{dom} f)$ and f is Gâteaux differentiable at x, then f is subdifferentiable at x and $\partial f(x) = \{\nabla f(x)\}$.
(b) If f is continuous at x and its subdifferential $\partial f(x)$ is a singleton, then f is Gâteaux differentiable at x and $\partial f(x) = \{\nabla f(x)\}$.

In the next results the convexity of a Gâteaux differentiable function is characterized.

Proposition 2.3.20. *Let $U \subseteq X$ be a nonempty, open and convex set and $f : U \to \mathbb{R}$ a Gâteaux differentiable function on U. Then the function f is convex on U if and only if $f(y) \geq f(x) + \langle \nabla f(x), y - x\rangle$ for all $x, y \in U$. This is further equivalent to $\langle \nabla f(y) - \nabla f(x), y - x\rangle \geq 0$ for all $x, y \in U$.*

2.4 Minimal and maximal elements of sets

It is characteristic for *vector optimization problems* that, different to *scalar programming problems*, more than one conflicting objectives have to be taken into consideration. Thus, the objective values may be considered as vectors in a finite or even infinite dimensional vector space. As an example let us mention the *Markowitz portfolio optimization problem* which aims at finding an optimal portfolio of several risky securities, e.g. stocks and shares of companies. There are two reasonable objectives, the *expected return* which has to be maximized, and the *risk* (measured via the variance of the expected return or any other risk measure) that has to be minimized. These both objectives are conflicting because in general the risk grows if the expected return is increasing. This conflict must be reflected by a corresponding partial ordering relation in the two dimensional objective space. Such an ordering relation allows to compare different vector objective values in, at least, a partial sense. Based on the considered partial ordering one can define distinct types of solutions in connection to a vector optimization problem. Partial orderings in the sense considered in this book are defined by convex cones as we have already done in section 2.1. The basic terminology for such solutions is that of *efficiency*, i.e. we consider different types of so-called efficient solutions. For the first time efficient solutions have been considered by Edgeworth in [60] and Pareto in [148].

In this section we present different notions of minimality (maximality) for sets in vector spaces. In the next section these notions will be employed when introducing different efficiency solution concepts for vector optimization problems.

2.4.1 Minimality

Unless otherwise mentioned, in the following we consider V to be a vector space partially ordered by a convex cone $K \subseteq V$.

First of all let us define the usual notion of minimality for a nonempty set $M \subseteq V$ with respect to the partial ordering "$\leqq_K$" induced by K. Initially, we confine ourselves to the case where the ordering cone K is pointed, i.e. $l(K) = \{0\}$, in which case "$\leqq_K$" is antisymmetric, since this is the situation mostly encountered within this book and also in the majority of practical applications of vector optimization.

Definition 2.4.1. *An element $\bar{v} \in M$ is said to be a minimal element of M (regarding the partial ordering induced by K) if there is no $v \in M$ satisfying $v \leq_K \bar{v}$. The set of all minimal elements of M is denoted by* $\text{Min}(M, K)$ *and it is called the minimal set of M (regarding the partial ordering defined by K).*

Remark 2.4.1. There are several obviously equivalent formulations for an element $\bar{v} \in M$ to be a minimal element of M. We list some of them in the following:

(i) there is no $v \in M$ such that $\bar{v} - v \in K \backslash \{0\}$;
(ii) from $v \leqq_K \bar{v}$, $v \in M$, follows $v = \bar{v}$;
(iii) from $v \leqq_K \bar{v}$, $v \in M$, follows $v \geqq_K \bar{v}$;
(iv) $(\bar{v} - K) \cap M = \{\bar{v}\}$;
(v) $(M - \bar{v}) \cap (-K) = \{0\}$;
(vi) for all $v \in M$ there is $v \nleq_K \bar{v}$.

We would like to underline that for the equivalence $(ii) \Leftrightarrow (iii)$ the pointedness of K is indispensable.

Example 2.4.1. An important case which occurs in practice is when $V = \mathbb{R}^k$ and $K = \mathbb{R}^k_+$. Let $M \subseteq \mathbb{R}^k$. Then $\bar{v} = (\bar{v}_1, \ldots, \bar{v}_k)^T \in M$ is a minimal element of M if there is no $v = (v_1, \ldots, v_k)^T \in M$ such that $v \neq \bar{v}$ and $v_i \leq \bar{v}_i$ for all $i = 1, \ldots, k$, i.e. there is no $v = (v_1, \ldots, v_k)^T \in M$ fulfilling $v_i \leq \bar{v}_i$ for all $i = 1, \ldots, k$ and $v_j < \bar{v}_j$ for at least one $j \in \{1, \ldots, k\}$.

In an analogous way one can define the notion of *maximal element* of a set M.

Definition 2.4.2. *An element $\bar{v} \in M$ is said to be a maximal element of M (regarding the partial ordering induced by K) if there is no $v \in M$ satisfying $v \geq_K \bar{v}$. The set of all maximal elements of M is denoted by* $\text{Max}(M, K)$ *and it is called the maximal set of M (regarding the partial ordering defined by K).*

Remark 2.4.2. As in Remark 2.4.1 one can give the following equivalent formulations for the maximality of an element $\bar{v} \in M$ in M:

(i) there is no $v \in M$ such that $v - \bar{v} \in K \backslash \{0\}$;
(ii) from $v \geqq_K \bar{v}$, $v \in M$, follows $v = \bar{v}$;
(iii) from $v \geqq_K \bar{v}$, $v \in M$, follows $v \leqq_K \bar{v}$;
(iv) $(\bar{v} + K) \cap M = \{\bar{v}\}$;

(v) $(M - \bar{v}) \cap K = \{0\}$;
(vi) for all $v \in M$ there is $v \not\geqq_K \bar{v}$.

Remark 2.4.3. The problem of finding the maximal elements of the set M regarding the cone K may be reformulated as the problem of finding the minimal elements of the set $(-M)$ regarding K or, equivalently, as the problem of finding the minimal elements of the set M regarding the partial ordering induced by the cone $(-K)$. It holds $\operatorname{Max}(M, K) = \operatorname{Min}(M, -K) = -\operatorname{Min}(-M, K)$.

Although we mostly confine ourselves within this book to the most important framework of partial orderings induced by pointed convex cones, for the sake of completeness we present also the definition of minimality regarding a partial ordering induced by a convex but not pointed cone K. As noted in subsection 2.1.1, in this situation $l(K) = K \cap (-K)$ is a linear subspace of V not equal to $\{0\}$. If this situation occurs, then Definition 2.4.1 is not always suitable for defining minimal elements, and this because it may happen to have a $v \in M$, $v \neq \bar{v}$, such that $v \leqq_K \bar{v} \leqq_K v$. More precisely, if $\bar{v} - v \in l(K) \subseteq K$ then $v - \bar{v} \in l(K) \subseteq K$, too, and now it is clear that Definition 2.4.1 cannot be used if the ordering cone K is not pointed. This situation can be avoided if instead of Definition 2.4.1 one uses the following definition due to Borwein [19] (see also Remark 2.4.1(*iii*)).

Definition 2.4.3. *Let $K \subseteq V$ be an arbitrary ordering cone. An element $\bar{v} \in M$ is said to be a minimal element of M (regarding the partial ordering induced by K), if from $v \leqq_K \bar{v}$, $v \in M$, follows $v \geqq_K \bar{v}$.*

From this definition immediately follows that if $\bar{v} \in M$ is a minimal element of M then any $\tilde{v} \in M$ such that $\tilde{v} \leqq_K \bar{v}$ is also a minimal element of M. To see this take an arbitrary $v \in M$ such that $v \leqq_K \tilde{v}$. Since $\bar{v} \in \operatorname{Min}(M, K)$, it holds $v \geqq_K \bar{v} \geqq_K \tilde{v}$ and so $\tilde{v} \in \operatorname{Min}(M, K)$.

We observe further that $\bar{v}$ being minimal means that for all $v \in M$ fulfilling $v \leqq_K \bar{v}$ it is binding to have $\bar{v} - v \in l(K)$. If K is a pointed cone then $l(K) = \{0\}$ and in this case we have $\bar{v} = v$. Therefore Definition 2.4.3 applies to pointed cones K, too, while Definition 2.4.1 can be seen as a particular case of it.

Next we give some equivalent formulations to the notion of minimality in case the cone K is not assumed to be pointed. More precisely, $\bar{v} \in M$ is a minimal element of M if and only if one of the following conditions is fulfilled:

(i) there is no $v \in M$ such that $\bar{v} - v \in K \setminus l(K)$;
(ii) $(\bar{v} - K) \cap M \subseteq \bar{v} + K$;
(iii) $(-K) \cap (M - \bar{v}) \subseteq K$.

The *maximal elements* of the set M (in case the cone K is not assumed pointed) can be defined following the same idea as in Definition 2.4.3. Analogously to Remark 2.4.3 one has $\operatorname{Max}(M, K) = \operatorname{Min}(M, -K) = -\operatorname{Min}(-M, K)$.

The next result describes the relation between the minimal elements of the sets M and $M + K$.

Lemma 2.4.1. *(a) It holds* $\mathrm{Min}(M, K) \subseteq \mathrm{Min}(M + K, K)$.
(b) If K *is pointed, then* $\mathrm{Min}(M, K) = \mathrm{Min}(M + K, K)$.

Proof. (a) Take an arbitrary $\bar{v} \in \mathrm{Min}(M, K)$. By definition $\bar{v} \in M \subseteq M + K$. Let us prove now that for $v \in M + K$ such that $v \leqq_K \bar{v}$ the relation $\bar{v} \leqq_K v$ holds, too. Since $v \in M + K$ we have $v = \tilde{v} + k$ for some $\tilde{v} \in M$ and $k \in K$. Obviously $\tilde{v} = v - k \leqq_K \bar{v} - k \leqq_K \bar{v}$. But the minimality of $\bar{v}$ secures $\bar{v} \leqq_K \tilde{v}$ and, since $\tilde{v} = v - k \leqq_K v$, the desired conclusion follows.

(b) Assuming now K pointed, let $\bar{v} \in \mathrm{Min}(M+K, K)$. We have $\bar{v} \in M+K$ and we show that actually $\bar{v} \in M$. Assuming the contrary implies $\bar{v} = \tilde{v} + k$ with $\tilde{v} \in M$ and $k \in K \backslash \{0\}$. This yields $\tilde{v} \leq_K \bar{v}$ and as $\tilde{v} \in M+K$, one would get a contradiction to the minimality of $\bar{v}$ in $M + K$. Following a similar reasoning one can prove that in fact $\bar{v} \in \mathrm{Min}(M, K)$. Now (a) yields the desired conclusion. □

The next minimality notion we introduce is the so-called *strong minimality*. We work in the same setting, with V a vector space partially ordered by the (not necessarily pointed) convex cone K and M a nonempty subset of V.

Definition 2.4.4. *An element* $\bar{v} \in M$ *is said to be a strongly minimal element of* M *(regarding the partial ordering induced by* K*) if* $\bar{v} \leqq_K v$ *for all* $v \in M$*, i.e.* $M \subseteq \bar{v} + K$.

For vector optimization this definition is of secondary importance because in the most practical cases strongly minimal elements do not exist. If we consider the classical situation when $V = \mathbb{R}^k$ and $K = \mathbb{R}^k_+$, then the strong minimality of $\bar{v} \in M \subseteq \mathbb{R}^k$ means $\bar{v}_i \leq v_i$, $i = 1, \ldots, k$, for all $v = (v_1, \ldots, v_k)^T \in M$. Thus, in case of a multiobjective optimization problem, this must imply that all the k components of its objective function attain their minima at the same point, i.e. the objectives are not conflicting as it is typical for vector optimization.

Obviously, every strongly minimal element is minimal. A *strongly maximal element* $\bar{v} \in M$ is defined in analogous manner, namely one must have $\bar{v} \geqq_K v$ for all $v \in M$.

2.4.2 Weak minimality

Although from the practical point of view not so important as the minimal elements, the so-called *weakly minimal elements* of a given set are of theoretical interest, one of the arguments sustaining this assertion being that they allow a complete characterization by linear scalarization in the convex case, which is not always possible with minimal elements. We consider in this subsection V to be a vector space partially ordered by the convex cone $K \subseteq V$ fulfilling $\mathrm{core}(K) \neq \emptyset$ and $M \subseteq V$ being a nonempty set.

Definition 2.4.5. *An element $\bar{v} \in M$ is said to be a weakly minimal element of M (regarding the partial ordering induced by K) if $(\bar{v} - \operatorname{core}(K)) \cap M = \emptyset$. The set of all weakly minimal elements of M is denoted by* $\operatorname{WMin}(M, K)$ *and it is called the weakly minimal set of M (regarding the partial ordering induced by K).*

The relation $(\bar{v} - \operatorname{core}(K)) \cap M = \emptyset$ in Definition 2.4.5 is obviously equivalent to $(M - \bar{v}) \cap (-\operatorname{core}(K)) = \emptyset$. From here follows that $\operatorname{WMin}(M, V) = \emptyset$. Whenever the cone K is nontrivial one may also notice that if we consider as ordering cone $\widehat{K} = \operatorname{core}(K) \cup \{0\}$, then $\bar{v} \in \operatorname{WMin}(M, K)$ if and only if $(\bar{v} - \widehat{K}) \cap M = \{\bar{v}\}$, or, equivalently, $\bar{v} \in \operatorname{Min}(M, \widehat{K})$ (see Remark 2.4.1(iv)). Of course, if $K = \widehat{K}$, then the minimal and weakly minimal elements of M regarding the partial ordering induced by K coincide. This, however, is not the case in general.

If $K \neq V$, any minimal element of M is also weakly minimal, since $(\bar{v} - K) \cap M \subseteq \bar{v} + K$ implies $(\bar{v} - \operatorname{core}(K)) \cap M = \emptyset$. Indeed, notice that $(\bar{v} - \operatorname{core}(K)) \cap (\bar{v} + K) = \emptyset$, as in this situation $-\operatorname{core}(K) \cap K = \emptyset$. This result is summarized in the following statement.

Proposition 2.4.2. *If $K \neq V$, then* $\operatorname{Min}(M, K) \subseteq \operatorname{WMin}(M, K)$.

Next we provide a result, similar to Lemma 2.4.1, for the weakly minimal elements of a set $M \subseteq V$, the proof of which being relinquished to the reader (see, for instance, [104, Lemma 4.13]).

Lemma 2.4.3. *It holds*

(a) $\operatorname{WMin}(M, K) \subseteq \operatorname{WMin}(M + K, K)$;
(b) $\operatorname{WMin}(M + K, K) \cap M \subseteq \operatorname{WMin}(M, K)$.

Remark 2.4.4. When V is taken to be a topological vector space, in the above assertions the algebraic interior $\operatorname{core}(K)$ can be replaced with the topological interior $\operatorname{int}(K)$ when the latter is nonempty.

Weakly maximal elements may be defined analogously, namely an element $\bar{v} \in M$ is called a weakly maximal element of M (regarding the partial ordering induced by K) if $(\bar{v} + \operatorname{core}(K)) \cap M = \emptyset$. The set of all weakly maximal elements of M is denoted by $\operatorname{WMax}(M, K)$ and it is called the *weakly maximal set* of M. Also here it holds $\operatorname{WMax}(M, K) = \operatorname{WMin}(M, -K) = -\operatorname{WMin}(-M, K)$.

Consequently, one can formulate obvious variants of Lemma 2.4.1, Proposition 2.4.2 and Lemma 2.4.3 for maximal and weakly maximal elements of the set M, respectively.

2.4.3 Proper minimality

There is another very important notion of minimality subsumed under the category of *properly minimal elements.* Properly minimal elements turn out to be minimal elements with additional properties. There are a lot of different kinds of properly minimal elements. In the following we present an overview of their distinct definitions and establish some relations between them. The majority of these notions have been introduced in connection to some vector optimization problems under the name *proper efficiency*. We introduce them in this subsection as proper minimality notions for a given set by extending to this general situation the corresponding notions for vector optimization problems. To this end one has only to take in this situation M to be the image set of the feasible set through the objective function. Let us mention also that here we work only with properly minimal elements, for considering properly maximal elements one needs only replace the cone K by $-K$.

The first notion we present concerns a nonempty set $M \subseteq \mathbb{R}^k$ when the space $\mathbb{R}^k$ is partially ordered by the cone $K = \mathbb{R}^k_+$, being inspired by Geoffrion's paper [71].

Definition 2.4.6. *An element $\bar{v} = (\bar{v}_1, \ldots, \bar{v}_k)^T \in M$ is said to be a properly minimal element of M in the sense of Geoffrion if $\bar{v} \in \text{Min}(M, \mathbb{R}^k_+)$ and if there exists a real number $N > 0$ such that for every $i \in \{1, \ldots, k\}$ and every $v = (v_1, \ldots, v_k)^T \in M$ satisfying $v_i < \bar{v}_i$ there exists at least one $j \in \{1, \ldots, k\}$ such that $\bar{v}_j < v_j$ and*

$$\frac{\bar{v}_i - v_i}{v_j - \bar{v}_j} \leq N.$$

The set of all properly minimal elements of M in the sense of Geoffrion is denoted by $\text{PMin}_{Ge}(M, \mathbb{R}^k_+)$.

The definition above can be interpreted as follows: a decrease in one component relative to $\bar{v}$ entails an increase in at least another component such that the ratio of the absolute values of those differences is bounded. In multiobjective optimization this means that the trade-offs among the different components of the vector objective function are bounded. In economics unbounded trade-offs are mostly undesirable. However, not only with respect to the practical applications but also from the theoretical point of view, properly minimal (or maximal) elements have nice and beneficial properties as we will see in the next subsection.

One can establish an analogous lemma to Lemma 2.4.1 and Lemma 2.4.3 by replacing M with $M + \mathbb{R}^k_+$.

Lemma 2.4.4. *There is* $\text{PMin}_{Ge}(M + \mathbb{R}^k_+, \mathbb{R}^k_+) = \text{PMin}_{Ge}(M, \mathbb{R}^k_+)$.

Proof. Take an arbitrary $\bar{v} \in \text{PMin}_{Ge}(M + \mathbb{R}^k_+, \mathbb{R}^k_+)$. By definition $\bar{v} \in \text{Min}(M + \mathbb{R}^k_+, \mathbb{R}^k_+)$ and, by Lemma 2.4.1, it holds $\bar{v} \in \text{Min}(M, \mathbb{R}^k_+)$. This implies, in particular, that $\bar{v} \in M$. Because $M \subseteq M + \mathbb{R}^k_+$, Definition 2.4.6 applies to any $v \in M$, i.e. $\bar{v} \in \text{PMin}_{Ge}(M, \mathbb{R}^k_+)$.

Now let $\bar{v} \in \text{PMin}_{Ge}(M, \mathbb{R}^k_+)$ and $N > 0$ be the positive constant provided by Definition 2.4.6. Clearly $\bar{v} \in M + \mathbb{R}^k_+$ and by Lemma 2.4.1, we get $\bar{v} \in \text{Min}(M, \mathbb{R}^k_+) = \text{Min}(M + \mathbb{R}^k_+, \mathbb{R}^k_+)$. Take an arbitrary $v = \tilde{v} + h \in M + \mathbb{R}^k_+$

where $\tilde{v} \in M$ and $h \in \mathbb{R}^k_+$ and $i \in \{1, \ldots, k\}$ such that $v_i = \tilde{v}_i + h_i < \bar{v}_i$. Obviously $\tilde{v}_i < \bar{v}_i$ and, according to Definition 2.4.6, there exists at least one $j \in \{1, \ldots, k\}$ such that $\bar{v}_j < \tilde{v}_j$ and

$$\frac{\bar{v}_i - \tilde{v}_i}{\tilde{v}_j - \bar{v}_j} \leq N.$$

But this implies $\bar{v}_j < \tilde{v}_j + h_j = v_j$ and

$$\frac{\bar{v}_i - v_i}{v_j - \bar{v}_j} = \frac{\bar{v}_i - \tilde{v}_i - h_i}{\tilde{v}_j + h_j - \bar{v}_j} \leq \frac{\bar{v}_i - \tilde{v}_i}{\tilde{v}_j - \bar{v}_j} \leq N.$$

Consequently, we get $\bar{v} \in \mathrm{PMin}_{Ge}(M + \mathbb{R}^k_+, \mathbb{R}^k_+)$. □

Already ten years earlier Hurwicz [93] has introduced a notion of proper efficiency for vector optimization problems which has been generalized in [83] to sets in partially ordered topological vector spaces.

Further we assume that V is a topological vector space partially ordered by the pointed convex cone K and $M \subseteq V$ is an arbitrary nonempty set.

Definition 2.4.7. *An element $\bar{v} \in M$ is said to be a properly minimal element of M in the sense of Hurwicz if* $\mathrm{cl}(\mathrm{coneco}((M - \bar{v}) \cup K)) \cap (-K) = \{0\}$. *The set of all properly minimal elements of M in the sense of Hurwicz is denoted by* $\mathrm{PMin}_{Hu}(M, K)$.

This definition seems to be in a certain manner natural if it is compared with the definition of minimality of $\bar{v} \in M$ in the equivalent formulation given in Remark 2.4.1(v) which states that $(M - \bar{v}) \cap (-K) = \{0\}$. Because $M - \bar{v} \subseteq \mathrm{cl}(\mathrm{coneco}((M - \bar{v}) \cup K))$ it is clear that $\mathrm{PMin}_{Hu}(M, K) \subseteq \mathrm{Min}(M, K)$.

Lemma 2.4.5. *There is* $\mathrm{PMin}_{Hu}(M + K, K) = \mathrm{PMin}_{Hu}(M, K)$.

Proof. Noting that $\mathrm{coneco}((M - \bar{v}) \cup K) = \mathrm{coneco}((M + K - \bar{v}) \cup K)$, we obtain the conclusion. □

Geoffrion's definition of proper efficiency is very illustrative concerning economical and geometrical aspects. But its drawback is the restriction to the ordering cone $K = \mathbb{R}^k_+$. To overcome this disadvantage Borwein proposed in [17] a notion of proper efficiency for vector maximization problems given in Hausdorff locally convex spaces partially ordered by a pointed convex closed cone which generalizes Geoffrion's definition. We employ Borwein's definition to sets and use the notion proper minimality instead of proper efficiency in accordance with our general context.

For the remaining part of this subsection we take V to be a Hausdorff locally convex space partially ordered by the pointed convex cone K and $M \subseteq V$ an arbitrary nonempty set.

Definition 2.4.8. *An element $\bar{v} \in M$ is said to be a properly minimal element of M in the sense of Borwein if* $\operatorname{cl}(T(M+K,\bar{v})) \cap (-K) = \{0\}$. *The set of all properly minimal elements of M in the sense of Borwein is denoted by* $\operatorname{PMin}_{Bo}(M,K)$.

Remark 2.4.5. The proper minimality in the sense of Borwein can be equivalently written as $0 \in \operatorname{Min}(\operatorname{cl}(T(M+K,\bar{v})),K)$. Observing that $T(M+K,\bar{v}) = T(M+K-\bar{v},0)$ one can see the affinity of this kind of proper minimality to the notion of minimality, via Remark 2.4.1(v) and Lemma 2.4.1. The element $\bar{v} \in M$ is minimal if and only if $(M+K-\bar{v}) \cap (-K) = \{\bar{v}\}$. Thus Borwein's definition of proper minimality is nothing else than additionally demanding $\operatorname{cl}(T(M+K-\bar{v},0)) \cap (-K) = \{0\}$. Moreover, if V is metrizable, then the tangent cone is closed and in this situation one may omit the closure operation within Definition 2.4.8.

Remark 2.4.6. (a) Let us mention here that in the original definition in [17] the ordering cone was not explicitly assumed to be pointed. But this has to be assumed, otherwise $\operatorname{PMin}_{Bo}(M,K)$ is the empty set. Indeed, if K is not pointed, then let $v \in K \cap (-K)$, $v \neq 0$, be arbitrarily chosen and $\bar{v} \in \operatorname{PMin}_{Bo}(M,K)$. Setting $v_l = \bar{v} + (1/l)v \in M+K$ for $l \geq 1$, we get $\lim_{l\to+\infty} v_l = \bar{v}$ and $\lim_{l\to\infty} l(v_l - \bar{v}) = v$. But this means nothing else than $v \in T(M+K,\bar{v})$. Therefore $\operatorname{cl}(T(M+K,\bar{v})) \cap (-K) \neq \{0\}$ and this means that in this situation $\operatorname{PMin}_{Bo}(M,K) = \emptyset$.

(b) A second observation in this context is that in the original definition for a vector maximization problem a properly efficient solution is, additionally, assumed to be an efficient solution. This would mean to require in our definition that $\bar{v} \in \operatorname{Min}(M,K)$. But this hypothesis is superfluous and turns out to be a consequence of the condition $\operatorname{cl}(T(M+K,\bar{v})) \cap (-K) = \{0\}$. The conclusion is obvious if $K = \{0\}$. Assume that $K \neq \{0\}$ and that for a $\bar{v} \in \operatorname{PMin}_{Bo}(M,K)$ it holds $\bar{v} \notin \operatorname{Min}(M,K)$. Then there exists $v \in M$ such that $\bar{v} - v \in K \backslash \{0\}$. We show that $v - \bar{v} \in T(M+K,\bar{v})$. Setting $v_l = \bar{v} + (1/l)(v-\bar{v})$, $l \geq 1$, we can easily see that $v_l = v + ((l-1)/l)(\bar{v} - v) \in M+K$ for $l \geq 1$. Even more, as $\lim_{l\to+\infty} v_l = \bar{v}$ and $\lim_{l\to+\infty} l(v_l - \bar{v}) = v - \bar{v}$, it follows $v - \bar{v} \in T(M+K,\bar{v})$. Finally, since $0 \neq v - \bar{v} \in T(M+K,\bar{v}) \cap (-K) \subseteq \operatorname{cl}(T(M+K,\bar{v})) \cap (-K)$, the equality $\operatorname{cl}(T(M+K,\bar{v})) \cap (-K) = \{0\}$ fails, and this contradicts the assumption $\bar{v} \in \operatorname{PMin}_{Bo}(M,K)$.

Since by the convexity of K it holds $(M+K)+K = M+K$, the following result follows easily via Lemma 2.4.1 and Remark 2.4.6(b).

Lemma 2.4.6. *There is* $\operatorname{PMin}_{Bo}(M+K,K) = \operatorname{PMin}_{Bo}(M,K)$.

Regarding the proper minimality in the sense of Geoffrion and in the sense of Borwein in case $V = \mathbb{R}^k$ and $K = \mathbb{R}^k_+$, we have that for a nonempty set $M \subseteq \mathbb{R}^k$ it holds $\operatorname{PMin}_{Ge}(M,\mathbb{R}^k_+) \subseteq \operatorname{PMin}_{Bo}(M,\mathbb{R}^k_+)$, which can be proven similarly as in [17, Proposition 1]. The following result, giving a sufficient condition for the coincidence of both notions, can also be proven similarly as for a corresponding assertion in [17] regarding a vector maximization problem under convexity assumptions.

Proposition 2.4.7. *If $M \subseteq \mathbb{R}^k$ is nonempty and $M + \mathbb{R}^k_+$ is convex, then* $\mathrm{PMin}_{Ge}(M, \mathbb{R}^k_+) = \mathrm{PMin}_{Bo}(M, \mathbb{R}^k_+)$.

The next proper minimality notion we consider here originates from Benson's paper [15] and it was introduced in order to extend Geoffrion's proper minimality.

Definition 2.4.9. *An element $\bar{v} \in M$ is said to be a properly minimal element of M in the sense of Benson if* $\mathrm{cl}(\mathrm{cone}(M + K - \bar{v})) \cap (-K) = \{0\}$. *The set of all properly minimal elements of M in the sense of Benson is denoted by* $\mathrm{PMin}_{Be}(M, K)$.

Remark 2.4.7. In [15] the notion introduced above was given as proper efficiency for vector maximum problems in finite dimensional spaces, with the efficiency of the elements in discussion additionally assumed. This means in our situation to supplementary impose the condition $\bar{v} \in \mathrm{Min}(M, K)$. But this is superfluous since $M - \bar{v} \subseteq \mathrm{cl}(\mathrm{cone}(M + K - \bar{v}))$ implies $\bar{v} \in \mathrm{Min}(M, K)$ if $\mathrm{cl}(\mathrm{cone}(M + K - \bar{v})) \cap (-K) = \{0\}$, i.e. $(M - \bar{v}) \cap (-K) = \{0\}$, too.

The next result is a consequence of the convexity of K along with Lemma 2.4.1 and Remark 2.4.7.

Lemma 2.4.8. *There is* $\mathrm{PMin}_{Be}(M + K, K) = \mathrm{PMin}_{Be}(M, K)$.

As mentioned above, for $V = \mathbb{R}^k$ and $K = \mathbb{R}^k_+$, when $M \subseteq \mathbb{R}^k$ is an arbitrary nonempty set, it holds (cf. [15]) $\mathrm{PMin}_{Ge}(M, \mathbb{R}^k_+) = \mathrm{PMin}_{Be}(M, \mathbb{R}^k_+)$. By taking into consideration the way Borwein's and Benson's proper minimalities are defined, one has that $\mathrm{PMin}_{Be}(M, K) \subseteq \mathrm{PMin}_{Bo}(M, K)$ is always fulfilled. Further, let us notice that for $\bar{v} \in M$ it holds $\mathrm{cone}(M + K - \bar{v}) \subseteq \mathrm{coneco}((M - \bar{v}) \cup K)$, the two sets being equal if $M + K$ is convex. Thus we have in general that $\mathrm{PMin}_{Hu}(M, K) \subseteq \mathrm{PMin}_{Be}(M, K)$, while when $M + K$ is convex it follows that $\mathrm{PMin}_{Hu}(M, K) = \mathrm{PMin}_{Be}(M, K) = \mathrm{PMin}_{Bo}(M, K)$.

In the following we introduce another proper minimality concept due to Borwein (cf. [18]), which is similar to Definition 2.4.9.

Definition 2.4.10. *An element $\bar{v} \in M$ is said to be a properly minimal element of M in the global sense of Borwein if* $\mathrm{cl}(\mathrm{cone}(M - \bar{v})) \cap (-K) = \{0\}$. *The set of all properly minimal elements of M in the global sense of Borwein is denoted by* $\mathrm{PMin}_{GBo}(M, K)$.

If for $\bar{v} \in M$ Definition 2.4.10 is satisfied, then also $(M - \bar{v}) \cap (-K) = \{0\}$ and, due to Remark 2.4.1(v), we have $\bar{v} \in \mathrm{Min}(M, K)$. Furthermore, it is always true that $\mathrm{PMin}_{Be}(M, K) = \mathrm{PMin}_{GBo}(M + K, K)$.

The next result relates $\mathrm{PMin}_{GBo}(M + K, K)$ to $\mathrm{PMin}_{GBo}(M, K)$.

Proposition 2.4.9. *There is* $\mathrm{PMin}_{GBo}(M+K,K) \subseteq \mathrm{PMin}_{GBo}(M,K)$.

Proof. Take an arbitrary $\bar{v} \in \mathrm{PMin}_{GBo}(M+K,K)$. Thus $\bar{v} \in M$ and $\mathrm{cl}(\mathrm{cone}(M+K-\bar{v})) \cap (-K) = \{0\}$. Since $M - \bar{v} \subseteq M + K - \bar{v}$ it follows that $\mathrm{cl}(\mathrm{cone}(M-\bar{v})) \cap (-K) \subseteq \mathrm{cl}(\mathrm{cone}(M+K-\bar{v})) \cap (-K) = \{0\}$ and, consequently, $\mathrm{cl}(\mathrm{cone}(M-\bar{v})) \cap (-K) = \{0\}$, which completes the proof. □

Obviously, we have that $\mathrm{PMin}_{Be}(M,K) \subseteq \mathrm{PMin}_{GBo}(M,K)$. On the other hand, no relation of inclusion between $\mathrm{PMin}_{Bo}(M,K)$ and $\mathrm{PMin}_{GBo}(M,K)$ can be given in general. Obviously, when $M+K$ is convex, then $\mathrm{PMin}_{Bo}(M,K) \subseteq \mathrm{PMin}_{GBo}(M,K)$.

A formally different minimality approach is the one introduced by Henig [88] and Lampe [122] by employing a nontrivial convex cone K' containing in its interior the given ordering cone K.

Definition 2.4.11. *An element $\bar{v} \in M$ is said to be a properly minimal element of M in the sense of Henig and Lampe if there exists a nontrivial convex cone $K' \subseteq X$ with $K \setminus \{0\} \subseteq \mathrm{int}(K')$ such that $(M-\bar{v}) \cap (-K') = \{0\}$. The set of all properly minimal elements of M in the sense of Henig and Lampe is denoted by* $\mathrm{PMin}_{He-La}(M,K)$.

If the cone K' is assumed also pointed, instead of $(M-\bar{v}) \cap (-K') = \{0\}$ one can write $\bar{v} \in \mathrm{Min}(M,K')$. It is an immediate consequence of this definition that $\bar{v} \in \mathrm{PMin}_{He-La}(M,K)$ implies $\bar{v} \in \mathrm{Min}(M,K)$.

Lemma 2.4.10. *There is* $\mathrm{PMin}_{He-La}(M+K,K) = \mathrm{PMin}_{He-La}(M,K)$.

Proof. Let $\bar{v} \in \mathrm{PMin}_{He-La}(M+K,K)$ be arbitrarily taken. Then one can easily show that $\bar{v} \in M$. Moreover, there exists a nontrivial convex cone K' such that $K \setminus \{0\} \subseteq \mathrm{int}(K')$ and $(M+K-\bar{v}) \cap (-K') = \{0\}$. Thus $(M-\bar{v}) \cap (-K') = \{0\}$ and so $\bar{v} \in \mathrm{PMin}_{He-La}(M,K)$.

Vice versa, take $\bar{v} \in \mathrm{PMin}_{He-La}(M,K)$. Then $\bar{v} \in M \subseteq M+K$ and there exists a nontrivial convex cone K' such that $K \setminus \{0\} \subseteq \mathrm{int}(K')$ and $(M-\bar{v}) \cap (-K') = \{0\}$. We prove that $(M+K-\bar{v}) \cap (-K') = \{0\}$. If we assume the contrary there would exist $v \in M$, $k \in K$ and $k' \in K' \setminus \{0\}$ such that $v + k - \bar{v} = -k'$. Then $v - \bar{v} = -(k+k') \in -(K + (K' \setminus \{0\})) \subseteq -(K' + (K' \setminus \{0\})) \subseteq -K'$ and so $k + k' = 0$. Consequently, $-k \in K' \cap (-\mathrm{int}(K'))$, which leads to a contradiction. Therefore $\bar{v} \in \mathrm{PMin}_{He-La}(M+K,K)$ and the proof is complete. □

The following statement reveals the relation between the proper minimality notions in the sense of Benson and in the sense of Henig and Lampe.

Proposition 2.4.11. *There is* $\mathrm{PMin}_{He-La}(M,K) \subseteq \mathrm{PMin}_{Be}(M,K)$.

Proof. If $K = \{0\}$ the inclusion follows automatically. Assume that $K \neq \{0\}$. Let $\bar{v} \in \mathrm{PMin}_{He-La}(M,K)$. Then $\bar{v} \in M$ and there exists a nontrivial convex cone K' such that $K \setminus \{0\} \subseteq \mathrm{int}(K')$ and $(M-\bar{v}) \cap (-K') = \{0\}$. We prove that $\text{cl}(\text{cone}(M+K-\bar{v}))\cap(-K)=\{0\}$. To this end we assume the contrary, namely that there exists a $k\in K\backslash\{0\}$ such that $-k\in\text{cl}(\text{cone}(M+K-\bar{v}))$. Thus $-k\in\text{int}(-K')$ and consequently there exist $\tilde{v}\in M$, $\tilde{k}\in K$ and $\tilde{\lambda}\geq 0$ such that $\tilde{\lambda}(\tilde{v}+\tilde{k}-\bar{v})\in\text{int}(-K')$. Obviously, $\tilde{\lambda}\neq 0$ and so $\tilde{v}-\bar{v}\in -\text{int}(K')-K'\subseteq -\text{int}(K')$. This yields that $\tilde{v}-\bar{v}\neq 0$, contradicting the fact that $(M-\bar{v})\cap(-K')=\{0\}$. □

The following proper minimality notion, based on linear scalarization, allows the treatment of minimal elements as solutions of scalar optimization problems.

Definition 2.4.12. *An element $\bar{v}\in M$ is said to be a properly minimal element of M in the sense of linear scalarization if there exists a $v^*\in K^{*0}$ such that $\langle v^*,\bar{v}\rangle\leq\langle v^*,v\rangle$ for all $v\in M$. The set of properly minimal elements of M in the sense of linear scalarization is denoted by $\text{PMin}_{LSc}(M,K)$.*

The properly minimal elements of M in the sense of linear scalarization are also minimal, as the next result shows.

Proposition 2.4.12. *There is $\text{PMin}_{LSc}(M,K)\subseteq\text{Min}(M,K)$.*

Proof. Take $\bar{v}\in\text{PMin}_{LSc}(M,K)$ with the corresponding $v^*\in K^{*0}$. If $\bar{v}\notin\text{Min}(M,K)$, then there exists $v\in M$ satisfying $v\leq_K\bar{v}$. As $v^*\in K^{*0}$ and $\bar{v}-v\in K\backslash\{0\}$, there is $\langle v^*,\bar{v}-v\rangle>0$, contradicting Definition 2.4.12. □

Simple examples illustrating that the opposite inclusion is in general not fulfilled can be found in [80]. Without any additional assumption on M, the properly minimal elements in the sense of linear scalarization of M and $M+K$ coincide. The simple proof of this assertion is left to the reader.

Lemma 2.4.13. *There is $\text{PMin}_{LSc}(M+K,K)=\text{PMin}_{LSc}(M,K)$.*

The connection between the properly minimal elements in the sense of linear scalarization and the properly minimal elements in the sense of Hurwicz and Henig and Lampe, respectively, is outlined in the following statements.

Proposition 2.4.14. *There is $\text{PMin}_{LSc}(M,K)\subseteq\text{PMin}_{Hu}(M,K)$.*

Proof. If $K=\{0\}$ there is nothing to be proven. Assume that $K\neq\{0\}$ and take $\bar{v}\in\text{PMin}_{LSc}(M,K)$. Then $\bar{v}\in M$ and there exists $v^*\in K^{*0}$ such that $\langle v^*,v\rangle\geq 0$ for all $v\in M-\bar{v}$. This means that for all $v\in(M-\bar{v})\cup K$ it holds $\langle v^*,v\rangle\geq 0$ and, consequently, $\langle v^*,v\rangle\geq 0$ for all $v\in\text{cl}(\text{coneco}(M-\bar{v})\cup K)$.

Assuming that there exists a $k\in K\setminus\{0\}$ such that $-k\in\text{cl}(\text{coneco}(M-\bar{v})\cup K)$, we get $\langle v^*,k\rangle\leq 0$. On the other hand, since $v^*\in K^{*0}$ there is $\langle v^*,k\rangle>0$. □

Proposition 2.4.15. *There is $\text{PMin}_{LSc}(M,K)\subseteq\text{PMin}_{He-La}(M,K)$.*

Proof. Take an abitrary $\bar{v} \in \text{PMin}_{LSc}(M,K)$. Then $\bar{v} \in M$ and there exists $v^* \in K^{*0}$ such that $\langle v^*, v \rangle \geq 0$ for all $v \in M - \bar{v}$. Let be $K' := \{v \in X : \langle v^*, v \rangle > 0\} \cup \{0\}$. Obviously, K' is a nontrivial convex cone and, since $v^* \in K^{*0}$, $K \backslash \{0\} \subseteq \text{int}(K') = \{v \in X : \langle v^*, v \rangle > 0\}$.

We prove that $(M - \bar{v}) \cap (-K') = \{0\}$. Assuming the contrary yields that there exists $v \in M$ such that $\bar{v} - v \in K' \backslash \{0\}$, or, equivalently, $\langle v^*, \bar{v} - v \rangle > 0$. This contradicts the fact that $\langle v^*, \bar{v} \rangle \leq \langle v^*, v \rangle$ for all $v \in M$. □

Remark 2.4.8. In case the set $M + K$ is convex Lemma 2.4.13 allows to characterize the properly minimal elements $\bar{v} \in \text{PMin}_{LSc}(M,K)$ as solutions of the scalar convex optimization problem

$$\min_{v \in M+K} \langle v^*, v \rangle$$

with an appropriate $v^* \in K^{*0}$. This again makes it possible to derive necessary and sufficient optimality conditions via scalar duality and also to construct vector dual problems in particular in the case when M is the image set of a feasible set through the objective function of a given vector optimization problem.

Summarizing the results proven above, we come to the following general scheme for the proper minimal sets introduced in this section. First we consider the general situation of an underlying Hausdorff locally convex space V partially ordered by the pointed convex cone K and let $M \subseteq V$ be a nonempty set.

Proposition 2.4.16. *There holds*

$$\text{PMin}_{LSc}(M,K) \subseteq \begin{matrix} \text{PMin}_{Hu}(M,K) \\ \text{PMin}_{He-La}(M,K) \end{matrix} \subseteq \text{PMin}_{Be}(M,K) \subseteq \begin{matrix} \text{PMin}_{GBo}(M,K) \\ \text{PMin}_{Bo}(M,K). \end{matrix}$$

If $M + K$ is convex, then

$$\text{PMin}_{LSc}(M,K) \subseteq \text{PMin}_{He-La}(M,K) \subseteq \text{PMin}_{Hu}(M,K)$$

$$= \text{PMin}_{Be}(M,K) = \text{PMin}_{Bo}(M,K) \subseteq \text{PMin}_{GBo}(M,K).$$

Under additional hypotheses some opposite inclusions hold, too. We begin with a statement that can be proven by considering some results from [83].

Proposition 2.4.17. *(a) If the ordering cone K is closed and it has a compact base, then* $\text{PMin}_{LSc}(M,K) = \text{PMin}_{Hu}(M,K)$.

(b) If V is normed, the ordering cone K is closed and it has a weakly compact base, then $\text{PMin}_{He-La}(M,K) = \text{PMin}_{Be}(M,K) = \text{PMin}_{GBo}(M,K)$.

Whenever $V = \mathbb{R}^k$ and $K = \mathbb{R}^k_+$ more inclusions turn into equalities in the scheme considered in Proposition 2.4.16 one can be include the properly minimal elements in the sense of Geoffrion, too.

Proposition 2.4.18. *Let $V = \mathbb{R}^k$, $K = \mathbb{R}^k_+$ and $M \subseteq \mathbb{R}^k$ a nonempty set.*

(a) Then it holds

$$\text{PMin}_{LSc}(M, \mathbb{R}^k_+) = \text{PMin}_{Hu}(M, \mathbb{R}^k_+) \subseteq \text{PMin}_{He-La}(M, \mathbb{R}^k_+) =$$

$$\text{PMin}_{Be}(M, \mathbb{R}^k_+) = \text{PMin}_{Ge}(M, \mathbb{R}^k_+) = \text{PMin}_{GBo}(M, \mathbb{R}^k_+) \subseteq \text{PMin}_{Bo}(M, \mathbb{R}^k_+).$$

(b) If, additionally, $M + \mathbb{R}^k_+$ is convex, then all the inclusions in (a) turn into equalities.

Remark 2.4.9. In section 4.4 we consider other minimality notions with respect to general increasing scalarization functions used only there, while in chapter 7 some minimality notions introduced in this section are extended for sets $M \subseteq \overline{V}$.

2.4.4 Linear scalarization

In this subsection we turn our attention to linear scalarization and its connections to the different minimality concepts introduced before. Scalarization in general allows us to associate a *scalar optimization problem* to a given *vector optimization problem*. This is very closely related to the monotonicity properties of the scalarizing function. In this context the dual cone and the quasi interior of the dual cone of the underlying ordering cone plays a crucial role. As noticed in Remark 2.4.8, characterizing minimal, weakly minimal or properly minimal elements of a given set by monotone scalarization, in particular linear scalarization, offers the possibility to investigate these notions by means of techniques which come from the scalar optimization.

Unless otherwise mentioned, in this subsection we consider V to be a vector space partially ordered by a convex cone $K \subseteq V$. If the pointedness of the cone K is required in some particular result, it will be explicitly mentioned. Moreover, the set $M \subseteq V$ is assumed to be nonempty.

Lemma 2.4.19. *Let $f : V \to \overline{\mathbb{R}}$ be a given function.*

(a) If f is K-increasing on M and there exists an uniquely determined element $\bar{v} \in M$ satisfying $f(\bar{v}) \leq f(v)$ for all $v \in M$, then $\bar{v} \in \text{Min}(M, K)$.
(b) If f is strongly K-increasing on M and there exists $\bar{v} \in M$ satisfying $f(\bar{v}) \leq f(v)$ for all $v \in M$, then $\bar{v} \in \text{Min}(M, K)$.

Proof. (a) Assuming $\bar{v} \notin \text{Min}(M, K)$ yields the existence of $v \in M$ such that $v \leq_K \bar{v}$. Taking into consideration the fact that f is K-increasing we get $f(v) \leq f(\bar{v})$. Thus $f(v) = f(\bar{v})$ and this contradicts the uniqueness of $\bar{v}$ as solution of the problem $\min_{v \in M} f(v)$.

(b) Arguing as in part (a) in case $\bar{v} \notin \text{Min}(M, K)$ one can find an element $v \in M$ such that $f(v) < f(\bar{v})$, but this contradicts the minimality of $f(\bar{v})$. □

For weakly minimal elements one has the following analogous characterization. Note that no difficulties arise if the ordering cone K is not pointed.

Lemma 2.4.20. *Suppose that* $\operatorname{core}(K) \neq \emptyset$ *and consider a function* $f : V \to \overline{\mathbb{R}}$ *which is strictly* K*-increasing on* M*. If there is an element* $\bar{v} \in M$ *fulfilling* $f(\bar{v}) \leq f(v)$ *for all* $v \in M$*, then* $\bar{v} \in \operatorname{WMin}(M, K)$*.*

Proof. If $\bar{v} \notin \operatorname{WMin}(M, K)$ then there exists $v \in (\bar{v} - \operatorname{core}(K)) \cap M$. This implies $f(v) < f(\bar{v})$, contradicting the assumption. □

The next scalarization result provides a necessary optimality condition for the minimal elements of M. One can notice the usefulness of the assumption of convexity for $M + K$, which allows giving such characterizations even if M is not a convex set. We refer to the previous subsections for the connections between the minimality properties of the sets M and $M + K$.

Theorem 2.4.21. *Assume that the ordering cone* K *is nontrivial and pointed,* $M + K$ *is convex and* $\operatorname{core}(M + K) \neq \emptyset$*. If* $\bar{v} \in \operatorname{Min}(M, K)$*, then there exists some* $v^{\#} \in K^{\#}\backslash\{0\}$ *such that* $\langle v^{\#}, \bar{v}\rangle \leq \langle v^{\#}, v\rangle$ *for all* $v \in M$*.*

Proof. If $\bar{v} \in \operatorname{Min}(M, K)$, then according to Lemma 2.4.1(*a*) we have $\bar{v} \in \operatorname{Min}(M + K, K)$, too, and this can be equivalently rewritten as $(M + K - \bar{v}) \cap (-K) = \{0\}$. Even more, as $M + K - \bar{v}$ and $(-K)$ are convex sets, $\operatorname{core}(M + K - \bar{v}) \neq \emptyset$ and $\operatorname{core}(M + K - \bar{v}) \cap (-K) = \emptyset$, Theorem 2.1.3 can be applied. Thus there exist $\bar{v}^{\#} \in V^{\#}\backslash\{0\}$ and $\lambda \in \mathbb{R}$ such that

$$\langle \bar{v}^{\#}, v + k_1 - \bar{v}\rangle \leq \lambda \leq \langle \bar{v}^{\#}, -k_2\rangle \ \forall v \in M \ \forall k_1, k_2 \in K. \tag{2.9}$$

If there exists $\bar{k} \in K\backslash\{0\}$ such that $\langle \bar{v}^{\#}, \bar{k}\rangle > 0$, then choosing $k_1 = \alpha\bar{k}$ for $\alpha > 0$, we obtain a contradiction to (2.9), as the left-hand side is unbounded for $\alpha \to +\infty$. Thus $\langle \bar{v}^{\#}, k\rangle \leq 0$ for all $k \in K\backslash\{0\}$ and this actually means that $\bar{v}^{\#} \in -K^{\#}$. Taking $k_1 = k_2 = 0$ and setting $v^{\#} := -\bar{v}^{\#} \in K^{\#}$ we get $\langle v^{\#}, \bar{v}\rangle \leq \langle v^{\#}, v\rangle$ for all $v \in M$, which completes the proof. □

It is clear that by means of the topological version of Eidelheit's separation theorem one can state an analogous scalarization result for the minimal elements of a subset M of a topological vector space.

Corollary 2.4.22. *Let* V *be a topological vector space partially ordered by a nontrivial pointed convex cone* K*. Moreover, assume that* $M + K$ *is convex and* $\operatorname{int}(M + K) \neq \emptyset$*. If* $\bar{v} \in \operatorname{Min}(M, K)$*, then there exists* $v^* \in K^*\backslash\{0\}$ *such that* $\langle v^*, \bar{v}\rangle \leq \langle v^*, v\rangle$ *for all* $v \in M$*.*

Now we present, again in the vector space setting, sufficient conditions for minimality which are immediate consequences of Lemma 2.4.19 and Example 2.2.3.

Theorem 2.4.23. *(a) If there exists* $v^{\#} \in K^{\#}$ *and* $\bar{v} \in M$ *such that* $\langle v^{\#}, \bar{v}\rangle < \langle v^{\#}, v\rangle$ *for all* $v \in M$*,* $v \neq \bar{v}$*, then* $\bar{v} \in \operatorname{Min}(M, K)$*.*
(b) If there exist $v^{\#} \in K^{\#0}$ *and* $\bar{v} \in M$ *such that* $\langle v^{\#}, \bar{v}\rangle \leq \langle v^{\#}, v\rangle$ *for all* $v \in M$*, then* $\bar{v} \in \operatorname{Min}(M, K)$*.*

Remark 2.4.10. (*a*) In Theorem 2.4.23(*b*) it is not necessary to impose the pointedness of K, because otherwise $K^{\#0} = \emptyset$.

(*b*) The necessary condition in Theorem 2.4.21 is not also sufficient because, as follows from Theorem 2.4.23(*a*), for this we need a strict inequality. Indeed, if $\langle v^{\#}, \bar{v} \rangle \leq \langle v^{\#}, v \rangle$ is for all $v \in M$ fulfilled, then $\bar{v}$ is weakly minimal to M, (see Theorem 2.4.25 below), but not necessarily minimal.

We would like to mention that in locally convex spaces partially ordered by a convex closed cone the strongly minimal elements can be as well equivalently characterized via linear scalarization by using linear continuous functionals from K^* (see [104, Theorem 5.6]). We omit giving this statement here, since strongly minimal elements are not interesting from the viewpoint of vector optimization and do not play any role in this book.

Next we turn our attention to necessary and sufficient optimality conditions characterizing via linear scalarization the weakly minimal elements of a nonempty subset of a vector space.

Theorem 2.4.24. *Let $K \subseteq V$ be such that* $\operatorname{core}(K) \neq \emptyset$ *and $M+K$ is convex. If $\bar{v} \in \operatorname{WMin}(M, K)$ then there exists $v^{\#} \in K^{\#} \backslash \{0\}$ such that $\langle v^{\#}, \bar{v} \rangle \leq \langle v^{\#}, v \rangle$ for all $v \in M$.*

Proof. The proof follows the lines of the proof of Theorem 2.4.21 using again the algebraic version of Eidelheit's separation theorem. □

Theorem 2.4.25. *Suppose that* $\operatorname{core}(K) \neq \emptyset$*. If there exist $v^{\#} \in K^{\#} \backslash \{0\}$ and $\bar{v} \in M$ such that for all $v \in M$ it holds $\langle v^{\#}, \bar{v} \rangle \leq \langle v^{\#}, v \rangle$, then $\bar{v} \in$* $\operatorname{WMin}(M, K)$.

Proof. The assertion is a straightforward conclusion of Lemma 2.4.20 and Example 2.2.3. □

Combining Theorem 2.4.24 and Theorem 2.4.25 one obtains an equivalent characterization via linear scalarization for weakly minimal elements.

Corollary 2.4.26. *Let $K \subseteq V$ be such that* $\operatorname{core}(K) \neq \emptyset$ *and $M+K$ is convex. Then $\bar{v} \in \operatorname{WMin}(M, K)$ if and only if there exists $v^{\#} \in K^{\#} \backslash \{0\}$ satisfying $\langle v^{\#}, \bar{v} \rangle \leq \langle v^{\#}, v \rangle$ for all $v \in M$.*

The following remark plays an important role when dealing with vector duality with respect to weakly minimal elements.

Remark 2.4.11. Assuming that V is a topological vector space partially ordered by the convex cone K with $\operatorname{int}(K) \neq \emptyset$ and $M \subseteq V$ is a nonempty set with $M + K$ convex, then by using the topological version of Eidelheit's separation theorem and the analog of Lemma 2.4.19 and Example 2.2.3 for topological vector spaces, one gets that $\bar{v} \in \operatorname{WMin}(M, K)$ if and only if there exists a $v^* \in K^* \backslash \{0\}$ such that $\langle v^*, \bar{v} \rangle \leq \langle v^*, v \rangle$ for all $v \in M$. Theorem 2.4.24 and Theorem 2.4.25 remain valid when formulated in a corresponding topological framework.

After characterizing minimal and weakly minimal elements of a set $M \subseteq V$ regarding the partial ordering induced by the convex cone $K \subseteq V$ via linear scalarization it is natural to ask whether it is possible to give analogous characterizations also for the properly minimal elements. First of all let us take a closer look at Definition 2.4.12, where we introduced $\text{PMin}_{LSc}(M, K)$, the set of properly minimal elements of M with respect to linear scalarization. This definition itself is already based on linear scalarization. If we look at Proposition 2.4.17(a) we see that under some additional hypotheses $\text{PMin}_{LSc}(M, K) = \text{PMin}_{Hu}(M, K)$, i.e. the properly minimal elements of M in the sense of Hurwicz may be characterized by linear scalarization using a functional $v^* \in K^{*0}$. Even more, as follows from Proposition 2.4.18(b), if $V = \mathbb{R}^k$, $K = \mathbb{R}^k_+$ and $M + K$ is a convex set, then all the properly minimal elements introduced in this section may be characterized in an equivalent manner by linear scalarization. But as far as properly minimal elements in the sense of Borwein are concerned, there exists a more general linear scalarization result, which can be proven like [104, Theorem 5.11 and Theorem 5.21].

Theorem 2.4.27. *Let V be a Hausdorff locally convex space partially ordered by the pointed convex closed cone K with $\text{int}_{w(V^*,V)}(K^*) \neq \emptyset$ and the nonempty set $M \subseteq V$ for which we assume that $M + K$ is convex. Then $\bar{v} \in \text{PMin}_{Bo}(M, K)$ if and only if there exists $v^* \in K^{*0}$ such that $\langle v^*, \bar{v}\rangle \leq \langle v^*, v\rangle$ for all $v \in M$.*

Remark 2.4.12. One should notice that for $V = \mathbb{R}^k$, $K = \mathbb{R}^k_+$ and $M \subseteq \mathbb{R}^k$ with $M + \mathbb{R}^k_+$ convex, when the hypotheses of Theorem 2.4.27 are fulfilled, there is

$$\text{PMin}_{LSc}(M, \mathbb{R}^k_+) = \text{PMin}_{Hu}(M, \mathbb{R}^k_+) = \text{PMin}_{He-La}(M, \mathbb{R}^k_+) =$$

$$\text{PMin}_{Be}(M, \mathbb{R}^k_+) = \text{PMin}_{Ge}(M, \mathbb{R}^k_+) = \text{PMin}_{GBo}(M, \mathbb{R}^k_+) = \text{PMin}_{Bo}(M, \mathbb{R}^k_+),$$

which is nothing but the assertion of Proposition 2.4.18(b).

2.5 Vector optimization problems

An optimization problem consisting in the minimization or maximization of several objective functions is a particular case of a *vector optimization problem*, for which one can find in the literature also the denotations *multiobjective* (or *multicriteria*) *optimization* (or *programming*) *problem* as well as *multiple objective optimization* (or *programming*) *problem*. The characteristic feature is the occurrence of several conflicting objectives, i.e. not all objectives under consideration attain their minimal or maximal values at the same element of the feasible set, which is a subset of the space where the objective functions are defined on, sometimes called *decision space* (or *input space*). In most real life decisions it is much more realistic to take into account not only one objective but different ones. For instance, if we look at an investment decision on the capital market it is reasonable to consider at least two objectives, namely the expected return which has to be maximized and the risk of an investment in a security or a portfolio of securities which should be minimized. In other situations one wants to minimize the cost and to maximize different features of quality of a product or a production process or to minimize the production time and to maximize the production capacity etc. It is obvious that such objectives often appear in a conflicting manner or as conflicting interests between different persons, groups of people or within a single decision-maker itself.

A widely used way of assessing the multiple objectives is on the base of partial ordering relations induced by convex cones. This allows to compare different vector objective values in the sense that an objective value is preferred if it is less than (if we consider a minimization problem) or greater than (if case of a maximization problem) another one with respect to the considered partial ordering induced by the underlying convex cone. The solutions are defined by those objective values that cannot be improved by another one in the sense of this preference notion. Thus, one immediately sees that the notions of minimal elements for sets introduced in section 2.4 turn out to be natural solution concepts in vector optimization. Although in many practical applications the number of considered objectives is finite, from a mathematical point of view the *objective space* (or *image space*), sometimes also called *outcome space*, may be an infinite dimensional space. So, for the sake of generality, we will define the vector optimization problem initially by considering general vector spaces for the decision and outcome spaces, the latter partially ordered by a convex cone.

Let X and V be vector spaces and assume that V is partially ordered by the convex cone $K \subseteq V$. For a given proper function $h : X \to \overline{V} = V \cup \{\pm\infty_K\}$ we investigate the *vector optimization problem* formally denoted by

$$(PVG) \quad \operatorname*{Min}_{x \in X} h(x).$$

It consists in determining the minimal, weakly minimal or properly minimal elements of the image set of X through h, also called *outcome set* (or *image set*), $h(\operatorname{dom} h) = \{v \in V : \exists x \in \operatorname{dom} h, v = h(x)\}$. In other words, we are interested in determining the sets $\operatorname{Min}(h(\operatorname{dom} h), K)$, $\operatorname{WMin}(h(\operatorname{dom} h), K)$ or $\operatorname{PMin}(h(\operatorname{dom} h), K)$, where PMin is a generic notation for all sets of properly minimal elements. On the other hand, we are also interested in finding the so-called *efficient*, *weakly efficient* or *properly efficient solutions* to (PVG).

Definition 2.5.1. *An element $\bar{x} \in X$ is said to be*

(a) an efficient solution to the vector optimization problem (PVG) if $\bar{x} \in \operatorname{dom} h$ and $h(\bar{x}) \in \operatorname{Min}(h(\operatorname{dom} h), K)$;

(b) a weakly efficient solution to the vector optimization problem (PVG) if $\bar{x} \in \operatorname{dom} h$ and $h(\bar{x}) \in \operatorname{WMin}(h(\operatorname{dom} h), K)$;

(c) *a properly efficient solution to the vector optimization problem* (PVG) *if* $\bar{x} \in \operatorname{dom} h$ *and* $h(\bar{x}) \in \operatorname{PMin}(h(\operatorname{dom} h), K)$.

The set containing all the efficient solutions to (PVG) is called the *efficiency set* of (PVG), the one containing all the weakly efficient solutions to (PVG) is said to be the *weak efficiency set* of (PVG), while the name used for the one containing all the properly efficient solutions to (PVG) is the *proper efficiency set* of (PVG).

It is worth mentioning that in many cases in practice a decision-maker is only interested to have a subset or even a single element of one of these efficiency sets. This is a direct consequence of the practical requirements in applications.

Frequently, one looks for efficient elements in a nonempty subset $\mathcal{A} \subseteq X$, where the objective function is $h : \mathcal{A} \to V$. This problem can be reformulated in the form of (PVG) by considering as objective function $\tilde{h} : X \to \overline{V}$,

$$\tilde{h}(x) = \begin{cases} h(x), & \text{if } x \in \mathcal{A}, \\ +\infty_K, & \text{otherwise.} \end{cases}$$

Although we have just defined the efficient solutions via the minimality notions for the image set, for the sake of convenience let us state them in an explicit manner.

Definition 2.5.2. *An element* $\bar{x} \in X$ *is said to be an efficient solution to the vector optimization problem* (PVG) *if* $\bar{x} \in \operatorname{dom} h$ *and for all* $x \in \operatorname{dom} h$ *from* $h(x) \leqq_K h(\bar{x})$ *follows* $h(\bar{x}) \leqq_K h(x)$. *The set of efficient solutions to* (PVG) *is denoted by* $\operatorname{Eff}(PVG)$.

As pointed out in the previous section, there are several equivalent formulations for $\bar{x} \in Eff(PVG)$, like, for example, $(h(\bar{x})-K) \cap h(\operatorname{dom} h) \subseteq h(\bar{x})+K$ and, in case K is pointed, $(h(\bar{x}) - K) \cap h(\operatorname{dom} h) = \{h(\bar{x})\}$.

Definition 2.5.3. *Suppose that* $\operatorname{core}(K) \neq \emptyset$. *An element* $\bar{x} \in X$ *is said to be a weakly efficient solution to the vector optimization problem* (PVG) *if* $\bar{x} \in \operatorname{dom} h$ *and* $(h(\bar{x}) - \operatorname{core}(K)) \cap h(\operatorname{dom} h) = \emptyset$. *The set of weakly efficient solutions to* (PVG) *is denoted by* $\operatorname{WEff}(PVG)$.

One can see that $\bar{x} \in \operatorname{WEff}(PVG)$ if and only if $\bar{x} \in \operatorname{dom} h$ and there is no $x \in \operatorname{dom} h$ satisfying $h(x) <_K h(\bar{x})$.

Taking into consideration Proposition 2.4.2, whenever $\operatorname{core}(K) \neq \emptyset$ and $K \neq V$, we have $\operatorname{Eff}(PVG) \subseteq \operatorname{WEff}(PVG)$. In section 2.4 we have pointed out the close connection between the different types of minimal elements to the sets M and $M + K$, when $M \subseteq X$ is a nonempty set. These results are important in the context of scalarization since we have seen that the property of $M+K$ to be convex is sufficient for the characterization of minimal elements of M by means of linear scalarization. We may transfer this to the vector optimization problem in an obvious manner.

From section 2.2 we know that if $h : X \to \overline{V}$ is a proper function, the assumption that $h(\operatorname{dom} h) + K$ is convex is equivalent to the property that the function h is K-convexlike. This allows to establish scalarization results for vector optimization problems with K-convexlike and indirectly with K-convex objective functions.

In an analogous manner one can deliver explicitly definitions for the different notions of properly efficient solutions. We restrict ourselves here only to the properly efficient solutions with respect to linear scalarization based on Definition 2.4.12 because this type of properly efficient solutions will be later involved in different duality statements for vector optimization problems.

Let us suppose that V is a Hausdorff locally convex space partially ordered by a pointed convex cone K. One can alternatively define the properly efficient solutions in the sense of linear scalarization in the following manner.

Definition 2.5.4. *An element $\bar{x} \in X$ is said to be a properly efficient solution to (PVG) in the sense of linear scalarization if there exists $v^* \in K^{*0}$ such that $(v^*h)(\bar{x}) \leq (v^*h)(x)$ for all $x \in X$. The set of properly efficient solutions in the sense of linear scalarization to (PVG) is denoted by* $\mathrm{PEff}_{LSc}(PVG)$.

In other words, $\bar{x} \in \mathrm{PEff}_{LSc}(PVC)$ if and only if $\bar{x}$ is an optimal solution to the scalar optimization problem

$$\min_{x \in X} (v^*h)(x).$$

The results in this section can be used for providing corresponding characterizations for the properly efficient solutions to (PVG) by means of linear scalarization.

Bibliographical notes

Convex analysis established itself as a distinct area of mathematics after the publishing of Rockafellar's seminal book [157] where one can find most of the known results concerning convex sets and functions in finite dimensional spaces. Besides it, for our exposition on convex sets and functions we used mainly the books of Hiriart-Urruty and Lemarechal [90] and Borwein and Lewis [22] when working in finite dimensional spaces and the ones of Ekeland and Temam [67] and Zălinescu [207] for infinite dimensional spaces. Most of the statements we give here without proofs are demonstrated in these books. We also refer to the books [104, 125] for notions and results concerning vector functions and cones and to [20, 27, 207] for some results involving generalized interiors.

Conjugate functions have been introduced in finite dimensional spaces by Fenchel in [68]. The further development of this theory in finite dimensional spaces can be pursued in Rockafellar's book [157] and in topological vector spaces in the books of Ekeland and Temam [67] and Zălinescu [207]. A good and detailed treatment of the finite dimensional theory can be found also in the books [91,92] of Hiriart-Urruty and Lemaréchal. Our presentation contains many results, partially with some modified proofs and some extensions, the reader can find in the mentioned books.

As far as subdifferentiability is concerned we refer to the mentioned books, too. Subdifferentiability is closely related to conjugacy as we have conveyed in this chapter and as an appropriate reference for this notion we quote here the book [159] of Rockafellar. Concerning partially ordered vector spaces we would like to refer to the initiating paper [110] of Kantorovitch, which is more than seventy years old. Books in this field are due to Nachbin [141], Peressini [151], Jameson [105] and others. Fuchssteiner and Lusky investigated convex cones in [69].

In vector optimization one of the essential and basic notions is that of minimal elements of a set regarding the partial ordering induced by a convex cone. The first definition of proper minimal elements has been proposed by Kuhn and Tucker [119] for finite dimensional vector optimization problems under differentiability assumptions, followed by the one due to Hurwicz [93]. Geoffrion introduced his widely used and economically inspired definition in [71], also for finite dimensional vector maximum problems. Later, more general notions of properly efficient elements, also for infinite dimensional vector optimization problems, came into discussion in papers written by Borwein [17], Benson [15], etc.

Very beneficial regarding the connections to scalar optimization is the characterization of different types of properly efficient and weakly efficient solutions by linear scalarization. For contributions on the relations between different kinds of properly minimal elements of a set we refer to [83,87] and the book [81], while linear scalarization results for minimal, properly and weakly minimal (or efficient) solutions can be found in Jahn's book [104].

Linear scalarization results for properly efficient elements in the sense of Borwein can be found in [17]. In Jahn's book [104] one can find also so-called norm scalarization results. An overview about different linear and other scalarization methods for finite dimensional vector optimization problems including sensitivity results and numerical methods is given in [66]. Vector optimization has attracted the attention and the research activities of many scientists. Let us mention here only some textbooks and monographs which allow to get an introduction as well as a deeper insight into the field (cf. [49, 64, 65, 80, 81, 104, 108, 125, 130, 163, 181, 209]).

3

Conjugate duality in scalar optimization

The aim of this chapter is to describe the so-called *conjugate duality theory* for scalar optimization problems, which represents a cornerstone in the duality theory for vector optimization problems. The most duality concepts which can be found in the literature on vector optimization, this book being no exception, have the origin in well-developed duality theories for scalar problems. This is the reason why we intensively deal in this chapter with the scalar case, providing some results to which we relate later.

3.1 Perturbation theory and dual problems

In this section we describe a general approach for introducing a conjugate dual optimization problem to a scalar one. We treat first a general scalar optimization problem, followed by some important particular instances of it.

3.1.1 The general scalar optimization problem

Let X be a Hausdorff locally convex space and $F : X \to \overline{\mathbb{R}}$ a given function. In this section we assign to the general optimization problem

$$(PG) \quad \inf_{x \in X} F(x)$$

a *conjugate dual problem* introduced by making use of the so-called *perturbation approach*. To this end we consider another Hausdorff locally convex space Y and the function $\Phi : X \times Y \to \overline{\mathbb{R}}$ fulfilling $\Phi(x, 0) = F(x)$ for all $x \in X$. The function Φ is the so-called *perturbation function* of the problem (PG). In this way one can embed the problem (PG) into a family of so-called *perturbed problems* which looks like

$$(PG_y) \quad \inf_{x \in X} \Phi(x, y),$$

R.I. Boţ et al., *Duality in Vector Optimization*, Vector Optimization,
DOI 10.1007/978-3-642-02886-1_3,

where $y \in Y$. Obviously, the problem

$$(PG_0) \quad \inf_{x \in X} \Phi(x, 0)$$

is nothing else than the initial optimization problem (PG). A *conjugate dual problem* to (PG) can be now formulated as being

$$(DG) \quad \sup_{y^* \in Y^*} \{-\Phi^*(0, y^*)\},$$

where $\Phi^* : X^* \times Y^* \to \overline{\mathbb{R}}$ is the conjugate function of Φ. Throughout this chapter we assume that the topological dual spaces X^* and Y^* of the *space of the feasible variables* X and of the *space of the perturbation variables* Y, respectively, are both endowed with the corresponding weak* topology (denoted by $w(X^*, X)$ and $w(Y^*, Y)$, respectively).

Further, let us denote by $v(PG)$ and $v(DG)$ the optimal objective values of the problems (PG) and (DG), respectively. The next result shows that *weak duality* is a consequence of the way in which the dual problem was defined.

Theorem 3.1.1. *It holds*

$$-\infty \leq v(DG) \leq v(PG) \leq +\infty.$$

Proof. For all $x \in X$ and all $y^* \in Y^*$, by the Young-Fenchel inequality one has

$$\Phi(x, 0) + \Phi^*(0, y^*) \geq \langle 0, x \rangle + \langle y^*, 0 \rangle = 0 \Leftrightarrow \Phi(x, 0) \geq -\Phi^*(0, y^*),$$

which implies that $v(PG) \geq v(DG)$. □

In the following we characterize the existence of *strong duality*, namely of the situation when the gap between $v(PG)$ and $v(DG)$ disappears and the dual (DG) has an *optimal solution*. Here an important role is played by the infimal value function of Φ, $h : Y \to \overline{\mathbb{R}}$, $h(y) = \inf\{\Phi(x, y) : x \in X\}$. One can notice that $v(PG_y) = h(y)$ and $v(PG) = h(0)$. The next proposition connects the infimal value function h to the optimal objective valued of the dual problem.

Proposition 3.1.2. *If $h : Y \to \overline{\mathbb{R}}$ is the infimal value function of Φ, then one has $v(DG) = h^{**}(0)$.*

Proof. For all $y^* \in Y^*$, by the definition of the conjugate function, we get

$$h^*(y^*) = \sup_{y \in Y}\{\langle y^*, y \rangle - h(y)\} = \sup_{\substack{x \in X, \\ y \in Y}} \{\langle y^*, y \rangle - \Phi(x, y)\} = \Phi^*(0, y^*). \quad (3.1)$$

Thus

$$h^{**}(0) = \sup_{y^* \in Y^*} \{-h^*(y^*)\} = \sup_{y^* \in Y^*} \{-\Phi^*(0, y^*)\} = v(DG)$$

and this concludes the proof. □

Remark 3.1.1. One can easily notice that the relation which states the weak duality, i.e. $v(DG) \leq v(PG)$ can be equivalently written as $h^{**}(0) \leq h(0)$. By Lemma 2.3.3 we know that this inequality is always true.

Definition 3.1.1. *We say that the problem* (PG) *is normal if* $h(0) \in \mathbb{R}$ *and* h *is lower semicontinuous at 0.*

For the following result we refer to [67].

Theorem 3.1.3. *Assume that* $\Phi : X \times Y \to \overline{\mathbb{R}}$ *is a proper and convex function. Then the following statements are equivalent:*

(i) the problem (PG) *is normal;*
(ii) it holds $v(PG) = v(DG)$ *and this value is finite.*

Proof. $(i) \Rightarrow (ii)$. Let $\bar{h} : Y \to \overline{\mathbb{R}}$ be the lower semicontinuous hull of h. By Proposition 2.3.4(b) it holds

$$h^{**}(y) \leq \bar{h}(y) \leq h(y) \ \forall y \in Y. \tag{3.2}$$

Since Φ is convex one has that h is convex (cf. Theorem 2.2.6) and this means that $\bar{h}$ is convex, too (cf. Proposition 2.2.1 and Theorem 2.2.12(a)). The problem (PG) being normal, it follows that $\bar{h}(0) = h(0) \in \mathbb{R}$. Using that $\bar{h}$ is a convex and lower semicontinuous function, Proposition 2.2.14 implies that $\bar{h}(y) > -\infty$ for all $y \in Y$. This guarantees the properness of $\bar{h}$. Taking now Corollary 2.3.7 into consideration we obtain $\bar{h} = (\bar{h})^{**} = ((\bar{h})^*)^* = (h^*)^* = h^{**}$ and so $h^{**}(0) = \bar{h}(0) = h(0) \in \mathbb{R}$. Since $v(PG) = h(0)$ and $v(DG) = h^{**}(0)$, (ii) is valid.

$(ii) \Rightarrow (i)$. As we have seen, the statement (ii) can be equivalently written as $h^{**}(0) = h(0) \in \mathbb{R}$. Then, by (3.2), we get that $\bar{h}(0) = h(0) \in \mathbb{R}$, which means that (PG) is normal. □

The notion we introduce in the definition below characterizes the existence of strong duality for (PG) and (DG) (see also [67]).

Definition 3.1.2. *We say that the problem* (PG) *is stable if* $h(0) \in \mathbb{R}$ *and* h *is subdifferentiable at 0.*

Theorem 3.1.4. *Assume that* $\Phi : X \times Y \to \overline{\mathbb{R}}$ *is a proper and convex function. Then the following statements are equivalent:*

(i) the problem (PG) *is stable;*
(ii) the problem (PG) *is normal and the dual* (DG) *has an optimal solution.*

In this situation the set of optimal solutions to (DG) *is equal to* $\partial h(0)$.

Proof. $(i) \Rightarrow (ii)$. Assume that $h(0) \in \mathbb{R}$ and $\partial h(0) \neq \emptyset$. By Theorem 2.3.16$(a)$ we get that $h(0) = h^{**}(0)$ and this is nothing else than $v(PG) = v(DG) \in \mathbb{R}$. Then, by Theorem 3.1.3, (PG) is normal. Further, let us consider an element

$\bar{y}^* \in \partial h(0)$. We have $h(0) + h^*(\bar{y}^*) = 0$ (cf. Theorem 2.3.12) or, equivalently, $v(PG) = h(0) = -h^*(\bar{y}^*) = -\Phi^*(0, \bar{y}^*)$. By Theorem 3.1.1 follows that $\bar{y}^*$ is an optimal solution to (DG).

(ii) $\Rightarrow$ *(i)*. Assume now that (PG) is normal and that the dual (DG) has an optimal solution $\bar{y}^*$. Applying again Theorem 3.1.3 and relation (3.1) we get $h(0) = v(PG) = v(DG) = -\Phi^*(0, \bar{y}^*) = -h^*(\bar{y}^*) \in \mathbb{R}$, which is the same as $h(0) + h^*(\bar{y}^*) = 0 \Leftrightarrow \bar{y}^* \in \partial h(0)$. Consequently, we have proved that the set $\partial h(0)$ is nonempty and this guarantees the stability of (PG). □

Since for a given primal problem and its conjugate dual the stability completely characterizes the existence of strong duality, one of the main issues in the optimization theory is to give sufficient conditions, called *regularity conditions*, which guarantee that a problem is stable. An overview of this kind of conditions for different classes of optimization problems will be given in the next section.

In the following we construct by means of the general perturbation approach described above conjugate dual problems to different primal optimization problems and corresponding perturbation functions. As one can notice, in this way some classical duality concepts, like the ones due to Fenchel and Lagrange, can be provided.

3.1.2 Optimization problems having the composition with a linear continuous mapping in the objective function

Consider the following primal optimization problem

$$(P^A) \quad \inf_{x \in X} \{f(x) + g(Ax)\},$$

where X and Y are Hausdorff locally convex spaces, $A \in \mathcal{L}(X, Y)$ and $f : X \to \overline{\mathbb{R}}$ and $g : Y \to \overline{\mathbb{R}}$ are proper functions fulfilling $\operatorname{dom} f \cap A^{-1}(\operatorname{dom} g) \neq \emptyset$. The perturbation function we consider for assigning a dual problem to (P^A) is $\Phi^A : X \times Y \to \overline{\mathbb{R}}$, $\Phi^A(x, y) = f(x) + g(Ax + y)$, where $y \in Y$ is the perturbation variable. Obviously, $\Phi^A(x, 0) = f(x) + g(Ax)$ for all $x \in X$. The conjugate function of Φ^A, $(\Phi^A)^* : X^* \times Y^* \to \overline{\mathbb{R}}$, has for all $(x^*, y^*) \in X^* \times Y^*$ the following formulation

$$(\Phi^A)^*(x^*, y^*) = \sup_{\substack{x \in X, \\ y \in Y}} \{\langle x^*, x \rangle + \langle y^*, y \rangle - f(x) - g(Ax + y)\} = \sup_{\substack{x \in X, \\ r \in Y}} \{\langle x^*, x \rangle +$$

$$\langle y^*, r - Ax \rangle - f(x) - g(r)\} = \sup_{\substack{x \in X, \\ r \in Y}} \{\langle x^* - A^* y^*, x \rangle + \langle y^*, r \rangle - f(x) - g(r)\}$$

$$= f^*(x^* - A^* y^*) + g^*(y^*). \tag{3.3}$$

The conjugate dual to (P^A) obtained by means of the perturbation function Φ^A is

$$(D^A) \quad \sup_{y^* \in Y^*} \{-(\Phi^A)^*(0, y^*)\},$$

which is nothing else than

$$(D^A) \quad \sup_{y^* \in Y^*} \{-f^*(-A^* y^*) - g^*(y^*)\}.$$

The dual (D^A) is the so-called *Fenchel dual* problem to (P^A). By the weak duality assertion, Theorem 3.1.1, it follows $v(D^A) \leq v(P^A)$.

Next we consider a particular instance of (P^A), by taking $X = Y$ and $A = \mathrm{id}_X$. The primal problems becomes

$$(P^{\mathrm{id}}) \quad \inf_{x \in X} \{f(x) + g(x)\},$$

while the perturbation function turns out to be $\Phi^{\mathrm{id}} : X \times X \to \overline{\mathbb{R}}$, $\Phi^{\mathrm{id}}(x, y) = f(x) + g(x + y)$. Its conjugate function $(\Phi^{\mathrm{id}})^* : X^* \times X^* \to \overline{\mathbb{R}}$ is given by the following formula

$$(\Phi^{\mathrm{id}})^*(x^*, y^*) = f^*(x^* - y^*) + g^*(y^*) \ \forall (x^*, y^*) \in X^* \times X^*. \tag{3.4}$$

The dual problem of (P^{id}) is obtained as a special case of (D^A), namely being

$$(D^{\mathrm{id}}) \quad \sup_{y^* \in X^*} \{-f^*(-y^*) - g^*(y^*)\},$$

which is actually the classical *Fenchel dual* problem to (P^{id}). The existence of weak duality between (P^{id}) and (D^{id}) follows also from Theorem 3.1.1.

The next optimization problem to which we assign a conjugate dual is another special case of (P^A) obtained by choosing $f : X \to \mathbb{R}$, $f \equiv 0$. This leads to the primal problem

$$(P^{A_g}) \quad \inf_{x \in X} \{g(Ax)\}$$

and to the corresponding perturbation function $\Phi^{A_g} : X \times Y \to \overline{\mathbb{R}}$, $\Phi^{A_g}(x, y) = g(Ax + y)$. The conjugate of Φ^{A_g} is the function $(\Phi^{A_g})^* : X^* \times Y^* \to \overline{\mathbb{R}}$ defined for all $(x^*, y^*) \in X^* \times Y^*$ by

$$(\Phi^{A_g})^*(x^*, y^*) = f^*(x^* - A^* y^*) + g^*(y^*) = \delta_{\{A^* y^*\}}(x^*) + g^*(y^*) \tag{3.5}$$

and leads to the following dual problem to (P^{A_g})

$$(D^{A_g}) \quad \sup_{\substack{y^* \in Y^*, \\ A^* y^* = 0}} \{-g^*(y^*)\}.$$

By Theorem 3.1.1 it holds $v(D^{A_g}) \leq v(P^{A_g})$.

The next pair of primal-dual problems that we consider here will be introduced as a special instance of $(P^{A_g}) - (D^{A_g})$. Let $f_i : X \to \overline{\mathbb{R}}, i = 1, ..., m$, be proper functions and take $Y = X^m$, $g : X^m \to \overline{\mathbb{R}}$, $g(x^1, ..., x^m) = \sum_{i=1}^m f_i(x^i)$ and $A : X \to X^m, Ax = (x, ..., x)$. The primal optimization problem (P^{A_g}) has now the following formulation

$$(P^{\Sigma}) \quad \inf_{x \in X} \left\{ \sum_{i=1}^m f_i(x) \right\}.$$

Since for all $(x^{1*}, ..., x^{m*}) \in (X^*)^m$, $g^*(x^{1*}, ..., x^{m*}) = \sum_{i=1}^m f_i^*(x^{i*})$ (cf. Proposition 2.3.2(l)) and $A^*(x^{1*}, ..., x^{m*}) = \sum_{i=1}^m x^{i*}$, we get by means of (D^{A_g}) the following dual problem to (P^{Σ})

$$(D^{\Sigma}) \quad \sup_{\substack{x^{i*} \in X^*, i=1,...,m, \\ \sum_{i=1}^m x^{i*} = 0}} \left\{ - \sum_{i=1}^m f_i^*(x^{i*}) \right\}.$$

Also for this pair of primal-dual problems we have weak duality, i.e. $v(D^{\Sigma}) \leq v(P^{\Sigma})$.

3.1.3 Optimization problems with geometric and cone constraints

The primal problem we consider in this subsection is

$$(P^C) \quad \inf_{x \in \mathcal{A}} f(x),$$
$$\mathcal{A} = \{x \in S : g(x) \in -C\}$$

where X is a Hausdorff locally convex space, Z is another Hausdorff locally convex space partially ordered by the convex cone $C \subseteq Z$, $S \subseteq X$ is a given nonempty set, $f : X \to \overline{\mathbb{R}}$ a proper function and $g : X \to \overline{Z} = Z \cup \{\pm\infty_C\}$ a proper vector function fulfilling $\operatorname{dom} f \cap S \cap g^{-1}(-C) \neq \emptyset$.

By considering different perturbation functions we first assign in the following three dual problems to (P^C) and then we establish the relations between their optimal objective values. The first dual we get is the classical Lagrange dual.

To begin let take Z as the space of the perturbation variables and define $\Phi^{C_L} : X \times Z \to \overline{\mathbb{R}}$,

$$\Phi^{C_L}(x, z) = \begin{cases} f(x), & \text{if } x \in S, g(x) \in z - C, \\ +\infty, & \text{otherwise.} \end{cases}$$

For this choice of the perturbation function we have $\Phi^{C_L}(x, 0) = f(x) + \delta_{\mathcal{A}}(x)$ for all $x \in X$. The conjugate function of Φ^{C_L}, $(\Phi^{C_L})^* : X^* \times Z^* \to \overline{\mathbb{R}}$ has for all $(x^*, z^*) \in X^* \times Z^*$ the following form

$$(\Phi^{C_L})^*(x^*, z^*) = \sup_{\substack{x \in X, \\ z \in Z}} \{\langle x^*, x\rangle + \langle z^*, z\rangle - \Phi^{C_L}(x, z)\} = \sup_{\substack{x \in S, z \in Z, \\ g(x) - z \in -C}} \{\langle x^*, x\rangle +$$

$$\langle z^*, z\rangle - f(x)\} = \sup_{x\in S, s\in -C} \{\langle x^*, x\rangle + \langle z^*, g(x) - s\rangle - f(x)\} =$$

$$\sup_{s\in -C} \{\langle -z^*, s\rangle\} + \sup_{x\in S}\{\langle x^*, x\rangle + \langle z^*, g(x)\rangle - f(x)\}.$$

Since for all $z^* \in Z^*$

$$\sup_{s\in -C} \{\langle -z^*, s\rangle\} = \delta_{-C^*}(z^*),$$

we get further for all $(x^*, z^*) \in X^* \times Z^*$

$$(\Phi^{C_L})^*(x^*, z^*) = (f + (-z^* g))^*_S(x^*) + \delta_{-C^*}(z^*). \tag{3.6}$$

Thus the dual problem to (P^C) which we obtain by means of the perturbation function Φ^{C_L} is

$$(D^{C_L}) \quad \sup_{z^*\in Z^*} \left\{-(\Phi^{C_L})^*(0, z^*)\right\},$$

which becomes

$$(D^{C_L}) \quad \sup_{z^*\in C^*} \left\{-(f + (z^* g))^*_S(0)\right\},$$

or, equivalently,

$$(D^{C_L}) \quad \sup_{z^*\in C^*} \inf_{x\in S}\{f(x) + (z^* g)(x)\}.$$

The optimization problem (D^{C_L}) is the classical *Lagrange dual* problem to (P^C). By Theorem 3.1.1 we automatically have that $v(D^{C_L}) \leq v(P^C)$.

The second perturbation function we consider for (P^C) is $\Phi^{C_F} : X \times X \to \overline{\mathbb{R}}$,

$$\Phi^{C_F}(x, y) = \begin{cases} f(x+y), & \text{if } x \in S, g(x) \in -C, \\ +\infty, & \text{otherwise,} \end{cases}$$

with X being the space of the perturbation variables. Obviously, $\Phi^{C_F}(x, 0) = f(x) + \delta_{\mathcal{A}}(x)$ for all $x \in X$. Because of $\Phi^{C_F}(x, y) = \delta_{\mathcal{A}}(x) + f(x+y)$ for all $(x, y) \in X \times X$, the formula of its conjugate function $(\Phi^{C_F})^* : X^* \times X^* \to \overline{\mathbb{R}}$ can be directly provided via (3.4). Thus for all $(x^*, y^*) \in X^* \times X^*$ we get

$$(\Phi^{C_F})^*(x^*, y^*) = \sigma_{\mathcal{A}}(x^* - y^*) + f^*(y^*) \tag{3.7}$$

and this leads to the following conjugate dual problem to (P^C)

$$(D^{C_F}) \quad \sup_{y^*\in X^*} \{-f^*(y^*) - \sigma_{\mathcal{A}}(-y^*)\}.$$

Since the primal problem (P^C) can be written as

$$(P^C) \quad \inf_{x\in X} \left\{f(x) + \delta_{\mathcal{A}}(x)\right\},$$

one can notice that (D^{C_F}) is nothing else than its Fenchel dual problem (cf. subsection 3.1.2). This is the reason why we call (D^{C_F}) the *Fenchel dual* problem to (P^C). Also in this case the weak duality, i.e. $v(D^{C_F}) \leq v(P^C)$, is fulfilled.

The last conjugate dual problem we consider to (P^C) is obtained by perturbing both the argument of the objective function and the cone constraints. We take $X \times Z$ as the space of perturbation variables and define $\Phi^{C_{FL}} : X \times X \times Z \to \overline{\mathbb{R}}$ by

$$\Phi^{C_{FL}}(x, y, z) = \begin{cases} f(x+y), & \text{if } x \in S, g(x) \in z - C, \\ +\infty, & \text{otherwise.} \end{cases}$$

The equality $\Phi^{C_{FL}}(x, 0, 0) = f(x) + \delta_{\mathcal{A}}(x)$ is again for all $x \in X$ fulfilled. For $(x^*, y^*, z^*) \in X^* \times X^* \times Z^*$ the conjugate function of $\Phi^{C_{FL}}$, $(\Phi^{C_{FL}})^* : X^* \times X^* \times Z^* \to \overline{\mathbb{R}}$, looks like

$$\begin{aligned}(\Phi^{C_{FL}})^*(x^*, y^*, z^*) &= \sup_{\substack{x \in X, y \in X,\\ z \in Z}} \{\langle x^*, x\rangle + \langle y^*, y\rangle + \langle z^*, z\rangle - \Phi^{C_{FL}}(x, y, z)\} = \\ \sup_{\substack{x \in S, y \in X, z \in Z,\\ g(x) - z \in -C}} &\{\langle x^*, x\rangle + \langle y^*, y\rangle + \langle z^*, z\rangle - f(x+y)\} = \sup_{\substack{x \in S, r \in X,\\ s \in -C}} \{\langle x^*, x\rangle + \\ &\langle y^*, r - x\rangle + \langle z^*, g(x) - s\rangle - f(r)\} = \sup_{x \in S}\{\langle x^* - y^*, x\rangle + \langle z^*, g(x)\rangle\} + \\ &\sup_{r \in X}\{\langle y^*, r\rangle - f(r)\} + \sup_{s \in -C}\{\langle -z^*, s\rangle\} \\ &= f^*(y^*) + (-z^* g)^*_S(x^* - y^*) + \delta_{-C^*}(z^*). \end{aligned} \tag{3.8}$$

One can define now the following dual problem to (P^C)

$$(D^{C_{FL}}) \qquad \sup_{y^* \in X^*, z^* \in Z^*} \left\{-(\Phi^{C_{FL}})^*(0, y^*, z^*)\right\},$$

which is actually

$$(D^{C_{FL}}) \qquad \sup_{y^* \in X^*, z^* \in C^*} \left\{-f^*(y^*) - (z^* g)^*_S(-y^*)\right\}.$$

We call $(D^{C_{FL}})$ the *Fenchel-Lagrange dual* problem of (P^C), since it can be seen as a combination of the classical Fenchel and Lagrange duals. By the weak duality theorem we have $v(D^{C_{FL}}) \leq v(P^C)$.

The three dual problems considered here for the optimization problem with geometric and cone constraints have been introduced and studied for problems in finite dimensional spaces in [186] and for problems in infinite dimensional spaces in [39].

Remark 3.1.2. The name Fenchel-Lagrange for the dual problem $(D^{C_{FL}})$ is motivated by the fact that in the definition of $\Phi^{C_{FL}}$ we perturb both the cone constraints (like for (D^{C_L})) and the argument of the objective function (like for (D^{C_F})). Another motivation for the choice of this name can be found in the following considerations.

The Lagrange dual problem to (P^C)

$$(D^{C_L}) \quad \sup_{z^* \in C^*} \inf_{x \in S} \{f(x) + (z^* g)(x)\}$$

can also be formulated as

$$(D^{C_L}) \quad \sup_{z^* \in C^*} \inf_{x \in X} \{f(x) + ((z^* g) + \delta_S)(x)\}.$$

Consider for a fixed $z^* \in C^*$ the infimum problem in the objective function of the problem above

$$\inf_{x \in X} \{f(x) + ((z^* g) + \delta_S)(x)\}.$$

Its Fenchel dual (cf. subsection 3.1.2) is

$$\sup_{y^* \in X^*} \{-f^*(y^*) - (z^* g)_S^*(-y^*)\}.$$

By taking in the objective function of (D^{C_L}) instead of the infimum problem its Fenchel dual, one gets exactly the Fenchel-Lagrange dual to (P^C) (weak duality is automatically ensured).

Consider now the Fenchel dual problem of (P^C)

$$(D^{C_F}) \quad \sup_{y^* \in X^*} \{-f^*(y^*) - \sigma_{\mathcal{A}}(-y^*)\},$$

which is the same as

$$(D^{C_F}) \quad \sup_{y^* \in X^*} \{-f^*(y^*) + \inf_{x \in \mathcal{A}} \langle y^*, x\rangle\}.$$

We fix an element $y^* \in X^*$ and consider the infimum problem which appears in the objective function of (D^{C_F})

$$\inf_{x \in \mathcal{A}} \langle y^*, x\rangle.$$

Its Lagrange dual is

$$\sup_{z^* \in C^*} \inf_{x \in S} \{\langle y^*, x\rangle + (z^* g)(x)\} = \sup_{z^* \in C^*} \{-(z^* g)_S^*(-y^*)\}.$$

By taking in the objective function of (D^{C_F}) instead of the infimum problem its Lagrange dual, what we get is again exactly the Fenchel-Lagrange dual to (P^C) (weak duality is also in this situation automatically ensured).

Having now the three dual problems for the optimization problem with geometric and cone constraints, it is natural to try to find out which relations exist between their optimal objective values. The weak duality theorem Theorem 3.1.1 is guaranteeing that these values are less than or equal to the optimal objective value of the primal problem, but the following results offer a more precise answer to this question.

Proposition 3.1.5. *It holds* $v(D^{C_{FL}}) \leq v(D^{C_L})$.

Proof. Let $z^* \in C^*$ be fixed. Since the optimization problem

$$\sup_{y^* \in X^*} \{-f^*(y^*) - (z^*g)^*_S(-y^*)\}$$

is the Fenchel dual to

$$\inf_{x \in X} \{f(x) + ((z^*g) + \delta_S)(x)\},$$

we obtain that (cf. Theorem 3.1.1)

$$\sup_{y^* \in X^*} \{-f^*(y^*) - (z^*g)^*_S(-y^*)\} \leq \inf_{x \in X} \{f(x) + ((z^*g) + \delta_S)(x)\}.$$

Taking the supremum over $z^* \in C^*$ in both sides of the equality above, the inequality $v(D^{C_{FL}}) \leq v(D^{C_L})$ follows automatically. □

Proposition 3.1.6. *It holds* $v(D^{C_{FL}}) \leq v(D^{C_F})$.

Proof. Let be $y^* \in X^*$ fixed. We have seen in Remark 3.1.2 that

$$\sup_{z^* \in C^*} \{-(z^*g)^*_S(-y^*)\}$$

is the Lagrange dual problem to

$$\inf_{x \in \mathcal{A}} \langle y^*, x \rangle.$$

This implies that

$$-f^*(y^*) + \sup_{z^* \in C^*} \{-(z^*g)^*_S(-y^*)\} \leq -f^*(y^*) + \inf_{x \in \mathcal{A}} \langle y^*, x \rangle.$$

Taking the supremum over $y^* \in X^*$ in both sides of the inequality above, we get $v(D^{C_{FL}}) \leq v(D^{C_F})$. □

Combining the results of the last two propositions we obtain the following scheme for the relations between the optimal objective values of the primal problem (P^C) and of the three conjugate duals introduced in this subsection (see also [39, 186])

$$v(D^{C_{FL}}) \leq \begin{array}{c} v(D^{C_L}) \\ v(D^{C_F}) \end{array} \leq v(P^C). \tag{3.9}$$

Remark 3.1.3. In [186] one can find examples which show that the inequalities in (3.9) can be strict and, on the other hand, that in general between $v(D^{C_L})$ and $v(D^{C_F})$ no ordering relation can be established. In order to close the gap between the optimal objective values in (3.9) and to guarantee the existence of optimal solutions to the duals, one needs so-called *regularity conditions.* The next section is dedicated to the formulation of different regularity conditions and strong duality results for the primal-dual pairs considered in this section.

3.2 Regularity conditions and strong duality

The *regularity conditions* we give first are regarding the primal optimization problem (PG) and its conjugate dual (DG). They are expressed by means of the perturbation function Φ and guarantee the stability of (PG). Afterwards we derive from these general conditions corresponding regularity conditions for the different classes of primal-dual problems considered in section 3.1 and also state strong duality theorems.

3.2.1 Regularity conditions for the general scalar optimization problem

Throughout this subsection we assume that Φ is a proper and convex function with $0 \in \Pr_Y(\operatorname{dom} \Phi)$.

The first regularity conditions we give are so-called *generalized interior point* regularity conditions. We start with a classical regularity condition stated in a general framework, in which the space of the feasible variables X and the space of the perturbation variables Y are assumed to be Hausdorff locally convex spaces:

$$(RC_1^{\Phi}) \;\big|\; \exists x' \in X \text{ such that } (x',0) \in \operatorname{dom} \Phi \text{ and } \Phi(x',\cdot) \text{ is continuous at } 0.$$

For the next three regularity conditions we have to assume that X and Y are Fréchet spaces (cf. [158, 205]):

$$(RC_2^{\Phi}) \;\left|\; \begin{array}{l} X \text{ and } Y \text{ are Fréchet spaces, } \Phi \text{ is lower semicontinuous and} \\ 0 \in \operatorname{sqri}(\Pr_Y(\operatorname{dom} \Phi)). \end{array}\right.$$

The regularity condition (RC_2^{Φ}) is the weakest one in relation to the following regularity conditions which involve further generalized interiority notions (cf. [158]):

$$(RC_{2'}^{\Phi}) \;\left|\; \begin{array}{l} X \text{ and } Y \text{ are Fréchet spaces, } \Phi \text{ is lower semicontinuous and} \\ 0 \in \operatorname{core}(\Pr_Y(\operatorname{dom} \Phi)), \end{array}\right.$$

respectively,

$$(RC_{2''}^{\Phi}) \;\left|\; \begin{array}{l} X \text{ and } Y \text{ are Fréchet spaces, } \Phi \text{ is lower semicontinuous and} \\ 0 \in \operatorname{int}(\Pr_Y(\operatorname{dom} \Phi)). \end{array}\right.$$

Regarding the last two conditions we want to make the following comment. In case Φ is convex and lower semicontinuous its infimal value function $h : Y \to \overline{\mathbb{R}}$, $h(y) = \inf_{x \in X} \Phi(x,y)$, is convex but not necessarily lower semicontinuous, fulfilling $\operatorname{dom} h = \Pr_Y(\operatorname{dom} \Phi)$. Nevertheless, when X and Y are Fréchet spaces, by [207, Proposition 2.2.18] the function h is li-convex. Using [207, Theorem 2.2.20] it follows that $\operatorname{core}(\operatorname{dom} h) = \operatorname{int}(\operatorname{dom} h)$, which has as consequence the equivalence of the regularity conditions $(RC_{2'}^{\Phi})$ and $(RC_{2''}^{\Phi})$. Thus, $(RC_{2''}^{\Phi}) \Leftrightarrow (RC_{2'}^{\Phi}) \Rightarrow (RC_2^{\Phi})$, all these conditions being implied by (RC_1^{Φ}) when X and Y are Fréchet spaces and Φ is lower semicontinuous.

Another generalized interior point regularity condition, we consider of interest especially when dealing with convex optimization problems in finite dimensional spaces, is the following (cf. [157, 207]):

$$(RC_3^{\Phi}) \mid \dim(\operatorname{lin}(\operatorname{Pr}_Y(\operatorname{dom}\Phi))) < +\infty \text{ and } 0 \in \operatorname{ri}(\operatorname{Pr}_Y(\operatorname{dom}\Phi)).$$

Now we state the strong duality theorem for the pair of primal-dual problems $(PG) - (DG)$ (for the proof, see [67, 157, 158, 207]).

Theorem 3.2.1. *Let $\Phi : X \times Y \to \overline{\mathbb{R}}$ be a proper and convex function such that $0 \in \operatorname{Pr}_Y(\operatorname{dom}\Phi)$. If one of the regularity conditions (RC_i^{Φ}), $i \in \{1,2,3\}$, is fulfilled, then $v(PG) = v(DG)$ and the dual has an optimal solution.*

Let us come now to a different class of regularity conditions, called *closedness type* regularity conditions. They are expressed by means of the epigraph of the conjugate of the perturbation function Φ, provided that X and Y are Hausdorff locally convex spaces and Φ is a proper, convex and lower semicontinuous function.

We prove first the following important result (see also [155]).

Theorem 3.2.2. *Let $\Phi : X \times Y \to \overline{\mathbb{R}}$ be a proper, convex and lower semicontinuous function fulfilling that $0 \in \operatorname{Pr}_Y(\operatorname{dom}\Phi)$. Then the following statements are equivalent:*

(i) $\sup\limits_{x \in X}\{\langle x^*, x\rangle - \Phi(x,0)\} = \min\limits_{y^* \in Y^*}\{\Phi^*(x^*, y^*)\}\ \forall x^* \in X^*$;
(ii) $\operatorname{Pr}_{X^* \times \mathbb{R}}(\operatorname{epi}\Phi^*)$ *is closed in the topology* $w(X^*, X) \times \mathbb{R}$.

Proof. Let $\eta : X^* \to \overline{\mathbb{R}}$, $\eta(x^*) = \inf\{\Phi^*(x^*, y^*) : y^* \in Y^*\}$ be the infimal value function of the conjugate function of Φ. Then, as pointed out in Remark 2.2.7, we have that

$$\operatorname{Pr}_{X^* \times \mathbb{R}}(\operatorname{epi}\Phi^*) \subseteq \operatorname{epi}\eta \subseteq \operatorname{cl}(\operatorname{Pr}_{X^* \times \mathbb{R}}(\operatorname{epi}\Phi^*)). \tag{3.10}$$

By using the Fenchel-Moreau theorem (Theorem 2.3.6) we obtain the following sequence of equalities which hold for all $x \in X$

$$\eta^*(x) = \sup_{x^* \in X^*}\{\langle x^*, x\rangle - \eta(x^*)\} = \sup_{x^* \in X^*}\left\{\langle x^*, x\rangle - \inf_{y^* \in Y^*}\{\Phi^*(x^*, y^*)\}\right\}$$

$$= \sup_{x^* \in X^*, y^* \in Y^*}\{\langle x^*, x\rangle - \Phi^*(x^*, y^*)\} = \Phi^{**}(x,0) = \Phi(x,0).$$

As η is convex (cf. Theorem 2.2.6), it follows that $\bar{\eta}$ is also convex. More than that, the latter is not identical $+\infty$. Otherwise would follow that $(\bar{\eta})^*(x) = \eta^*(x) = -\infty$ for all $x \in X$, but this would contradict the properness of Φ.

Because of $0 \in \operatorname{Pr}_Y(\operatorname{dom}\Phi)$, there exists $x_0 \in X$ such that $\Phi(x_0, 0) < +\infty$. Then by the Young-Fenchel inequality we get

$$\eta(x^*) = \inf_{y^* \in Y^*}\{\Phi^*(x^*, y^*)\} \geq \langle x^*, x_0\rangle - \Phi(x_0, 0)\ \forall x^* \in X^*$$

and this yields $\overline{\eta}(x^*) \geq \langle x^*, x_0 \rangle - \Phi(x_0, 0) > -\infty$ for all $x^* \in X^*$.

This means that $\overline{\eta}$ is also proper and by using Corollary 2.3.7 we obtain for all $x^* \in X^*$

$$\sup_{x \in X}\{\langle x^*, x \rangle - \Phi(x, 0)\} = \sup_{x \in X}\{\langle x^*, x \rangle - \eta^*(x)\}$$

$$= \eta^{**}(x^*) = \overline{\eta}(x^*) \leq \eta(x^*) = \inf_{y^* \in Y^*}\{\Phi^*(x^*, y^*)\}. \tag{3.11}$$

$(i) \Rightarrow (ii)$. In case relation (i) is fulfilled, from (3.11) follows that η is lower semicontinuous and that for all $x^* \in X^*$ there exists $y^* \in Y^*$ such that $\eta(x^*) = \Phi^*(x^*, y^*)$. From here we have that $\operatorname{epi} \eta \subseteq \operatorname{Pr}_{X^* \times \mathbb{R}}(\operatorname{epi} \Phi^*)$. Taking now into consideration (3.10) it follows that $\operatorname{epi} \eta = \operatorname{Pr}_{X^* \times \mathbb{R}}(\operatorname{epi} \Phi^*)$ and this proves that $\operatorname{Pr}_{X^* \times \mathbb{R}}(\operatorname{epi} \Phi^*)$ is closed in the topology $w(X^*, X) \times \mathbb{R}$.

$(ii) \Rightarrow (i)$. Assuming that $\operatorname{Pr}_{X^* \times \mathbb{R}}(\operatorname{epi} \Phi^*)$ is closed in the topology $w(X^*, X) \times \mathbb{R}$, relation (3.10) implies that η is a lower semicontinuous function and $\operatorname{epi} \eta = \operatorname{Pr}_{X^* \times \mathbb{R}}(\operatorname{epi} \Phi^*)$. This is nothing else than that for all $x^* \in X^*$ $\eta(x^*) = \min_{y^* \in Y^*}\{\Phi^*(x^*, y^*)\}$. Considering again (3.11) one can easily conclude that for all $x^* \in X^*$ the equality

$$\sup_{x \in X}\{\langle x^*, x \rangle - \Phi(x, 0)\} = \min_{y^* \in Y^*}\{\Phi^*(x^*, y^*)\}$$

holds. □

Inspired by Theorem 3.2.2 we can state the following closedness type regularity condition:

(RC_4^Φ) | Φ is lower semicontinuous and $\operatorname{Pr}_{X^* \times \mathbb{R}}(\operatorname{epi} \Phi^*)$ is closed in the topology $w(X^*, X) \times \mathbb{R}$.

Theorem 3.2.3. *Let $\Phi : X \times Y \to \overline{\mathbb{R}}$ be a proper and convex function such that $0 \in \operatorname{Pr}_Y(\operatorname{dom} \Phi)$. If the regularity condition (RC_4^Φ) is fulfilled, then $v(PG) = v(DG)$ and the dual has an optimal solution.*

Proof. The conclusion follows from Theorem 3.2.2$(ii) \Rightarrow (i)$, by considering (i) for $x^* = 0$. □

Remark 3.2.1. Whenever Φ is proper, convex and lower semicontinuous, then the conditions $(RC_i^\Phi), i \in \{2, 2', 2''\}$, are sufficient for having (RC_4^Φ) fulfilled.

In what follows we particularize the general regularity conditions as well as the strong duality theorems introduced above for the primal-dual pairs of optimization problems given in the previous section.

3.2.2 Regularity conditions for problems having the composition with a linear continuous mapping in the objective function

For X and Y Hausdorff locally convex spaces, $A \in \mathcal{L}(X,Y)$, $f : X \to \overline{\mathbb{R}}$ and $g : Y \to \overline{\mathbb{R}}$ proper and convex functions fulfilling $\operatorname{dom} f \cap A^{-1}(\operatorname{dom} g) \neq \emptyset$ we consider the optimization problem from subsection 3.1.2

$$(P^A) \quad \inf_{x \in X} \{f(x) + g(Ax)\},$$

along with the perturbation function $\Phi^A : X \times Y \to \overline{\mathbb{R}}$, $\Phi^A(x,y) = f(x) + g(Ax+y)$.

The regularity condition (RC_1^Φ) becomes in this particular case

$$(RC_1^A) \left| \begin{array}{l} \exists x' \in X \text{ such that } f(x') + g(Ax') < +\infty \text{ and the function} \\ y \mapsto f(x') + g(Ax'+y) \text{ is continuous at } 0 \end{array} \right.$$

or, equivalently,

$$(RC_1^A) \mid \exists x' \in \operatorname{dom} f \cap A^{-1}(\operatorname{dom} g) \text{ such that } g \text{ is continuous at } Ax'.$$

In Fréchet spaces, since $\Pr_Y(\operatorname{dom} \Phi^A) = \{y \in Y : \exists x \in \operatorname{dom} f \text{ such that } y \in \operatorname{dom} g - Ax\} = \operatorname{dom} g - A(\operatorname{dom} f)$, we can state the following regularity condition for $(P^A) - (D^A)$

$$(RC_2^A) \left| \begin{array}{l} X \text{ and } Y \text{ are Fréchet spaces, } f \text{ and } g \text{ are lower semicontinuous} \\ \text{and } 0 \in \operatorname{sqri}(\operatorname{dom} g - A(\operatorname{dom} f)) \end{array} \right.$$

along with its stronger versions

$$(RC_{2'}^A) \left| \begin{array}{l} X \text{ and } Y \text{ are Fréchet spaces, } f \text{ and } g \text{ are lower semicontinuous} \\ \text{and } 0 \in \operatorname{core}(\operatorname{dom} g - A(\operatorname{dom} f)) \end{array} \right.$$

and

$$(RC_{2''}^A) \left| \begin{array}{l} X \text{ and } Y \text{ are Fréchet spaces, } f \text{ and } g \text{ are lower semicontinuous} \\ \text{and } 0 \in \operatorname{int}(\operatorname{dom} g - A(\operatorname{dom} f)), \end{array} \right.$$

which are in fact equivalent.

The condition (RC_2^A) was introduced in [160], while $(RC_{2'}^A)$ was given for the first time in Banach spaces in [158]. In the finite dimensional case one has for the the pair of problems $(P^A) - (D^A)$ the following regularity condition

$$(RC_3^A) \mid \dim(\operatorname{lin}(\operatorname{dom} g - A(\operatorname{dom} f))) < +\infty \text{ and } \operatorname{ri}(A(\operatorname{dom} f)) \cap \operatorname{ri}(\operatorname{dom} g) \neq \emptyset.$$

The next strong duality theorem follows by Theorem 3.2.1.

Theorem 3.2.4. *Let $f : X \to \overline{\mathbb{R}}, g : Y \to \overline{\mathbb{R}}$ be proper and convex functions and $A \in \mathcal{L}(X,Y)$ such that* $\operatorname{dom} f \cap A^{-1}(\operatorname{dom} g) \neq \emptyset$. *If one of the regularity conditions (RC_i^A), $i \in \{1,2,3\}$, is fulfilled, then $v(P^A) = v(D^A)$ and the dual has an optimal solution.*

For further regularity conditions, expressed my means of the quasi interior and quasi relative interior of the domains of the functions involved, which also guarantee strong duality for $(P^A) - (D^A)$, we refer to [27]. In order to derive an appropriate closedness type condition for this primal-dual pair we use the formula of the conjugate of Φ^A (cf. (3.3)), which states that for all $(x^*, y^*) \in X^* \times Y^*$ it holds $(\Phi^A)^*(x^*, y^*) = f^*(x^* - A^*y^*) + g^*(y^*)$. Thus

$$(x^*, r) \in \Pr_{X^* \times \mathbb{R}}(\operatorname{epi}(\Phi^A)^*) \Leftrightarrow \exists y^* \in Y^* \text{ such that } f^*(x^* - A^*y^*)$$
$$+g^*(y^*) \leq r \Leftrightarrow \exists y^* \in Y^* \text{ such that } (x^* - A^*y^*, r - g^*(y^*)) \in \operatorname{epi} f^*$$
$$\Leftrightarrow \exists y^* \in Y^* \text{ such that } (x^*, r) \in \operatorname{epi} f^* + (A^*y^*, g^*(y^*))$$
$$\Leftrightarrow (x^*, r) \in \operatorname{epi} f^* + (A^* \times \mathrm{id}_{\mathbb{R}})(\operatorname{epi} g^*).$$

Here $A^* \times \mathrm{id}_{\mathbb{R}} : Y^* \times \mathbb{R} \to X^* \times \mathbb{R}$ is defined as $(A^* \times \mathrm{id}_{\mathbb{R}})(y^*, r) = (A^*y^*, r)$. This leads to the following regularity condition

(RC_4^A) | f and g are lower semicontinuous and $\operatorname{epi} f^* + (A^* \times \mathrm{id}_{\mathbb{R}})(\operatorname{epi} g^*)$ is closed in the topology $w(X^*, X) \times \mathbb{R}$.

From Theorem 3.2.3 one can deduce the following strong duality theorem.

Theorem 3.2.5. *Let $f : X \to \overline{\mathbb{R}}$ and $g : Y \to \overline{\mathbb{R}}$ be proper and convex functions and $A \in \mathcal{L}(X, Y)$ such that* $\operatorname{dom} f \cap A^{-1}(\operatorname{dom} g) \neq \emptyset$. *If the regularity condition (RC_4^A) is fulfilled, then $v(P^A) = v(D^A)$ and the dual has an optimal solution.*

Remark 3.2.2. The regularity condition (RC_4^A) has been introduced by Boţ and Wanka in [38]. In case the functions f and g are lower semicontinuous (RC_4^A) is proven to be implied by the generalized interior point regularity conditions given in the literature for the pair of problems $(P^A) - (D^A)$. In [38] it is shown by means of some examples that (RC_4^A) is indeed weaker than the other regularity conditions considered for this pair of primal-dual problems.

In case $X = Y$ and $A = \mathrm{id}_X$ the regularity conditions enunciated for (P^A) become

(RC_1^{id}) | $\exists x' \in \operatorname{dom} f \cap \operatorname{dom} g$ such that g is continuous at x',

in case X is a Fréchet space

(RC_2^{id}) | X is a Fréchet space, f and g are lower semicontinuous and $0 \in \operatorname{sqri}(\operatorname{dom} g - \operatorname{dom} f)$,

along with its stronger versions

$(RC_{2'}^{\mathrm{id}})$ | X is a Fréchet space, f and g are lower semicontinuous and $0 \in \operatorname{core}(\operatorname{dom} g - \operatorname{dom} f)$

and

$(RC_{2''}^{\mathrm{id}})$ | X is a Fréchet space, f and g are lower semicontinuous and $0 \in \operatorname{int}(\operatorname{dom} g - \operatorname{dom} f)$, which are in fact equivalent, in the finite dimensional case

$$(RC_3^{\text{id}}) \mid \dim(\text{lin}(\text{dom}\, g - \text{dom}\, f)) < +\infty \text{ and } \text{ri}(\text{dom}\, f) \cap \text{ri}(\text{dom}\, g) \neq \emptyset,$$

while the closedness type regularity condition states

$$(RC_4^{\text{id}}) \left| \begin{array}{l} f \text{ and } g \text{ are lower semicontinuous and } \text{epi}\, f^* + \text{epi}\, g^* \\ \text{is closed in the topology } w(X^*, X) \times \mathbb{R}, \end{array} \right.$$

respectively. Condition (RC_2^{id}) was introduced by Attouch and Breézis in [7] (and bears their names), while (RC_4^{id}) is due to Burachik and Jeyakumar (cf. [41]; see also the paper of Boţ and Wanka [38]). The strong duality theorem follows automatically from Theorem 3.2.4 and Theorem 3.2.5.

Theorem 3.2.6. *Let $f, g : X \to \overline{\mathbb{R}}$ be proper and convex functions such that $\text{dom}\, f \cap \text{dom}\, g \neq \emptyset$. If one of the regularity conditions (RC_i^{id}), $i \in \{1, 2, 3, 4\}$, is fulfilled, then $v(P^{\text{id}}) = v(D^{\text{id}})$ and the dual has an optimal solution.*

We come now to a second special case, namely the one where $f \equiv 0$. In this case it obviously holds $\text{epi}\, f^* = \{0\} \times \mathbb{R}_+$. The regularity conditions introduced for (P^A) give rise to the following formulations

$$(RC_1^{A_g}) \mid \exists x' \in A^{-1}(\text{dom}\, g) \text{ such that } g \text{ is continuous at } Ax',$$

in case X and Y are Fréchet spaces

$$(RC_2^{A_g}) \left| \begin{array}{l} X \text{ and } Y \text{ are Fréchet spaces, } g \text{ is lower semicontinuous} \\ \text{and } 0 \in \text{sqri}(\text{dom}\, g - A(X)), \end{array} \right.$$

along with its stronger versions

$$(RC_{2'}^{A_g}) \left| \begin{array}{l} X \text{ and } Y \text{ are Fréchet spaces, } g \text{ is lower semicontinuous} \\ \text{and } 0 \in \text{core}(\text{dom}\, g - A(X)) \end{array} \right.$$

and

$$(RC_{2''}^{A_g}) \left| \begin{array}{l} X \text{ and } Y \text{ are Fréchet spaces, } g \text{ is lower semicontinuous} \\ \text{and } 0 \in \text{int}(\text{dom}\, g - A(X)), \end{array} \right.$$

which are in fact equivalent, in the finite dimensional case

$$(RC_3^{A_g}) \mid \dim(\text{lin}(\text{dom}\, g - A(X))) < +\infty \text{ and } \text{ri}(A(X)) \cap \text{ri}(\text{dom}\, g) \neq \emptyset,$$

while the closedness type regularity condition states that

$$(RC_4^{A_g}) \left| \begin{array}{l} g \text{ is lower semicontinuous and } (A^* \times \text{id}_{\mathbb{R}})(\text{epi}\, g^*) \\ \text{is closed in the topology } w(X^*, X) \times \mathbb{R}. \end{array} \right.$$

Theorem 3.2.7. *Let $g : Y \to \overline{\mathbb{R}}$ be a proper and convex function and $A \in \mathcal{L}(X, Y)$ such that $A^{-1}(\text{dom}\, g) \neq \emptyset$. If one of the regularity conditions $(RC_i^{A_g})$, $i \in \{1, 2, 3, 4\}$, is fulfilled, then $v(P^{A_g}) = v(D^{A_g})$ and the dual has an optimal solution.*

Next we particularize the problem (P^{A_g}) in a similar way like in section 3.1. Let $f_i : X \to \overline{\mathbb{R}}, i = 1, ..., m$, be proper and convex functions, $Y = X^m$, $g : X^m \to \overline{\mathbb{R}}$, $g(x^1, ..., x^m) = \sum_{i=1}^m f_i(x^i)$ and $A : X \to X^m$, $Ax = (x, ..., x)$. Obviously, it holds $\operatorname{dom} g = \prod_{i=1}^m \operatorname{dom} f_i$ and $A^{-1}(\operatorname{dom} g) = \bigcap_{i=1}^m \operatorname{dom} f_i$. In this situation the condition $(RC_1^{A_g})$ states that there exists $x' \in \bigcap\limits_{i=1}^m \operatorname{dom} f_i$ such that f_i is continuous at $x', i = 1, ..., m$. Nevertheless, we state here a weaker condition, asking (only) that

$$(RC_1^\Sigma) \left| \begin{array}{l} \exists x' \in \bigcap\limits_{i=1}^m \operatorname{dom} f_i \text{ such that } m-1 \text{ of the functions} \\ f_i, i = 1, ..., m, \text{ are continuous at } x'. \end{array} \right.$$

That (RC_1^Σ) guarantees strong duality for the primal-dual pair $(P^\Sigma) - (D^\Sigma)$ follows, for instance, by applying (RC_1^{id}) for $m-1$ times.

Coming back to the special choice of the function g and of the mapping A from above, one has $\operatorname{dom} g - A(X) = \prod_{i=1}^m \operatorname{dom} f_i - \Delta_{X^m}$. Thus $(RC_i^{A_g})$, $i \in \{2, 2', 2'', 3\}$, lead to the following regularity conditions (if X is a Fréchet space, then X^m is a Fréchet space, too):

$$(RC_2^\Sigma) \left| \begin{array}{l} X \text{ is a Fréchet space, } f_i \text{ is lower semicontinuous, } i = 1, ..., m, \\ \text{and } 0 \in \operatorname{sqri}\left(\prod\limits_{i=1}^m \operatorname{dom} f_i - \Delta_{X^m}\right), \end{array} \right.$$

along with its stronger versions

$$(RC_{2'}^\Sigma) \left| \begin{array}{l} X \text{ is a Fréchet space, } f_i \text{ is lower semicontinuous, } i = 1, ..., m, \\ \text{and } 0 \in \operatorname{core}\left(\prod\limits_{i=1}^m \operatorname{dom} f_i - \Delta_{X^m}\right) \end{array} \right.$$

and

$$(RC_{2''}^\Sigma) \left| \begin{array}{l} X \text{ is a Fréchet space, } f_i \text{ is lower semicontinuous, } i = 1, ..., m, \\ \text{and } 0 \in \operatorname{int}\left(\prod\limits_{i=1}^m \operatorname{dom} f_i - \Delta_{X^m}\right), \end{array} \right.$$

which are in fact equivalent, while in the finite dimensional case we have

$$(RC_3^\Sigma) \left| \dim\left(\operatorname{lin}\left(\prod_{i=1}^m \operatorname{dom} f_i - \Delta_{X^m}\right)\right) < +\infty \text{ and } \bigcap_{i=1}^m \operatorname{ri}(\operatorname{dom} f_i) \neq \emptyset. \right.$$

For stating the closedness type regularity condition we need to establish the shape of the set $(A^* \times \mathrm{id}_{\mathbb{R}})(\operatorname{epi} g^*)$. One has

$$(x^*, r) \in (A^* \times \mathrm{id}_{\mathbb{R}})(\operatorname{epi} g^*) \Leftrightarrow \exists (x^{1*}, ..., x^{m*}) \in (X^*)^m \text{ such that}$$

$$g^*(x^{1*}, ..., x^{m*}) \leq r \text{ and } A^*(x^{1*}, ..., x^{m*}) = x^* \Leftrightarrow \exists (x^{1*}, ..., x^{m*}) \in (X^*)^m$$

$$\text{such that} \sum_{i=1}^m f_i^*(x^{i*}) \leq r \text{ and } \sum_{i=1}^m x^{i*} = x^* \Leftrightarrow (x^*, r) \in \sum_{i=1}^m \operatorname{epi} f_i^*.$$

Thus $(RC_4^{A_g})$ is nothing else than

(RC_4^{Σ}) | f_i is lower semicontinuous, $i = 1, ..., m$, and $\sum_{i=1}^{m} \operatorname{epi} f_i^*$ is closed in the topology $w(X^*, X) \times \mathbb{R}$.

Remark 3.2.3. The regularity condition (RC_1^{Σ}) can also be obtained as a particular instance of (RC_1^{Φ}) when considering an appropriate perturbation function. This should have $m-1$ perturbation variables, perturbing the arguments of $m-1$ of the functions $f_i, i = 1, ..., m$. One can prove that, when employing the regularity conditions $(RC_i^{\Phi}), i \in \{2, 2', 2'', 3, 4\}$, for this perturbation function one obtains nothing else than equivalent formulations for the regularity conditions $(RC_i^{\Sigma}), i \in \{2, 2', 2'', 3, 4\}$, respectively.

We state the strong duality theorem for the primal-dual pair $(P^{\Sigma}) - (D^{\Sigma})$.

Theorem 3.2.8. *Let $f_i : X \to \overline{\mathbb{R}}, i = 1, ..., m$, be proper and convex functions such that $\bigcap_{i=1}^{m} \operatorname{dom} f_i \neq \emptyset$. If one of the regularity conditions (RC_i^{Σ}), $i \in \{1, 2, 3, 4\}$, is fulfilled, then $v(P^{\Sigma}) = v(D^{\Sigma})$ and the dual has an optimal solution.*

3.2.3 Regularity conditions for problems with geometric and cone constraints

In this subsection we consider again the primal problem (P^C) (cf. subsection 3.1.3)

$$(P^C) \quad \inf_{x \in \mathcal{A}} f(x),$$
$$\mathcal{A} = \{x \in S : g(x) \in -C\}$$

where X is a Hausdorff locally convex space, Z is another Hausdorff locally convex space partially ordered by the convex cone $C \subseteq Z$, $S \subseteq X$ is a given nonempty convex set, $f : X \to \overline{\mathbb{R}}$ a proper and convex function and $g : X \to \overline{Z}$ a proper and C-convex function fulfilling $\operatorname{dom} f \cap S \cap g^{-1}(-C) \neq \emptyset$.

We deal first with regularity conditions for (P^C) and its Lagrange dual (D^{C_L}), which we derive from the general ones by considering as perturbation function $\Phi^{C_L} : X \times Z \to \overline{\mathbb{R}}$,

$$\Phi^{C_L}(x, z) = \begin{cases} f(x), & \text{if } x \in S, g(x) \in z - C, \\ +\infty, & \text{otherwise.} \end{cases}$$

The first condition we introduce is the well-known *Slater constraint qualification*

$$(RC_1^{C_L}) \mid \exists x' \in \operatorname{dom} f \cap S \text{ such that } g(x') \in -\operatorname{int}(C).$$

Indeed, having for $x' \in X$ that $(x', 0) \in \operatorname{dom} \Phi^{C_L}$ and $\Phi^{C_L}(x', \cdot)$ is continuous at 0, this is nothing else than supposing that $x' \in \operatorname{dom} f \cap S$ and δ_{-C} is continuous at $g(x')$ or, equivalently, $x' \in \operatorname{dom} f \cap S$ and $g(x') \in -\operatorname{int}(C)$.

We come now to the class of regularity conditions which assumes that X and Z are Fréchet spaces. One has $\operatorname{Pr}_Z(\operatorname{dom} \Phi^{C_L}) = g(\operatorname{dom} f \cap S \cap \operatorname{dom} g) + C$.

In order to guarantee the lower semicontinuity of Φ^{C_L} we additionally assume that S is a closed set, f is a lower semicontinuous function and g is a C-epi closed function. As under these assumptions the epigraph of the perturbation function

$$\operatorname{epi}\Phi^{C_L} = \{(x,z,r) \in X \times Z \times \mathbb{R} : (x,r) \in \operatorname{epi} f\} \cap (S \times Z \times \mathbb{R}) \cap (\operatorname{epi}_C g \times \mathbb{R})$$

is a closed set, the perturbation function Φ^{C_L} is lower semicontinuous. These considerations lead to the following regularity condition

$$(RC_2^{C_L}) \left| \begin{array}{l} X \text{ and } Z \text{ are Fréchet spaces, } S \text{ is closed, } f \text{ is lower semicontinuous,} \\ g \text{ is } C\text{-epi closed and } 0 \in \operatorname{sqri}(g(\operatorname{dom} f \cap S \cap \operatorname{dom} g) + C), \end{array} \right.$$

along with its stronger versions

$$(RC_{2'}^{C_L}) \left| \begin{array}{l} X \text{ and } Z \text{ are Fréchet spaces, } S \text{ is closed, } f \text{ is lower semicontinuous,} \\ g \text{ is } C\text{-epi closed and } 0 \in \operatorname{core}(g(\operatorname{dom} f \cap S \cap \operatorname{dom} g) + C) \end{array} \right.$$

and

$$(RC_{2''}^{C_L}) \left| \begin{array}{l} X \text{ and } Z \text{ are Fréchet spaces, } S \text{ is closed, } f \text{ is lower semicontinuous,} \\ g \text{ is } C\text{-epi closed and } 0 \in \operatorname{int}(g(\operatorname{dom} f \cap S \cap \operatorname{dom} g) + C), \end{array} \right.$$

which are in fact equivalent. Mentioning that in the finite dimensional case the regularity condition looks like

$$(RC_3^{C_L}) \left| \begin{array}{l} \dim(\operatorname{lin}(g(\operatorname{dom} f \cap S \cap \operatorname{dom} g) + C)) < +\infty \text{ and} \\ 0 \in \operatorname{ri}(g(\operatorname{dom} f \cap S \cap \operatorname{dom} g) + C), \end{array} \right.$$

we can give the following strong duality theorem for the pair $(P^C) - (D^{C_L})$, which is a consequence of Theorem 3.2.1.

Theorem 3.2.9. *Let Z be partially ordered by the convex cone $C \subseteq Z$, $S \subseteq X$ be a nonempty convex set, $f : X \to \overline{\mathbb{R}}$ a proper and convex function and $g : X \to \overline{Z}$ a proper and C-convex function such that* $\operatorname{dom} f \cap S \cap g^{-1}(-C) \neq \emptyset$. *If one of the regularity conditions $(RC_i^{C_L})$, $i \in \{1,2,3\}$, is fulfilled, then $v(P^C) = v(D^{C_L})$ and the dual has an optimal solution.*

We come now to the formulation of the closedness type regularity condition and to this aim we determine the set $\operatorname{Pr}_{X^* \times \mathbb{R}}(\operatorname{epi}(\Phi^{C_L})^*)$. By (3.6) one has

$$(x^*,r) \in \operatorname{Pr}_{X^* \times \mathbb{R}}(\operatorname{epi}(\Phi^{C_L})^*) \Leftrightarrow \exists z^* \in -C^* \text{ such that } (f + (-z^*g))_S^*(x^*) \leq r$$

$$\Leftrightarrow (x^*,r) \in \bigcup_{z^* \in C^*} \operatorname{epi}(f + (z^*g) + \delta_S)^*$$

and this provides the following regularity condition

$(RC_4^{C_L})$ | S is closed, f is lower semicontinuous, g is C-epi closed and $\bigcup_{z^* \in C^*} \operatorname{epi}(f + (z^*g) + \delta_S)^*$ is closed in the topology $w(X^*, X) \times \mathbb{R}$.

From Theorem 3.2.3 we get the following result.

Theorem 3.2.10. *Let Z be partially ordered by the convex cone $C \subseteq Z$, $S \subseteq X$ be a nonempty convex set, $f : X \to \overline{\mathbb{R}}$ a proper and convex function and $g : X \to \overline{Z}$ a proper and C-convex function such that* $\operatorname{dom} f \cap S \cap g^{-1}(-C) \neq \emptyset$. *If the regularity condition $(RC_4^{C_L})$ is fulfilled, then $v(P^C) = v(D^{C_L})$ and the dual has an optimal solution.*

Remark 3.2.4. The regularity condition $(RC_4^{C_L})$ was introduced by Boţ, Grad and Wanka in [32] (see also [31]). There has been shown that, concerning the function g, it is enough to assume that g is C-epi closed, different to other regularity conditions given in the literature, which assume for g the stronger hypotheses of C-lower semicontinuity or star C-lower semicontinuity, respectively. In case S is closed, f is lower semicontinuous and g is C-epi closed, $(RC_4^{C_L})$ has been proven to be weaker than the generalized interior point regularity conditions given with respect to Lagrange duality in the literature.

We turn our attention now to the regularity conditions which guarantee strong duality for (P^C) and its dual (D^{C_F}), the perturbation function used in this context being $\Phi^{C_F} : X \times X \to \overline{\mathbb{R}}, \Phi^{C_F}(x, y) = \delta_{\mathcal{A}}(x) + f(x + y)$. They can be derived from the ones given in subsection 3.2.2 in the general framework of Fenchel duality. The assumptions we made in the beginning of this subsection imply that $\mathcal{A}$ is a nonempty convex set. Then $(RC_i^{\text{id}}), i \in \{1, 2, 2', 2'', 3, 4\}$, lead to the following conditions

$(RC_1^{C_F})$ | $\exists x' \in \operatorname{dom} f \cap \mathcal{A}$ such that f is continuous at x',

in case X is a Fréchet space

$(RC_2^{C_F})$ | X is a Fréchet space, $\mathcal{A}$ is closed, f is lower semicontinuous and $0 \in \operatorname{sqri}(\operatorname{dom} f - \mathcal{A})$,

along with its stronger versions

$(RC_{2'}^{C_F})$ | X is a Fréchet space, $\mathcal{A}$ is closed, f is lower semicontinuous and $0 \in \operatorname{core}(\operatorname{dom} f - \mathcal{A})$

and

$(RC_{2''}^{C_F})$ | X is a Fréchet space, $\mathcal{A}$ is closed, f is lower semicontinuous and $0 \in \operatorname{int}(\operatorname{dom} f - \mathcal{A})$,

which are in fact equivalent, in the finite dimensional case

$(RC_3^{C_F})$ | $\dim(\operatorname{lin}(\operatorname{dom} f - \mathcal{A})) < +\infty$ and $\operatorname{ri}(\operatorname{dom} f) \cap \operatorname{ri}(\mathcal{A}) \neq \emptyset$,

while the closedness type regularity condition states that

$(RC_4^{C_F})$	$\mathcal{A}$ is closed, f is lower semicontinuous and $\operatorname{epi} f^* + \operatorname{epi} \sigma_{\mathcal{A}}$ is closed in the topology $w(X^*, X) \times \mathbb{R}$,

respectively. From Theorem 3.2.6 we get the following result.

Theorem 3.2.11. *Let Z be partially ordered by the convex cone $C \subseteq Z$, $S \subseteq X$ be a nonempty convex set, $f : X \to \overline{\mathbb{R}}$ a proper and convex function and $g : X \to \overline{Z}$ a proper and C-convex function such that* $\operatorname{dom} f \cap S \cap g^{-1}(-C) \neq \emptyset$. *If one of the regularity conditions $(RC_i^{C_F})$, $i \in \{1, 2, 3, 4\}$, is fulfilled, then $v(P^C) = v(D^{C_F})$ and the dual has an optimal solution.*

The third conjugate dual problem we introduced for (P^C) is the so-called Fenchel-Lagrange dual obtained via the perturbation function $\Phi^{C_{FL}} : X \times X \times Z \to \overline{\mathbb{R}}$,

$$\Phi^{C_{FL}}(x, y, z) = \begin{cases} f(x + y), & \text{if } x \in S, g(x) \in z - C, \\ +\infty, & \text{otherwise.} \end{cases}$$

The first regularity condition we state for the pair of problems $(P^C) - (D^{C_{FL}})$ is given in the general framework regarding the spaces X and Z

$(RC_1^{C_{FL}})$	$\exists x' \in \operatorname{dom} f \cap S$ such that f is continuous at x' and $g(x') \in -\operatorname{int}(C)$.

One can easily see that $(RC_1^{C_{FL}})$ is exactly the reformulation of (RC_1^{Φ}) when considering as perturbation function $\Phi^{C_{FL}}$. We show in the following that

$$\operatorname{Pr}_{X \times Z}(\operatorname{dom} \Phi^{C_{FL}}) = \operatorname{dom} f \times C - \operatorname{epi}_{(-C)}(-g) \cap (S \times Z),$$

where we consider the projection on the product of the last two spaces in the domain of definition of $\Phi^{C_{FL}}$. Indeed,

$$(y, z) \in \operatorname{Pr}_{X \times Z}(\operatorname{dom} \Phi^{C_{FL}}) \Leftrightarrow \exists x \in X \text{ such that } \Phi^{C_{FL}}(x, y, z) < +\infty$$

$$\Leftrightarrow \exists x \in S \text{ such that } x + y \in \operatorname{dom} f \text{ and } g(x) \in z - C \Leftrightarrow \exists x \in S \text{ such that}$$
$$(y, z) \in (\operatorname{dom} f - x) \times (g(x) + C) \Leftrightarrow \exists x \in S \text{ such that } (y, z) \in \operatorname{dom} f \times C$$
$$-(x, -g(x)) \Leftrightarrow (y, z) \in \operatorname{dom} f \times C - \operatorname{epi}_{(-C)}(-g) \cap (S \times Z)$$

and this leads to the desired formula.

One can formulate the following generalized interior point regularity condition

$(RC_2^{C_{FL}})$	X and Z are Fréchet spaces, S is closed, f is lower semicontinuous, g is C-epi closed and $0 \in \operatorname{sqri}\Big(\operatorname{dom} f \times C - \operatorname{epi}_{(-C)}(-g) \cap (S \times Z)\Big)$,

along with its stronger versions

$$(RC_{2'}^{C_{FL}}) \quad \begin{array}{|l} X \text{ and } Z \text{ are Fréchet spaces, } S \text{ is closed, } f \text{ is lower} \\ \text{semicontinuous, } g \text{ is } C\text{-epi closed and} \\ 0 \in \operatorname{core}\Big(\operatorname{dom} f \times C - \operatorname{epi}_{(-C)}(-g) \cap (S \times Z)\Big) \end{array}$$

and

$$(RC_{2''}^{C_{FL}}) \quad \begin{array}{|l} X \text{ and } Z \text{ are Fréchet spaces, } S \text{ is closed, } f \text{ is lower} \\ \text{semicontinuous, } g \text{ is } C\text{-epi closed and} \\ 0 \in \operatorname{int}\Big(\operatorname{dom} f \times C - \operatorname{epi}_{(-C)}(-g) \cap (S \times Z)\Big), \end{array}$$

which are in fact equivalent, while in the finite dimensional case one can state

$$(RC_3^{C_{FL}}) \quad \begin{array}{|l} \dim\Big(\operatorname{lin}\Big(\operatorname{dom} f \times C - \operatorname{epi}_{(-C)}(-g) \cap (S \times Z)\Big)\Big) < +\infty \text{ and} \\ 0 \in \operatorname{ri}\Big(\operatorname{dom} f \times C - \operatorname{epi}_{(-C)}(-g) \cap (S \times Z)\Big). \end{array}$$

The strong duality theorem for the pair $(P^C)-(D^{C_{FL}})$ follows from Theorem 3.2.1.

Theorem 3.2.12. *Let Z be partially ordered by the convex cone $C \subseteq Z$, $S \subseteq X$ be a nonempty convex set, $f : X \to \overline{\mathbb{R}}$ a proper and convex function and $g : X \to \overline{Z}$ a proper and C-convex function such that* $\operatorname{dom} f \cap S \cap g^{-1}(-C) \neq \emptyset$. *If one of the regularity conditions $(RC_i^{C_{FL}})$, $i \in \{1, 2, 3\}$, is fulfilled, then $v(P^C) = v(D^{C_{FL}})$ and the dual has an optimal solution.*

Before introducing the closedness type condition we need to determine first the set $\operatorname{Pr}_{X^* \times \mathbb{R}}(\operatorname{epi}(\Phi^{C_{FL}})^*)$. One has, by (3.8),

$$(x^*, r) \in \operatorname{Pr}_{X^* \times \mathbb{R}}(\operatorname{epi}(\Phi^{C_{FL}})^*) \Leftrightarrow \exists (y^*, z^*) \in X^* \times Z^* \text{ such that}$$

$$(\Phi^{C_{FL}})^*(x^*, y^*, z^*) \leq r \Leftrightarrow \exists (y^*, z^*) \in X^* \times C^* \text{ such that } f^*(y^*)$$

$$+(z^*g)_S^*(x^* - y^*) \leq r \Leftrightarrow \exists (y^*, z^*) \in X^* \times C^* \text{ such that } (x^* - y^*, r - f^*(y^*))$$

$$\in \operatorname{epi}(z^*g)_S^* \Leftrightarrow \exists (y^*, z^*) \in X^* \times C^* \text{ such that } (x^*, r) \in (y^*, f^*(y^*)) + \operatorname{epi}(z^*g)_S^*$$

$$\Leftrightarrow (x^*, r) \in \operatorname{epi} f^* + \bigcup_{z^* \in C^*} \operatorname{epi}((z^*g) + \delta_S)^*.$$

In conclusion, $\operatorname{Pr}_{X^* \times \mathbb{R}}(\operatorname{epi}(\Phi^{C_{FL}})^*) = \operatorname{epi} f^* + \bigcup_{z^* \in C^*} \operatorname{epi}((z^*g) + \delta_S)^*$ and thus the closedness type regularity condition looks like

$$(RC_4^{C_{FL}}) \quad \begin{array}{|l} S \text{ is closed, } f \text{ is lower semicontinuous, } g \text{ is } C\text{-epi closed} \\ \text{and } \operatorname{epi} f^* + \bigcup\limits_{z^* \in C^*} \operatorname{epi}((z^*g) + \delta_S)^* \text{ is closed in the topology} \\ w(X^*, X) \times \mathbb{R}, \end{array}$$

while the strong duality theorem can be enunciated as follows.

Theorem 3.2.13. *Let Z be partially ordered by the convex cone $C \subseteq Z$, $S \subseteq X$ be a nonempty convex set, $f : X \to \overline{\mathbb{R}}$ a proper and convex function and $g : X \to \overline{Z}$ a proper and C-convex function such that* $\operatorname{dom} f \cap S \cap g^{-1}(-C) \neq \emptyset$. *If the regularity condition $(RC_4^{C_{FL}})$ is fulfilled, then $v(P^C) = v(D^{C_{FL}})$ and the dual has an optimal solution.*

Remark 3.2.5. Taking into consideration the relations that exist between the optimal objective values of the three conjugate duals to (P^C) (cf. (3.9)), one has that if between (P^C) and $(D^{C_{FL}})$ strong duality holds, then $v(P^C) = v(D^{C_L}) = v(D^{C_F}) = v(D^{C_{FL}})$. Moreover, if $(\bar{y}^*, \bar{z}^*) \in X^* \times C^*$ is an optimal solution to $(D^{C_{FL}})$, then $\bar{y}^* \in X^*$ is an optimal solution to (D^{C_F}) and $\bar{z}^* \in C^*$ is an optimal solution to (D^{C_L}). This means that for the pairs $(P^C) - (D^{C_L})$ and $(P^C) - (D^{C_F})$ strong duality holds, too.

We close the section by considering a particular instance of the primal problem (P^C) for which we give some weak regularity conditions which ensure strong duality between it and the three conjugate dual problems assigned to it. Consider $S \subseteq \mathbb{R}^n$ to be a nonempty convex set, $f : \mathbb{R}^n \to \overline{\mathbb{R}}$ a proper and convex function and $g : \mathbb{R}^n \to \mathbb{R}^m$, $g(x) = (g_1(x), ..., g_m(x))^T$ a vector function having each component $g_i, i = 1, ..., m$, convex. We consider as primal problem

$$(\widetilde{P}^C) \quad \inf_{x \in \mathcal{A}} f(x).$$
$$\mathcal{A} = \{x \in S : g(x) \leqq 0\}$$

Assume that $\operatorname{dom} f \cap S \cap g^{-1}(-\mathbb{R}^m_+) \neq \emptyset$. Further, let be $L = \{i \in \{1, ..., m\} : g_i \text{ is affine}\}$ and $N = \{1, ..., m\} \setminus L$. Consider the following regularity condition (cf. [186])

$$(\widetilde{RC}^{C_{FL}}) \left| \begin{array}{l} \exists x' \in \operatorname{ri}(\operatorname{dom} f) \cap \operatorname{ri}(S) \text{ such that } g_i(x') \leq 0, i \in L, \text{ and} \\ g_i(x') < 0, i \in N. \end{array} \right.$$

In [186] the following strong duality theorem has been formulated.

Theorem 3.2.14. *Let $S \subseteq \mathbb{R}^n$ be a nonempty convex set, $f : \mathbb{R}^n \to \overline{\mathbb{R}}$ a proper and convex function and $g : \mathbb{R}^n \to \mathbb{R}^m$, $g(x) = (g_1(x), ..., g_m(x))^T$ a vector function having each component $g_i, i = 1, ..., m$, convex such that $\operatorname{dom} f \cap S \cap g^{-1}(-\mathbb{R}^m_+) \neq \emptyset$. If the regularity condition $(\widetilde{RC}^{C_{FL}})$ is fulfilled, then for (P^C) and its Fenchel-Lagrange dual strong duality holds, namely $v(\widetilde{P}^C) = v(\widetilde{D}^{C_{FL}})$ and the dual has an optimal solution.*

Remark 3.2.6. The condition $(\widetilde{RC}^{C_{FL}})$ is ensuring strong duality also for the Fenchel and Lagrange dual problems. Nevertheless, for having strong duality between $(\widetilde{P}^C)$ and its Fenchel dual it is enough to assume that (cf. [157, 186])

$$(\widetilde{RC}^{C_F}) \mid \operatorname{ri}(\operatorname{dom} f) \cap \operatorname{ri}(\mathcal{A}) \neq \emptyset$$

holds, while for having strong duality between $(\widetilde{P}^C)$ and its Lagrange dual it is enough to assume that (cf. [157, 186])

$$(\widetilde{RC}^{C_L}) \left| \begin{array}{l} \exists x' \in \operatorname{ri}(\operatorname{dom} f \cap S) \text{ such that } g_i(x') \leq 0, i \in L, \text{ and} \\ g_i(x') < 0, i \in N \end{array} \right.$$

holds.

3.3 Optimality conditions and saddle points

In this section we derive by using the already given strong duality theorems necessary and sufficient optimality conditions for the pairs of primal-dual problems considered until now in this chapter. For these pairs we also define, taking into consideration the corresponding perturbation functions, so-called *Lagrangian functions* (cf. [67]) and give some characterizations of their *saddle points* with respect to the optimal solutions to the primal and dual problems. All this will be first done in the general case (for the primal-dual pair $(PG) - (DG)$), followed by a particularization to the different classes of optimization problems treated above.

3.3.1 The general scalar optimization problem

Assume that the perturbation function $\Phi : X \times Y \to \overline{\mathbb{R}}$ is a proper function fulfilling $0 \in \Pr_Y(\operatorname{dom} \Phi)$. The following theorem (cf. [67]) formulates necessary and sufficient optimality conditions for the primal-dual pair $(PG) - (DG)$.

Theorem 3.3.1. *(a) Assume that $\Phi : X \times Y \to \overline{\mathbb{R}}$ is a proper and convex function such that $0 \in \Pr_Y(\operatorname{dom} \Phi)$. Let $\bar{x} \in X$ be an optimal solution to (PG) and assume that one of the regularity conditions $(RC_i^{\Phi}), i \in \{1, 2, 3, 4\}$, is fulfilled. Then there exists $\bar{y}^* \in Y^*$, an optimal solution to (DG), such that*

$$\Phi(\bar{x}, 0) + \Phi^*(0, \bar{y}^*) = 0 \tag{3.12}$$

or, equivalently,

$$(0, \bar{y}^*) \in \partial\Phi(\bar{x}, 0). \tag{3.13}$$

(b) Assume that $\bar{x} \in X$ and $\bar{y}^* \in Y^*$ fulfill one of the relations above. Then $\bar{x}$ is an optimal solution to (PG), $\bar{y}^*$ is an optimal solution to (DG) and $v(PG) = v(DG)$.

Proof. (a) By Theorem 3.2.1 and Theorem 3.2.3 follows that there exists $\bar{y}^* \in Y^*$ such that

$$\Phi(\bar{x}, 0) = v(PG) = v(DG) = -\Phi^*(0, \bar{y}^*) \in \mathbb{R}.$$

Thus $\Phi(\bar{x}, 0) + \Phi^*(0, \bar{y}^*) = 0$ or, equivalently, by Theorem 2.3.12, $(0, \bar{y}^*) \in \partial\Phi(\bar{x}, 0)$.

(b) The assumption that (3.12) or (3.13) are fulfilled automatically guarantees that $\Phi(\bar{x}, 0) \in \mathbb{R}$. Since, by the weak duality theorem $-\Phi^*(0, y^*) \leq \Phi(x, 0)$ for all $x \in X$ and all $y^* \in Y^*$, we get that $-\Phi^*(0, \bar{y}^*) = \sup_{y^* \in Y^*}\{-\Phi^*(0, y^*)\} = v(DG)$, $\Phi(\bar{x}, 0) = \inf_{x \in X}\{\Phi(x, 0)\} = v(PG)$ and that these values are equal. □

Remark 3.3.1. We want to underline the fact that the statement (b) in the theorem above is true in the most general case without any assumption for Φ regarding convexity or the fulfillment of any regularity condition.

We come now to the definition of the Lagrangian function for the pair of primal-dual problems $(PG) - (DG)$ (cf. [67]). This will be done by means of the perturbation function Φ.

Definition 3.3.1. *The function $L^{\Phi} : X \times Y^* \to \overline{\mathbb{R}}$ defined by*

$$L^{\Phi}(x, y^*) = \inf_{y \in Y} \{\Phi(x, y) - \langle y^*, y \rangle\}$$

is called the Lagrangian function of the pair of primal-dual problems $(PG) - (DG)$ relative to the perturbation function Φ.

One can easily see that for all $x \in X$ it holds $L^{\Phi}(x, y^*) = -\Phi_x^*(y^*)$ for all $y^* \in Y^*$, where $\Phi_x : Y \to \overline{\mathbb{R}}$ is defined by $\Phi_x(y) = \Phi(x, y)$. Thus for all $x \in X$ the function $L^{\Phi}(x, \cdot)$ is concave and upper semicontinuous. On the other hand, assuming that Φ is convex, for all $y^* \in Y^*$ the function $L^{\Phi}(\cdot, y^*)$ is also convex (cf. Theorem 2.2.6).

In the following we express the problems (PG) and (DG) in terms of the Lagrangian function L^{Φ}. For all $(x^*, y^*) \in X^* \times Y^*$ we have

$$\Phi^*(x^*, y^*) = \sup_{x \in X, y \in Y} \{\langle x^*, x \rangle + \langle y^*, y \rangle - \Phi(x, y)\} = \sup_{x \in X} \{\langle x^*, x \rangle -$$

$$\inf_{y \in Y} \{\Phi(x, y) - \langle y^*, y \rangle\}\} = \sup_{x \in X} \left\{\langle x^*, x \rangle - L^{\Phi}(x, y^*)\right\}$$

and so

$$-\Phi^*(0, y^*) = \inf_{x \in X} L^{\Phi}(x, y^*) \ \forall y^* \in Y^*. \tag{3.14}$$

Then the dual (DG) can be reformulated as

$$(DG) \quad \sup_{y^* \in Y^*} \inf_{x \in X} L^{\Phi}(x, y^*).$$

Assume in the following that for all $x \in X$ the function Φ_x is convex and lower semicontinuous and it fulfills $\Phi_x(y) > -\infty$ for all $y \in Y$. In case Φ is a proper, convex and lower semicontinuous function, the assumption required above is fulfilled for all $x \in X$.

Thus for a given $x \in X$ one has

$$\Phi_x(y) = \Phi_x^{**}(y) = \sup_{y^* \in Y^*} \{\langle y^*, y \rangle - \Phi_x^*(y^*)\}$$

$$= \sup_{y^* \in Y^*} \left\{\langle y^*, y \rangle + L^{\Phi}(x, y^*)\right\} \ \forall y \in Y.$$

For $y = 0$ one gets

$$\Phi(x, 0) = \Phi_x(0) = \sup_{y^* \in Y^*} L^{\Phi}(x, y^*) \ \forall x \in X \tag{3.15}$$

and the problem (PG) can be reformulated as

$$(PG) \quad \inf_{x \in X} \sup_{y^* \in Y^*} L^{\Phi}(x, y^*).$$

By these reformulations of the primal and dual problems the weak duality theorem is nothing else than the well-known "minmax"-inequality

$$\sup_{y^* \in Y^*} \inf_{x \in X} L^{\Phi}(x, y^*) \leq \inf_{x \in X} \sup_{y^* \in Y^*} L^{\Phi}(x, y^*). \tag{3.16}$$

Definition 3.3.2. *We say that $(\bar{x}, \bar{y}^*) \in X \times Y^*$ is a saddle point of the Lagrangian function L^{Φ} if*

$$L^{\Phi}(\bar{x}, y^*) \leq L^{\Phi}(\bar{x}, \bar{y}^*) \leq L^{\Phi}(x, \bar{y}^*) \ \forall x \in X \ \forall y^* \in Y^*. \tag{3.17}$$

Next we relate the saddle points of the Lagrangian function L^{Φ} to the optimal solutions to the problems (PG) and (DG).

Theorem 3.3.2. *Assume that for all $x \in X$ the function $\Phi_x : Y \to \overline{\mathbb{R}}$ is a convex and lower semicontinuous function fulfilling $\Phi_x(y) > -\infty$ for all $y \in Y$. Then the following statements are equivalent:*

(i) $(\bar{x}, \bar{y}^*)$ is a saddle point of L^{Φ};
(ii) $\bar{x} \in X$ is an optimal solution to (PG), $\bar{y}^* \in Y^*$ is an optimal solution to (DG) and $v(PG) = v(DG)$.

Proof. $(i) \Rightarrow (ii)$. Since $(\bar{x}, \bar{y}^*)$ is a saddle point of L^{Φ}, by (3.14) and (3.15) follows that

$$-\Phi^*(0, \bar{y}^*) = \inf_{x \in X} L^{\Phi}(x, \bar{y}^*) = L^{\Phi}(\bar{x}, \bar{y}^*) = \sup_{y^* \in Y^*} L^{\Phi}(\bar{x}, y^*) = \Phi(\bar{x}, 0).$$

Thus $\Phi(\bar{x}, 0) + \Phi^*(0, \bar{y}^*) = 0$ and the conclusion follows by Theorem 3.3.1(b).

$(ii) \Rightarrow (i)$. Using again (3.14) and (3.15) we have

$$-\Phi^*(0, \bar{y}^*) = \inf_{x \in X} L^{\Phi}(x, \bar{y}^*) \leq L^{\Phi}(\bar{x}, \bar{y}^*)$$

and

$$\Phi(\bar{x}, 0) = \sup_{y^* \in Y^*} L^{\Phi}(\bar{x}, y^*) \geq L^{\Phi}(\bar{x}, \bar{y}^*),$$

respectively. Since $v(PG) = v(DG)$ is equivalent to $\Phi(\bar{x}, 0) = -\Phi^*(0, \bar{y}^*)$, it holds

$$\sup_{y^* \in Y^*} L^{\Phi}(\bar{x}, y^*) = L^{\Phi}(\bar{x}, \bar{y}^*) = \inf_{x \in X} L^{\Phi}(x, \bar{y}^*),$$

which means that $(\bar{x}, \bar{y}^*)$ is a saddle point of L^{Φ}. □

Remark 3.3.2. For all $x \in X$ the inequality $\Phi(x, 0) \geq \sup_{y^* \in Y^*} L^{\Phi}(x, y^*)$ is always true. This means that the implication $(ii) \Rightarrow (i)$ in the theorem above is true in the most general case without any assumption for Φ.

Theorem 3.3.1 and Theorem 3.3.2 lead to the following result.

Corollary 3.3.3. *Assume that* $\Phi : X \times Y \to \overline{\mathbb{R}}$ *is a proper and convex function such that* $0 \in \Pr_Y(\operatorname{dom}\Phi)$ *and* Φ_x *is lower semicontinuous for all* $x \in X$. *Let one of the regularity conditions* $(RC_i^{\Phi}), i \in \{1,2,3,4\}$, *be fulfilled. Then* $\bar{x} \in X$ *is an optimal solution to* (PG) *if and only if there exists* $\bar{y}^* \in Y^*$ *such that* $(\bar{x}, \bar{y}^*)$ *is a saddle point of* L^{Φ}. *In this case* $\bar{y}^*$ *is an optimal solution to the dual* (DG).

In what follows we give necessary and sufficient optimality conditions and, respectively, introduce corresponding Lagrangian functions for the different particular pairs of primal-dual problems derived from the general one $(PG)-(DG)$. The connection between the saddle points of these Lagrangian functions and the optimal solutions to the corresponding problems is also established.

3.3.2 Problems having the composition with a linear continuous mapping in the objective function

Let X and Y be Hausdorff locally convex spaces, $f : X \to \overline{\mathbb{R}}$ and $g : Y \to \overline{\mathbb{R}}$ be proper functions and $A \in \mathcal{L}(X,Y)$ fulfilling $\operatorname{dom} f \cap A^{-1}(\operatorname{dom} g) \neq \emptyset$ and consider the primal problem

$$(P^A) \quad \inf_{x \in X} \{f(x) + g(Ax)\}$$

and its Fenchel dual

$$(D^A) \quad \sup_{y^* \in Y^*} \{-f^*(-A^*y^*) - g^*(y^*)\}.$$

We can state the following result.

Theorem 3.3.4. *(a) Let* $f : X \to \overline{\mathbb{R}}, g : Y \to \overline{\mathbb{R}}$ *be proper and convex functions and* $A \in \mathcal{L}(X,Y)$ *such that* $\operatorname{dom} f \cap A^{-1}(\operatorname{dom} g) \neq \emptyset$. *Let* $\bar{x} \in X$ *be an optimal solution to* (P^A) *and assume that one of the regularity conditions* $(RC_i^A), i \in \{1,2,3,4\}$, *is fulfilled. Then there exists* $\bar{y}^* \in Y^*$, *an optimal solution to* (D^A), *such that*

(i) $f(\bar{x}) + f^*(-A^*\bar{y}^*) = \langle -\bar{y}^*, A\bar{x}\rangle$;

(ii) $g(A\bar{x}) + g^*(\bar{y}^*) = \langle \bar{y}^*, A\bar{x}\rangle$.

(b) Assume that $\bar{x} \in X$ *and* $\bar{y}^* \in Y^*$ *fulfill the relations* $(i)-(ii)$. *Then* $\bar{x}$ *is an optimal solution to* (P^A), $\bar{y}^*$ *is an optimal solution to* (D^A) *and* $v(P^A) = v(D^A)$.

Proof. The result follows from Theorem 3.3.1. What we have to prove is just that in this particular case the relation (3.12) is equivalent to $(i)-(ii)$. Indeed, (3.12) can be rewritten as

$$\Phi^A(\bar{x},0) + (\Phi^A)^*(0,\bar{y}^*) = 0 \Leftrightarrow f(\bar{x}) + g(A\bar{x}) + f^*(-A^*\bar{y}^*) + g^*(\bar{y}^*) = 0$$

$$\Leftrightarrow [f(\bar{x}) + f^*(-A^*\bar{y}^*) - \langle -A^*\bar{y}^*, \bar{x}\rangle] + [g(A\bar{x}) + g^*(\bar{y}^*) - \langle \bar{y}^*, A\bar{x}\rangle] = 0.$$

On the other hand, the Young-Fenchel inequality implies $f(\bar{x}) + f^*(-A^*\bar{y}^*) - \langle -A^*\bar{y}^*, \bar{x}\rangle \geq 0$ and $g(A\bar{x}) + g^*(\bar{y}^*) - \langle \bar{y}^*, A\bar{x}\rangle \geq 0$, which means that these inequalities must be fulfilled as equalities. This concludes the proof. □

Remark 3.3.3. The optimality conditions $(i) - (ii)$ in Theorem 3.3.4 can be equivalently written as

$$-A^*\bar{y}^* \in \partial f(\bar{x}) \text{ and } \bar{y}^* \in \partial g(A\bar{x}).$$

The Lagrangian function assigned to the pair of primal-dual problems $(P^A) - (D^A)$, $L^A : X \times Y^* \to \overline{\mathbb{R}}$, is defined by

$$L^A(x, y^*) = \inf_{y \in Y} [\Phi^A(x, y) - \langle y^*, y\rangle] = \inf_{y \in Y} [f(x) + g(Ax + y) - \langle y^*, y\rangle]$$

$$= f(x) + \inf_{r \in Y} [g(r) - \langle y^*, r - Ax\rangle] = f(x) + \langle A^*y^*, x\rangle - g^*(y^*).$$

Thus

$$\sup_{y^* \in Y^*} \inf_{x \in X} L^A(x, y^*) = \sup_{y^* \in Y^*} \{-f^*(-A^*y^*) - g^*(y^*)\}$$

and

$$\inf_{x \in X} \sup_{y^* \in Y^*} L^A(x, y^*) = \inf_{x \in X} \{f(x) + g^{**}(Ax)\}.$$

The assumption that for all $x \in X$ the mapping $\Phi_x^A : Y \to \overline{\mathbb{R}}$ is convex, lower semicontinuous and fulfills $\Phi_x^A(y) > -\infty$ for all $y \in Y$ is nothing else than assuming that g is a convex and lower semicontinuous function such that $g(y) > -\infty$ for all $y \in Y$. This is what we do in the following. By Theorem 3.3.2 we get the next result which holds without any assumption regarding f.

Theorem 3.3.5. *Assume that g is a convex and lower semicontinuous function fulfilling $g(y) > -\infty$ for all $y \in Y$. Then the following statements are equivalent:*

(i) $(\bar{x}, \bar{y}^*)$ is a saddle point of L^A;
(ii) $\bar{x} \in X$ is an optimal solution to (P^A), $\bar{y}^* \in Y^*$ is an optimal solution to (D^A) and $v(P^A) = v(D^A)$.

Remark 3.3.4. The hypotheses of Theorem 3.3.5 are natural, since in this case, $g(Ax) = g^{**}(Ax)$ for all $x \in X$, which implies that $\inf_{x \in X} \sup_{y^* \in Y^*} L^A(x, y^*) = \inf_{x \in X}\{f(x) + g(Ax)\}$. As in general $g^{**}(Ax) \leq g(Ax)$ for all $x \in X$, the implication $(ii) \Rightarrow (i)$ in the theorem above is true without any further assumption.

Corollary 3.3.3 is providing the following characterization of the saddle points of L^A.

Corollary 3.3.6. *Assume that $f : X \to \overline{\mathbb{R}}$ is a proper and convex function, $g : Y \to \overline{\mathbb{R}}$ a proper, convex and lower semicontinuous function and $A \in \mathcal{L}(X,Y)$ such that $\operatorname{dom} f \cap A^{-1}(\operatorname{dom} g) \neq \emptyset$. Let one of the regularity conditions $(RC_i^A), i \in \{1,2,3,4\}$, be fulfilled. Then $\bar{x} \in X$ is an optimal solution to (P^A) if and only if there exists $\bar{y}^* \in Y^*$ such that $(\bar{x}, \bar{y}^*)$ is a saddle point of L^A. In this case $\bar{y}^*$ is an optimal solution to the dual (D^A).*

In case $X = Y$ and $A = \mathrm{id}_X$, under the assumption that $f, g : X \to \overline{\mathbb{R}}$ are proper functions fulfilling $\operatorname{dom} f \cap \operatorname{dom} g \neq \emptyset$, we get the following necessary and sufficient optimality conditions for the primal problem

$$(P^{\mathrm{id}}) \quad \inf_{x \in X} \{f(x) + g(x)\}$$

and its Fenchel dual

$$(D^{\mathrm{id}}) \quad \sup_{y^* \in X^*} \{-f^*(-y^*) - g^*(y^*)\}.$$

Theorem 3.3.7. *(a) Let $f, g : X \to \overline{\mathbb{R}}$ be proper and convex functions such that $\operatorname{dom} f \cap \operatorname{dom} g \neq \emptyset$. Let $\bar{x} \in X$ be an optimal solution to (P^{id}) and assume that one of the regularity conditions $(RC_i^{\mathrm{id}}), i \in \{1,2,3,4\}$, is fulfilled. Then there exists $\bar{y}^* \in X^*$, an optimal solution to (D^{id}), such that*
(i) $f(\bar{x}) + f^*(-\bar{y}^*) = \langle -\bar{y}^*, \bar{x} \rangle$;
(ii) $g(\bar{x}) + g^*(\bar{y}^*) = \langle \bar{y}^*, \bar{x} \rangle$.
(b) Assume that $\bar{x} \in X$ and $\bar{y}^* \in X^*$ fulfill the relations $(i) - (ii)$. Then $\bar{x}$ is an optimal solution to (P^{id}), $\bar{y}^*$ is an optimal solution to (D^{id}) and $v(P^{\mathrm{id}}) = v(D^{\mathrm{id}})$.

Remark 3.3.5. The optimality conditions $(i) - (ii)$ in Theorem 3.3.7 can be equivalently written as

$$\bar{y}^* \in (-\partial f(\bar{x})) \cap \partial g(\bar{x}).$$

The Lagrangian function assigned to the pair of primal-dual problems $(P^{\mathrm{id}}) - (D^{\mathrm{id}})$, $L^{\mathrm{id}} : X \times X^* \to \overline{\mathbb{R}}$, looks like $L^{\mathrm{id}}(x, y^*) = f(x) + \langle y^*, x \rangle - g^*(y^*)$ and one has

$$\sup_{y^* \in X^*} \inf_{x \in X} L^{\mathrm{id}}(x, y^*) = \sup_{y^* \in X^*} \{-f^*(-y^*) - g^*(y^*)\}$$

and

$$\inf_{x \in X} \sup_{y^* \in X^*} L^{\mathrm{id}}(x, y^*) = \inf_{x \in X} \{f(x) + g^{**}(x)\}.$$

For the following result we omit asking that g is proper. The following two results are consequences of Theorem 3.3.5 and Corollary 3.3.6, respectively.

Theorem 3.3.8. *Assume that g is a convex and lower semicontinuous function fulfilling $g(x) > -\infty$ for all $x \in X$. Then the following statements are equivalent:*

(i) $(\bar{x}, \bar{y}^*)$ *is a saddle point of* L^{id}*;*
(ii) $\bar{x} \in X$ *is an optimal solution to* (P^{id}), $\bar{y}^* \in X^*$ *is an optimal solution to* (D^{id}) *and* $v(P^{\mathrm{id}}) = v(D^{\mathrm{id}})$.

Corollary 3.3.9. *Assume that $f, g : X \to \overline{\mathbb{R}}$ are proper and convex function such that g is lower semicontinuous and $\operatorname{dom} f \cap \operatorname{dom} g \neq \emptyset$. Let one of the regularity conditions $(RC_i^{\mathrm{id}}), i \in \{1, 2, 3, 4\}$, be fulfilled. Then $\bar{x} \in X$ is an optimal solution to (P^{id}) if and only if there exists $\bar{y}^* \in X^*$ such that $(\bar{x}, \bar{y}^*)$ is a saddle point of L^{id}. In this case $\bar{y}^*$ is an optimal solution to the dual (D^{id}).*

Now let $f : X \to \mathbb{R}$ be such that $f \equiv 0$, $g : Y \to \overline{\mathbb{R}}$ be a proper function, $A \in \mathcal{L}(X, Y)$ fulfilling $A^{-1}(\operatorname{dom} g) \neq \emptyset$ and consider the following pair of primal-dual problems (cf. 3.1.2)

$$(P^{A_g}) \quad \inf_{x \in X} \{g(Ax)\}$$

and

$$(D^{A_g}) \quad \sup_{\substack{y^* \in Y^*,\\ A^* y^* = 0}} \{-g^*(y^*)\},$$

respectively. Since $f^*(x^*) = \delta_{\{0\}}(x^*)$ for all $x^* \in X^*$, from Theorem 3.3.4 one can derive the following result.

Theorem 3.3.10. *(a) Let $g : Y \to \overline{\mathbb{R}}$ be a proper and convex function and $A \in \mathcal{L}(X, Y)$ such that $A^{-1}(\operatorname{dom} g) \neq \emptyset$. Let $\bar{x} \in X$ be an optimal solution to (P^{A_g}) and assume that one of the regularity conditions $(RC_i^{A_g}), i \in \{1, 2, 3, 4\}$, is fulfilled. Then there exists $\bar{y}^* \in Y^*$, an optimal solution to (D^{A_g}), such that*

(i) $A^* \bar{y}^* = 0$;
(ii) $g(A\bar{x}) + g^*(\bar{y}^*) = \langle \bar{y}^*, A\bar{x} \rangle$.

(b) Assume that $\bar{x} \in X$ and $\bar{y}^* \in Y^*$ fulfill the relations $(i) - (ii)$. Then $\bar{x}$ is an optimal solution to (P^{A_g}), $\bar{y}^*$ is an optimal solution to (D^{A_g}) and $v(P^{A_g}) = v(D^{A_g})$.

Remark 3.3.6. The optimality conditions $(i) - (ii)$ in Theorem 3.3.10 can be equivalently written as

$$A^* \bar{y}^* = 0 \text{ and } \bar{y}^* \in \partial g(A\bar{x}).$$

The Lagrangian function of the pair of primal-dual problems $(P^{A_g}) - (D^{A_g})$, $L^{A_g} : X \times Y^* \to \overline{\mathbb{R}}$, $L^{A_g}(x, y^*) = \langle A^* y^*, x \rangle - g^*(y^*)$ verifies

$$\sup_{y^* \in Y^*} \inf_{x \in X} L^{A_g}(x, y^*) = \sup_{\substack{y^* \in Y^*, \\ A^* y^* = 0}} \{-g^*(y^*)\}$$

and

$$\inf_{x \in X} \sup_{y^* \in Y^*} L^{A_g}(x, y^*) = \inf_{x \in X} \{g^{**}(Ax)\}.$$

For the following result we omit asking that g is proper. Theorem 3.3.5 and Corollary 3.3.6 lead to the following results, respectively.

Theorem 3.3.11. *Assume that g is a convex and lower semicontinuous function fulfilling $g(y) > -\infty$ for all $y \in Y$. Then the following statements are equivalent:*

(i) $(\bar{x}, \bar{y}^*)$ is a saddle point of L^{A_g};
(ii) $\bar{x} \in X$ is an optimal solution to (P^{A_g}), $\bar{y}^* \in Y^*$ is an optimal solution to (D^{A_g}) and $v(P^{A_g}) = v(D^{A_g})$.

Corollary 3.3.12. *Assume that $g : Y \to \overline{\mathbb{R}}$ is a proper, convex and lower semicontinuous function and $A \in \mathcal{L}(X, Y)$ such that $A^{-1}(\operatorname{dom} g) \neq \emptyset$. Let one of the regularity conditions $(RC_i^{A_g}), i \in \{1, 2, 3, 4\}$, be fulfilled. Then $\bar{x} \in X$ is an optimal solution to (P^{A_g}) if and only if there exists $\bar{y}^* \in Y^*$ such that $(\bar{x}, \bar{y}^*)$ is a saddle point of L^{A_g}. In this case $\bar{y}^*$ is an optimal solution to the dual (D^{A_g}).*

We close this subsection by considering $f_i : X \to \overline{\mathbb{R}}, i = 1, \dots, m$, proper functions such that $\bigcap_{i=1}^m \operatorname{dom} f_i \neq \emptyset$. For the primal optimization problem

$$(P^{\Sigma}) \quad \inf_{x \in X} \left\{ \sum_{i=1}^m f_i(x) \right\}$$

we obtained in subsection 3.1.2 the following dual

$$(D^{\Sigma}) \quad \sup_{\substack{x^{i*} \in X^*, i=1,\dots,m, \\ \sum_{i=1}^m x^{i*} = 0}} \left\{ -\sum_{i=1}^m f_i^*(x^{i*}) \right\}.$$

Using Theorem 3.3.10, one can easily prove the following result.

Theorem 3.3.13. *(a) Let $f_i : X \to \overline{\mathbb{R}}, i = 1, \dots, m$, be proper and convex functions such that $\bigcap_{i=1}^m \operatorname{dom} f_i \neq \emptyset$. Let $\bar{x} \in X$ be an optimal solution to (P^{Σ}) and assume that one of the regularity conditions $(RC_i^{\Sigma}), i \in \{1, 2, 3, 4\}$, is fulfilled. Then there exists $(\bar{x}^{1*}, \dots, \bar{x}^{m*}) \in (X^*)^m$, an optimal solution to (D^{Σ}), such that*

(i) $\sum_{i=1}^m \bar{x}^{i*} = 0$;
(ii) $f_i(\bar{x}) + f_i^*(\bar{x}^{i*}) = \langle \bar{x}^{i*}, \bar{x} \rangle, i = 1, \dots, m$.

(b) *Assume that* $\bar{x} \in X$ *and* $(\bar{x}^{1*}, ..., \bar{x}^{m*}) \in (X^*)^m$ *fulfill the relations* (i) – (ii). *Then* $\bar{x}$ *is an optimal solution to* (P^Σ), $(\bar{x}^{1*}, ..., \bar{x}^{m*})$ *is an optimal solution to* (D^Σ) *and* $v(P^\Sigma) = v(D^\Sigma)$.

Proof. The result is a direct consequence of Theorem 3.3.10, where (i) and (ii) look like $\sum_{i=1}^m \bar{x}^{i*} = 0$ and $\sum_{i=1}^m f_i(\bar{x}) + \sum_{i=1}^m f_i^*(\bar{x}^{i*}) = \langle \sum_{i=1}^m \bar{x}^{i*}, \bar{x} \rangle$, respectively. Since for all $i = 1, ..., m$, $f_i(\bar{x}) + f_i^*(\bar{x}^{i*}) \geq \langle \bar{x}^{i*}, \bar{x} \rangle$, the latter is further equivalent to $f_i(\bar{x}) + f_i^*(\bar{x}^{i*}) = \langle \bar{x}^{i*}, \bar{x} \rangle$ for all $i = 1, ..., m$. □

Remark 3.3.7. The optimality conditions (i) – (ii) in Theorem 3.3.13 can be equivalently written as

$$\sum_{i=1}^m \bar{x}^{i*} = 0 \text{ and } \bar{x}^{i*} \in \partial f_i(\bar{x}), i = 1, ..., m.$$

The Lagrangian function of the pair of primal-dual problems $(P^\Sigma) - (D^\Sigma)$ is $L^\Sigma : X \times (X^*)^m \to \overline{\mathbb{R}}$, $L^\Sigma(x, x^{1*}, ..., x^{m*}) = \sum_{i=1}^m \langle x^{i*}, x \rangle - \sum_{i=1}^m f_i^*(x^{i*})$ and it verifies

$$\sup_{(x^{1*},...,x^{m*}) \in (X^*)^m} \inf_{x \in X} L^\Sigma(x, x^{1*}, ..., x^{m*}) = \sup_{\substack{x^{i*} \in X^*, i=1,...,m, \\ \sum_{i=1}^m x^{i*} = 0}} \left\{ -\sum_{i=1}^m f_i^*(x^{i*}) \right\}$$

and

$$\inf_{x \in X} \sup_{(x^{1*},...,x^{m*}) \in (X^*)^m} L^\Sigma(x, x^{1*}, ..., x^{m*}) = \inf_{x \in X} \left\{ \sum_{i=1}^m f_i^{**}(x) \right\}.$$

For the next result we omit asking that the functions f_i, $i = 1, ..., m$, are proper. Theorem 3.3.11 and Corollary 3.3.12 lead to the following results, respectively.

Theorem 3.3.14. *Assume for all* $i = 1, ..., m$, *that* $f_i : X \to \overline{\mathbb{R}}$ *are convex and lower semicontinuous functions fulfilling* $f_i(x) > -\infty$ *for all* $x \in X$. *Then the following statements are equivalent:*

(i) $(\bar{x}, \bar{x}^{1*}, ..., \bar{x}^{m*})$ *is a saddle point of* L^Σ;
(ii) $\bar{x} \in X$ *is an optimal solution to* (P^Σ), $(\bar{x}^{1*}, ..., \bar{x}^{m*}) \in (X^*)^m$ *is an optimal solution to* (D^Σ) *and* $v(P^\Sigma) = v(D^\Sigma)$.

Corollary 3.3.15. *Assume that* $f_i : X \to \overline{\mathbb{R}}, i = 1, ..., m$, *are proper, convex and lower semicontinuous functions such that* $\bigcap_{i=1}^m \operatorname{dom} f_i \neq \emptyset$. *Let one of the regularity conditions* $(RC_i^\Sigma), i \in \{1, 2, 3, 4\}$, *be fulfilled. Then* $\bar{x} \in X$ *is an optimal solution to* (P^Σ) *if and only if there exists* $(\bar{x}^{1*}, ..., \bar{x}^{m*}) \in (X^*)^m$ *such that* $(\bar{x}, \bar{x}^{1*}, ..., \bar{x}^{m*})$ *is a saddle point of* L^Σ. *In this case* $(\bar{x}^{1*}, ..., \bar{x}^{m*})$ *is an optimal solution to the dual* (D^Σ).

3.3.3 Problems with geometric and cone constraints

Consider now the Hausdorff locally convex spaces X and Z, the latter being partially ordered by the convex cone $C \subseteq Z$, S a nonempty subset of X and $f : X \to \overline{\mathbb{R}}$ and $g : X \to \overline{Z}$ proper functions fulfilling $\operatorname{dom} f \cap S \cap g^{-1}(-C) \neq \emptyset$. For the optimization problem

$$(P^C) \quad \inf_{x \in \mathcal{A}} f(x),$$
$$\mathcal{A} = \{x \in S : g(x) \in -C\}$$

and its three conjugate duals introduced in subsection 3.1.3 we derive necessary and sufficient optimality conditions, introduce corresponding Lagrangian functions and characterize the existence of saddle points for the latter.

We consider first the Lagrange dual problem to (P^C)

$$(D^{C_L}) \quad \sup_{z^* \in C^*} \inf_{x \in S} \{f(x) + (z^* g)(x)\}$$

and formulate the optimality conditions for the pair $(P^C) - (D^{C_L})$.

Theorem 3.3.16. *(a) Let Z be partially ordered by the convex cone $C \subseteq Z$, $S \subseteq X$ be a nonempty convex set, $f : X \to \overline{\mathbb{R}}$ a proper and convex function and $g : X \to \overline{Z}$ a proper and C-convex function such that* $\operatorname{dom} f \cap S \cap g^{-1}(-C) \neq \emptyset$. *If $\bar{x} \in \mathcal{A}$ is an optimal solution to (P^C) and one of the regularity conditions $(RC_i^{C_L})$, $i \in \{1, 2, 3, 4\}$, is fulfilled, then there exists $\bar{z}^* \in C^*$, an optimal solution to (D^{C_L}), such that*

(i) $\min\limits_{x \in S}\{f(x) + (\bar{z}^* g)(x)\} = f(\bar{x})$;

(ii) $(\bar{z}^* g)(\bar{x}) = 0$.

(b) Assume that $\bar{x} \in \mathcal{A}$ and $\bar{z}^* \in C^*$ fulfill the relations $(i) - (ii)$. Then $\bar{x}$ is an optimal solution to (P^C), $\bar{z}^*$ is an optimal solution to (D^{C_L}) and $v(P^C) = v(D^{C_L})$.

Proof. The result follows by Theorem 3.3.1, since $(i) - (ii)$ are nothing else than an equivalent formulation of (3.12). Indeed, $\Phi^{C_L}(\bar{x}, 0) + (\Phi^{C_L})^*(0, \bar{z}^*) = 0$ is nothing else than $f(\bar{x}) - \inf_{x \in S}\{f(x) + (\bar{z}^* g)(x)\} = 0$. But, as $\inf_{x \in S}\{f(x) + (\bar{z}^* g)(x)\} \leq f(\bar{x}) + (\bar{z}^* g)(\bar{x})$, this is true only if $(\bar{z}^* g)(\bar{x}) \geq 0$, which the same with $(\bar{z}^* g)(\bar{x}) = 0$ (since $g(\bar{x}) \in -C$ and $\bar{z}^* \in C^*$). This leads to the desired conclusion. □

Remark 3.3.8. The optimality conditions $(i) - (ii)$ in Theorem 3.3.16 can be equivalently written as

$$0 \in \partial(f + (\bar{z}^* g) + \delta_S)(\bar{x}) \text{ and } (\bar{z}^* g)(\bar{x}) = 0.$$

The Lagrangian function introduced by the perturbation function Φ^{C_L}, $L^{C_L} : X \times Z^* \to \overline{\mathbb{R}}$, has the following formulation

$$L^{C_L}(x, z^*) = \inf_{z \in Z} \left\{\Phi^{C_L}(x, z) - \langle z^*, z \rangle\right\}$$

$$= \begin{cases} \inf\limits_{z \in g(x)+C} \{f(x) - \langle z^*, z\rangle\}, & \text{if } x \in S, \\ +\infty, & \text{otherwise}, \end{cases}$$

$$= \begin{cases} f(x) + \inf\limits_{s \in C} \{-\langle z^*, g(x) + s\rangle\}, & \text{if } x \in S, \\ +\infty, & \text{otherwise}, \end{cases}$$

$$= \begin{cases} f(x) + (-z^*g)(x) + \inf\limits_{s \in C} \langle -z^*, s\rangle, & \text{if } x \in S, \\ +\infty, & \text{otherwise}, \end{cases}$$

$$= \begin{cases} f(x) + (-z^*g)(x), & \text{if } x \in S, z^* \in -C^*, \\ -\infty, & \text{if } x \in S, z^* \notin -C^*, \\ +\infty, & \text{otherwise}. \end{cases}$$

Thus

$$\sup_{z^* \in Z^*} \inf_{x \in X} L^{C_L}(x, z^*) = \sup_{z^* \in Z^*} \inf_{x \in S} L^{C_L}(x, z^*)$$

$$= \sup_{z^* \in -C^*} \inf_{x \in S} \{f(x) + (-z^*g)(x)\} = \sup_{z^* \in C^*} \inf_{x \in S} \{f(x) + (z^*g)(x)\}$$

and

$$\inf_{x \in X} \sup_{z^* \in Z^*} L^{C_L}(x, z^*) = \inf_{x \in S} \sup_{z^* \in -C^*} \{f(x) + (-z^*g)(x)\}$$

$$= \inf_{x \in S} \left\{ f(x) + \sup_{z^* \in C^*} \langle z^*, g(x)\rangle \right\}.$$

Since for all $x \in S$, $\sup\limits_{z^* \in C^*} \langle z^*, g(x)\rangle = \delta_{-C^{**}}(g(x))$, we obtain

$$\inf_{x \in X} \sup_{z^* \in Z^*} L^{C_L}(x, z^*) = \inf_{\substack{x \in S, \\ g(x) \in -C^{**}}} f(x) = \inf_{\substack{x \in S, \\ g(x) \in -\operatorname{cl}(C)}} f(x).$$

If C is closed, then one has that

$$\inf_{x \in X} \sup_{z^* \in Z^*} L^{C_L}(x, z^*) = \inf_{\substack{x \in S, \\ g(x) \in -C}} f(x).$$

For the next result, which is a consequence of Theorem 3.3.2, excepting the additional closedness for the convex cone C, no other assumption regarding the sets and functions involved is made. This is because of the fact that for all $x \in X$ the function $\Phi_x^{C_L} : Z \to \overline{\mathbb{R}}$, $\Phi_x^{C_L}(z) = f(x) + \delta_S(x) + \delta_{g(x)+C}(z)$, is convex and lower semicontinuous and fulfills $\Phi_x^{C_L}(z) > -\infty$ for all $z \in Z$.

Theorem 3.3.17. *Assume that C is closed. The following statements are equivalent:*

(i) $(\bar{x}, -\bar{z}^*)$ *is a saddle point of* L^{C_L}*;*
(ii) $\bar{x} \in \mathcal{A}$ *is an optimal solution to* (P^C)*,* $\bar{z}^* \in C^*$ *is an optimal solution to* (D^{C_L}) *and* $v(P^C) = v(D^{C_L})$*.*

The following assertion is a consequence of Corollary 3.3.3.

Corollary 3.3.18. *Let Z be partially ordered by the convex closed cone $C \subseteq Z$, $S \subseteq X$ be a nonempty convex set, $f : X \to \overline{\mathbb{R}}$ a proper and convex function and $g : X \to \overline{Z}$ a proper and C-convex function such that $\operatorname{dom} f \cap S \cap g^{-1}(-C) \neq \emptyset$. Assume that one of the regularity conditions $(RC_i^{C_L})$, $i \in \{1,2,3,4\}$, is fulfilled. Then $\bar{x} \in \mathcal{A}$ is an optimal solution to (P^C) if and only if there exists $\bar{z}^* \in -C^*$ such that $(\bar{x}, \bar{z}^*)$ is a saddle point of L^{C_L}. In this case $-\bar{z}^*$ is an optimal solution to the dual (D^{C_L}).*

Coming now to the Fenchel dual problem to (P^C) (cf. subsection 3.1.3)

$$(D^{C_F}) \quad \sup_{y^* \in X^*} \{-f^*(y^*) - \sigma_{\mathcal{A}}(-y^*)\}$$

the necessary and sufficient optimality conditions can be obtained also in this case by particularizing the general ones.

Theorem 3.3.19. *(a) Let Z be partially ordered by the convex cone $C \subseteq Z$, $S \subseteq X$ be a nonempty convex set, $f : X \to \overline{\mathbb{R}}$ a proper and convex function and $g : X \to \overline{Z}$ a proper and C-convex function such that $\operatorname{dom} f \cap S \cap g^{-1}(-C) \neq \emptyset$. If $\bar{x} \in \mathcal{A}$ is an optimal solution to (P^C) and one of the regularity conditions $(RC_i^{C_F})$, $i \in \{1,2,3,4\}$, is fulfilled, then there exists $\bar{y}^* \in X^*$, an optimal solution to (D^{C_F}), such that*

(i) $\min_{x \in \mathcal{A}} \langle \bar{y}^*, x \rangle = \langle \bar{y}^*, \bar{x} \rangle$;

(ii) $f(\bar{x}) + f^*(\bar{y}^*) = \langle \bar{y}^*, \bar{x} \rangle$.

(b) Assume that $\bar{x} \in \mathcal{A}$ and $\bar{y}^* \in X^*$ fulfill the relations (i) – (ii). Then $\bar{x}$ is an optimal solution to (P^C), $\bar{y}^*$ is an optimal solution to (D^{C_F}) and $v(P^C) = v(D^{C_F})$.

Proof. The result follows from Theorem 3.3.7 by taking into consideration that $\delta_{\mathcal{A}}^*(y^*) = \sup_{x \in \mathcal{A}} \langle y^*, x \rangle$ for all $y^* \in X^*$. □

Remark 3.3.9. The optimality conditions (i) – (ii) in Theorem 3.3.19 can be equivalently written as

$$\bar{y}^* \in \partial f(\bar{x}) \cap (-N(\mathcal{A}, \bar{x})).$$

The Lagrangian function for (P^C) – (D^{C_F}) is $L^{C_F} : X \times X^* \to \overline{\mathbb{R}}$, $L^{C_F}(x, y^*) = \delta_{\mathcal{A}}(x) + \langle y^*, x \rangle - f^*(y^*)$, and one has

$$\sup_{y^* \in X^*} \inf_{x \in X} L^{C_F}(x, y^*) = \sup_{y^* \in X^*} \{-f^*(y^*) - \sigma_{\mathcal{A}}(-y^*)\}$$

and

$$\inf_{x \in X} \sup_{y^* \in Y^*} L^{C_F}(x, y^*) = \inf_{x \in X} \{f^{**}(x) + \delta_{\mathcal{A}}(x)\}.$$

For the next result we omit asking that f is proper. Theorem 3.3.8 and Corollary 3.3.9 lead to the following assertions, respectively.

Theorem 3.3.20. *Assume that f is a convex and lower semicontinuous function fulfilling $f(x) > -\infty$ for all $x \in X$. Then the following statements are equivalent:*

(i) $(\bar{x}, \bar{y}^*)$ *is a saddle point of* L^{C_F};
(ii) $\bar{x} \in \mathcal{A}$ *is an optimal solution to* (P^C), $\bar{y}^* \in X^*$ *is an optimal solution to* (D^{C_F}) *and* $v(P^C) = v(D^{C_F})$.

Corollary 3.3.21. *Let Z be partially ordered by the convex cone $C \subseteq Z$, $S \subseteq X$ be a nonempty convex set, $f : X \rightarrow \overline{\mathbb{R}}$ a proper, convex and lower semicontinuous function and $g : X \rightarrow \overline{Z}$ a proper and C-convex function such that $\operatorname{dom} f \cap S \cap g^{-1}(-C) \neq \emptyset$. Assume that one of the regularity conditions $(RC_i^{C_F})$, $i \in \{1,2,3,4\}$, is fulfilled. Then $\bar{x} \in \mathcal{A}$ is an optimal solution to (P^C) if and only if there exists $\bar{y}^* \in X^*$ such that $(\bar{x}, \bar{y}^*)$ is a saddle point of L^{C_F}. In this case $\bar{y}^*$ is an optimal solution to the dual (D^{C_F}).*

Concerning the Fenchel-Lagrange dual problem to (P^C)

$$(D^{C_{FL}}) \qquad \sup_{y^* \in X^*, z^* \in C^*} \left\{ -f^*(y^*) - (z^*g)^*_S(-y^*) \right\},$$

we can derive by means of Theorem 3.3.1 the following necessary and sufficient optimality conditions.

Theorem 3.3.22. *(a) Let Z be partially ordered by the convex cone $C \subseteq Z$, $S \subseteq X$ be a nonempty convex set, $f : X \rightarrow \overline{\mathbb{R}}$ a proper and convex function and $g : X \rightarrow \overline{Z}$ a proper and C-convex function such that $\operatorname{dom} f \cap S \cap g^{-1}(-C) \neq \emptyset$. If $\bar{x} \in \mathcal{A}$ is an optimal solution to (P^C) and one of the regularity conditions $(RC_i^{C_{FL}})$, $i \in \{1,2,3,4\}$, is fulfilled, then there exists $(\bar{y}^*, \bar{z}^*) \in X^* \times C^*$, an optimal solution to $(D^{C_{FL}})$, such that*
(i) $(\bar{z}^*g)^*_S(-\bar{y}^*) = -\langle \bar{y}^*, \bar{x} \rangle$;
(ii) $(\bar{z}^*g)(\bar{x}) = 0$;
(iii) $f(\bar{x}) + f^*(\bar{y}^*) = \langle \bar{y}^*, \bar{x} \rangle$.
(b) Assume that $\bar{x} \in \mathcal{A}$ and $(\bar{y}^*, \bar{z}^*) \in X^* \times C^*$ fulfill the relations $(i) - (iii)$. Then $\bar{x}$ is an optimal solution to (P^C), $(\bar{y}^*, \bar{z}^*)$ is an optimal solution to $(D^{C_{FL}})$ and $v(P^C) = v(D^{C_{FL}})$.

Proof. We prove that for this pair of primal-dual problems relation (3.12) is equivalent to the fact that $(i) - (iii)$ are fulfilled. Indeed, for $\bar{x} \in \mathcal{A}$ and $(\bar{y}^*, \bar{z}^*) \in X^* \times C^*$ one has

$$\begin{aligned} \Phi^{C_{FL}}(\bar{x},0,0) + (\Phi^{C_{FL}})^*(0,\bar{y}^*,\bar{z}^*) = 0 &\Leftrightarrow f(\bar{x}) + f^*(\bar{y}^*) + (\bar{z}^*g)^*_S(-\bar{y}^*) = 0 \\ &\Leftrightarrow [(\bar{z}^*g)^*_S(-\bar{y}^*) + (\bar{z}^*g)(\bar{x}) - \langle -\bar{y}^*, \bar{x} \rangle] + [-(\bar{z}^*g)(\bar{x})] \\ &\qquad + [f(\bar{x}) + f^*(\bar{y}^*) - \langle \bar{y}^*, \bar{x} \rangle] = 0. \end{aligned}$$

Since $(\bar{z}^*g)^*_S(-\bar{y}^*) + (\bar{z}^*g)(\bar{x}) - \langle -\bar{y}^*, \bar{x} \rangle \geq 0$, $-(\bar{z}^*g)(\bar{x}) \geq 0$ and $f(\bar{x}) + f^*(\bar{y}^*) - \langle \bar{y}^*, \bar{x} \rangle \geq 0$, all the inequalities must be fulfilled as equalities and the conclusion follows. □

Remark 3.3.10. The optimality conditions $(i)-(iii)$ in Theorem 3.3.19 can be equivalently written as

$$\bar{y}^* \in \partial f(\bar{x}) \cap (-\partial((\bar{z}^* g) + \delta_S)(\bar{x})) \text{ and } (\bar{z}^* g)(\bar{x}) = 0.$$

The Lagrangian function $L^{C_{FL}} : X \times X^* \times Z^* \to \overline{\mathbb{R}}$ looks like

$$L^{C_{FL}}(x, y^*, z^*) = \inf_{\substack{y\in X,\\ z\in Z}} \{\Phi^{C_{FL}}(x,y,z) - \langle y^*, y\rangle - \langle z^*, z\rangle\}$$

$$= \begin{cases} \inf\limits_{\substack{y\in X,\\ z\in g(x)+C}} \{f(x+y) - \langle y^*, y\rangle - \langle z^*, z\rangle\}, & \text{if } x \in S,\\ +\infty, & \text{otherwise,}\end{cases}$$

$$= \begin{cases} \inf\limits_{\substack{r\in X,\\ s\in C}} \{f(r) - \langle y^*, r-x\rangle - \langle z^*, g(x)+s\rangle\}, & \text{if } x \in S,\\ +\infty, & \text{otherwise,}\end{cases}$$

$$= \begin{cases} \langle y^*, x\rangle - \langle z^*, g(x)\rangle - f^*(y^*) + \inf\limits_{s\in C}\langle -z^*, s\rangle, & \text{if } x \in S,\\ +\infty, & \text{otherwise,}\end{cases}$$

$$= \begin{cases} \langle y^*, x\rangle - \langle z^*, g(x)\rangle - f^*(y^*), & \text{if } x \in S, z^* \in -C^*,\\ -\infty, & \text{if } x \in S, z^* \notin -C^*,\\ +\infty, & \text{otherwise.}\end{cases}$$

In this way we get that

$$\sup_{(y^*,z^*)\in X^*\times Z^*} \inf_{x\in X} L^{C_{FL}}(x,y^*,z^*) = \sup_{(y^*,z^*)\in X^*\times Z^*} \inf_{x\in S} L^{C_{FL}}(x,y^*,z^*)$$

$$= \sup_{y^*\in X^*, -z^*\in C^*} \left\{ -f^*(y^*) + \inf_{x\in S}\{\langle y^*, x\rangle + \langle -z^*, g(x)\rangle\}\right\}$$

$$= \sup_{y^*\in X^*, z^*\in C^*} \{-f^*(y^*) - (z^* g)_S^*(-y^*)\}$$

and

$$\inf_{x\in X} \sup_{(y^*,z^*)\in X^*\times Z^*} L^{C_{FL}}(x,y^*,z^*)$$

$$= \inf_{x\in S}\left\{ \sup_{y^*\in X^*}\{\langle y^*, x\rangle - f^*(y^*)\} + \sup_{-z^*\in C^*}\langle -z^*, g(x)\rangle\right\}$$

$$= \inf_{x\in S}\{f^{**}(x) + \delta_{\{y\in X: g(y)\in -C^{**}\}}(x)\} = \inf_{\substack{x\in S,\\ g(x)\in -\operatorname{cl}(C)}} f^{**}(x).$$

The characterization of the saddle points of $L^{C_{FL}}$ via the optimal solutions of the pair $(P^C)-(D^{C_{FL}})$ follows. As above we can again weaken the properness assumption for f.

Theorem 3.3.23. *Assume that C is closed and that f is a convex and lower semicontinuous function fulfilling $f(x) > -\infty$ for all $x \in X$. Then the following statements are equivalent:*

(i) $(\bar{x}, \bar{y}^*, \bar{z}^*)$ *is a saddle point of* $L^{C_{FL}}$;
(ii) $\bar{x} \in \mathcal{A}$ *is an optimal solution to* (P^C), $(\bar{y}^*, -\bar{z}^*) \in X^* \times C^*$ *is an optimal solution to* $(D^{C_{FL}})$ *and* $v(P^C) = v(D^{C_{FL}})$.

Corollary 3.3.24. *Let Z be partially ordered by the convex closed cone $C \subseteq Z$, $S \subseteq X$ be a nonempty convex set, $f : X \to \overline{\mathbb{R}}$ a proper, convex and lower semicontinuous function and $g : X \to \overline{Z}$ a proper and C-convex function such that $\operatorname{dom} f \cap S \cap g^{-1}(-C) \neq \emptyset$. Assume that one of the regularity conditions $(RC_i^{C_{FL}})$, $i \in \{1, 2, 3, 4\}$, is fulfilled. Then $\bar{x} \in \mathcal{A}$ is an optimal solution to (P^C) if and only if there exists $(\bar{y}^*, \bar{z}^*) \in X^* \times -C^*$ such that $(\bar{x}, \bar{y}^*, \bar{z}^*)$ is a saddle point of $L^{C_{FL}}$. In this case $(\bar{y}^*, -\bar{z}^*)$ is an optimal solution to the dual $(D^{C_{FL}})$.*

3.4 The composed convex optimization problem

In this section we construct by means of the perturbation approach described in section 3.1 two conjugate dual problems to the *composed convex optimization problem.* We give also regularity conditions which guarantee the existence of strong duality and derive from the latter necessary and sufficient optimality conditions. Lagrangian functions for each pair of primal-dual problems are also introduced.

Let X and Z be Hausdorff locally convex spaces, where Z is assumed to be partially ordered by the convex cone $C \subseteq Z$. Consider $f : X \to \overline{\mathbb{R}}$ a proper and convex function, $h : X \to \overline{Z}$ a proper and C-convex function and $g : Z \cup \{+\infty_C\} \to \overline{\mathbb{R}}$ a proper, convex and C-increasing function fulfilling $g(+\infty_C) = +\infty$ and $h(\operatorname{dom} f \cap \operatorname{dom} h) \cap \operatorname{dom} g \neq \emptyset$. The primal problem we deal with in this section is

$$(P^{CC}) \quad \inf_{x \in X} \{f(x) + g \circ h(x)\}.$$

3.4.1 A first dual problem to (P^{CC})

Let Z be the space of perturbation variables and $\Phi^{CC_1} : X \times Z \to \overline{\mathbb{R}}$ be the perturbation function defined by $\Phi^{CC_1}(x, z) = f(x) + g(h(x) + z)$. Obviously, $\Phi^{CC_1}(x, 0) = f(x) + g(h(x))$ for all $x \in X$. The conjugate function $(\Phi^{CC_1})^* : X^* \times Z^* \to \overline{\mathbb{R}}$ of Φ^{CC_1} has for all $(x^*, z^*) \in X^* \times Z^*$ the following form

$$(\Phi^{CC_1})^*(x^*, z^*) = \sup_{\substack{x \in X, \\ z \in Z}} \{\langle x^*, x\rangle + \langle z^*, z\rangle - f(x) - g(h(x) + z)\}$$

$$= \sup_{\substack{x\in X,\\ s\in Z}} \{\langle x^*, x\rangle + \langle z^*, s - h(x)\rangle - f(x) - g(s)\}$$

$$= \sup_{x\in X}\{\langle x^*, x\rangle - \langle z^*, h(x)\rangle - f(x)\} + \sup_{s\in Z}\{\langle z^*, s\rangle - g(s)\}$$

$$= (f + (z^*h))^*(x^*) + g^*(z^*).$$

Since g is C-increasing we have by Proposition 2.3.11 that $g^*(z^*) = +\infty$ if $z^* \notin C^*$. Thus

$$(\Phi^{CC_1})^*(x^*, z^*) = (f + (z^*h))^*(x^*) + g^*(z^*) + \delta_{C^*}(z^*). \tag{3.18}$$

The dual problem we get by means of Φ^{CC_1} is

$$(D^{CC_1}) \quad \sup_{z^*\in C^*} \{-(\Phi^{CC_1})^*(0, z^*)\},$$

which is nothing else than

$$(D^{CC_1}) \quad \sup_{z^*\in C^*} \{-g^*(z^*) - (f + (z^*h))^*(0)\}.$$

By Theorem 3.1.1 it holds $v(D^{CC_1}) \leq v(P^{CC})$. In the following we introduce some regularity conditions which close the gap between these optimal objective values and ensure that the dual has an optimal solution. First, let us mention that the assumptions we made at the beginning of the section guarantee the properness and convexity of Φ^{CC_1}. The regularity conditions which follows are derived from the general ones given in section 3.2.

The regularity condition (RC_1^{Φ}) assumes in general that there exists $x' \in X$ such that $(x', 0) \in \operatorname{dom} \Phi^{CC_1}$ and $\Phi^{CC_1}(x', \cdot)$ is continuous at 0, and becomes in this special case

$$(RC_1^{CC_1}) \left| \begin{array}{l} \exists x' \in \operatorname{dom} f \cap \operatorname{dom} h \cap h^{-1}(\operatorname{dom} g) \text{ such that} \\ g \text{ is continuous at } h(x'). \end{array}\right.$$

In order to provide regularity conditions in case X and Z are Fréchet spaces we have to establish the set $\Pr_Z(\operatorname{dom} \Phi^{CC_1})$ and to guarantee lower semicontinuity for Φ^{CC_1}. First notice that

$$z \in \Pr_Z(\operatorname{dom} \Phi^{CC_1}) \Leftrightarrow \exists x \in X \text{ such that } \Phi^{CC_1}(x, z) < +\infty$$

$$\Leftrightarrow \exists x \in \operatorname{dom} f \cap \operatorname{dom} h \text{ such that } z \in \operatorname{dom} g - h(x)$$

$$\Leftrightarrow z \in \operatorname{dom} g - h(\operatorname{dom} f \cap \operatorname{dom} h),$$

and so $\Pr_Z(\operatorname{dom} \Phi^{CC_1}) = \operatorname{dom} g - h(\operatorname{dom} f \cap \operatorname{dom} h)$.

Next we show that if f and g are lower semicontinuous and h is star C-lower semicontinuous, i.e. (z^*h) is lower semicontinuous for all $z^* \in C^*$, then Φ^{CC_1} is lower semicontinuous. To this end we calculate the biconjugate of Φ^{CC_1}. We have for all $(x, z) \in X \times Z$

$$(\Phi^{CC_1})^{**}(x,z) = \sup_{\substack{x^* \in X^*,\\ z^* \in Z^*}} \left\{ \langle x^*, x\rangle + \langle z^*, z\rangle - (\Phi^{CC_1})^*(x^*, z^*) \right\}$$

$$= \sup_{\substack{x^* \in X^*,\\ z^* \in C^*}} \{ \langle x^*, x\rangle + \langle z^*, z\rangle - (f + (z^*h))^*(x^*) - g^*(z^*) \}$$

$$= \sup_{z^* \in C^*} \left\{ \langle z^*, z\rangle - g^*(z^*) + \sup_{x^* \in X^*} \{ \langle x^*, x\rangle - (f + (z^*h))^*(x^*) \} \right\}$$

$$= \sup_{z^* \in C^*} \{ \langle z^*, z\rangle - g^*(z^*) + (f + (z^*h))^{**}(x) \}.$$

Since for all $z^* \in C^*$, $f + (z^*h)$ is proper, convex and lower semicontinuous and g is proper, convex and lower semicontinuous, we get by Theorem 2.3.5 that

$$(\Phi^{CC_1})^{**}(x,z) = \sup_{z^* \in C^*} \{ \langle z^*, z\rangle - g^*(z^*) + f(x) + (z^*h)(x) \}$$

$$= f(x) + \sup_{z^* \in C^*} \{ \langle z^*, h(x) + z\rangle - g^*(z^*) \}$$

$$= f(x) + \sup_{z^* \in Z^*} \{ \langle z^*, h(x) + z\rangle - g^*(z^*) \}$$

$$= f(x) + g^{**}(h(x) + z) = f(x) + g(h(x) + z) = \Phi^{CC_1}(x,z),$$

which proves, via Theorem 2.3.6, that Φ^{CC_1} is lower semicontinuous. We can state now the following regularity condition

$(RC_2^{CC_1})$ | X and Z are Fréchet spaces, f and g are lower semicontinuous, h is star C-lower semicontinuous and $0 \in \operatorname{sqri}(\operatorname{dom} g - h(\operatorname{dom} f \cap \operatorname{dom} h))$,

along with its stronger variants

$(RC_{2'}^{CC_1})$ | X and Z are Fréchet spaces, f and g are lower semicontinuous, h is star C-lower semicontinuous and $0 \in \operatorname{core}(\operatorname{dom} g - h(\operatorname{dom} f \cap \operatorname{dom} h))$

and

$(RC_{2''}^{CC_1})$ | X and Z are Fréchet spaces, f and g are lower semicontinuous, h is star C-lower semicontinuous and $0 \in \operatorname{int}(\operatorname{dom} g - h(\operatorname{dom} f \cap \operatorname{dom} h))$,

which are in fact equivalent. In the finite dimensional case one can consider as regularity condition

$(RC_3^{CC_1})$ | $\dim(\operatorname{lin}(\operatorname{dom} g - h(\operatorname{dom} f \cap \operatorname{dom} h))) < +\infty$ and $\operatorname{ri}(\operatorname{dom} g) \cap \operatorname{ri}(h(\operatorname{dom} f \cap \operatorname{dom} h)) \neq \emptyset$.

The conditions $(RC_2^{CC_1})$ and $(RC_{2'}^{CC_1})$ have been introduced in [50] but under the assumption that h is a C-lower semicontinuous function. The condition $(RC_1^{CC_1})$ is a classical one, while $(RC_3^{CC_1})$ has been stated for the first time in [204]. A refinement of $(RC_3^{CC_1})$ for X and Z finite dimensional spaces has been given by the authors in [30].

Before enunciating the strong duality theorem we formulate a closedness type regularity condition for the composed convex optimization problem. To this end, along the lower semicontinuity of Φ^{CC_1}, we have to ensure that $\Pr_{X^*\times\mathbb{R}}(\operatorname{epi}(\Phi^{CC_1})^*)$ is closed in the topology $w(X^*,X)\times\mathbb{R}$. One has

$$(x^*,r)\in\Pr_{X^*\times\mathbb{R}}(\operatorname{epi}(\Phi^{CC_1})^*)\Leftrightarrow \exists z^*\in Z^* \text{ such that } (\Phi^{CC_1})^*(x^*,z^*)\le r$$

$$\Leftrightarrow \exists z^*\in C^* \text{ such that } (f+(z^*h))^*(x^*)+g^*(z^*)\le r \Leftrightarrow \exists z^*\in C^* \text{ such that}$$
$$(x^*,r)\in \operatorname{epi}(f+(z^*h))^*+(0,g^*(z^*)),$$

which means that

$$\Pr_{X^*\times\mathbb{R}}(\operatorname{epi}(\Phi^{CC_1})^*)=\bigcup_{z^*\in C^*}\left(\operatorname{epi}(f+(z^*h))^*+(0,g^*(z^*))\right).$$

Thus the closedness type regularity condition looks like (cf. [33])

$(RC_4^{CC_1})$ | f and g are lower semicontinuous, h is star C-lower semicontinuous and $\bigcup_{z^*\in C^*}\left(\operatorname{epi}(f+(z^*h))^*+(0,g^*(z^*))\right)$ is closed in the topology $w(X^*,X)\times\mathbb{R}$.

The strong duality theorem follows as a consequence of Theorem 3.2.1 and Theorem 3.2.3.

Theorem 3.4.1. *Let Z be partially ordered by the convex cone $C\subseteq Z$, $f:X\to\overline{\mathbb{R}}$ be a proper and convex function, $h:X\to Z\cup\{+\infty_C\}$ a proper and C-convex function and $g:Z\cup\{+\infty_C\}\to\overline{\mathbb{R}}$ a proper, convex and C-increasing function such that $g(+\infty_C)=+\infty$ and $h(\operatorname{dom} f\cap\operatorname{dom} h)\cap\operatorname{dom} g\neq\emptyset$. If one of the regularity conditions $(RC_i^{CC_1})$, $i\in\{1,2,3,4\}$, is fulfilled, then $v(P^{CC})=v(D^{CC_1})$ and the dual has an optimal solution.*

The necessary and sufficient optimality conditions for the pair of primal-dual problems $(P^{CC})-(D^{CC_1})$ are a consequence of Theorem 3.3.1.

Theorem 3.4.2. *(a) Let Z be partially ordered by the convex cone $C\subseteq Z$, $f:X\to\overline{\mathbb{R}}$ be a proper and convex function, $h:X\to Z\cup\{+\infty_C\}$ a proper and C-convex function and $g:Z\cup\{+\infty_C\}\to\overline{\mathbb{R}}$ a proper, convex and C-increasing function such that $g(+\infty_C)=+\infty$ and $h(\operatorname{dom} f\cap\operatorname{dom} h)\cap\operatorname{dom} g\neq\emptyset$. Let $\bar{x}\in X$ be an optimal solution to (P^{CC}) and assume that one of the regularity conditions $(RC_i^{CC_1})$, $i\in\{1,2,3,4\}$, is fulfilled. Then there exists $\bar{z}^*\in C^*$, an optimal solution to (D^{CC_1}), such that*

(i) $\min_{x\in X}\{f(x)+(\bar{z}^*h)(x)\}=f(\bar{x})+(\bar{z}^*h)(\bar{x})$;

(ii) $g^*(\bar{z}^*) + g(h(\bar{x})) = (\bar{z}^* h)(\bar{x})$.

(b) Assume that $\bar{x} \in X$ and $\bar{z}^* \in C^*$ fulfill the relations $(i) - (ii)$. Then $\bar{x}$ is an optimal solution to (P^{CC}), $\bar{z}^*$ is an optimal solution to (D^{CC_1}) and $v(P^{CC}) = v(D^{CC_1})$.

Proof. We show that the relation (3.12) becomes in this special case $(i) - (ii)$ and in this way the result follows by Theorem 3.3.1. We have that $\Phi^{CC_1}(\bar{x}, 0) + (\Phi^{CC_1})^*(0, \bar{z}^*) = 0$ is equivalent to $\bar{z}^* \in C^*$ and

$$f(\bar{x}) + g(h(\bar{x})) + g^*(\bar{z}^*) + (f + (\bar{z}^* h))^*(0) = 0 \Leftrightarrow \bar{z}^* \in C^* \text{ and}$$

$$[f(\bar{x}) + (\bar{z}^* h)(\bar{x}) + (f + (\bar{z}^* h))^*(0)] + [g^*(\bar{z}^*) + g(h(\bar{x})) - \langle \bar{z}^*, h(\bar{x}) \rangle] = 0.$$

Since $f(\bar{x}) + (\bar{z}^* h)(\bar{x}) + (f + (\bar{z}^* h))^*(0) \geq 0$ and $g^*(\bar{z}^*) + g(h(\bar{x})) - \langle \bar{z}^*, h(\bar{x}) \rangle \geq 0$, the inequalities must be satisfied as equalities and so the desired conclusion follows. □

Remark 3.4.1. The optimality conditions $(i) - (ii)$ in Theorem 3.4.2 can be equivalently written as

$$0 \in \partial(f + (\bar{z}^* h))(\bar{x}) \text{ and } \bar{z}^* \in \partial g(h(\bar{x})).$$

The Lagrangian function assigned to $(P^{CC}) - (D^{CC_1})$ is denoted by $L^{CC_1} : X \times Z^* \to \overline{\mathbb{R}}$, being defined for all $(x, z^*) \in X \times Z^*$ by

$$L^{CC_1}(x, z^*) = \inf_{z \in Z} \{\Phi^{CC_1}(x, z) - \langle z^*, z \rangle\} = \inf_{z \in Z} \{f(x) + g(h(x) + z) - \langle z^*, z \rangle\}$$

$$= f(x) + \inf_{s \in Z} \{g(s) - \langle z^*, s - h(x) \rangle\} = f(x) + (z^* h)(x) + \inf_{s \in Z} \{g(s) - \langle z^*, s \rangle\}$$

$$= f(x) + (z^* h)(x) - g^*(z^*) = \begin{cases} f(x) + (z^* h)(x) - g^*(z^*), & \text{if } z^* \in C^*, \\ -\infty, & \text{otherwise.} \end{cases}$$

Thus

$$\sup_{z^* \in Z^*} \inf_{x \in X} L^{CC_1}(x, z^*) = \sup_{z^* \in C^*} \{-g^*(z^*) - (f + (z^* h))^*(0)\}$$

and

$$\inf_{x \in X} \sup_{z^* \in Z^*} L^{CC_1}(x, z^*) = \inf_{x \in X} \{f(x) + g^{**}(h(x))\}.$$

We have the following characterization for the saddle points of the Lagrangian L^{CC_1}, where the properness assumption for g is slightly weakened.

Theorem 3.4.3. *Assume that g is a convex and lower semicontinuous function fulfilling $g(z) > -\infty$ for all $z \in Z$. Then the following statements are equivalent:*

(i) $(\bar{x}, \bar{z}^*)$ is a saddle point of L^{CC_1};

(ii) $\bar{x} \in X$ is an optimal solution to (P^{CC}), $\bar{z}^* \in C^*$ is an optimal solution to (D^{CC_1}) and $v(P^{CC}) = v(D^{CC_1})$.

Remark 3.4.2. Since for all $x \in X$ $g^{**}(h(x)) \leq g(h(x))$, the implication $(ii) \Rightarrow (i)$ in Theorem 3.4.3 holds in the most general case without any assumption.

The following statement is a consequence of Corollary 3.3.3.

Corollary 3.4.4. *Let Z be partially ordered by the convex cone $C \subseteq Z$, $S \subseteq X$ be a nonempty convex set, $f : X \to \overline{\mathbb{R}}$ be a proper and convex function, $h : X \to Z \cup \{+\infty_C\}$ a proper and C-convex function and $g : Z \cup \{+\infty_C\} \to \overline{\mathbb{R}}$ a proper, convex, lower semicontinuous and C-increasing function fulfilling $g(+\infty_C) = +\infty$ and $h(\operatorname{dom} f \cap \operatorname{dom} h) \cap \operatorname{dom} g \neq \emptyset$. Assume that one of the regularity conditions $(RC_i^{CC_1})$, $i \in \{1,2,3,4\}$, is fulfilled. Then $\bar{x} \in X$ is an optimal solution to (P^{CC}) if and only if there exists $\bar{z}^* \in C^*$ such that $(\bar{x}, \bar{z}^*)$ is a saddle point of L^{CC_1}. In this case $\bar{z}^*$ is an optimal solution to the dual (D^{CC_1}).*

3.4.2 A second dual problem to (P^{CC})

Let now $X \times Z$ be the space of perturbation variables and $\Phi^{CC_2} : X \times X \times Z \to \overline{\mathbb{R}}$ the perturbation function defined by $\Phi^{CC_2}(x, y, z) = f(x+y) + g(h(x)+z)$. Obviously, $\Phi^{CC_2}(x, 0, 0) = f(x) + g(h(x))$ for all $x \in X$. The conjugate function $(\Phi^{CC_2})^* : X^* \times X^* \times Z^* \to \overline{\mathbb{R}}$ of Φ^{CC_2} looks for all $(x^*, y^*, z^*) \in X^* \times X^* \times Z^*$ like

$$(\Phi^{CC_2})^*(x^*, y^*, z^*) = \sup_{\substack{x,y \in X,\\ z \in Z}} \{\langle x^*, x\rangle + \langle y^*, y\rangle + \langle z^*, z\rangle - f(x+y)$$

$$-g(h(x)+z)\} = \sup_{\substack{x,r \in X,\\ s \in Z}} \{\langle x^*, x\rangle + \langle y^*, r-x\rangle + \langle z^*, s-h(x)\rangle - f(r) - g(s)\}$$

$$= \sup_{x \in X}\{\langle x^* - y^*, x\rangle - \langle z^*, h(x)\rangle\} + \sup_{s \in Z}\{\langle z^*, s\rangle - g(s)\} + \sup_{r \in Y}\{\langle y^*, r\rangle - f(r)\}$$

$$= (z^*h)^*(x^* - y^*) + f^*(y^*) + g^*(z^*).$$

Taking again Proposition 2.3.11 into consideration, we get

$$(\Phi^{CC_2})^*(x^*, y^*, z^*) = (z^*h)^*(x^* - y^*) + f^*(y^*) + g^*(z^*) + \delta_{C^*}(z^*). \quad (3.19)$$

The dual problem of (P^{CC}) obtained via the perturbation function Φ^{CC_2} is

$$(D^{CC_2}) \qquad \sup_{y^* \in X^*, z^* \in Z^*} \{-(\Phi^{CC_2})^*(0, y^*, z^*)\}$$

or, equivalently,

$$(D^{CC_2}) \qquad \sup_{y^* \in X^*, z^* \in C^*} \{-g^*(z^*) - f^*(y^*) - (z^*h)^*(-y^*)\}.$$

Since the existence of weak duality is always ensured by Theorem 3.1.1, we look now for some regularity conditions for having also in this case strong duality. The properness and convexity of Φ^{CC_2} follow by the assumptions considered in this section and thus one can derive these regularity conditions from the general ones given in section 3.2. The condition (RC_1^{Φ}) leads in this special case to

$$(RC_1^{CC_2}) \left| \begin{array}{l} \exists x' \in \operatorname{dom} f \cap \operatorname{dom} h \cap h^{-1}(\operatorname{dom} g) \text{ such that } f \text{ is continuous} \\ \text{at } x' \text{ and } g \text{ is continuous at } h(x'). \end{array} \right.$$

Assuming that X and Z are Fréchet spaces as in (RC_2^{Φ}), we have to guarantee that Φ^{CC_2} is lower semicontinuous and $0 \in \operatorname{sqri}(\operatorname{Pr}_{X \times Z}(\operatorname{dom} \Phi^{CC_2}))$. Similar to the considerations in the previous section one can show that if f and g are lower semicontinuous and h is star C-lower semicontinuous, then Φ^{CC_2} is lower semicontinuous, too. Further, one can notice that

$$\operatorname{Pr}_{X \times Z}(\operatorname{dom} \Phi^{CC_2}) = \operatorname{dom} f \times \operatorname{dom} g - \operatorname{epi}_C h,$$

where the projection on the product of the last two spaces in the domain of definition of Φ^{CC_2} is considered. Indeed, this is a consequence of the following sequence of equivalences

$$\begin{aligned} (y, z) \in \operatorname{Pr}_{X \times Z}(\operatorname{dom} \Phi^{CC_2}) &\Leftrightarrow \exists x \in X \text{ such that } \Phi^{CC_2}(x, y, z) < +\infty \\ &\Leftrightarrow \exists x \in \operatorname{dom} h \text{ such that } x + y \in \operatorname{dom} f \text{ and } h(x) + z \in \operatorname{dom} g \\ &\Leftrightarrow \exists x \in \operatorname{dom} h \text{ such that } (y, z) \in \operatorname{dom} f \times \operatorname{dom} g - (x, h(x)) \\ &\Leftrightarrow (y, z) \in \operatorname{dom} f \times \operatorname{dom} g - \operatorname{epi}_C h. \end{aligned}$$

For the last equivalence the assumption that g is C-increasing is here determinant.

When X and Z are Fréchet spaces we can formulate the following condition

$$(RC_2^{CC_2}) \left| \begin{array}{l} X \text{ and } Z \text{ are Fréchet spaces, } f \text{ and } g \text{ are lower semicontinuous,} \\ h \text{ is star } C\text{-lower semicontinuous and} \\ 0 \in \operatorname{sqri}(\operatorname{dom} f \times \operatorname{dom} g - \operatorname{epi}_C h) \end{array} \right.$$

along with its stronger variants

$$(RC_{2'}^{CC_2}) \left| \begin{array}{l} X \text{ and } Z \text{ are Fréchet spaces, } f \text{ and } g \text{ are lower semicontinuous,} \\ h \text{ is star } C\text{-lower semicontinuous and} \\ 0 \in \operatorname{core}(\operatorname{dom} f \times \operatorname{dom} g - \operatorname{epi}_C h) \end{array} \right.$$

and

$$(RC_{2''}^{CC_2}) \left| \begin{array}{l} X \text{ and } Z \text{ are Fréchet spaces, } f \text{ and } g \text{ are lower semicontinuous,} \\ h \text{ is star } C\text{-lower semicontinuous and} \\ 0 \in \operatorname{int}(\operatorname{dom} f \times \operatorname{dom} g - \operatorname{epi}_C h), \end{array} \right.$$

which are in fact equivalent, while in the finite dimensional case one has

$$(RC_3^{CC_2}) \left| \begin{array}{l} \dim(\operatorname{lin}(\operatorname{dom} f \times \operatorname{dom} g - \operatorname{epi}_C h)) < +\infty \text{ and} \\ \operatorname{ri}(\operatorname{dom} f \times \operatorname{dom} g) \cap \operatorname{ri}(\operatorname{epi}_C h) \neq \emptyset. \end{array} \right.$$

Let us determine now the set $\Pr_{X^*\times\mathbb{R}}(\operatorname{epi}(\Phi^{CC_2})^*)$. It holds

$$(x^*,r)\in \Pr_{X^*\times\mathbb{R}}(\operatorname{epi}(\Phi^{CC_2})^*) \Leftrightarrow \exists y^*\in X^* \text{ and } \exists z^*\in Z^* \text{ such that}$$

$$(\Phi^{CC_2})^*(x^*,y^*,z^*)\le r \Leftrightarrow \exists y^*\in X^* \text{ and } \exists z^*\in C^* \text{ such that } (z^*h)^*(x^*-y^*)$$
$$+f^*(y^*)+g^*(z^*)\le r \Leftrightarrow \exists (y^*,z^*)\in X^*\times C^* \text{ such that } (x^*,r)\in (y^*,f^*(y^*))$$
$$+\operatorname{epi}(z^*h)^*+(0,g^*(z^*)) \Leftrightarrow \exists z^*\in C^* \text{ such that } (x^*,r)\in \operatorname{epi} f^*+\operatorname{epi}(z^*h)^*$$
$$+(0,g^*(z^*)) \Leftrightarrow (x^*,r)\in \operatorname{epi} f^* + \bigcup_{z^*\in C^*}(\operatorname{epi}(z^*h)^*+(0,g^*(z^*))).$$

This leads to the following closedness type regularity condition (cf. [33])

$(RC_4^{CC_2})$ | f and g are lower semicontinuous, h is star C-lower semicontinuous and $\operatorname{epi} f^* + \bigcup_{z^*\in C^*}(\operatorname{epi}(z^*h)^*+(0,g^*(z^*)))$ is closed in the topology $w(X^*,X)\times\mathbb{R}$.

We can formulate the following strong duality theorem.

Theorem 3.4.5. *Let Z be partially ordered by the convex cone $C\subseteq Z$, $f: X\to\overline{\mathbb{R}}$ be a proper and convex function, $h: X\to Z\cup\{+\infty_C\}$ a proper and C-convex function and $g: Z\cup\{+\infty_C\}\to\overline{\mathbb{R}}$ a proper, convex and C-increasing function such that $g(+\infty_C)=+\infty$ and $h(\operatorname{dom} f\cap\operatorname{dom} h)\cap\operatorname{dom} g\neq\emptyset$. If one of the regularity conditions $(RC_i^{CC_2})$, $i\in\{1,2,3,4\}$, is fulfilled, then $v(P^{CC})=v(D^{CC_2})$ and the dual has an optimal solution.*

Remark 3.4.3. (a) Since for all $z^*\in C^*$

$$(f+(z^*h))^*(0)\le \inf_{y^*\in X^*}\{f^*(y^*)+(z^*h)^*(-y^*)\}, \tag{3.20}$$

it is obvious that in general one has $v(D^{CC_2})\le v(D^{CC_1})\le v(P^{CC})$. This means that if for (P^{CC}) and (D^{CC_2}) strong duality holds, then we also have $v(P^{CC})=v(D^{CC_1})$. Moreover, if $(\bar y^*,\bar z^*)\in X^*\times C^*$ is an optimal solution to (D^{CC_2}), then $\bar z^*\in C^*$ is an optimal solution to (D^{CC_1}). Thus for (P^{CC}) and (D^{CC_1}) strong duality holds, too.

(b) The authors have given in [31] closedness type regularity conditions for strong duality between (P^{CC}) and the duals (D^{CC_1}) and (D^{CC_2}), respectively, also in case h is only C-epi closed. Those regularity conditions are stronger than $(RC_4^{CC_1})$ and $(RC_4^{CC_2})$, respectively, which is quite natural since we assume less for the function h.

The necessary and sufficient optimality conditions for the pair of primal-dual problems (P^{CC}) – (D^{CC_2}) follow.

Theorem 3.4.6. *(a) Let the assumptions of Theorem 3.4.5 be fulfilled and assume that $\bar x\in X$ is an optimal solution to (P^{CC}) and that one of the regularity conditions $(RC_i^{CC_2})$, $i\in\{1,2,3,4\}$, is fulfilled. Then there exists $(\bar y^*,\bar z^*)\in X^*\times C^*$, an optimal solution to (D^{CC_2}), such that*

(i) $\min_{x \in X}\{\langle \bar{y}^*, x\rangle + (\bar{z}^* h)(x)\} = \langle \bar{y}^*, \bar{x}\rangle + (\bar{z}^* h)(\bar{x})$;
(ii) $f^*(\bar{y}^*) + f(\bar{x}) = \langle \bar{y}^*, \bar{x}\rangle$;
(iii) $g^*(\bar{z}^*) + g(h(\bar{x})) = (\bar{z}^* h)(\bar{x})$.

(b) Assume that $\bar{x} \in X$ *and* $(\bar{y}^*, \bar{z}^*) \in X^* \times C^*$ *fulfill the relations (i) – (iii). Then* $\bar{x}$ *is an optimal solution to* (P^{CC}), $(\bar{y}^*, \bar{z}^*)$ *is an optimal solution to* (D^{CC_2}) *and* $v(P^{CC}) = v(D^{CC_2})$.

Proof. We have just to evaluate the relation (3.12) and to show that it is equivalent to *(i)* – *(iii)*. By Theorem 3.3.1 the conclusion will follow automatically. Indeed, $\Phi^{CC_2}(\bar{x}, 0, 0) + (\Phi^{CC_2})^*(0, \bar{y}^*, \bar{z}^*) = 0$ is equivalent to $(\bar{y}^*, \bar{z}^*) \in X^* \times C^*$ and

$$f(\bar{x}) + g(h(\bar{x})) + f^*(\bar{y}^*) + g^*(\bar{z}^*) + (\bar{z}^* h)^*(-\bar{y}^*) = 0,$$

which is the same as $(\bar{y}^*, \bar{z}^*) \in X^* \times C^*$ and

$$\{f^*(\bar{y}^*) + f(\bar{x}) - \langle \bar{y}^*, \bar{x}\rangle\} + \{g^*(\bar{z}^*) + g(h(\bar{x})) - (\bar{z}^* h)(\bar{x})\}$$

$$+ \left\{\langle \bar{y}^*, \bar{x}\rangle + (\bar{z}^* h)(\bar{x}) - \inf_{x \in X}\{\langle \bar{y}^*, x\rangle + (\bar{z}^* h)(x)\}\right\} = 0.$$

Since the summands in the braces are nonnegative, they must be equal to zero and this leads to the desired conclusion. □

Remark 3.4.4. The optimality conditions *(i)* – *(iii)* in Theorem 3.4.6 can be equivalently written as

$$\bar{y}^* \in \partial f(\bar{x}) \cap (-\partial(\bar{z}^* g)(\bar{x})) \text{ and } \bar{z}^* \in \partial g(h(\bar{x})).$$

The Lagrangian function of the pair of primal-dual problems (P^{CC}) – (D^{CC_2}) is denoted by $L^{CC_2} : X \times X^* \times Z^* \to \overline{\mathbb{R}}$ and is defined for all $(x, y^*, z^*) \in X \times X^* \times Z^*$ as being

$$L^{CC_2}(x, y^*, z^*) = \inf_{\substack{y \in X,\\ z \in Z}} \{\Phi^{CC_2}(x, y, z) - \langle y^*, y\rangle - \langle z^*, z\rangle\}$$

$$= \inf_{\substack{y \in X,\\ z \in Z}} \{f(x+y) + g(h(x)+z) - \langle y^*, y\rangle - \langle z^*, z\rangle\} = \inf_{\substack{r \in X,\\ s \in Z}} \{f(r) + g(s) - \langle y^*, r - x\rangle$$

$$-\langle z^*, s - h(x)\rangle\} = \langle y^*, x\rangle + (z^* h)(x) + \inf_{r \in Y}\{f(r) - \langle y^*, r\rangle\}$$

$$+ \inf_{s \in Z}\{g(s) - \langle z^*, s\rangle\} = \langle y^*, x\rangle + (z^* h)(x) - f^*(y^*) - g^*(z^*)$$

$$= \begin{cases} \langle y^*, x\rangle + (z^* h)(x) - f^*(y^*) - g^*(z^*), & \text{if } z^* \in C^*, \\ -\infty, & \text{otherwise.} \end{cases}$$

It holds

$$\sup_{y^* \in X^*, z^* \in Z^*} \inf_{x \in X} L^{CC_2}(x, y^* z^*)$$

$$= \sup_{y^* \in X^*, z^* \in C^*} \left\{ -f^*(y^*) - g^*(z^*) + \inf_{x \in X} \{\langle y^*, x\rangle + (z^*h)(x)\} \right\}$$
$$= \sup_{y^* \in X^*, z^* \in Z^*} \{ -f^*(y^*) - g^*(z^*) - (z^*h)^*(-y^*)\}$$

and

$$\inf_{x \in X} \sup_{y^* \in X^*, z^* \in C^*} L^{CC_2}(x, y^*, z^*)$$
$$= \inf_{x \in X} \left\{ \sup_{y^* \in Y^*} \{\langle y^*, x\rangle - f^*(y^*)\} + \sup_{z^* \in Z^*} \{\langle z^*, h(x)\rangle - g^*(z^*)\} \right\}$$
$$= \inf_{x \in X} \{ f^{**}(x) + g^{**}(h(x))\}.$$

By Theorem 3.3.2 we obtain the following result, where we omit assuming properness for f and g.

Theorem 3.4.7. *Assume that f and g are convex and lower semicontinuous functions fulfilling $f(x) > -\infty$ for all $x \in X$ and $g(z) > -\infty$ for all $z \in Z$, respectively. Then the following statements are equivalent:*

(i) $(\bar{x}, \bar{y}^*, \bar{z}^*)$ is a saddle point of L^{CC_2};
(ii) $\bar{x} \in X$ is an optimal solution to (P^{CC}), $(\bar{y}^*, \bar{z}^*) \in X^* \times C^*$ is an optimal solution to (D^{CC_2}) and $v(P^{CC}) = v(D^{CC_2})$.

Remark 3.4.5. Since $f^{**}(x) \leq f(x)$ and $g^{**}(h(x)) \leq g(h(x))$ for all $x \in X$, the implication $(ii) \Rightarrow (i)$ in Theorem 3.4.7 holds always without any assumption on the functions involved.

Combining Theorem 3.4.5 and Theorem 3.4.7 one can state the following result.

Corollary 3.4.8. *Let Z be partially ordered by the convex cone $C \subseteq Z$, $S \subseteq X$ be a nonempty convex set, $f : X \to \overline{\mathbb{R}}$ a proper, convex and lower semicontinuous function, $h : X \to Z \cup \{+\infty_C\}$ a proper and C-convex function and $g : Z \cup \{+\infty_C\} \to \overline{\mathbb{R}}$ a proper, convex, lower semicontinuous and C-increasing function fulfilling $g(+\infty_C) = +\infty$ and $h(\operatorname{dom} f \cap \operatorname{dom} h) \cap \operatorname{dom} g \neq \emptyset$. Assume that one of the regularity conditions $(RC_i^{CC_2})$, $i \in \{1, 2, 3, 4\}$, is fulfilled. Then $\bar{x} \in X$ is an optimal solution to (P^{CC}) if and only if there exists $(\bar{y}^*, \bar{z}^*) \in X^* \times C^*$ such that $(\bar{x}, \bar{y}^*, \bar{z}^*)$ is a saddle point of L^{CC_2}. In this case $(\bar{y}^*, \bar{z}^*)$ is an optimal solution to the dual (D^{CC_2}).*

3.5 Stable strong duality and formulae for conjugate functions and subdifferentials

The aim of this section is to introduce the concept of *stable strong duality*, to prove that the regularity conditions introduced in section 3.2 ensure it and to derive as a consequence of it different formulae for conjugate functions and subdifferentials.

3.5.1 Stable strong duality for the general scalar optimization problem

Consider again the general optimization problem

$$(PG) \quad \inf_{x \in X} \Phi(x, 0),$$

where $\Phi : X \times Y \to \overline{\mathbb{R}}$ is a perturbation function and X and Y are Hausdorff locally convex spaces. By means of Φ we assigned in section 3.1 to (PG) the following dual problem

$$(DG) \quad \sup_{y^* \in Y^*} \{-\Phi^*(0, y^*)\}.$$

For every $x^* \in X^*$ one can consider the following *extended primal optimization problem*

$$(PG^{x^*}) \quad \inf_{x \in X} \{\Phi(x, 0) - \langle x^*, x \rangle\}.$$

The function $\Phi^{x^*} : X \times Y \to \overline{\mathbb{R}}$, $\Phi^{x^*}(x, y) = \Phi(x, y) - \langle x^*, x \rangle$ is a perturbation function for (PG^{x^*}) and it introduces the following conjugate dual

$$(DG^{x^*}) \quad \sup_{y^* \in Y^*} \{-(\Phi^{x^*})^*(0, y^*)\}.$$

Since

$$(\Phi^{x^*})^*(u^*, y^*) = \sup_{\substack{x \in X, \\ y \in Y}} \left\{\langle u^*, x \rangle + \langle y^*, y \rangle - \Phi^{x^*}(x, y)\right\}$$

$$= \sup_{\substack{x \in X, \\ y \in Y}} \{\langle u^* + x^*, x \rangle + \langle y^*, y \rangle - \Phi(x, y)\} = \Phi^*(u^* + x^*, y^*),$$

we get for every $x^* \in X^*$ the following formulation for the conjugate dual of (PG^{x^*})

$$(DG^{x^*}) \quad \sup_{y^* \in Y^*} \{-\Phi^*(x^*, y^*)\}.$$

The following definition introduces the notion of stable strong duality.

Definition 3.5.1. *We say that between the optimization problems (PG) and (DG) stable strong duality holds, if for all $x^* \in X^*$ for (PG^{x^*}) and (DG^{x^*}) strong duality holds, i.e. $v(PG^{x^*}) = v(DG^{x^*})$ and the dual (DG^{x^*}) has an optimal solution.*

Next we show that the regularity conditions we introduced in order to guarantee strong duality for (PG) and (DG) are also guaranteeing the existence of stable strong duality for (PG) and (DG).

Assume that Φ is a proper and convex function with $0 \in \Pr_Y(\operatorname{dom} \Phi)$. For all $x^* \in X^*$ we have $\operatorname{dom} \Phi^{x^*} = \operatorname{dom} \Phi$. Obviously, the generalized interior point regularity conditions (RC_i^{Φ}), $i \in \{1,2,3\}$, are sufficient for strong duality for (PG^{x^*}) and (DG^{x^*}). Coming now to the closedness type regularity condition (RC_4^{Φ}), if this is fulfilled, then by Theorem 3.2.2 follows that

$$\sup_{x \in X} \{\langle x^*, x\rangle - \Phi(x,0)\} = \min_{y^* \in Y^*} \{\Phi^*(x^*, y^*)\} \ \forall x^* \in X^*,$$

which is the same with $v(PG^{x^*}) = v(DG^{x^*})$ and the dual (DG^{x^*}) has an optimal solution, for all $x^* \in X^*$. Thus one can state the following stable strong duality result (see [25]).

Theorem 3.5.1. *Let $\Phi : X \times Y \to \overline{\mathbb{R}}$ be a proper and convex function such that $0 \in \Pr_Y(\operatorname{dom} \Phi)$. If one of the regularity conditions (RC_i^{Φ}), $i \in \{1,2,3,4\}$, is fulfilled, then for (PG) and (DG) stable strong duality holds, which is nothing else than*

$$\sup_{x \in X} \{\langle x^*, x\rangle - \Phi(x,0)\} = \min_{y^* \in Y^*} \{\Phi^*(x^*, y^*)\} \ \forall x^* \in X^*. \tag{3.21}$$

Next we state for the different pairs of scalar primal-dual problems investigated in this chapter stable strong duality theorems. We also show that the existence of stable strong duality is a sufficient condition for deriving different subdifferential formulae.

The approach we choose is the following: we start with the optimization problem with a composed convex function as objective function and treat the other classes of optimization problems as special cases of it.

3.5.2 The composed convex optimization problem

Let X and Z be Hausdorff locally convex spaces, where Z is assumed to be partially ordered by the convex cone $C \subseteq Z$. Consider $f : X \to \overline{\mathbb{R}}$ a proper and convex function, $h : X \to Z \cup \{+\infty_C\}$ a proper and C-convex function and $g : Z \cup \{+\infty_C\} \to \overline{\mathbb{R}}$ a proper, convex and C-increasing function fulfilling $g(+\infty_C) = +\infty$ and $h(\operatorname{dom} f \cap \operatorname{dom} h) \cap \operatorname{dom} g \neq \emptyset$. For the optimization problem

$$(P^{CC}) \quad \inf_{x \in X} \{f(x) + g \circ h(x)\},$$

by using as perturbation function $\Phi^{CC_1} : X \times Z \to \overline{\mathbb{R}}$, $\Phi^{CC_1}(x,z) = f(x) + g(h(x) + z)$, we introduced in section 3.4 the following dual problem

$$(D^{CC_1}) \quad \sup_{z^* \in C^*} \{-g^*(z^*) - (f + (z^*h))^*(0)\}.$$

Next one can state the following result.

Theorem 3.5.2. *(a) Let Z be partially ordered by the convex cone $C \subseteq Z$, $f : X \to \overline{\mathbb{R}}$ be a proper and convex function, $h : X \to Z \cup \{+\infty_C\}$ a proper and C-convex function and $g : Z \cup \{+\infty_C\} \to \overline{\mathbb{R}}$ a proper, convex and C-increasing function such that $g(+\infty_C) = +\infty$ and $h(\operatorname{dom} f \cap \operatorname{dom} h) \cap$*

$\operatorname{dom} g \neq \emptyset$. If one of the regularity conditions $(RC_i^{CC_1})$, $i \in \{1,2,3,4\}$, is fulfilled, then for (P^{CC}) and (D^{CC_1}) stable strong duality holds, that is

$$(f + g \circ h)^*(x^*) = \min_{z^* \in C^*} \{g^*(z^*) + (f + (z^*h))^*(x^*)\} \ \forall x^* \in X^*. \quad (3.22)$$

(b) If for (P^{CC}) and (D^{CC_1}) stable strong duality holds, then for all $x \in X$ one has

$$\partial(f + g \circ h)(x) = \bigcup_{z^* \in \partial g(h(x))} \partial(f + (z^*h))(x). \quad (3.23)$$

Proof. (*a*) The assertion follows as a consequence of Theorem 3.4.1, Theorem 3.5.1 and relation (3.18).

(*b*) In case $x \notin \operatorname{dom} f \cap \operatorname{dom} h \cap h^{-1}(\operatorname{dom} g)$ the conclusion follows via the conventions made for the subdifferential. Let be $x \in \operatorname{dom} f \cap \operatorname{dom} h \cap h^{-1}(\operatorname{dom} g)$.

"$\supseteq$" Consider $z^* \in \partial g(h(x))$ and $x^* \in \partial(f + (z^*h))(x)$. We have

$$\langle z^*, z - h(x)\rangle \leq g(z) - g(h(x)) \ \forall z \in Z$$

and

$$\langle x^*, t - x\rangle \leq (f + (z^*h))(t) - (f + (z^*h))(x) \ \forall t \in X.$$

Take an arbitrary $t \in X$. If $h(t) = +\infty_C$, then $g(h(t)) = +\infty$ and $(f + g \circ h)(t) = +\infty$. Thus $\langle x^*, t - x\rangle \leq (f + g \circ h)(t) - (f + g \circ h)(x)$.

In case $h(t) \in Z$ we have

$$\langle x^*, t - x\rangle \leq f(t) - f(x) + \langle z^*, h(t) - h(x)\rangle \leq f(t) - f(x)$$
$$+ g(h(t)) - g(h(x)) = (f + g \circ h)(t) - (f + g \circ h)(x).$$

Since in both situations $\langle x^*, t - x\rangle \leq (f + g \circ h)(t) - (f + g \circ h)(x)$ for all $t \in X$, it follows that $x^* \in \partial(f + g \circ h)(x)$.

"$\subseteq$" For proving the opposite inclusion we take an arbitrary $x^* \in \partial(f + g \circ h)(x)$. By Theorem 2.3.12 we get $(f + g \circ h)^*(x^*) + (f + g \circ h)(x) = \langle x^*, x\rangle$. Since (3.22) holds, there exists $\bar{z}^* \in C^*$ such that $(f + g \circ h)^*(x^*) = g^*(\bar{z}^*) + (f + (\bar{z}^*h))^*(x^*)$ and by means of the Young-Fenchel inequality we obtain further

$$\langle x^*, x\rangle = g^*(\bar{z}^*) + g(h(x)) + (f + (\bar{z}^*h))^*(x^*) + f(x)$$
$$\geq (\bar{z}^*h)(x) + f(x) + (f + (\bar{z}^*h))^*(x^*) \geq \langle x^*, x\rangle.$$

The inequalities in the relation above must be fulfilled as equalities and this means that

$$g^*(\bar{z}^*) + g(h(x)) = \langle \bar{z}^*, h(x)\rangle \Leftrightarrow \bar{z}^* \in \partial g(h(x))$$

and

$$(f + (\bar{z}^*h))^*(x^*) + (f + \bar{z}^*h)(x) = \langle x^*, x\rangle \Leftrightarrow x^* \in \partial(f + \bar{z}^*h)(x),$$

respectively. In conclusion, $x^* \in \bigcup_{z^* \in \partial g(h(x))} \partial(f + (z^*h))(x)$ and this delivers the desired result. □

Remark 3.5.1. From the proof of the previous theorem one can easily deduce that the inclusion

$$\bigcup_{z^* \in \partial g(h(x))} \partial(f + (z^*h))(x) \subseteq \partial(f + g \circ h)(x)$$

holds for all $x \in X$ without any other assumption.

Considering $\Phi^{CC_2} : X \times X \times Z \to \overline{\mathbb{R}}$, $\Phi^{CC_2}(x, y, z) = f(x+y)+g(h(x)+z)$, as perturbation function, in subsection 3.4.2 we introduced to (P^{CC}) another dual problem, which can be seen as a refinement of (D^{CC_1}), namely

$$(D^{CC_2}) \qquad \sup_{y^* \in X^*, z^* \in C^*} \{-g^*(z^*) - f^*(y^*) - (z^*h)^*(-y^*)\}.$$

Like for (D^{CC_1}), one can formulate also for (D^{CC_2}) a stable strong duality theorem.

Theorem 3.5.3. *(a) Let Z be partially ordered by the convex cone $C \subseteq Z$, $f : X \to \overline{\mathbb{R}}$ be a proper and convex function, $h : X \to Z \cup \{+\infty_C\}$ a proper and C-convex function and $g : Z \cup \{+\infty_C\} \to \overline{\mathbb{R}}$ a proper, convex and C-increasing function such that $g(+\infty_C) = +\infty$ and $h(\operatorname{dom} f \cap \operatorname{dom} h) \cap \operatorname{dom} g \neq \emptyset$. If one of the regularity conditions $(RC_i^{CC_2})$, $i \in \{1, 2, 3, 4\}$, is fulfilled, then for (P^{CC}) and (D^{CC_2}) stable strong duality holds, that is*

$$(f + g \circ h)^*(x^*) = \min_{\substack{y^* \in X^*, \\ z^* \in C^*}} \{g^*(z^*) + f^*(y^*) + (z^*h)^*(x^* - y^*)\} \;\; \forall x^* \in X^*. \tag{3.24}$$

(b) If for (P^{CC}) and (D^{CC_2}) stable strong duality holds, then for all $x \in X$ one has

$$\partial(f + g \circ h)(x) = \partial f(x) + \bigcup_{z^* \in \partial g(h(x))} \partial(z^*h)(x). \tag{3.25}$$

Proof. (a) The assertion follows as a consequence of Theorem 3.4.1, Theorem 3.5.1 and relation (3.19).

(b) We omit giving the proof of (3.25) as it is similar to the one given for (3.23). □

Remark 3.5.2. (a) For all $x^* \in X^*$ the following inequalities are fulfilled (see also Remark 3.4.3)

$$(f + g \circ h)^*(x^*) \leq \inf_{z^* \in C^*} \{g^*(z^*) + (f + (z^*h))^*(x^*)\}$$

$$\leq \inf_{\substack{y^* \in X^*, \\ z^* \in C^*}} \{g^*(z^*) + f^*(y^*) + (z^*h)^*(x^* - y^*)\}.$$

Thus, if (3.24) holds, then (3.22) holds, too.

(b) From the proof of Theorem 3.5.3 follows that for all $x \in X$

$$\partial f(x) + \bigcup_{z^* \in \partial g(h(x))} \partial(z^* h)(x) \subseteq \bigcup_{z^* \in \partial g(h(x))} \partial(f + (z^* h))(x)$$

$$\subseteq \partial(f + g \circ h)(x),$$

which is true without any other assumption. This means that if (3.25) holds, then (3.23) holds, too.

In the following we consider different classes of optimization problems and show how they can be treated as special cases of the composed convex optimization problem (P^{CC}). Formulae for conjugate functions and subdifferentials are also derived from the general ones given in Theorem 3.5.2 and Theorem 3.5.3, respectively.

3.5.3 Problems having the composition with a linear continuous mapping in the objective function

Consider X and Y Hausdorff locally convex spaces, $f : X \to \overline{\mathbb{R}}$ and $g : Y \to \overline{\mathbb{R}}$ proper and convex functions and $A \in \mathcal{L}(X, Y)$ fulfilling $\operatorname{dom} f \cap A^{-1}(\operatorname{dom} g) \neq \emptyset$. To the primal problem

$$(P^A) \quad \inf_{x \in X} \{f(x) + g(Ax)\}$$

we assigned the following dual

$$(D^A) \quad \sup_{y^* \in Y^*} \{-f^*(-A^* y^*) - g^*(y^*)\}.$$

From Theorem 3.5.1 and (3.3) we get the following result (see also Theorem 3.2.4 and Theorem 3.2.5).

Theorem 3.5.4. *Let $f : X \to \overline{\mathbb{R}}$ and $g : Y \to \overline{\mathbb{R}}$ be proper and convex functions and $A \in \mathcal{L}(X, Y)$ such that $\operatorname{dom} f \cap A^{-1}(\operatorname{dom} g) \neq \emptyset$. If one of the regularity conditions (RC_i^A), $i \in \{1, 2, 3, 4\}$, is fulfilled, then for (P^A) and (D^A) stable strong duality holds, that is*

$$(f + g \circ A)^*(x^*) = (f^* \Box A^* g^*)(x^*) = \min_{y^* \in Y^*} \{f^*(x^* - A^* y^*) + g^*(y^*)\} \ \forall x^* \in X^*. \tag{3.26}$$

For deriving a formula for $\partial(f + g \circ A)(x)$, when $x \in X$, we reformulate (P^A) in the framework of (P^{CC}). Let $h : X \to Y$, $h(x) = Ax$ and take as ordering cone of Y $C = \{0\}$, which is convex and closed. For this choice h is proper and C-convex and g is C-increasing. Moreover, $C^* = Y^*$ and for all $z^* \in Y^*$ it holds (cf. Proposition 2.3.2(h)) $(f + (z^* h))^*(x^*) = (f + \langle A^* z^*, \cdot \rangle)^*(x^*) = f^*(x^* - A^* z^*)$ for all $x^* \in X^*$. Notice that we also have $\operatorname{dom} f \cap \operatorname{dom} h \cap h^{-1}(\operatorname{dom} g) = \operatorname{dom} f \cap A^{-1}(\operatorname{dom} g)$ Thus the formula (3.26) is nothing else than (3.22). But, as we have seen in the proof of Theorem

3.5.2, if (3.22) holds, then for all $x \in X$, one has (cf. (3.23) and Proposition 2.3.13(*a*))

$$\partial(f + g \circ A)(x) = \partial(f + g \circ h)(x) = \bigcup_{z^* \in \partial g(h(x))} \partial(f + (z^* h))(x)$$

$$= \bigcup_{z^* \in \partial g(Ax)} \partial(f + \langle A^* z^*, \cdot \rangle)(x) = \partial f(x) + \bigcup_{z^* \in \partial g(Ax)} A^* z^* = \partial f(x) + A^* \partial g(Ax).$$

This leads to the following statement.

Theorem 3.5.5. *If for (P^A) and (D^A) stable strong duality holds, then for all $x \in X$ one has*

$$\partial(f + g \circ A)(x) = \partial f(x) + A^* \partial g(Ax). \tag{3.27}$$

Remark 3.5.3. One can notice that for this special choice of h the formula (3.24) turns out to be (3.26), too. Indeed,

$$(f + g \circ h)^*(x^*) = \min_{\substack{y^* \in X^*, \\ z^* \in C^*}} \{g^*(z^*) + f^*(y^*) + (z^* h)^*(x^* - y^*)\} \ \forall x^* \in X^*$$

is nothing else than

$$(f + g \circ A)^*(x^*) = \min_{y^*, z^* \in X^*} \{g^*(z^*) + f^*(y^*) + (\langle A^* z^*, \cdot \rangle)^*(x^* - y^*)\} \ \forall x^* \in X^*$$

$$\Leftrightarrow (f + g \circ A)^*(x^*) = \min_{\substack{y^*, z^* \in X^*, \\ x^* = y^* + A^* z^*}} \{g^*(z^*) + f^*(y^*)\} \ \forall x^* \in X^*$$

$$\Leftrightarrow (f + g \circ A)^*(x^*) = \min_{y^* \in X^*} \{f^*(x^* - A^* y^*) + g^*(y^*)\} \ \forall x^* \in X^*.$$

More than that, one can easily see that (3.25) is in this case equivalent to (3.27).

In case $X = Y$ and $A = \mathrm{id}_X$, by Theorem 3.5.4 and Theorem 3.5.5 we get the following results concerning the primal problem

$$(P^{\mathrm{id}}) \quad \inf_{x \in X} \{f(x) + g(x)\}$$

and its Fenchel dual

$$(D^{\mathrm{id}}) \quad \sup_{y^* \in X^*} \{-f^*(-y^*) - g^*(y^*)\}.$$

Theorem 3.5.6. *(a) Let $f, g : X \to \overline{\mathbb{R}}$ be proper and convex functions such that $\operatorname{dom} f \cap \operatorname{dom} g \neq \emptyset$. If one of the regularity conditions (RC_i^{id}), $i \in \{1, 2, 3, 4\}$, is fulfilled, then for (P^{id}) and (D^{id}) stable strong duality holds, that is*

$$(f + g)^*(x^*) = (f^* \square g^*)(x^*) = \min_{y^* \in X^*} \{f^*(x^* - y^*) + g^*(y^*)\} \ \forall x^* \in X^*. \tag{3.28}$$

(b) If for (P^{id}) *and* (D^{id}) *stable strong duality holds, then for all* $x \in X$ *one has*

$$\partial(f+g)(x) = \partial f(x) + \partial g(x). \tag{3.29}$$

Another consequence of Theorem 3.5.4 and Theorem 3.5.5 can be delivered whenever $f : X \to \mathbb{R}$, $f \equiv 0$, by considering the primal problem

$$(P^{A_g}) \quad \inf_{x \in X} \{g(Ax)\}$$

along with its conjugate dual

$$(D^{A_g}) \quad \sup_{\substack{y^* \in Y^*, \\ A^*y^*=0}} \{-g^*(y^*)\}.$$

Theorem 3.5.7. *(a) Let* $g : Y \to \overline{\mathbb{R}}$ *be a proper and convex function and* $A \in \mathcal{L}(X,Y)$ *such that* $A^{-1}(\operatorname{dom} g) \neq \emptyset$. *If one of the regularity conditions* $(RC_i^{A_g})$, $i \in \{1,2,3,4\}$, *is fulfilled, then for* (P^{A_g}) *and* (D^{A_g}) *stable strong duality holds, that is*

$$(g \circ A)^*(x^*) = (A^*g^*)(x^*) = \min\{g^*(y^*) : x^* = A^*y^*\} \ \forall x^* \in X^*. \tag{3.30}$$

(b) If for (P^{A_g}) *and* (D^{A_g}) *stable strong duality holds, then for all* $x \in X$ *one has*

$$\partial(g \circ A)(x) = A^*\partial g(Ax). \tag{3.31}$$

Further, consider $f_i : X \to \overline{\mathbb{R}}, i = 1,...,m$, proper and convex functions such that $\bigcap_{i=1}^m \operatorname{dom} f_i \neq \emptyset$. We take $g : X^m \to \overline{\mathbb{R}}$, $g(x^1,...,x^m) = \sum_{i=1}^m f_i(x^i)$ and $A : X \to X^m, Ax = (x,...,x)$. To the primal optimization problem

$$(P^{\Sigma}) \quad \inf_{x \in X} \left\{\sum_{i=1}^m f_i(x)\right\}$$

we assigned the following conjugate dual

$$(D^{\Sigma}) \quad \sup_{\substack{x^{i*} \in X^*, i=1,...,m, \\ \sum_{i=1}^m x^{i*}=0}} \left\{-\sum_{i=1}^m f_i^*(x^{i*})\right\}.$$

Theorem 3.5.7 and Proposition 2.3.13(b) lead to the following result (see also Theorem 3.2.8), noticing that for all $(x^{1*},...,x^{m*}) \in (X^*)^m$, $g^*(x^{1*},...,x^{m*}) = \sum_{i=1}^m f_i^*(x^{i*})$ and $A^*(x^{1*},...,x^{m*}) = \sum_{i=1}^m x^{i*}$.

Theorem 3.5.8. *(a) Let* $f_i : X \to \overline{\mathbb{R}}, i = 1,...,m$, *be proper and convex functions such that* $\bigcap_{i=1}^m \operatorname{dom} f_i \neq \emptyset$. *If one of the regularity conditions* (RC_i^{Σ}), $i \in \{1,2,3,4\}$, *is fulfilled, then for* (P^{Σ}) *and* (D^{Σ}) *stable strong duality holds, that is*

$$\left(\sum_{i=1}^{m} f_i\right)^*(x^*) = (f_1^* \Box \ldots \Box f_m^*)(x^*) =$$

$$\min\left\{\sum_{i=1}^{m} f_i^*(x^{i*}) : x^* = \sum_{i=1}^{m} x^{i*}\right\} \ \forall x^* \in X^*. \tag{3.32}$$

(b) If for (P^{Σ}) and (D^{Σ}) stable strong duality holds, then for all $x \in X$ one has

$$\partial\left(\sum_{i=1}^{m} f_i\right)(x) = \sum_{i=1}^{m} \partial f_i(x). \tag{3.33}$$

3.5.4 Problems with geometric and cone constraints

Consider the Hausdorff locally convex spaces X and Z, the latter partially ordered by the convex cone $C \subseteq Z$, S a nonempty and convex subset of X, $f : X \to \overline{\mathbb{R}}$ a proper and convex function and $g : X \to \overline{Z}$ a proper and C-convex function fulfilling $\operatorname{dom} f \cap S \cap g^{-1}(-C) \neq \emptyset$. For the optimization problem

$$(P^C) \quad \inf_{x \in \mathcal{A}} f(x),$$
$$\mathcal{A} = \{x \in S : g(x) \in -C\}$$

and its Lagrange dual problem

$$(D^{C_L}) \quad \sup_{z^* \in C^*} \inf_{x \in S} \{f(x) + (z^*g)(x)\}$$

we formulate in the following a stable strong duality result by using Theorem 3.5.1 and relation (3.6) (see also Theorem 3.2.9 and Theorem 3.2.10).

Theorem 3.5.9. *Let Z be partially ordered by the convex cone $C \subseteq Z$, $S \subseteq X$ be a nonempty convex set, $f : X \to \overline{\mathbb{R}}$ a proper and convex function and $g : X \to \overline{Z}$ a proper and C-convex function such that* $\operatorname{dom} f \cap S \cap g^{-1}(-C) \neq \emptyset$. *If one of the regularity conditions $(RC_i^{C_L})$, $i \in \{1, 2, 3, 4\}$, is fulfilled, then for (P^C) and (D^{C_L}) stable strong duality holds, that is*

$$(f + \delta_{\mathcal{A}})^*(x^*) = \min_{z^* \in C^*} (f + (z^*g))_S^*(x^*) \ \forall x^* \in X^*. \tag{3.34}$$

Next we rewrite (P^C) as a particularization of the composed convex problem (P^{CC}). Define $\tilde{h} : X \to Z \cup \{+\infty_C\}$ by

$$\tilde{h}(x) = \begin{cases} g(x), & \text{if } x \in S, \\ +\infty_C, & \text{otherwise,} \end{cases}$$

and $\tilde{g} : Z \cup \{+\infty_C\} \to \overline{\mathbb{R}}$, which fulfills $\tilde{g} = \delta_{-C}$. We also suppose that $\tilde{g}(+\infty_C) = +\infty$. For all $x \in X$ and $x^* \in X^*$ it holds

$$(f + \tilde{g} \circ \tilde{h})(x) = \begin{cases} f(x), & \text{if } x \in S, g(x) \in -C, \\ +\infty, & \text{otherwise,} \end{cases}$$

and

$$(f + \tilde{g} \circ \tilde{h})^*(x^*) = (f + \delta_{\mathcal{A}})^*(x^*).$$

Further, for $z^* \in Z^*$ we have

$$\tilde{g}^*(z^*) = \begin{cases} 0, & \text{if } z^* \in C^*, \\ +\infty, & \text{otherwise,} \end{cases}$$

while if $z^* \in C^*$ it holds

$$(z^* \tilde{h})(x) = \begin{cases} (z^* g)(x), & \text{if } x \in S, \\ +\infty, & \text{otherwise} \end{cases} = ((z^* g) + \delta_S)(x) \ \forall x \in X.$$

This means that (3.34) is nothing else than relation (3.22) applied for $f, \tilde{g}$ and $\tilde{h}$, since the latter asserts that

$$(f + \tilde{g} \circ \tilde{h})^*(x^*) = \min_{z^* \in C^*} \left\{ \tilde{g}^*(z^*) + (f + (z^* \tilde{h}))^*(x^*) \right\} \ \forall x^* \in X^*$$

$$\Leftrightarrow (f + \delta_{\mathcal{A}})^*(x^*) = \min_{z^* \in C^*} \{ (f + (z^* g) + \delta_S)^*(x^*) \} \ \forall x^* \in X^*$$

$$\Leftrightarrow (f + \delta_{\mathcal{A}})^*(x^*) = \min_{z^* \in C^*} \{ (f + (z^* g))_S^*(x^*) \} \ \forall x^* \in X^*.$$

On the other hand, by Theorem 3.5.2 relation (3.22) guarantees that for all $x \in X$ one has

$$\partial(f + \tilde{g} \circ \tilde{h})(x) = \bigcup_{z^* \in \partial \tilde{g}(\tilde{h}(x))} \partial(f + (z^* \tilde{h}))(x). \tag{3.35}$$

One can easily notice that $\operatorname{dom} f \cap \operatorname{dom} \tilde{h} \cap \tilde{h}^{-1}(\operatorname{dom} \tilde{g}) = \operatorname{dom} f \cap S \cap g^{-1}(-C)$, $(f + \tilde{g} \circ \tilde{h})(x) = (f + \delta_{\mathcal{A}})(x)$ for all $x \in X$ and, for $z^* \in C^*$, $(f + z^* \tilde{h})(x) = (f + (z^* g) + \delta_S)(x)$ for all $x \in X$. Moreover, for $x \in \mathcal{A}$ it holds

$$z^* \in \partial \tilde{g}(\tilde{h}(x)) \Leftrightarrow \tilde{g}^*(z^*) + \tilde{g}(\tilde{h}(x)) = \langle z^*, \tilde{h}(x) \rangle$$

$$\Leftrightarrow z^* \in C^* \text{ and } (z^* g)(x) = 0.$$

Thus, in this special case, relation (3.35) can be equivalently written as

$$\partial(f + \delta_{\mathcal{A}})(x) = \bigcup_{\substack{z^* \in C^*, \\ (z^* g)(x) = 0}} \partial(f + (z^* g) + \delta_S)(x) \ \forall x \in X$$

and this leads to the following result.

Theorem 3.5.10. *If for (P^C) and (D^{C_L}) stable strong duality holds, then for all $x \in X$ one has*

$$\partial(f + \delta_{\mathcal{A}})(x) = \bigcup_{\substack{z^* \in C^*, \\ (z^* g)(x) = 0}} \partial(f + (z^* g) + \delta_S)(x). \tag{3.36}$$

Considering now the Fenchel dual to (P^C)

$$(D^{C_F}) \quad \sup_{y^* \in X^*} \{-f^*(y^*) - \sigma_{\mathcal{A}}(-y^*)\},$$

by Theorem 3.5.1, Theorem 3.2.11 and relations (3.7) and (3.29) we get the following result.

Theorem 3.5.11. *(a) Let Z be partially ordered by the convex cone $C \subseteq Z$, $S \subseteq X$ be a nonempty convex set, $f : X \to \overline{\mathbb{R}}$ a proper and convex function and $g : X \to \overline{Z}$ a proper and C-convex function such that* $\operatorname{dom} f \cap S \cap g^{-1}(-C) \neq \emptyset$*. If one of the regularity conditions $(RC_i^{C_F})$, $i \in \{1,2,3,4\}$, is fulfilled, then for (P^C) and (D^{C_F}) stable strong duality holds, that is*

$$(f + \delta_{\mathcal{A}})^*(x^*) = \min_{y^* \in X^*} \{f^*(y^*) + \sigma_{\mathcal{A}}(x^* - y^*)\} \ \forall x^* \in X^*. \tag{3.37}$$

(b) If for (P^C) and (D^{C_F}) stable strong duality holds, then for all $x \in X$ one has

$$\partial(f + \delta_{\mathcal{A}})(x) = \partial f(x) + N(\mathcal{A}, x). \tag{3.38}$$

Remark 3.5.4. Theorem 3.5.11 indirectly provides regularity conditions under which the so-called *Pshenichnyi-Rockafellar formula*, which is nothing else than relation (3.38) in case $g \equiv 0$, is fulfilled.

The third dual to (P^C) we consider here is the Fenchel-Lagrange dual

$$(D^{C_{FL}}) \qquad \sup_{y^* \in X^*, z^* \in C^*} \{-f^*(y^*) - (z^*g)_S^*(-y^*)\}.$$

First we characterize the stable strong duality for (P^C) and $(D^{C_{FL}})$ (cf. Theorem 3.5.1, Theorem 3.2.12, Theorem 3.2.13 and relation (3.8)).

Theorem 3.5.12. *Let Z be partially ordered by the convex cone $C \subseteq Z$, $S \subseteq X$ be a nonempty convex set, $f : X \to \overline{\mathbb{R}}$ a proper and convex function and $g : X \to \overline{Z}$ a proper and C-convex function such that* $\operatorname{dom} f \cap S \cap g^{-1}(-C) \neq \emptyset$*. If one of the regularity conditions $(RC_i^{C_{FL}})$, $i \in \{1,2,3,4\}$, is fulfilled, then for (P^C) and $(D^{C_{FL}})$ stable strong duality holds, that is*

$$(f + \delta_{\mathcal{A}})^*(x^*) = \min_{y^* \in X^*, z^* \in C^*} \{f^*(y^*) + (z^*g)_S^*(x^* - y^*)\} \ \forall x^* \in X^*. \tag{3.39}$$

By using the notations we made above, relation (3.39) is nothing else than

$$(f + \tilde{g} \circ \tilde{h})^*(x^*) = \min_{y^* \in X^*, z^* \in C^*} \left\{\tilde{g}^*(z^*) + f^*(y^*) + (z^*\tilde{h})^*(x^* - y^*)\right\} \ \forall x^* \in X^*,$$

which is relation (3.24) in Theorem 3.5.3. By this theorem one has that for all $x \in X$ (cf. (3.25))

$$\partial(f + \tilde{g} \circ \tilde{h})(x) = \partial f(x) + \bigcup_{z^* \in \partial \tilde{g}(\tilde{h}(x))} \partial(z^* \tilde{h})(x)$$

or, equivalently, for all $x \in X$

$$\partial(f + \delta_{\mathcal{A}})(x) = \partial f(x) + \bigcup_{\substack{z^* \in C^*, \\ (z^* g)(x) = 0}} \partial((z^* g) + \delta_S)(x).$$

The following result closes the section.

Theorem 3.5.13. *If for (P^C) and $(D^{C_{FL}})$ stable strong duality holds, then for all $x \in X$ one has*

$$\partial(f + \delta_{\mathcal{A}})(x) = \partial f(x) + \bigcup_{\substack{z^* \in C^*, \\ (z^* g)(x) = 0}} \partial((z^* g) + \delta_S)(x). \tag{3.40}$$

Bibliographical notes

The perturbation approach for constructing a dual problem to a convex optimization problem via the theory of conjugate functions was first used by Rockafellar in [158] and Ekeland and Temam in [67]. The derivation of the classical Fenchel and Lagrange dual problems by means of the aforementioned approach is classical, while the Fenchel-Lagrange dual problem for an optimization problem with geometric and cone constraints has been recently introduced by Boţ and Wanka (cf. [35,39,186], etc.). The relations between the Lagrange, Fenchel and Fenchel-Lagrange dual problems have been studied in [186] in finite dimensional spaces and in [39] in infinite dimensional spaces.

A good reference for regularity conditions guaranteeing the existence of strong duality for a general optimization problem and its conjugate dual is the book of Zălinescu [207]. There are collected different, mostly generalized interior point, regularity conditions introduced in the last thirty years in the literature. The regularity condition (RC_1^{Φ}) is classical (see [67,158]), the condition $(RC_{2'}^{\Phi})$ is due to Rockafellar (cf. [158]), while its generalization (RC_2^{Φ}) was first given in [205]. Regularity conditions for problems given in finite dimensional spaces can be found in Rockafellar's book [157]. Theorem 3.2.2, which is the starting point for different closedness type regularity conditions is due to Precupanu (cf. [155]). Coming now to the optimization problem having as objective function the sum of a convex function with the composition of another convex function with a linear continuous mapping, let us mention that $(RC_{2'}^A)$ (see also $(RC_{2'}^{\mathrm{id}})$) is known as the *Attouch-Brézis regularity condition* (cf. [7]). Its weaker version (RC_2^A) is due to Rodrigues (cf. [160]), while the closedness type regularity condition (RC_4^A) was first formulated in [38] and in the special case when $X = Y$ and $A = \mathrm{id}_X$ in [41]. Interior point regularity conditions for the optimization problem with geometric and cone constraints were given by Rockafellar in [158] in infinite dimensional spaces, while the closedness type regularity conditions $(RC_4^{C_L})$ and $(RC_4^{C_{FL}})$ have been recently introduced in [32]. It is worth mentioning that $(\widetilde{RC}^{C_L})$ is the weakest interior point condition which exists in the literature for $(\widetilde{P}^C)$ and was first given in the monograph of Rockafellar on convex analysis [157].

Theorem 3.3.1, which is providing necessary and sufficient optimality conditions for the pair of primal-dual optimization problems $(PG) - (DG)$ was given in [67]. In the same book one can also find the definition of the notion of a Lagrangian function. The optimality conditions and the Lagrangian function for the problem with geometric and cone constraints and its Lagrange dual are classical results, which can be found in almost every optimization book.

The composed convex optimization problem was considered in case $f \equiv 0$ by different authors (see, for example, [89, 120, 204]). Some of the generalized interior point regularity conditions listed in subsection 3.4.1 were given in [50, 51] under stronger assumptions for h. A collection of regularity conditions for this kind of optimization problems can be found in [207]. Closedness type regularity conditions for both dual problems introduced in section 3.4 have been given in [33] in case h is star C-lower semicontinuous. For the case h is C-epi closed the reader can consult our paper [31].

An overview over generalized interior point regularity conditions which ensure the existence of stable strong duality can be found in [206, 207]. Some of the formulae for the subdifferential were given also there, while some recent results in this direction have been published in [31–33]. The connection between the stable strong duality and formulae for the subdifferential has been pointed out in [34, 89].

4

Conjugate vector duality via scalarization

In this chapter we introduce to different vector optimization problems corresponding vector dual problems by using some duality concepts having as starting point the scalar duality theory. Since there is a certain similarity between its definition and the one of the duals introduced via scalarization, we also investigate the classical geometric vector duality concept. More than that, we give a general approach for treating duality in vector optimization independently from the nature of the scalarization functions considered. We also provide a first look at the duality theory for linear vector optimization problems in Hausdorff locally convex spaces.

4.1 Fenchel type vector duality

Let X, Y and V be Hausdorff locally convex spaces and assume that V is partially ordered by the nontrivial pointed convex cone $K \subseteq V$. Further, let $f : X \to \overline{V} = V \cup \{\pm\infty_K\}$ and $g : Y \to \overline{V}$ be given proper and K-convex functions and $A \in \mathcal{L}(X, Y)$ such that $\operatorname{dom} f \cap A^{-1}(\operatorname{dom} g) \neq \emptyset$.

To the primal vector optimization problem

$$(PV^A) \quad \operatorname*{Min}_{x \in X} \{f(x) + g(Ax)\}$$

we introduce dual vector optimization problems with respect to both properly efficient solutions in the sense of linear scalarization and weakly efficient solutions and prove weak, strong and converse duality theorems.

4.1.1 Duality with respect to properly efficient solutions

In this subsection we investigate a duality approach to (PV^A) with respect to properly efficient solutions in the sense of linear scalarization. Since we do not have to differentiate between different classes of such solutions we call them simply properly efficient solutions. We say that $\bar{x} \in X$ is a properly efficient

R.I. Boţ et al., *Duality in Vector Optimization*, Vector Optimization,
DOI 10.1007/978-3-642-02886-1_4,

solution to (PV^A) if $\bar{x} \in \operatorname{dom} f \cap A^{-1}(\operatorname{dom} g)$ and $f(\bar{x}) \in \mathrm{PMin}_{LSc}((f + g \circ A)(\operatorname{dom} f \cap A^{-1}(\operatorname{dom} g)), K)$. This means that there exists $v^* \in K^{*0}$ such that $\langle v^*, (f + g \circ A)(\bar{x}) \rangle \leq \langle v^*, (f + g \circ A)(x) \rangle$ for all $x \in X$.

The vector dual problem to (PV^A) we investigate in this subsection is

$$(DV^A) \qquad \underset{(v^*, y^*, v) \in \mathcal{B}^A}{\mathrm{Max}} h^A(v^*, y^*, v),$$

where

$$\mathcal{B}^A = \{(v^*, y^*, v) \in K^{*0} \times Y^* \times V : \langle v^*, v \rangle \leq -(v^* f)^*(-A^* y^*) - (v^* g)^*(y^*)\}$$

and

$$h^A(v^*, y^*, v) = v.$$

We prove first the existence of *weak duality* for (PV^A) and (DV^A).

Theorem 4.1.1. *There is no $x \in X$ and no $(v^*, y^*, v) \in \mathcal{B}^A$ such that $(f + g \circ A)(x) \leq_K h^A(v^*, y^*, v)$.*

Proof. We assume the contrary, namely that there exist $x \in X$ and $(v^*, y^*, v) \in \mathcal{B}^A$ such that $v - (f + g \circ A)(x) = h^A(v^*, y^*, v) - (f + g \circ A)(x) \geq_K 0$. It is obvious that $x \in \operatorname{dom} f \cap A^{-1}(\operatorname{dom} g)$ and that $\langle v^*, v \rangle > \langle v^*, f(x) \rangle + \langle v^*, g(Ax) \rangle$. On the other hand, we have

$$\langle v^*, f(x) \rangle + \langle v^*, g(Ax) \rangle \geq \inf_{y \in X} \{\langle v^*, f(y) \rangle + \langle v^*, g(Ay) \rangle\}.$$

Since the infimum on the right-hand side of the relation above is greater than or equal to the optimal objective value of its corresponding scalar Fenchel dual problem (cf. subsection 3.1.2), it holds

$$\langle v^*, v \rangle > \inf_{y \in X} \{(v^* f)(y) + (v^* g)(Ay)\}$$

$$\geq \sup_{z^* \in Y^*} \{-(v^* f)^*(-A^* z^*) - (v^* g)^*(z^*)\} \geq -(v^* f)^*(-A^* y^*) - (v^* g)^*(y^*).$$

As this contradicts the fact that $(v^*, y^*, v) \in \mathcal{B}^A$, the conclusion follows. □

In order to be able to prove the following *strong duality theorem* for the primal-dual vector pair $(PV^A) - (DV^A)$ we have to impose a regularity condition which actually ensures the existence of strong duality for the scalar problem

$$\inf_{x \in X} \{(v^* f)(x) + (v^* g)(Ax)\}$$

and its Fenchel dual

$$\sup_{y^* \in Y^*} \{-(v^* f)^*(-A^* y^*) - (v^* g)^*(y^*)\},$$

for all $v^* \in K^{*0}$. This means that we are looking for sufficient conditions which are independent from the choice of $v^* \in K^{*0}$ and therefore we consider the following regularity condition

$(RCV^A) \mid \exists x' \in \operatorname{dom} f \cap A^{-1}(\operatorname{dom} g)$ such that g is continuous at Ax'.

One can notice that (RCV^A) as well as the generalized interior point regularity conditions which can be considered here are in this situation preferable to a closedness type condition, as they are formulated independently of the choice of $v^* \in K^{*0}$.

Theorem 4.1.2. *Assume that the regularity condition (RCV^A) is fulfilled. If $\bar{x} \in X$ is a properly efficient solution to (PV^A), then there exists $(\bar{v}^*, \bar{y}^*, \bar{v})$, an efficient solution to (DV^A), such that $(f + g \circ A)(\bar{x}) = h^A(\bar{v}^*, \bar{y}^*, \bar{v}) = \bar{v}$.*

Proof. Having $\bar{x} \in X$ a properly efficient solution to (PV^A), it follows that $\bar{x} \in \operatorname{dom} f \cap A^{-1}(\operatorname{dom} g)$ and that there exists $\bar{v}^* \in K^{*0}$, which fulfills

$$\langle \bar{v}^*, (f + g \circ A)(\bar{x}) \rangle = \inf_{x \in X} \{(\bar{v}^* f)(x) + (\bar{v}^* g)(Ax)\}.$$

The functions $(\bar{v}^* f) : X \to \overline{\mathbb{R}}$ and $(\bar{v}^* g) : Y \to \overline{\mathbb{R}}$ are proper and convex functions with $\operatorname{dom}(\bar{v}^* f) = \operatorname{dom} f$ and $\operatorname{dom}(\bar{v}^* g) = \operatorname{dom} g$. The regularity condition (RCV^A) yields that $(\bar{v}^* g)$ is continuous at Ax' and so, by Theorem 3.2.4, there exists $\bar{y}^* \in Y^*$ such that

$$\langle \bar{v}^*, (f + g \circ A)(\bar{x}) \rangle = \inf_{x \in X} \{(\bar{v}^* f)(x) + (\bar{v}^* g)(Ax)\}$$

$$= \sup_{y^* \in Y^*} \{-(\bar{v}^* f)^*(-A^* y^*) - (\bar{v}^* g)^*(y^*)\} = -(\bar{v}^* f)^*(-A^* \bar{y}^*) - (\bar{v}^* g)^*(\bar{y}^*).$$

Defining $\bar{v} := (f + g \circ A)(\bar{x}) \in V$ one has $(\bar{v}^*, \bar{y}^*, \bar{v}) \in \mathcal{B}^A$. Assuming that $(\bar{v}^*, \bar{y}^*, \bar{v})$ is not an efficient solution to (DV^A), there must exist an element (v^*, y^*, v) in $\mathcal{B}^A$ such that $(f + g \circ A)(\bar{x}) = \bar{v} \leq_K v = h^A(v^*, y^*, v)$. But this contradicts Theorem 4.1.1 and in this way the conclusion follows. □

Remark 4.1.1. In case X and Y are Fréchet spaces and the functions f and g are star K-lower semicontinuous, instead of assuming (RCV^A) fulfilled, for having strong duality for (PV^A) and (DV^A) it is enough to assume that $0 \in \operatorname{sqri}(\operatorname{dom} g - A(\operatorname{dom} f))$. On the other hand, if $\operatorname{lin}(\operatorname{dom} g - A(\operatorname{dom} f))$ is a finite dimensional linear subspace one can ask that $\operatorname{ri}(A(\operatorname{dom} f)) \cap \operatorname{ri}(\operatorname{dom} g) \neq \emptyset$. In both situations Theorem 3.2.4 guarantees strong duality for the scalar problem

$$\inf_{x \in X} \{(v^* f)(x) + (v^* g)(Ax)\}$$

and its Fenchel dual problem for all $v^* \in K^{*0}$.

The next result plays an important role in proving the converse duality theorem.

Theorem 4.1.3. *Assume that $\mathcal{B}^A$ is nonempty and that the regularity condition (RCV^A) is fulfilled. Then*

$$V \setminus \operatorname{cl}\big((f + g \circ A)(\operatorname{dom} f \cap A^{-1}(\operatorname{dom} g)) + K\big) \subseteq \operatorname{core}(h^A(\mathcal{B}^A)).$$

Proof. Let $\bar{v} \in V \setminus \operatorname{cl}\big((f + g \circ A)(\operatorname{dom} f \cap A^{-1}(\operatorname{dom} g)) + K\big)$ be arbitrarily chosen. Since $\operatorname{cl}\big((f + g \circ A)(\operatorname{dom} f \cap A^{-1}(\operatorname{dom} g)) + K\big) \subseteq V$ is a convex and closed set, by Theorem 2.1.5 there exist $\bar{v}^* \in V^* \setminus \{0\}$ and $\alpha \in \mathbb{R}$ such that

$$\langle \bar{v}^*, \bar{v} \rangle < \alpha < \langle \bar{v}^*, v \rangle \ \forall v \in \operatorname{cl}\big((f + g \circ A)(\operatorname{dom} f \cap A^{-1}(\operatorname{dom} g)) + K\big). \quad (4.1)$$

Obviously, $\bar{v}^* \in K^* \setminus \{0\}$. Further, since $\mathcal{B}^A \neq \emptyset$, there exists $(\tilde{v}^*, \tilde{y}^*, \tilde{v}) \in K^{*0} \times Y^* \times V$ fulfilling

$$\langle \tilde{v}^*, \tilde{v} \rangle \leq -(\tilde{v}^* f)^*(-A^* \tilde{y}^*) - (\tilde{v}^* g)^*(\tilde{y}^*) \leq \inf_{x \in X} \langle \tilde{v}^*, (f + g \circ A)(x) \rangle. \quad (4.2)$$

Denote by $\gamma := \alpha - \langle \bar{v}^*, \bar{v} \rangle > 0$. For all $s \in (0, 1)$ we have

$$\langle s\tilde{v}^* + (1-s)\bar{v}^*, \bar{v} \rangle = \langle \bar{v}^*, \bar{v} \rangle + s(\langle \tilde{v}^*, \bar{v} \rangle - \langle \bar{v}^*, \bar{v} \rangle) = \alpha - \gamma + s(\langle \tilde{v}^*, \bar{v} \rangle - \alpha + \gamma),$$

while, by (4.1) and (4.2), for all $v \in (f + g \circ A)(\operatorname{dom} f \cap A^{-1}(\operatorname{dom} g))$ it holds

$$\langle s\tilde{v}^* + (1-s)\bar{v}^*, v \rangle > s\langle \tilde{v}^*, \tilde{v} \rangle + (1-s)\alpha = \alpha + s(\langle \tilde{v}^*, \tilde{v} \rangle - \alpha).$$

Now we choose $\bar{s} \in (0, 1)$ close enough to 0 such that $\bar{s}(\langle \tilde{v}^*, \bar{v} \rangle - \alpha + \gamma) < \gamma/2$ and $\bar{s}(\langle \tilde{v}^*, \tilde{v} \rangle - \alpha) > -\gamma/2$. For $v_{\bar{s}}^* := \bar{s}\tilde{v}^* + (1-\bar{s})\bar{v}^* \in K^{*0}$ it holds

$$\langle v_{\bar{s}}^*, \bar{v} \rangle < \alpha - \frac{\gamma}{2} < \langle v_{\bar{s}}^*, v \rangle \ \forall v \in (f + g \circ A)(\operatorname{dom} f \cap A^{-1}(\operatorname{dom} g)),$$

which implies that

$$\langle v_{\bar{s}}^*, \bar{v} \rangle < \inf_{x \in X} \langle v_{\bar{s}}^*, (f + g \circ A)(x) \rangle.$$

Using the fact that the regularity assumption (RCV^A) is fulfilled, by Theorem 3.2.4 there exists $y_{\bar{s}}^* \in Y^*$ such that

$$\begin{aligned} \langle v_{\bar{s}}^*, \bar{v} \rangle &< \inf_{x \in X} \langle v_{\bar{s}}^*, (f + g \circ A)(x) \rangle \\ = \sup_{y^* \in Y^*} \{-(v_{\bar{s}}^* f)^*(-A^* y^*) - (v_{\bar{s}}^* g)^*(y^*)\} &= -(v_{\bar{s}}^* f)^*(-A^* y_{\bar{s}}^*) - (v_{\bar{s}}^* g)^*(y_{\bar{s}}^*). \end{aligned} \quad (4.3)$$

Let $\varepsilon > 0$ be such that

$$\langle v_{\bar{s}}^*, \bar{v} \rangle + \varepsilon < -(v_{\bar{s}}^* f)^*(-A^* y_{\bar{s}}^*) - (v_{\bar{s}}^* g)^*(y_{\bar{s}}^*).$$

For all $v \in V$ there exists $\delta_v > 0$ such that it holds

$$\langle v_{\bar{s}}^*, \bar{v} + \lambda v \rangle \leq \langle v_{\bar{s}}^*, \bar{v} \rangle + \varepsilon < -(v_{\bar{s}}^* f)^*(-A^* y_{\bar{s}}^*) - (v_{\bar{s}}^* g)^*(y_{\bar{s}}^*) \ \forall \lambda \in [0, \delta_v].$$

This means that for all $\lambda \in [0, \delta_v]$, $(v_{\bar{s}}^*, y_{\bar{s}}^*, \bar{v} + \lambda v) \in \mathcal{B}^A$ and, further, $\bar{v} + \lambda v \in h^A(\mathcal{B}^A)$. In conclusion, $\bar{v} \in \operatorname{core}(h^A(\mathcal{B}^A))$. □

We come now to the proof of the *converse duality theorem.*

Theorem 4.1.4. *Assume that the regularity condition* (RCV^A) *is fulfilled and that the set* $(f + g \circ A)(\operatorname{dom} f \cap A^{-1}(\operatorname{dom} g)) + K$ *is closed. Then for every efficient solution* $(\bar{v}^*, \bar{y}^*, \bar{v})$ *to* (DV^A) *there exists* $\bar{x} \in X$*, a properly efficient solution to* (PV^A)*, such that* $(f + g \circ A)(\bar{x}) = h^A(\bar{v}^*, \bar{y}^*, \bar{v}) = \bar{v}$.

Proof. Assume that $\bar{v} \notin (f + g \circ A)(\operatorname{dom} f \cap A^{-1}(\operatorname{dom} g)) + K$. By Theorem 4.1.3 follows that $\bar{v} \in \operatorname{core}(h^A(\mathcal{B}^A))$. Thus for $k \in K \setminus \{0\}$ there exists $\lambda > 0$ such that $v_\lambda := \bar{v} + \lambda k \geq_K \bar{v}$ and $v_\lambda \in h^A(\mathcal{B}^A)$. Since this contradicts the fact that $(\bar{v}^*, \bar{y}^*, \bar{v})$ is an efficient solution to (DV^A), we must have $\bar{v} \in (f + g \circ A)(\operatorname{dom} f \cap A^{-1}(\operatorname{dom} g)) + K$. This means that there exist $\bar{x} \in \operatorname{dom} f \cap A^{-1}(\operatorname{dom} g)$ and $\bar{k} \in K$ fulfilling $\bar{v} = (f + g \circ A)(\bar{x}) + \bar{k}$. By Theorem 4.1.1 follows that $\bar{k} = 0$ and, consequently, $\bar{v} = (f + g \circ A)(\bar{x})$. Since

$$\langle \bar{v}^*, (f + g \circ A)(\bar{x})\rangle = \langle \bar{v}^*, \bar{v}\rangle$$

$$\leq -(\bar{v}^* f)^*(-A^*\bar{y}^*) - (\bar{v}^* g)^*(\bar{y}^*) \leq \inf_{x \in X} \langle \bar{v}^*, (f + g \circ A)(x)\rangle,$$

$\bar{x}$ is a properly efficient solution to (PV^A). □

Remark 4.1.2. In the following we want to point out that in Theorem 4.1.3 and, consequently, in Theorem 4.1.4 the regularity condition (RCV^A) can be replaced with a weaker sufficient condition. As we have seen, in case (RCV^A) is fulfilled, for all $v^* \in K^{*0}$ one has that

$$\inf_{x \in X} \langle v^*, (f + g \circ A)(x)\rangle = \sup_{y^* \in Y^*} \{-(v^* f)^*(-A^* y^*) - (v^* g)^*(y^*)\}$$

and the supremum is attained. This means that for all $v^* \in K^{*0}$ the scalar optimization problem

$$\inf_{x \in X} \langle v^*, (f + g \circ A)(x)\rangle$$

is stable (cf. section 3.1). But for the purposes of the last two theorems it is enough to assume that (see also [101]) for all $v^* \in K^{*0}$ the problem

$$\inf_{x \in X} \langle v^*, (f + g \circ A)(x)\rangle$$

is normal with respect to its Fenchel dual. This means that the optimal objective values of the infimum problem from above coincide with the optimal objective value of its Fenchel dual even if the existence of an optimal solution to the dual is not guaranteed. Nevertheless, this is enough to ensure in the proof of Theorem 4.1.3 the existence of $y_{\bar{s}}^* \in Y^*$ such that (see relation (4.3))

$$\langle v_{\bar{s}}^*, \bar{v}\rangle < -(v_{\bar{s}}^* f)^*(-A^* y_{\bar{s}}^*) - (v_{\bar{s}}^* g)^*(y_{\bar{s}}^*).$$

Having this fulfilled, the conclusion of the theorem follows in an identical manner.

The scalar Fenchel duality was involved for the first time in the definition of a vector dual problem by Breckner and Kolumbán in [42,43] in a very general framework (for more details the reader can consult also [73]). Particularizing the approach introduced in these works to the primal problem (PV^A) one gets the following dual vector optimization problem

$$(DV^A_{BK}) \qquad \underset{(v^*,y^*,v)\in\mathcal{B}^A_{BK}}{\operatorname{Max}} \; h^A_{BK}(v^*,y^*,v),$$

where

$$\mathcal{B}^A_{BK} = \{(v^*,y^*,v) \in K^{*0} \times Y^* \times V : \langle v^*, v\rangle = -(v^*f)^*(-A^*y^*) - (v^*g)^*(y^*)\}$$

and

$$h^A_{BK}(v^*,y^*,v) = v.$$

It is not hard to see that $h^A_{BK}(\mathcal{B}^A_{BK}) \subseteq h^A(\mathcal{B}^A)$ and that this inclusion is in general strict. This fact together with Theorem 4.1.1 guarantees that between (PV^A) and (DV^A_{BK}) weak duality holds, too. Instead of proving the strong and converse duality theorems for this primal-dual pair we show that (see also [28]) the sets of maximal elements of $h^A_{BK}(\mathcal{B}^A_{BK})$ and $h^A(\mathcal{B}^A)$ coincide. In this way the mentioned duality results will follow automatically from Theorem 4.1.2 and Theorem 4.1.4, respectively. According to the notations in section 2.4 the sets of maximal elements of $h^A_{BK}(\mathcal{B}^A_{BK})$ and $h^A(\mathcal{B}^A)$ regarding the partial ordering induced by the cone K will be denoted by $\operatorname{Max}(h^A_{BK}(\mathcal{B}^A_{BK}),K)$ and $\operatorname{Max}(h^A(\mathcal{B}^A),K)$, respectively.

Theorem 4.1.5. *It holds* $\operatorname{Max}(h^A_{BK}(\mathcal{B}^A_{BK}),K) = \operatorname{Max}(h^A(\mathcal{B}^A),K)$.

Proof. "$\subseteq$" Let be $(\bar v^*,\bar y^*,\bar v) \in \mathcal{B}^A_{BK}$ such that $\bar v \in \operatorname{Max}(h^A_{BK}(\mathcal{B}^A_{BK}),K)$. Then it holds $\bar v \in h^A_{BK}(\mathcal{B}^A_{BK}) \subseteq h^A(\mathcal{B}^A)$. Assuming that $\bar v \notin \operatorname{Max}(h^A(\mathcal{B}^A),K)$, there exists $(v^*,y^*,v) \in \mathcal{B}^A$ fulfilling $v \geq_K \bar v$. It is obvious that $(v^*,y^*,v) \notin \mathcal{B}^A_{BK}$, which means that

$$\langle v^*, v\rangle < -(v^*f)^*(-A^*y^*) - (v^*g)^*(y^*).$$

Consequently, there exists $\tilde v \in v + K \setminus \{0\}$ such that

$$\langle v^*, \tilde v\rangle = -(v^*f)^*(-A^*y^*) - (v^*g)^*(y^*),$$

which yields $(v^*,y^*,\tilde v) \in \mathcal{B}^A_{BK}$. Further, we have $\tilde v \geq_K \bar v$ and this contradicts the maximality of $\tilde v$ in $h^A_{BK}(\mathcal{B}^A_{BK})$. Hence we must have $\operatorname{Max}(h^A_{BK}(\mathcal{B}^A_{BK}),K) \subseteq \operatorname{Max}(h^A(\mathcal{B}^A),K)$.

"$\supseteq$" We take an element $(\bar v^*,\bar y^*,\bar v) \in \mathcal{B}^A$ such that $\bar v \in \operatorname{Max}(h^A(\mathcal{B}^A),K)$ and prove first that $(\bar v^*,\bar y^*,\bar v) \in \mathcal{B}^A_{BK}$. Assuming the contrary, one has

$$\langle \bar v^*, \bar v\rangle < -(\bar v^*f)^*(-A^*\bar y^*) - (\bar v^*g)^*(\bar y^*).$$

In this situation it is easy to find $\tilde v \in \bar v + K \setminus \{0\}$ fulfilling

$$\langle \bar{v}^*, \bar{v} \rangle < \langle \bar{v}^*, \tilde{v} \rangle = -(\bar{v}^* f)^*(-A^* \bar{y}^*) - (\bar{v}^* g)^*(\bar{y}^*).$$

As $(\bar{v}^*, \bar{y}^*, \tilde{v}) \in \mathcal{B}^A$ and $\tilde{v} \geq_K \bar{v}$ this leads to a contradiction to the fact that $\bar{v}$ belongs to $\operatorname{Max}(h^A(\mathcal{B}^A), K)$. Consequently, $\bar{v} \in h^A_{BK}(\mathcal{B}^A_{BK})$. We suppose further that $\bar{v} \notin \operatorname{Max}(h^A_{BK}(\mathcal{B}^A_{BK}), K)$. Then there exists $v \in h^A_{BK}(\mathcal{B}^A_{BK})$ such that $v \geq_K \bar{v}$. Since $h^A_{BK}(\mathcal{B}^A_{BK}) \subseteq h^A(\mathcal{B}^A)$, it yields $v \in h^A(\mathcal{B}^A)$, but this contradicts the maximality of $\bar{v}$ in $h^A(\mathcal{B}^A)$. Thus the opposite inclusion $\operatorname{Max}(h^A(\mathcal{B}^A), K) \subseteq \operatorname{Max}(h^A_{BK}(\mathcal{B}^A_{BK}), K)$ is also shown. □

Remark 4.1.3. We emphasize the fact that in the proof of the previous theorem no assumptions regarding the functions and sets involved in the formulation of (PV^A) have been used. This means that the maximal sets of $h^A(\mathcal{B}^A)$ and $h^A_{BK}(\mathcal{B}^A_{BK})$ are always identical.

In case $V = \mathbb{R}$ and $K = \mathbb{R}_+$ one can identify $\overline{V}$ with $\overline{\mathbb{R}} = \mathbb{R} \cup \{\pm\infty\}$ and, assuming that $f : X \to \overline{\mathbb{R}}$ and $g : Y \to \overline{\mathbb{R}}$ are proper and convex functions, the primal problem (PV^A) becomes in this particular case

$$(P^A) \quad \inf_{x \in X} \{f(x) + g(Ax)\}.$$

In this situation finding the properly efficient solutions to (PV^A) means in fact establishing which are the optimal solutions to (P^A). An element (v^*, y^*, v) belongs to $\mathcal{B}^A$ if and only if $v^* > 0$, $y^* \in Y^*$ and $v \in \mathbb{R}$ fulfill

$$v^* v \leq -(v^* f)^*(-A^* y^*) - (v^* g)^*(y^*)$$

or, equivalently,

$$v^* v \leq -v^* f^* \left(-\frac{1}{v^*} A^* y^*\right) - v^* g^* \left(\frac{1}{v^*} y^*\right)$$

$$\Leftrightarrow v \leq -f^* \left(-\frac{1}{v^*} A^* y^*\right) - g^* \left(\frac{1}{v^*} y^*\right).$$

The vector dual problem looks in this case like

$$(D^A) \quad \sup_{v^* > 0, y^* \in Y^*} \left\{-f^* \left(-\frac{1}{v^*} A^* y^*\right) - g^* \left(\frac{1}{v^*} y^*\right)\right\}$$

or, equivalently,

$$(D^A) \quad \sup_{y^* \in Y^*} \{-f^*(-A^* y^*) - g^*(y^*)\},$$

which is nothing else than the classical scalar Fenchel dual problem to (P^A) (cf. subsection 3.1.2). The same conclusion can be drawn when particularizing in an analogous way the vector dual problem (DV^A_{BK}).

4.1.2 Duality with respect to weakly efficient solutions

Next we assume that the ordering cone K has a nonempty interior and we introduce a dual vector problem to

$$(PV_w^A) \quad \underset{x\in X}{\mathrm{WMin}}\{f(x) + g\circ A(x)\},$$

which puts in relation the weakly efficient solutions to the primal vector and the dual vector problems. We say that $\bar{x} \in X$ is a weakly efficient solution to (PV_w^A) if $\bar{x} \in \operatorname{dom} f \cap A^{-1}(\operatorname{dom} g)$ and $f(\bar{x}) \in \mathrm{WMin}((f+g\circ A)(\operatorname{dom} f \cap A^{-1}(\operatorname{dom} g)), K)$. The dual vector problem, which we investigate in the following, is defined by slightly modifying the formulation of (DV^A)

$$(DV_w^A) \quad \underset{(v^*,y^*,v)\in\mathcal{B}_w^A}{\mathrm{WMax}} h_w^A(v^*,y^*,v),$$

where

$$\mathcal{B}_w^A = \{(v^*,y^*,v) \in (K^* \setminus \{0\}) \times Y^* \times V : \\ \langle v^*, v\rangle \leq -(v^*f)^*(-A^*y^*) - (v^*g)^*(y^*)\}$$

and

$$h_w^A(v^*,y^*,v) = v.$$

Next we prove the *weak duality* theorem for (PV_w^A) and (DV_w^A).

Theorem 4.1.6. *There is no $x \in X$ and no $(v^*,y^*,v) \in \mathcal{B}_w^A$ such that $(f+g\circ A)(x) <_K h_w^A(v^*,y^*,v)$.*

Proof. Assume that there exist $x \in X$ and $(v^*,y^*,v) \in \mathcal{B}_w^A$ such that $v-(f+g\circ A)(x) = h_w^A(v^*,y^*,v)-(f+g\circ A)(x) >_K 0$. Thus $x \in \operatorname{dom} f\cap A^{-1}(\operatorname{dom} g)$ and since $v^* \in K^*\setminus\{0\}$ one has

$$\langle v^*, (f+g\circ A)(x)\rangle < \langle v^*, v\rangle \leq -(v^*f)^*(-A^*y^*) - (v^*g)^*(y^*).$$

Like in the proof of Theorem 4.1.1, this leads to a contradiction. □

We come now to the proof of the *strong duality* theorem.

Theorem 4.1.7. *Assume that the regularity condition (RCV^A) is fulfilled. If $\bar{x} \in X$ is a weakly efficient solution to (PV_w^A), then there exists $(\bar{v}^*,\bar{y}^*,\bar{v})$, a weakly efficient solution to (DV_w^A), such that $(f+g\circ A)(\bar{x}) = h_w^A(\bar{v}^*,\bar{y}^*,\bar{v}) = \bar{v}$.*

Proof. If $\bar{x} \in X$ is a weakly efficient solution to (PV_w^A), then $\bar{x} \in \operatorname{dom} f\cap A^{-1}(\operatorname{dom} g)$ and $(f+g\circ A)(\bar{x})$ is a weakly minimal element of the set $(f+g\circ A)(\operatorname{dom} f\cap A^{-1}(\operatorname{dom} g)) \subseteq V$. As $(f+g\circ A)(\operatorname{dom} f\cap A^{-1}(\operatorname{dom} g))+K$ is a nonempty convex set, by Corollary 2.4.26 (see Remark 2.4.11), there exists $\bar{v}^* \in K^*\setminus\{0\}$, such that

$$\langle \bar{v}^*, (f+g\circ A)(\bar{x})\rangle = \inf_{x\in X}\{(\bar{v}^*f)(x) + (\bar{v}^*g)(Ax)\}.$$

Like in the proof of Theorem 4.1.2, one can provide a $\bar{y}^* \in Y^*$ such that $(\bar{v}^*, \bar{y}^*, \bar{v}) \in \mathcal{B}_w^A$, where $\bar{v} := (f + g \circ A)(\bar{x}) \in V$. Further, by Theorem 4.1.6 there exists no (v^*, y^*, v) in $\mathcal{B}_w^A$ such that $(f + g \circ A)(\bar{x}) = \bar{v} <_K v = h^A(v^*, y^*, v)$, which means that in fact $(\bar{v}^*, \bar{y}^*, \bar{v})$ is a weakly efficient solution to (DV_w^A). □

A converse duality theorem for the vector primal-dual pair $(PV_w^A)-(DV_w^A)$ can be also given. To this end we prove first the following preliminary result.

Theorem 4.1.8. *Assume that the regularity condition (RCV^A) is fulfilled. Then*

$$V \setminus \operatorname{cl}\big((f + g \circ A)(\operatorname{dom} f \cap A^{-1}(\operatorname{dom} g)) + K\big) \subseteq \operatorname{core}(h_w^A(\mathcal{B}_w^A)).$$

Proof. Let $\bar{v} \in V \setminus \operatorname{cl}\big((f + g \circ A)(\operatorname{dom} f \cap A^{-1}(\operatorname{dom} g)) + K\big)$ be arbitrarily chosen. Since $\operatorname{cl}\big((f + g \circ A)(\operatorname{dom} f \cap A^{-1}(\operatorname{dom} g)) + K\big) \subseteq V$ is a convex and closed set, there exist $\bar{v}^* \in K^* \setminus \{0\}$ and $\alpha \in \mathbb{R}$ such that

$$\langle \bar{v}^*, \bar{v}\rangle < \alpha \leq \inf_{x \in X} \{(\bar{v}^* f)(x) + (\bar{v}^* g)(Ax)\}.$$

Having (RCV^A) fulfilled and one obtains a $\bar{y}^* \in Y^*$ such that

$$\langle \bar{v}^*, \bar{v}\rangle < -(\bar{v}^* f)^*(-A^* \bar{y}^*) - (\bar{v}^* g)^*(\bar{y}^*).$$

As in the proof of Theorem 4.1.3 one can conclude that $\bar{v} \in \operatorname{core}(h_w^A(\mathcal{B}_w^A))$. □

Now we are able to prove the *converse duality* result.

Theorem 4.1.9. *Assume that the regularity condition (RCV^A) is fulfilled and that the set $(f + g \circ A)(\operatorname{dom} f \cap A^{-1}(\operatorname{dom} g)) + K$ is closed. Then for every weakly efficient solution $(\bar{v}^*, \bar{y}^*, \bar{v})$ to (DV_w^A) one has that $\bar{v}$ is a weakly minimal element of the set $(f + g \circ A)(\operatorname{dom} f \cap A^{-1}(\operatorname{dom} g)) + K$.*

Proof. We assume that $\bar{v} \notin (f + g \circ A)(\operatorname{dom} f \cap A^{-1}(\operatorname{dom} g)) + K$. By Theorem 4.1.8, one has $\bar{v} \in \operatorname{core}(h_w^A(\mathcal{B}_w^A))$. Considering an element $k \in \operatorname{int}(K)$ there exists $\lambda > 0$ such that $v_\lambda := \bar{v} + \lambda k >_K \bar{v}$ and $v_\lambda \in h_w^A(\mathcal{B}_w^A)$. This contradicts the fact that $(\bar{v}^*, \bar{y}^*, \bar{v})$ is a weakly efficient solution to (DV_w^A) and, consequently, we must have that $\bar{v} \in (f + g \circ A)(\operatorname{dom} f \cap A^{-1}(\operatorname{dom} g)) + K$. Supposing that $\bar{v}$ is not a weakly minimal element of this set, there exist $x \in X$ and $k \in K$ such that $(f + g \circ A)(x) \leqq_K (f + g \circ A)(x) + k <_K \bar{v}$. Theorem 4.1.6 leads to a contradiction and this provides the desired conclusion. □

Remark 4.1.4. (a) The observations made in Remark 4.1.1 and Remark 4.1.2 apply also for the strong and converse duality theorems, respectively, given for the primal-dual vector pair (PV_w^A) - (DV_w^A).

(b) Another dual problem to (PV_w^A) can be defined in analogy to (DV_{BK}^A) as being

$$(DV^A_{BKw}) \quad \underset{(v^*,y^*,v)\in\mathcal{B}^A_{BKw}}{\text{WMax}} \quad h^A_{BKw}(v^*,y^*,v),$$

where

$$\mathcal{B}^A_{BKw} = \{(v^*,y^*,v) \in (K^* \setminus \{0\}) \times Y^* \times V : \\ \langle v^*, v\rangle = -(v^*f)^*(-A^*y^*) - (v^*g)^*(y^*)\}$$

and

$$h^A_{BKw}(v^*,y^*,v) = v.$$

Also in this case one has in general that $h^A_{BKw}(\mathcal{B}^A_{BKw}) \subsetneq h^A_w(\mathcal{B}^A_w)$ and it can be shown, like in the proof of Theorem 4.1.5 that the weakly maximal elements of these sets coincide.

(*c*) Particularizing (DV^A_w) and (DV^A_{BKw}) for $V = \mathbb{R}$ and $K = \mathbb{R}_+$ they both turn out to be the classical scalar Fenchel dual optimization problem.

4.2 Constrained vector optimization: a geometric approach

In this section we consider as primal problem a vector optimization problem with geometric and cone constraints. With respect to both properly and weakly efficient solutions to the primal we define a corresponding dual vector optimization problem by means of a so-called *geometric approach* which was considered for the first time by Nakayama for vector problems in finite dimensional spaces (cf. [142–144]).

Let X, Z and V be Hausdorff locally convex spaces and assume that Z is partially ordered by the convex cone $C \subseteq Z$, while V is partially ordered by the nontrivial pointed convex cone $K \subseteq V$. Further, let $S \subseteq X$ be a nonempty convex set, $f : X \to \overline{V} = V \cup \{\pm\infty_K\}$ a proper and K-convex function and $g : X \to \overline{Z} = Z \cup \{\pm\infty_C\}$ a proper and C-convex function such that $\operatorname{dom} f \cap S \cap g^{-1}(-C) \neq \emptyset$. The primal vector optimization problem with geometric and cone constraints we deal here with is

$$(PV^C) \quad \underset{x\in\mathcal{A}}{\text{Min}}\, f(x). \\ \mathcal{A} = \{x \in S : g(x) \in -C\}$$

4.2.1 Duality with respect to properly efficient solutions

The dual vector problem we construct in this part is with respect to the properly efficient solutions to (PV^C) in the sense of linear scalarization, which we simply shall call properly efficient solutions. We say that $\bar{x} \in \mathcal{A}$ is a properly efficient solution to (PV^C) if $\bar{x} \in \operatorname{dom} f$ and $f(\bar{x}) \in \text{PMin}_{LSc}(f(\operatorname{dom} f \cap \mathcal{A}), K)$. This means that there exists $v^* \in K^{*0}$ such that $\langle v^*, f(\bar{x})\rangle \leq \langle v^*, f(x)\rangle$ for all $x \in \mathcal{A}$. Consider the following dual vector problem with respect to the class of efficient solutions

$$(DV^{C_N}) \quad \underset{(U,v)\in\mathcal{B}^{C_N}}{\text{Max}} \; h^{C_N}(U,v),$$

where

$$\begin{aligned}\mathcal{B}^{C_N} = \{&(U,v) \in \mathcal{L}_+(Z,V) \times V : \\ &\nexists x \in S \cap \operatorname{dom} g \text{ such that } v \geqq_K f(x) + U(g(x))\},\end{aligned}$$

$$\mathcal{L}_+(Z,V) = \{U \in \mathcal{L}(Z,V) : U(C) \subseteq K\}$$

and

$$h^{C_N}(U,v) = v.$$

The set $\mathcal{L}_+(Z,V)$ is known in the literature as the set of *positive mappings* (cf. [52,156]). We first prove that for (PV^C) and (DV^{C_N}) weak duality holds.

Theorem 4.2.1. *There is no $x \in \mathcal{A}$ and no $(U,v) \in \mathcal{B}^{C_N}$ such that $f(x) \leq_K h^{C_N}(U,v)$.*

Proof. We assume the contrary, namely that there exist $x \in \mathcal{A}$ and $(U,v) \in \mathcal{B}^{C_N}$ such that $v = h^{C_N}(U,v) \geq_K f(x)$. Since $g(x) \in -C$ and $U \in \mathcal{L}_+(Z,V)$ we have that $U(g(x)) \in -K$, which yields $f(x)+U(g(x)) \leqq_K f(x) \leq_K v$. But this contradicts the fact that (U,v) is a feasible element to the dual (DV^{C_N}) and so the proof is done. □

The next theorem proves that under the fulfilment of the regularity condition

$$(RCV^{C_L}) \mid \exists x' \in \operatorname{dom} f \cap S \text{ such that } g(x') \in -\operatorname{int}(C)$$

the existence of strong duality for (PV^C) and (DV^{C_N}) is guaranteed.

Theorem 4.2.2. *Assume that the regularity condition (RCV^{C_L}) is fulfilled. If $\bar{x} \in \mathcal{A}$ is a properly efficient solution to (PV^C), then there exists $(\overline{U},\bar{v})$, an efficient solution to (DV^{C_N}), such that $f(\bar{x}) = h^{C_N}(\overline{U},\bar{v}) = \bar{v}$.*

Proof. Since $\bar{x} \in X$ is a properly efficient solution to (PV^C), there exists $\bar{v}^* \in K^{*0}$ such that $\bar{x}$ is an optimal solution to the scalar optimization problem

$$\inf_{x\in\mathcal{A}} \langle \bar{v}^*, f(x)\rangle.$$

Using that (RCV^{C_L}) is fulfilled and taking into consideration the fact that $\operatorname{dom}(\bar{v}^* f) = \operatorname{dom} f$, by Theorem 3.3.16 follows that there exists $\bar{z}^* \in C^*$ such that

$$\langle \bar{v}^*, f(\bar{x})\rangle = \inf_{x\in S}\{\langle \bar{v}^*, f(x)\rangle + \langle \bar{z}^*, g(x)\rangle\} \tag{4.4}$$

and

$$\langle \bar{z}^*, g(\bar{x})\rangle = 0. \tag{4.5}$$

Next we show that there exists a positive operator $\overline{U} \in \mathcal{L}_+(Z,V)$ such that $\overline{U}^* \bar{v}^* = \bar{z}^*$.

As $\bar{v}^* \in K^{*0}$ there exists $\bar{\mu} \in K$ such that $\langle \bar{v}^*, \bar{\mu} \rangle = 1$. Define $\overline{U} : Z \to V$ by $\overline{U}z = \langle \bar{z}^*, z \rangle \bar{\mu}$. It is obvious that $\overline{U}$ is linear and continuous. Further, take an arbitrary $c \in C$. As $\bar{z}^* \in C^*$ one has that $\langle \bar{z}^*, c \rangle \geq 0$, which yields $\langle \bar{z}^*, c \rangle \bar{\mu} \in K$. This is nothing else than $\overline{U}(C) \subseteq K$ and so $\overline{U} \in \mathcal{L}_+(Z, V)$. More than this, for all $z \in Z$ it holds

$$\langle \overline{U}^* \bar{v}^*, z \rangle = \langle \bar{v}^*, \overline{U}z \rangle = \langle \bar{z}^*, z \rangle \langle \bar{v}^*, \bar{\mu} \rangle = \langle \bar{z}^*, z \rangle,$$

which means that $\overline{U}^* \bar{v}^* = \bar{z}^*$.

Taking $\bar{v} := f(\bar{x})$ one can see that $(\overline{U}, \bar{v}) \in \mathcal{B}^{C_N}$. Indeed, a ssuming the contrary, one would have that there exists $x \in S \cap \operatorname{dom} g$ such that $f(\bar{x}) = \bar{v} \geq_K f(x) + \overline{U}(g(x))$. Thus

$$\langle \bar{v}^*, f(\bar{x}) \rangle > \langle \bar{v}^*, f(x) \rangle + \langle \bar{v}^*, \overline{U}(g(x)) \rangle$$

$$= \langle \bar{v}^*, f(x) \rangle + \langle \overline{U}^* \bar{v}^*, g(x) \rangle = \langle \bar{v}^*, f(x) \rangle + \langle \bar{z}^*, g(x) \rangle.$$

But this contradicts relation (4.4) and this means that $(\overline{U}, \bar{v})$ is a feasible solution to the dual problem (DV^{C_N}). In order to get the desired conclusion one has only to show that $(\overline{U}, \bar{v})$ is an efficient solution to (DV^{C_N}). If this were not the case, then there would exist a feasible element (U, v) to (DV^{C_N}) such that $v \geq_K \bar{v} = f(\bar{x})$. In this way we obtain a contradiction to Theorem 4.2.1 and the desired conclusion follows. □

Remark 4.2.1. As follows from the proof of the previous theorem, the regularity condition (RCV^{C_L}) is used in order to ensure the existence of strong duality for the scalar optimization problem

$$\inf_{x \in \mathcal{A}} \langle \bar{v}^*, f(x) \rangle$$

and its Lagrange dual problem

$$\sup_{z^* \in C^*} \inf_{x \in S} \{ \langle \bar{v}^*, f(x) \rangle + \langle z^*, g(x) \rangle \}.$$

This condition assumes implicitly that the cone C has a nonempty interior, an assumption which can fail in a lot of situations. In case X and Z are Fréchet spaces, S is closed, f is star K-lower semicontinuous and g is C-epi closed one can suppose instead, that $0 \in \operatorname{sqri}(g(\operatorname{dom} f \cap S \cap \operatorname{dom} g) + C)$ (cf. subsection 3.2.3). On the other hand, if $\operatorname{lin}(g(\operatorname{dom} f \cap S \cap \operatorname{dom} g) + C)$ is a finite dimensional linear subspace, (RCV^{C_L}) can be replaced with the assumption $0 \in \operatorname{ri}(g(\operatorname{dom} f \cap S \cap \operatorname{dom} g) + C)$.

Remark 4.2.2. In Nakayama's papers [142–144], where the concept of geometric duality for vector optimization problems in finite dimensional spaces has been introduced, the properly efficient solutions are also defined in the sense of linear scalarization, but by considering $\bar{v}^*$ from $\operatorname{int}(K^*)$. Working with the quasi interior of K^* we are able to cover a broader class of optimization problems, namely the ones for which one has $K^{*0} \neq \emptyset$, even if the interior of the dual cone is empty.

Before coming to the converse duality theorem we deliver some inclusion relations involving the set $h^{C_N}(\mathcal{B}^{C_N})$ which extend [142, Proposition 3.1].

Proposition 4.2.3. *It holds*

$$h^{C_N}(\mathcal{B}^{C_N}) - K = h^{C_N}(\mathcal{B}^{C_N}) \subseteq \operatorname{cl}(V \setminus (f(\operatorname{dom} f \cap \mathcal{A}) + K)). \tag{4.6}$$

Proof. That $h^{C_N}(\mathcal{B}^{C_N}) - K \supseteq h^{C_N}(\mathcal{B}^{C_N})$ is obvious. Further, consider an arbitrary element $v \in h^{C_N}(\mathcal{B}^{C_N}) - K$. This means that there exist $k \in K$ and $U \in \mathcal{L}_+(Z, V)$ such that $(U, v + k) \in \mathcal{B}^{C_N}$ or, equivalently, $v + k \ngeq_K f(x) + U(g(x))$ for all $x \in S \cap \operatorname{dom} g$. From here follows that $v \ngeq_K f(x) + U(g(x))$ for all $x \in S \cap \operatorname{dom} g$, which is nothing else than $(U, v) \in \mathcal{B}^{C_N}$. Thus $v \in h^{C_N}(\mathcal{B}^{C_N})$ and the equality in (4.6) is proven.

In order to show that $h^{C_N}(\mathcal{B}^{C_N}) \subseteq \operatorname{cl}(V \setminus (f(\operatorname{dom} f \cap \mathcal{A}) + K))$, it is enough to prove that the sets $h^{C_N}(\mathcal{B}^{C_N})$ and $\operatorname{int}(f(\operatorname{dom} f \cap \mathcal{A}) + K)$ have no point in common. Assume that this is not the case and that $\tilde{v}$ is a common element of these sets. Choosing $\tilde{k} \in K \setminus \{0\}$ one has that there exists $\tilde{\lambda} > 0$ such that $\tilde{v} - \tilde{\lambda}\tilde{k} \in \operatorname{int}(f(\operatorname{dom} f \cap \mathcal{A}) + K)$. But this implies that there exists $\tilde{x} \in \operatorname{dom} f \cap \mathcal{A}$ with the property that $\tilde{v} \geq_K f(\tilde{x})$, which is a contradiction to Theorem 4.2.1. □

The next result will play a crucial role in proving the converse duality theorem.

Theorem 4.2.4. *Assume that* $\operatorname{qi}(K) \neq \emptyset$ *and that the regularity condition* (RCV^{C_L}) *is fulfilled. Then it holds*

$$V \setminus \operatorname{cl}(f(\operatorname{dom} f \cap \mathcal{A}) + K) \subseteq h^{C_N}(\mathcal{B}^{C_N}) - (K \setminus \{0\}). \tag{4.7}$$

Proof. Take an arbitrary element $\bar{v} \in V \setminus \operatorname{cl}(f(\operatorname{dom} f \cap \mathcal{A}) + K)$. As $\operatorname{dom} f \cap \mathcal{A} \neq \emptyset$ and f is a K-convex function one has that $\operatorname{cl}(f(\operatorname{dom} f \cap \mathcal{A}) + K)$ is a nonempty convex and closed subset of V. Theorem 2.1.5 guarantees the existence of $\bar{v}^* \in Y^* \setminus \{0\}$ and $\alpha \in \mathbb{R}$ such that

$$\langle \bar{v}^*, \bar{v} \rangle < \alpha < \langle \bar{v}^*, v \rangle \ \forall v \in \operatorname{cl}(f(\operatorname{dom} f \cap \mathcal{A}) + K).$$

That $\bar{v}^* \in K^* \setminus \{0\}$ follows automatically. More than that, the relation above implies

$$\langle \bar{v}^*, \bar{v} \rangle < \inf_{x \in \mathcal{A}} \langle \bar{v}^*, f(x) \rangle.$$

Thus, by Theorem 3.2.9 there exists $\bar{z}^* \in C^*$ fulfilling

$$\langle \bar{v}^*, \bar{v} \rangle < \inf_{x \in \mathcal{A}} \langle \bar{v}^*, f(x) \rangle = \inf_{x \in S} \{ \langle \bar{v}^*, f(x) \rangle + \langle \bar{z}^*, g(x) \rangle \}. \tag{4.8}$$

As in the proof of Proposition 2.1.1 one can show that $\operatorname{qi}(K) \subseteq \{v \in K : \langle v^*, v \rangle > 0 \text{ for all } v^* \in K^* \setminus \{0\}\}$. On the other hand, the assumption of the nonemptiness for $\operatorname{qi}(K)$ guarantees the existence of a nonzero element in this set. Thus there exists $\bar{\mu} \in K$ such that $\langle \bar{v}^*, \bar{\mu} \rangle = 1$. Like in the proof of Theorem 4.2.2 one can construct a linear continuous mapping $\overline{U} \in \mathcal{L}_+(Z, V)$ such that $\overline{U}^* \bar{v}^* = \bar{z}^*$ and so relation (4.8) yields that

$$\langle \bar{v}^*, \bar{v} \rangle < \inf_{x \in S} \{ \langle \bar{v}^*, f(x) \rangle + \langle \bar{v}^*, \overline{U}(g(x)) \rangle \} = \inf_{x \in S} \langle \bar{v}^*, f(x) + \overline{U}(g(x)) \rangle.$$

Then there exists $\tilde{v} \in \bar{v} + (K \setminus \{0\})$ such that

$$\langle \bar{v}^*, \bar{v} \rangle < \langle \bar{v}^*, \tilde{v} \rangle < \inf_{x \in S} \langle \bar{v}^*, f(x) + \overline{U}(g(x)) \rangle. \tag{4.9}$$

Since $\bar{v} \in \tilde{v} - (K \setminus \{0\})$, in order to get the desired conclusion it is enough to prove that $\tilde{v} \in h^{C_N}(\mathcal{B}^{C_N})$. We claim that $(\overline{U}, \tilde{v}) \in \mathcal{B}^{C_N}$. Were this not the case, one could find an element $\bar{x} \in S \cap \operatorname{dom} g$ such that $\tilde{v} \geq_K f(\bar{x}) + \overline{U}(g(\bar{x}))$, which would imply $\langle \bar{v}^*, \tilde{v} \rangle \geq \langle \bar{v}^*, f(\bar{x}) + \overline{U}(g(\bar{x})) \rangle$. Since this would contradict relation (4.9), we must have $\bar{v} \in \tilde{v} - (K \setminus \{0\}) \subseteq h^{C_N}(\mathcal{B}^{C_N}) - (K \setminus \{0\})$. □

Remark 4.2.3. The regularity condition (RCV^{C_L}) is used in the proof of Theorem 4.2.4 in order to guarantee the existence of strong duality for the scalar optimization problem

$$\inf_{x \in \mathcal{A}} \langle \bar{v}^*, f(x) \rangle$$

and its Lagrange dual problem, more precisely, to ensure the existence of an element $\bar{z}^* \in C^*$ such that relation (4.8) is true. In fact, it is enough to assume that for all $v^* \in K^* \setminus \{0\}$ the optimization problem

$$\inf_{x \in \mathcal{A}} \langle v^*, f(x) \rangle$$

is normal with respect to its Lagrange dual. This means that

$$\inf_{x \in \mathcal{A}} \langle v^*, f(x) \rangle = \sup_{z^* \in C^*} \inf_{x \in S} \{ \langle v^*, f(x) \rangle + \langle z^*, g(x) \rangle \},$$

whereby the existence of an optimal solution to the dual is no further guaranteed. Nevertheless, this assumption is enough for getting an element $\bar{z}^* \in C^*$ which fulfills

$$\langle v^*, v \rangle < \inf_{x \in S} \{ \langle v^*, f(x) \rangle + \langle \bar{z}^*, g(x) \rangle \}.$$

Remark 4.2.4. Combining the last two results one has that, in case $\operatorname{qi}(K) \neq \emptyset$ and (RCV^{C_L}) is fulfilled, the following relations of inclusion hold

$$\begin{aligned} V \setminus \operatorname{cl}(f(\operatorname{dom} f \cap \mathcal{A}) + K) &\subseteq h^{C_N}(\mathcal{B}^{C_N}) - (K \setminus \{0\}) \\ \subseteq h^{C_N}(\mathcal{B}^{C_N}) - K = h^{C_N}(\mathcal{B}^{C_N}) &\subseteq \operatorname{cl}(V \setminus (f(\operatorname{dom} f \cap \mathcal{A}) + K)). \end{aligned} \tag{4.10}$$

Relation (4.10) generalizes [142, Proposition 3.1], providing a refinement of this result as well as an extension of it to infinite dimensional spaces. It is also worth mentioning that the assumptions we consider in this section are weaker than the ones in the original work.

We come now to the converse duality theorem for (PV^C) and (DV^{C_N}).

Theorem 4.2.5. *Assume that* $\text{qi}(K) \neq \emptyset$*, the regularity condition* (RCV^{C_L}) *is fulfilled and the set* $f(\text{dom} f \cap \mathcal{A}) + K$ *is closed. Then for every efficient solution* $(\overline{U}, \bar{v})$ *to* (DV^{C_N}) *there exists* $\bar{x} \in \mathcal{A}$*, an efficient solution to* (PV^C)*, such that* $f(\bar{x}) = h^{C_N}(\overline{U}, \bar{v}) = \bar{v}$*.*

Proof. We start by showing that $\bar{v} \in f(\text{dom} f \cap \mathcal{A}) + K$. Assuming the contrary, by Theorem 4.2.4 one has $\bar{v} \in h^{C_N}(\mathcal{B}^{C_N}) - (K \setminus \{0\})$, which means that there exists $(U, v) \in \mathcal{B}^{C_N}$ fulfilling $v \geq_K \bar{v}$. But this contradicts the fact that $(\overline{U}, \bar{v})$ is an efficient solution to (DV^{C_N}). Thus $\bar{v} \in f(\text{dom} f \cap \mathcal{A}) + K$ and so there exist $\bar{x} \in \text{dom} f \cap \mathcal{A}$ and $\bar{k} \in K$ for which $\bar{v} = f(\bar{x}) + \bar{k}$. By the weak duality result (see Theorem 4.2.1) follows that $\bar{k} = 0$, which yields $\bar{v} = f(\bar{x})$. That $\bar{x}$ is an efficient solution to (PV^C) follows also by Theorem 4.2.1. □

Remark 4.2.5. (*a*) As pointed out in Remark 4.2.3 the converse duality theorem remains valid even if one supposes that for all $v^* \in K^* \setminus \{0\}$ the optimization problem

$$\inf_{x \in \mathcal{A}} \langle v^*, f(x) \rangle$$

is normal with respect to its Lagrange dual.

(*b*) Working in finite dimensional spaces, Nakayama has given in [142–144] a converse duality result for the vector primal-dual pair $(PV^C) - (DV^{C_N})$ in a more particular framework, namely by considering $C \subseteq \mathbb{R}^m$ and $K \subseteq \mathbb{R}^k$ convex closed cones with nonempty interiors, $S \subseteq \mathbb{R}^n$ a nonempty convex set, $f : \mathbb{R}^n \to \mathbb{R}^k$ a K-convex function and $g : \mathbb{R}^n \to \mathbb{R}^m$ a C-convex function. Assuming that a Slater type condition is fulfilled, that the set $(f, g)(S) + K \times C$ is closed and that there exists at least one properly efficient solution to (PV^C), Nakayama proves that for (PV^C) and (DV^{C_N}) converse duality holds. As one can see, the assumption regarding the existence of a properly efficient solution to the primal is not necessary, while instead of asking that $(f, g)(S) + K \times C$ is closed one can consider in this particular situation the weaker hypothesis that $f(\mathcal{A}) + K$ is closed.

4.2.2 Duality with respect to weakly efficient solutions

In this second part of the section 4.2 we suppose that $\text{int}(K) \neq \emptyset$ and provide a vector dual problem to

$$(PV_w^C) \quad \underset{x \in \mathcal{A}}{\text{WMin}} f(x),$$
$$\mathcal{A} = \{x \in S : g(x) \in -C\}$$

which this time relates the weakly efficient solutions to the primal and dual problem. We say that $\bar{x} \in \mathcal{A}$ is a weakly efficient solution to (PV_w^C) if $\bar{x} \in \text{dom} f$ and $f(\bar{x}) \in \text{WMin}(f(\text{dom} f \cap \mathcal{A}), K)$. The vector dual is defined as follows

$$(DV_w^{C_N}) \quad \underset{(U,v)\in\mathcal{B}_w^{C_N}}{\text{WMax}} \; h_w^{C_N}(U,v),$$

where

$$\mathcal{B}_w^{C_N} = \{(U,v) \in \mathcal{L}_+(Z,V) \times V : \\ \nexists x \in S \cap \operatorname{dom} g \text{ such that } v >_K f(x) + U(g(x))\},$$

and

$$h_w^{C_N}(U,v) = v.$$

Since $\operatorname{int}(K) \subsetneq K \setminus \{0\}$ one has that $\mathcal{B}^{C_N} \subseteq \mathcal{B}_w^{C_N}$ and, consequently, $h^{C_N}(\mathcal{B}^{C_N}) \subseteq h_w^{C_N}(\mathcal{B}_w^{C_N})$. The weak and strong duality statements follow.

Theorem 4.2.6. *There is no $x \in \mathcal{A}$ and no $(U,v) \in \mathcal{B}_w^{C_N}$ such that $f(x) <_K h_w^{C_N}(U,v)$.*

We omit the proof of Theorem 4.2.6 as it follows in the lines of Theorem 4.2.1.

Theorem 4.2.7. *Assume that the regularity condition (RCV^{C_L}) is fulfilled. If $\bar{x} \in \mathcal{A}$ is a weakly efficient solution to (PV_w^C), then there exists $(\overline{U},\bar{v})$, a weakly efficient solution to $(DV_w^{C_N})$, such that $f(\bar{x}) = h_w^{C_N}(\overline{U},\bar{v}) = \bar{v}$.*

Proof. If $\bar{x} \in X$ is a weakly efficient efficient solution to (PV_w^C), then $\bar{x} \in \operatorname{dom} f \cap \mathcal{A}$ and $f(\bar{x})$ is a weakly minimal element of the set $f(\operatorname{dom} f \cap \mathcal{A}) \subseteq V$. Using that $f(\operatorname{dom} f \cap \mathcal{A}) + K$ is a nonempty convex set, by Corollary 2.4.26 (see also Remark 2.4.11) there exists $\bar{v}^* \in K^* \setminus \{0\}$ such that $\bar{x}$ is an optimal solution to the scalar optimization problem

$$\inf_{x\in\mathcal{A}} \langle \bar{v}^*, f(x)\rangle.$$

Like in the proof of Theorem 4.2.2 one can construct an element $\overline{U} \in \mathcal{L}_+(Z,V)$ such that for $\bar{v} = f(\bar{x})$ it holds $(\overline{U},\bar{v}) \in \mathcal{B}^{C_N} \subseteq \mathcal{B}_w^{C_N}$. By Theorem 4.2.6 no $(U,v) \in \mathcal{B}_w^{C_N}$, fulfilling $v >_K f(x)$, exists and this means that $(\overline{U},\bar{v})$ is a weakly efficient solution to the dual $(DV_w^{C_N})$. □

In analogy to Proposition 4.2.3 and Theorem 4.2.4 one can prove the following results, respectively. One can notice that since $\operatorname{int}(K)$ was assumed nonempty, the assumption of the nonemptiness for the quasi interior of the cone K becomes superfluous, thus it is omitted.

Proposition 4.2.8. *It holds*

$$h_w^{C_N}(\mathcal{B}_w^{C_N}) - K = h_w^{C_N}(\mathcal{B}_w^{C_N}) \subseteq \operatorname{cl}\left(V \setminus (f(\operatorname{dom} f \cap \mathcal{A}) + K)\right). \tag{4.11}$$

Theorem 4.2.9. *Assume that the regularity condition (RCV^{C_L}) is fulfilled. Then it holds*

$$V \setminus \operatorname{cl}\left(f(\operatorname{dom} f \cap \mathcal{A}) + K\right) \subseteq h^{C_N}(\mathcal{B}^{C_N}) - \operatorname{int}(K) \subseteq h_w^{C_N}(\mathcal{B}_w^{C_N}) - \operatorname{int}(K). \tag{4.12}$$

Combining (4.11) and (4.12), under the hypothesis that (RCV^{C_L}) is fulfilled, one gets the following relations of inclusion

$$\begin{aligned} &V \setminus \mathrm{cl}\,(f(\mathrm{dom}\, f \cap \mathcal{A}) + K) \subseteq h_w^{C_N}(\mathcal{B}_w^{C_N}) - \mathrm{int}(K) \\ &\subseteq h_w^{C_N}(\mathcal{B}_w^{C_N}) - K = h_w^{C_N}(\mathcal{B}_w^{C_N}) \subseteq \mathrm{cl}\,(V \setminus (f(\mathrm{dom}\, f \cap \mathcal{A}) + K)), \end{aligned} \tag{4.13}$$

which are useful when proving the converse duality theorem for (PV_w^C) and $(DV_w^{C_N})$. This is what we do next.

Theorem 4.2.10. *Assume that the regularity condition (RCV^{C_L}) is fulfilled and that the set $f(\mathrm{dom}\, f \cap \mathcal{A}) + K$ is closed. Then for every weakly efficient solution $(\overline{U}, \bar{v})$ to $(DV_w^{C_N})$ one has that $\bar{v}$ is a weakly minimal element of the set $f(\mathrm{dom}\, f \cap \mathcal{A}) + K$.*

Proof. Assuming that $\bar{v} \notin f(\mathrm{dom}\, f \cap \mathcal{A}) + K$, by (4.13) follows that $\bar{v} \in h_w^{C_N}(\mathcal{B}_w^{C_N}) - \mathrm{int}(K)$. Thus there exists $(U, v) \in \mathcal{B}_w^{C_N}$ such that $v >_K \bar{v}$, which contradicts the fact that $(\overline{U}, \bar{v})$ is a weakly efficient solution to $(DV_w^{C_N})$. Consequently, $\bar{v} \in f(\mathrm{dom}\, f \cap \mathcal{A}) + K$ and since there is no $x \in \mathrm{dom}\, f \cap \mathcal{A}$ with $\bar{v} >_K f(x)$, $\bar{v}$ turns out to be a weakly minimal element of $f(\mathrm{dom}\, f \cap \mathcal{A}) + K$. □

Remark 4.2.6. (*a*) The observation made in Remark 4.2.1 applies also for the strong duality theorem given for (PV_w^C) and $(DV_w^{C_L})$. In other words, also for this primal-dual vector pair the regularity condition (RCV^{C_L}) can be replaced with alternative regularity conditions if X and Z are Fréchet spaces and some topological assumptions for the sets and functions involved are fulfilled or, on the other hand, if $\mathrm{lin}(g(\mathrm{dom}\, f \cap S \cap \mathrm{dom}\, g) + C)$ is a finite dimensional linear subspace.

(*b*) In both Theorem 4.2.9 and Theorem 4.2.10 instead of assuming that (RCV^{C_L}) holds one can suppose the weaker assumption that for all $v^* \in K^* \setminus \{0\}$ the optimization problem

$$\inf_{x \in \mathcal{A}} \langle v^*, f(x) \rangle$$

is normal with respect to its Lagrange dual. The argumentation is the same as in Remark 4.2.3.

4.3 Constrained vector optimization: a linear scalarization approach

In the third section of this chapter we construct further dual problems to (PV^C) with respect to the properly efficient solutions, all these duals having in common the fact that in the formulation of their feasible sets scalar conjugate dual problems are involved. This duality scheme was used for the first time by Jahn in [101], where a vector dual problem to (PV^C) has been introduced having as starting point the classical scalar Lagrange duality. The relations between the maximal sets of the duals considered here are investigated and some considerations on the duality for (PV^C) with respect to weakly efficient solutions are made.

To begin we introduce a general approach for defining a vector dual problem based on linear scalarization which will provide as particular instances the above-mentioned vector dual problems to (PV^C).

4.3.1 A general approach for constructing a vector dual problem via linear scalarization

Let X and V be Hausdorff locally convex spaces and assume that V is partially ordered by the nontrivial pointed convex cone $K \subseteq V$. Let $F : X \to \overline{V} = V \cup \{\pm\infty_K\}$ be a proper and K-convex function and consider the general vector optimization problem

$$(PVG) \quad \operatorname*{Min}_{x \in X} F(x).$$

Take Y another Hausdorff locally convex space and $\Phi : X \times Y \to \overline{V}$ a proper and K-convex so-called *vector perturbation function* with $\Phi(x, 0) = F(x)$ for all $x \in X$. We say that $\bar{x} \in X$ is a properly efficient solution to (PVG) (here also considered in the sense of linear scalarization) if $\bar{x} \in \operatorname{dom} F$ and $F(\bar{x}) \in \operatorname{PMin}_{LSc}(F(\operatorname{dom} F), K)$. A vector dual problem to (PVG) with respect to the properly efficient solutions can be introduced in the following way (for a related approach, see [78])

$$(DVG) \quad \operatorname*{Max}_{(v^*, y^*, v) \in \mathcal{B}^G} h^G(v^*, y^*, v),$$

where

$$\mathcal{B}^G = \{(v^*, y^*, v) \in K^{*0} \times Y^* \times V : \langle v^*, v \rangle \leq -(v^*\Phi)^*(0, -y^*)\}$$

and

$$h^G(v^*, y^*, v) = v.$$

We prove that for the primal-dual pair $(PVG) - (DVG)$ weak duality is ensured.

Theorem 4.3.1. *There is no $x \in X$ and no $(v^*, y^*, v) \in \mathcal{B}^G$ such that $F(x) \leq_K h^G(v^*, y^*, v)$.*

Proof. We assume the contrary, namely that there exist $x \in X$ and $(v^*, y^*, v) \in \mathcal{B}^G$ such that $F(x) \leq_K h^G(v^*, y^*, v) = v$. It is obvious that $x \in \operatorname{dom} F$ and $\langle v^*, v \rangle > \langle v^*, F(x) \rangle$.

On the other hand, by applying the Young-Fenchel inequality, it holds

$$\langle v^*, v \rangle \leq -(v^*\Phi)^*(0, -y^*) \leq \langle v^*, F(x) \rangle,$$

which leads to a contradiction. □

For proving strong duality we consider the following regularity condition

$$(RCV^{\Phi}) \mid \exists x' \in X \text{ such that } (x',0) \in \operatorname{dom} \Phi \text{ and } \Phi(x',\cdot) \text{ is continuous at } 0.$$

Theorem 4.3.2. *Assume that the regularity condition (RCV^{Φ}) is fulfilled. If $\bar{x} \in X$ is a properly efficient solution to (PVG), then there exists $(\bar{v}^*, \bar{y}^*, \bar{v})$, an efficient solution to (DVG), such that $F(\bar{x}) = h^G(\bar{v}^*, \bar{y}^*, \bar{v}) = \bar{v}$.*

Proof. Since $\bar{x} \in X$ is a properly efficient solution to (PVG), it follows that $\bar{x} \in \operatorname{dom} F$ and there exists $\bar{v}^* \in K^{*0}$ fulfilling

$$\langle \bar{v}^*, F(\bar{x})\rangle = \inf_{x\in X} \langle \bar{v}^*, F(x)\rangle = \inf_{x\in X} (\bar{v}^*\Phi)(x,0).$$

The function $(x,y) \mapsto (\bar{v}^*\Phi)(x,y)$ is proper and convex and one has that there exists $x' \in X$ such that $(x',0) \in \operatorname{dom}(\bar{v}^*\Phi) = \operatorname{dom}\Phi$ and $(\bar{v}^*\Phi)(x',\cdot)$ is continuous at 0. By Theorem 3.2.1, there exists $\bar{y}^* \in Y^*$ such that

$$\langle \bar{v}^*, F(\bar{x})\rangle = \inf_{x\in X} (\bar{v}^*\Phi)(x,0) = \sup_{y^*\in Y^*} \{-(\bar{v}^*\Phi)^*(0,-y^*)\} = -(\bar{v}^*\Phi)^*(0,-\bar{y}^*).$$

This has as consequence the fact that for $\bar{v} = F(\bar{x})$ the element $(\bar{v}^*, \bar{y}^*, \bar{v})$ is a feasible solution to (DVG). By the weak duality statement (see Theorem 4.3.1) it follows that $(\bar{v}^*, \bar{y}^*, \bar{v})$ is an efficient solution to (DVG). □

Remark 4.3.1. For having strong duality it is enough to assume that for all $v^* \in K^{*0}$ the scalar optimization problem

$$\inf_{x\in X} (\bar{v}^*\Phi)(x,0)$$

is stable. In case X and Y are Fréchet spaces, Φ is star K-lower semicontinuous and $0 \in \operatorname{sqri}(\operatorname{Pr}_Y(\operatorname{dom}\Phi))$ this is guaranteed by Theorem 3.2.1. The same happens if $\operatorname{lin}(\operatorname{Pr}_Y(\operatorname{dom}\Phi))$ is a finite dimensional linear subspace and $0 \in \operatorname{ri}(\operatorname{Pr}_Y(\operatorname{dom}\Phi))$.

Before coming to the converse duality theorem we prove a preliminary result.

Theorem 4.3.3. *Assume that $\mathcal{B}^G$ is nonempty and that the regularity condition (RCV^{Φ}) is fulfilled. Then*

$$V \setminus \operatorname{cl}(F(\operatorname{dom} F) + K) \subseteq \operatorname{core}(h^G(\mathcal{B}^G)).$$

Proof. Consider an arbitrary element $\bar{v} \in V \setminus \operatorname{cl}(F(\operatorname{dom} F) + K)$. The set $\operatorname{cl}(F(\operatorname{dom} F) + K) \subseteq V$ is convex and closed and, by Theorem 2.1.5, there exists $\bar{v}^* \in V^* \setminus \{0\}$ and $\alpha \in \mathbb{R}$ fulfills

$$\langle \bar{v}^*, \bar{v}\rangle < \alpha < \langle \bar{v}^*, v\rangle \ \forall v \in \operatorname{cl}(F(\operatorname{dom} F) + K). \tag{4.14}$$

One can easily see that $\bar{v}^* \in K^* \setminus \{0\}$. Further, since $\mathcal{B}^G \neq \emptyset$, there exists $(\tilde{v}^*, \tilde{y}^*, \tilde{v}) \in K^{*0} \times Y^* \times V$ such that

$$\langle \tilde{v}^*, \tilde{v} \rangle \leq -(\tilde{v}^* \Phi)^*(0, -\tilde{y}^*) \leq \inf_{x \in X} \langle \tilde{v}^*, F(x) \rangle. \tag{4.15}$$

Denote by $\gamma := \alpha - \langle \bar{v}^*, \bar{v} \rangle > 0$. For all $s \in (0,1)$ it holds

$$\langle s\tilde{v}^* + (1-s)\bar{v}^*, \bar{v} \rangle = \langle \bar{v}^*, \bar{v} \rangle + s(\langle \tilde{v}^*, \bar{v} \rangle - \langle \bar{v}^*, \bar{v} \rangle) = \alpha - \gamma + s(\langle \tilde{v}^*, \bar{v} \rangle - \alpha + \gamma),$$

while, by (4.14) and (4.15), for all $v \in F(\operatorname{dom} F) + K$ it holds

$$\langle s\tilde{v}^* + (1-s)\bar{v}^*, v \rangle > s\langle \tilde{v}^*, \tilde{v} \rangle + (1-s)\alpha = \alpha + s(\langle \tilde{v}^*, \tilde{v} \rangle - \alpha).$$

Thus one can choose $\bar{s} \in (0,1)$ close enough to 0 such that $\bar{s}(\langle \tilde{v}^*, \bar{v} \rangle - \alpha + \gamma) < \gamma/2$ and $\bar{s}(\langle \tilde{v}^*, \tilde{v} \rangle - \alpha) > -\gamma/2$. For $v_{\bar{s}}^* := \bar{s}\tilde{v}^* + (1-\bar{s})\bar{v}^* \in K^{*0}$ it holds

$$\langle v_{\bar{s}}^*, \bar{v} \rangle < \alpha - \frac{\gamma}{2} < \langle v_{\bar{s}}^*, v \rangle \ \forall v \in F(\operatorname{dom} F),$$

which yields that

$$\langle v_{\bar{s}}^*, \bar{v} \rangle < \inf_{x \in X} \langle v_{\bar{s}}^*, F(x) \rangle. \tag{4.16}$$

Taking into account that (RCV^{Φ}) is fulfilled, there exists $y_{\bar{s}}^* \in Y^*$ such that

$$\langle v_{\bar{s}}^*, \bar{v} \rangle < \inf_{x \in X} (v_{\bar{s}}^* \Phi)(x, 0) = \sup_{y^* \in Y^*} \{-(v_{\bar{s}}^* \Phi)^*(0, -y^*)\} = -(v_{\bar{s}}^* \Phi)^*(0, -y_{\bar{s}}^*). \tag{4.17}$$

Obviously, $\bar{v} \in h^G(\mathcal{B}^G)$. Let $\varepsilon > 0$ be such that

$$\langle v_{\bar{s}}^*, \bar{v} \rangle + \varepsilon < -(v_{\bar{s}}^* \Phi)^*(0, -y_{\bar{s}}^*).$$

Then for all $v \in V$ there exists $\delta_v > 0$ such that

$$\langle v_{\bar{s}}^*, \bar{v} + \lambda v \rangle \leq \langle v_{\bar{s}}^*, \bar{v} \rangle + \varepsilon < -(v_{\bar{s}}^* \Phi)^*(0, -y_{\bar{s}}^*) \ \forall \lambda \in [0, \delta_v].$$

So, for all $\lambda \in [0, \delta_v]$, $(v_{\bar{s}}^*, y_{\bar{s}}^*, \bar{v} + \lambda v) \in \mathcal{B}^G$ and therefore $\bar{v} + \lambda v \in h^G(\mathcal{B}^G)$. In conclusion, $\bar{v} \in \operatorname{core}(h^G(\mathcal{B}^G))$. □

Next we state the converse duality theorem.

Theorem 4.3.4. *Assume that the regularity condition (RCV^{Φ}) is fulfilled and that the set $F(\operatorname{dom} F)+K$ is closed. Then for every efficient solution $(\bar{v}^*, \bar{y}^*, \bar{v})$ to (DVG) there exists $\bar{x} \in X$, a properly efficient solution to (PVG), such that $F(\bar{x}) = h^G(\bar{v}^*, \bar{y}^*, \bar{v}) = \bar{v}$.*

Proof. Assuming that $\bar{v} \notin F(\operatorname{dom} F) + K$, by the previous result follows that $\bar{v} \in \operatorname{core}(h^G(\mathcal{B}^G))$. Thus for $k \in K \setminus \{0\}$ there exists $\lambda > 0$ such that $v_\lambda := \bar{v} + \lambda k \geq_K \bar{v}$ and $v_\lambda \in h^G(\mathcal{B}^G)$. This contradicts the fact that $(\bar{v}^*, \bar{y}^*, \bar{v})$ is an efficient solution to (DVG). Thus we must have $\bar{v} \in F(\operatorname{dom} F) + K$ and this means that there exists $\bar{x} \in \operatorname{dom} F$ and $\bar{k} \in K$ fulfilling $\bar{v} = F(\bar{x}) + \bar{k}$. By Theorem 4.3.1 it follows that $\bar{k} = 0$ and, consequently, $\bar{v} = F(\bar{x})$. Since

$$\langle \bar{v}^*, F(\bar{x}) \rangle = \langle \bar{v}^*, \bar{v} \rangle \leq -(\bar{v}^* \Phi)^*(0, -\bar{y}^*) \leq \inf_{x \in X} \langle \bar{v}^*, F(x) \rangle,$$

$\bar{x}$ is a properly efficient solution to (PVG). □

Remark 4.3.2. In Theorem 4.3.3 and, consequently, in Theorem 4.3.4 the regularity condition (RCV^{Φ}) can be replaced with the weaker assumption that for all $v^* \in K^{*0}$ the problem

$$\inf_{x \in X} \langle v^*, F(x) \rangle$$

is normal. Normality, even if does not guarantee that the conjugate dual of this scalar problem has an optimal solution, ensures that

$$\inf_{x \in X} \langle v^*, F(x) \rangle = \sup_{y^* \in Y^*} \{-(v^* \Phi)^*(0, -y^*)\}.$$

This is enough to guarantee in the proof of Theorem 4.3.3, together with (4.16), the existence of $y_{\bar{s}}^* \in Y^*$ such that

$$\langle v_{\bar{s}}^*, \bar{v} \rangle < -(v_{\bar{s}}^* \Phi)(0, -y_{\bar{s}}^*).$$

Remark 4.3.3. Going back to the vector optimization (PV^A) treated in section 4.1 one can notice that for $\Phi : X \times Y \to \overline{V}$, $\Phi(x, y) = f(x) + g(Ax + y)$, which is in that setting a proper and K-convex function, (DVG) becomes exactly the vector dual problem (DV^A). The regularity condition (RCV^A) is nothing else than (RCV^{Φ}) and this means that the weak, strong and converse duality results stated for the primal-dual vector pair $(PV^A) - (DV^A)$ are particular instances of those introduced in this subsection.

Remark 4.3.4. In case $V = \mathbb{R}$ and $K = \mathbb{R}_+$ one can identify $\overline{V}$ with $\overline{\mathbb{R}}$ and for $F : X \to \overline{\mathbb{R}}$ a proper and convex function, the primal vector problem becomes

$$(PG) \quad \inf_{x \in X} F(x).$$

For $\Phi : X \times Y \to \overline{\mathbb{R}}$ fulfilling $\Phi(x, 0) = F(x)$ for all $x \in X$ one has that (v^*, y^*, v) belongs to $\mathcal{B}^G$ if and only if $v^* > 0$, $y^* \in Y^*$ and $v \in \mathbb{R}$ fulfill

$$v^* v \leq -(v^* \Phi)^*(0, -y^*) \Leftrightarrow v^* v \leq -v^* \Phi^* \left(0, -\frac{1}{v^*} y^*\right)$$

$$\Leftrightarrow v \leq -\Phi^* \left(0, -\frac{1}{v^*} y^*\right).$$

The dual vector problem has in this case the following formulation

$$(DG) \quad \sup_{v^* > 0, y^* \in Y^*} \left\{ -\Phi^* \left(0, -\frac{1}{v^*} y^*\right) \right\}$$

or, equivalently,

$$(DG) \quad \sup_{y^* \in Y^*} \{-\Phi^*(0, -y^*)\},$$

which is the general conjugate dual problem to (PG) investigated in section 3.1.

4.3.2 Vector dual problems to (PV^C) as particular instances of the general approach

For the primal vector optimization problem (PV^C) introduced in section 4.2 we construct some vector dual problems via the general approach described above by considering different vector perturbation functions taking at $(x,0)$ the value $f(x)+\delta_{\mathcal{A}}^V(x)$ for all $x \in X$. We do this in analogy to the investigations made in subsection 3.1.3 in the scalar case and assume to this end that the hypotheses stated for the sets and functions involved in the formulation of (PV^C) are fulfilled.

Consider first $\Phi^{C_L} : X \times Z \to \overline{V}$ defined by

$$\Phi^{C_L}(x,z) = \begin{cases} f(x), & \text{if } x \in S, g(x) \in z - C, \\ +\infty_K, & \text{otherwise.} \end{cases}$$

For $v^* \in K^{*0}$ the formula for the conjugate of $(v^*\Phi^{C_L})$ can be deduced from (3.6) and in this way one obtains the following *Lagrange type vector dual problem to* (PV^C)

$$(DV^{C_L}) \quad \underset{(v^*,z^*,v) \in \mathcal{B}^{C_L}}{\text{Max}} h^{C_L}(v^*,z^*,v),$$

where

$$\mathcal{B}^{C_L} = \left\{(v^*,z^*,v) \in K^{*0} \times C^* \times V : \langle v^*, v\rangle \leq \inf_{x \in S}\{(v^*f)(x) + (z^*g)(x)\}\right\}$$

and

$$h^{C_L}(v^*,z^*,v) = v.$$

The weak, strong and converse duality results given in the previous subsection in the general case lead to the following statement.

Theorem 4.3.5. *(a) There is no $x \in \mathcal{A}$ and no $(v^*,z^*,v) \in \mathcal{B}^{C_L}$ such that $f(x) \leq_K h^{C_L}(v^*,z^*,v)$.*

(b) If (RCV^{C_L}) is fulfilled and $\bar{x} \in \mathcal{A}$ is a properly efficient solution to (PV^C), then there exists $(\bar{v}^*,\bar{z}^*,\bar{v})$, an efficient solution to (DV^{C_L}), such that $f(\bar{x}) = h^{C_L}(\bar{v}^*,\bar{z}^*,\bar{v}) = \bar{v}$.

(c) If (RCV^{C_L}) is fulfilled, $f(\operatorname{dom} f \cap \mathcal{A}) + K$ is closed and $(\bar{v}^*,\bar{z}^*,\bar{v})$ is an efficient solution to (DV^{C_L}), then there exists $\bar{x} \in \mathcal{A}$, a properly efficient solution to (PV^C), such that $f(\bar{x}) = h^{C_L}(\bar{v}^*,\bar{z}^*,\bar{v}) = \bar{v}$.

Remark 4.3.5. (*a*) The dual problem (DV^{C_L}) is one of the classical vector duality concepts which exist in the literature. It was introduced by Jahn in [101] (see also [103,104]), where also corresponding weak, strong and converse duality results have been proven.

(*b*) In Theorem 4.3.5(*b*) the regularity condition (RCV^{C_L}) can be replaced in case X and Z are Fréchet spaces, S is closed, f is star K-lower semicontinuous and g is C-epi closed with the condition $0 \in \operatorname{sqri}(g(\operatorname{dom} f \cap S \cap \operatorname{dom} g) + C)$.

If $\operatorname{lin}(g(\operatorname{dom} f \cap S \cap \operatorname{dom} g) + C)$ is a finite dimensional linear subspace one can assume instead that $0 \in \operatorname{ri}(g(\operatorname{dom} f \cap S \cap \operatorname{dom} g) + C)$. As noticed in Remark 4.3.2 in the general case, in Theorem 4.3.5(*c*) the regularity condition can be replaced with the assumption that for all $v^* \in K^{*0}$ the optimization problem

$$\inf_{x \in \mathcal{A}} \langle v^*, f(x) \rangle$$

is normal with respect to its Lagrange dual (see also [101, Theorem 2.5]).

(*c*) If $V = \mathbb{R}$ and $K = \mathbb{R}_+$, then (PV^C) turns out to be the scalar optimization problem with geometric and cone constraints (P^C), while (DV^{C_L}) is nothing else than the classical scalar Lagrange dual problem to (P^C) (cf. subsection 3.1.3).

The second vector perturbation function we consider in this subsection is $\Phi^{C_F} : X \times X \to \overline{V}$,

$$\Phi^{C_F}(x, y) = \begin{cases} f(x + y), & \text{if } x \in \mathcal{A}, \\ +\infty_K, & \text{otherwise}, \end{cases}$$

which leads by taking into consideration (3.7) to the following so-called *Fenchel type vector dual problem to* (PV^C) (cf. [36, 37])

$$(DV^{C_F}) \qquad \operatorname*{Max}_{(v^*, y^*, v) \in \mathcal{B}^{C_F}} h^{C_F}(v^*, y^*, v),$$

where

$$\mathcal{B}^{C_F} = \left\{ (v^*, y^*, v) \in K^{*0} \times X^* \times V : \langle v^*, v \rangle \leq -(v^* f)^*(y^*) - \sigma_{\mathcal{A}}(-y^*) \right\}$$

and

$$h^{C_F}(v^*, y^*, v) = v.$$

Considering as regularity condition

$$(RCV^{C_F}) \mid \exists x' \in \operatorname{dom} f \cap \mathcal{A} \text{ such that } f \text{ is continuous at } x',$$

the results in the previous subsection can be summarized to the following theorem.

Theorem 4.3.6. *(a) There is no $x \in \mathcal{A}$ and no $(v^*, y^*, v) \in \mathcal{B}^{C_F}$ such that $f(x) \leq_K h^{C_F}(v^*, y^*, v)$.*

(b) If (RCV^{C_F}) is fulfilled and $\bar{x} \in \mathcal{A}$ is a properly efficient solution to (PV^C), then there exists $(\bar{v}^*, \bar{y}^*, \bar{v})$, an efficient solution to (DV^{C_F}), such that $f(\bar{x}) = h^{C_F}(\bar{v}^*, \bar{y}^*, \bar{v}) = \bar{v}$.

(c) If (RCV^{C_F}) is fulfilled, $f(\operatorname{dom} f \cap \mathcal{A}) + K$ is closed and $(\bar{v}^*, \bar{y}^*, \bar{v})$ is an efficient solution to (DV^{C_F}), then there exists $\bar{x} \in \mathcal{A}$, a properly efficient solution to (PV^C), such that $f(\bar{x}) = h^{C_F}(\bar{v}^*, \bar{y}^*, \bar{v}) = \bar{v}$.

Remark 4.3.6. (*a*) In statement (*b*) of Theorem 4.3.6 the regularity condition (RCV^{C_F}) can be replaced in case X is a Fréchet space, $\mathcal{A}$ is closed and f is star K-lower semicontinuous with the condition $0 \in \operatorname{sqri}(\operatorname{dom} f - \mathcal{A})$. If $\operatorname{lin}(\operatorname{dom} f - \mathcal{A})$ is a finite dimensional linear subspace one can assume instead that $0 \in \operatorname{ri}(\operatorname{dom} f - \mathcal{A})$. In Theorem 4.3.6(*c*) the regularity condition can be replaced with the assumption that for all $v^* \in K^{*0}$ the optimization problem

$$\inf_{x \in \mathcal{A}} \langle v^*, f(x) \rangle$$

is normal with respect to its Fenchel dual.

(*b*) If $V = \mathbb{R}$ and $K = \mathbb{R}_+$, then (DV^{C_F}) is nothing else than the scalar Fenchel dual problem to (P^C) (cf. subsection 3.1.3).

The third vector perturbation function we treat is $\Phi^{C_{FL}} : X \times X \times Z \to \overline{V}$,

$$\Phi^{C_{FL}}(x, y, z) = \begin{cases} f(x + y), & \text{if } x \in S, g(x) \in z - C, \\ +\infty_K, & \text{otherwise.} \end{cases}$$

The formula for the conjugate of $(v^* \Phi^{C_{FL}})$, when $v^* \in K^{*0}$, follows from (3.8) and it provides the following so-called *Fenchel-Lagrange type vector dual problem to* (PV^C) (cf. [36, 37])

$$(DV^{C_{FL}}) \qquad \operatorname*{Max}_{(v^*, y^*, z^*, v) \in \mathcal{B}^{C_{FL}}} h^{C_{FL}}(v^*, y^*, z^*, v),$$

where

$$\mathcal{B}^{C_{FL}} = \big\{(v^*, y^*, z^*, v) \in K^{*0} \times X^* \times C^* \times V : \\ \langle v^*, v \rangle \leq -(v^* f)^*(y^*) - (z^* g)^*_S(-y^*)\big\}$$

and

$$h^{C_{FL}}(v^*, y^*, z^*, v) = v.$$

We introduce the following regularity condition

$$(RCV^{C_{FL}}) \left| \begin{array}{l} \exists x' \in \operatorname{dom} f \cap S \text{ such that } f \text{ is continuous at } x' \\ \text{and } g(x') \in -\operatorname{int}(C) \end{array} \right.$$

and thus one can derive from the weak, strong and converse duality theorems in the previous subsection the following result.

Theorem 4.3.7. *(a) There is no $x \in \mathcal{A}$ and no $(v^*, y^*, z^*, v) \in \mathcal{B}^{C_{FL}}$ such that $f(x) \leq_K h^{C_{FL}}(v^*, y^*, z^*, v)$.*

(b) If $(RCV^{C_{FL}})$ is fulfilled and $\bar{x} \in \mathcal{A}$ is a properly efficient solution to (PV^C), then there exists $(\bar{v}^*, \bar{y}^*, \bar{z}^*, \bar{v})$, an efficient solution to $(DV^{C_{FL}})$, such that $f(\bar{x}) = h^{C_{FL}}(\bar{v}^*, \bar{y}^*, \bar{z}^*, \bar{v}) = \bar{v}$.

(c) If $(RCV^{C_{FL}})$ is fulfilled, $f(\operatorname{dom} f \cap \mathcal{A}) + K$ is closed and $(\bar{v}^*, \bar{y}^*, \bar{z}^*, \bar{v})$ is an efficient solution to $(DV^{C_{FL}})$, then there exists $\bar{x} \in \mathcal{A}$, a properly efficient solution to (PV^C), such that $f(\bar{x}) = h^{C_{FL}}(\bar{v}^*, \bar{y}^*, \bar{z}^*, \bar{v}) = \bar{v}$.

Remark 4.3.7. (*a*) If X and Z are Fréchet spaces, S is closed, f is star K-lower semicontinuous and g is C-epi closed, then the regularity condition in the statement (*b*) of Theorem 4.3.7 can be replaced with the condition $0 \in \operatorname{sqri}(\operatorname{dom} f \times C - \operatorname{epi}_{-C}(-g) \cap (S \times Z))$. If the linear subspace generated by $\operatorname{dom} f \times C - \operatorname{epi}_{-C}(-g) \cap (S \times Z)$ is finite dimensional, then one can assume instead, that 0 belongs to the relative interior of this set. More than that, in Theorem 4.3.7(*c*) $(RCV^{C_{FL}})$ can be weakened by assuming that for all $v^* \in K^{*0}$ the optimization problem

$$\inf_{x \in \mathcal{A}} \langle v^*, f(x) \rangle$$

is normal with respect to its Fenchel-Lagrange dual.

(*b*) In case $V = \mathbb{R}$ and $K = \mathbb{R}_+$, then $(DV^{C_{FL}})$ is exactly the scalar Fenchel-Lagrange dual problem to (P^C) (cf. subsection 3.1.3).

The last vector dual problem to (PV^C) investigated in this subsection is not related to any conjugate dual of the scalarized primal problem. What we do is in fact involving in the definition of its feasible set the scalarized problem itself. The vector dual looks like

$$(DV^{C_P}) \quad \operatorname*{Max}_{(v^*,v) \in \mathcal{B}^{C_P}} h^{C_P}(v^*, v),$$

where

$$\mathcal{B}^{C_P} = \left\{ (v^*, v) \in K^{*0} \times V : \langle v^*, v \rangle \leq \inf_{x \in \mathcal{A}} \langle v^*, f(x) \rangle \right\}$$

and

$$h^{C_P}(v^*, v) = v.$$

We omit the proofs of the weak and strong duality theorems since these assertions follow automatically.

Theorem 4.3.8. *There is no $x \in \mathcal{A}$ and no $(v^*, v) \in \mathcal{B}^{C_P}$ such that $f(x) \leq_K h^{C_P}(v^*, v)$.*

Theorem 4.3.9. *If $\bar{x} \in \mathcal{A}$ is a properly efficient solution to (PV^C), then there exists $(\bar{v}^*, \bar{v})$, an efficient solution to (DV^{C_P}), such that $f(\bar{x}) = h^{C_P}(\bar{v}^*, \bar{v}) = \bar{v}$.*

The proofs of the two statements in the next result can be made in analogy to the ones of Theorem 4.3.3 and Theorem 4.3.4, respectively.

Theorem 4.3.10. *(a) Assume that $\mathcal{B}^{C_P}$ is nonempty. Then*

$$V \setminus \operatorname{cl}(f(\operatorname{dom} f \cap \mathcal{A}) + K) \subseteq \operatorname{core}(h^{C_P}(\mathcal{B}^{C_P})).$$

(b) If $f(\operatorname{dom} f \cap \mathcal{A}) + K$ is closed and $(\bar{v}^*, \bar{v})$ is an efficient solution to (DV^{C_P}), then there exists $\bar{x} \in \mathcal{A}$, a properly efficient solution to (PV^C), such that $f(\bar{x}) = h^{C_P}(\bar{v}^*, \bar{v}) = \bar{v}$.

Remark 4.3.8. Different to the strong and converse duality theorems given in this subsection for Theorem 4.3.9 and Theorem 4.3.10 no regularity condition is needed.

4.3.3 The relations between the dual vector problems to (PV^C)

In this subsection we investigate some inclusion relations between the image sets of the feasible sets through the objective functions of the vector duals of (PV^C) introduced in this chapter. We start by proving two general results which are fulfilled without any further assumptions.

Proposition 4.3.11. *It holds*

(a) $h^{C_{FL}}(\mathcal{B}^{C_{FL}}) \subseteq h^{C_L}(\mathcal{B}^{C_L})$;
(b) $h^{C_{FL}}(\mathcal{B}^{C_{FL}}) \subseteq h^{C_F}(\mathcal{B}^{C_F})$.

Proof. Let $(v^*, y^*, z^*, v) \in \mathcal{B}^{C_{FL}}$ be arbitrarily chosen. We have (see also the proof of Proposition 3.1.5)

$$\langle v^*, v\rangle \leq -(v^*f)^*(y^*) - (z^*g)^*_S(-y^*)$$

$$\leq \inf_{x\in X}\{(v^*f)(x) + ((z^*g) + \delta_S)(x)\} = \inf_{x\in S}\{(v^*f)(x) + (z^*g)(x)\}$$

and so $v = h^{C_L}(v^*, z^*, v) \in h^{C_L}(\mathcal{B}^{C_L})$.

On the other hand, it holds (see also the proof of Proposition 3.1.6)

$$\langle v^*, v\rangle \leq -(v^*f)^*(y^*) - (z^*g)^*_S(-y^*) = -(v^*f)^*(y^*)$$

$$+ \inf_{x\in S}\{\langle y^*, x\rangle + \langle z^*, g(x)\rangle\} \leq -(v^*f)^*(y^*) + \inf_{x\in \mathcal{A}}\langle y^*, x\rangle$$

$$= -(v^*f)^*(y^*) - \sigma_{\mathcal{A}}(-y^*),$$

which means that $v = h^{C_F}(v^*, y^*, v) \in h^{C_F}(\mathcal{B}^{C_F})$. □

Proposition 4.3.12. *It holds*

(a) $h^{C_L}(\mathcal{B}^{C_L}) \subseteq h^{C_P}(\mathcal{B}^{C_P})$;
(b) $h^{C_F}(\mathcal{B}^{C_F}) \subseteq h^{C_P}(\mathcal{B}^{C_P})$.

Proof. (a) For $(v^*, z^*, v) \in \mathcal{B}^{C_L}$ one has

$$\langle v^*, v\rangle \leq \inf_{x\in S}\{(v^*f)(x) + (z^*g)(x)\} \leq \inf_{x\in \mathcal{A}}(v^*f)(x),$$

which guarantees that $v = h^{C_P}(v^*, v) \in h^{C_P}(\mathcal{B}^{C_P})$.

(b) For $(v^*, y^*, v) \in \mathcal{B}^{C_F}$ one has

$$\langle v^*, v\rangle \leq -(v^*f)^*(y^*) - \sigma_{\mathcal{A}}(-y^*) \leq \inf_{x\in X}\{(v^*f)(x) + \delta_{\mathcal{A}}(x)\} = \inf_{x\in \mathcal{A}}(v^*f)(x),$$

and so $v = h^{C_P}(v^*, v) \in h^{C_P}(\mathcal{B}^{C_P})$. □

Proposition 4.3.11 and Proposition 4.3.12 provide the following general scheme (cf. [36, 37])

$$h^{C_{FL}}(\mathcal{B}^{C_{FL}}) \subseteq \begin{matrix} h^{C_L}(\mathcal{B}^{C_L}) \\ h^{C_F}(\mathcal{B}^{C_F}) \end{matrix} \subseteq h^{C_P}(\mathcal{B}^{C_P}). \tag{4.18}$$

We invite the reader to consult [36, 37] for examples which show that in general the inclusion relations above can be strict. Further we give some sufficient conditions which close the "gaps" between the sets involved in (4.18).

Theorem 4.3.13. *Assume that there exists $x' \in \operatorname{dom} f \cap S \cap \operatorname{dom} g$ such that f is continuous at x'. Then $h^{C_{FL}}(\mathcal{B}^{C_{FL}}) = h^{C_L}(\mathcal{B}^{C_L})$.*

Proof. As follows from Proposition 4.3.11(a), it is enough to prove that for an arbitrary $v \in h^{C_L}(\mathcal{B}^{C_L})$ it holds $v \in h^{C_{FL}}(\mathcal{B}^{C_{FL}})$. Let be $v \in h^{C_L}(\mathcal{B}^{C_L})$ and $(v^*, z^*) \in K^{*0} \times C^*$ such that $(v^*, z^*, v) \in \mathcal{B}^{C_L}$ or, equivalently,

$$\langle v^*, v\rangle \leq \inf_{x\in S}\{(v^*f)(x) + (z^*g)(x)\} = \inf_{x\in X}\{(v^*f)(x) + ((z^*g) + \delta_S)(x)\}.$$

As $\operatorname{dom}((z^*g) + \delta_S) = S \cap \operatorname{dom} g$, by Theorem 3.2.6 follows that there exists $\bar{y}^* \in X^*$ fulfilling

$$\inf_{x\in X}\{(v^*f)(x) + ((z^*g) + \delta_S)(x)\} = \sup_{y^*\in X^*}\{-(v^*f)^*(y^*) - (z^*g)_S^*(-y^*)\}$$

$$= -(v^*f)^*(\bar{y}^*) - (z^*g)_S^*(-\bar{y}^*).$$

Thus $(v^*, \bar{y}^*, z^*, v) \in \mathcal{B}^{C_{FL}}$ and, consequently, $v \in h^{C_{FL}}(\mathcal{B}^{C_{FL}})$. □

Theorem 4.3.14. *Assume that there exists $x' \in \mathcal{A}$ such that $g(x') \in -\operatorname{int}(C)$. Then $h^{C_{FL}}(\mathcal{B}^{C_{FL}}) = h^{C_F}(\mathcal{B}^{C_F})$.*

Proof. As follows from Proposition 4.3.11(b), it is enough to prove that for an arbitrary $v \in h^{C_F}(\mathcal{B}^{C_F})$ if holds $v \in h^{C_{FL}}(\mathcal{B}^{C_{FL}})$. Let be $v \in h^{C_F}(\mathcal{B}^{C_F})$ and $(v^*, y^*) \in K^{*0} \times X^*$ such that $(v^*, y^*, v) \in \mathcal{B}^{C_F}$ or, equivalently,

$$\langle v^*, v\rangle \leq -(v^*f)^*(y^*) - \sigma_{\mathcal{A}}(-y^*).$$

By Theorem 3.2.9, there exists $\bar{z}^* \in C^*$ such that

$$\sigma_{\mathcal{A}}(-y^*) = -\inf_{x\in\mathcal{A}}\langle y^*, x\rangle = -\sup_{z^*\in C^*}\inf_{x\in S}\{\langle y^*, x\rangle + (z^*g)(x)\}$$

$$= -\inf_{x\in S}\{\langle y^*, x\rangle + (\bar{z}^*g)(x)\} = (\bar{z}^*g)_S^*(-y^*)$$

and this yields

$$\langle v^*, v\rangle \leq -(v^*f)^*(y^*) - (\bar{z}^*g)_S^*(-y^*).$$

Thus $(v^*, y^*, \bar{z}^*, v) \in \mathcal{B}^{C_{FL}}$ and, consequently, $v \in h^{C_{FL}}(\mathcal{B}^{C_{FL}})$. □

Combining the last two theorems it follows that in case $(RCV^{C_{FL}})$ is fulfilled, then $h^{C_{FL}}(\mathcal{B}^{C_{FL}}) = h^{C_F}(\mathcal{B}^{C_F}) = h^{C_L}(\mathcal{B}^{C_L})$. The next theorem shows that actually all inclusion relations in (4.18) are in fact equalities (see also [36, 37]).

Theorem 4.3.15. *Assume that the regularity condition* $(RCV^{C_{FL}})$ *is fulfilled. Then*

$$h^{C_{FL}}(\mathcal{B}^{C_{FL}}) = h^{C_F}(\mathcal{B}^{C_F}) = h^{C_L}(\mathcal{B}^{C_L}) = h^{C_P}(\mathcal{B}^{C_P}). \tag{4.19}$$

Consequently, under this hypothesis the maximal sets of the image sets of the feasible set through the objective functions of the vector dual problems $(DV^{C_{FL}}), (DV^{C_F}), (DV^{C_L})$ *and* (DV^{C_P}) *are identical.*

Proof. What we prove is that for an arbitrary $v \in h^{C_P}(\mathcal{B}^{C_P})$ it holds $v \in h^{C_{FL}}(\mathcal{B}^{C_{FL}})$. Let be $v \in h^{C_P}(\mathcal{B}^{C_P})$ and $v^* \in K^{*0}$ such that $(v^*, v) \in \mathcal{B}^{C_P}$ or, equivalently,

$$\langle v^*, v\rangle \leq \inf_{x \in \mathcal{A}} \langle v^*, f(x)\rangle.$$

By Theorem 3.2.12, there exist $\bar{y}^* \in X^*$ and $\bar{z}^* \in C^*$ such that

$$\begin{aligned}\inf_{x \in \mathcal{A}} \langle v^*, f(x)\rangle &= \sup_{y^* \in X^*, z^* \in C^*} \{-(v^*f)^*(y^*) - (z^*g)^*_S(-y^*)\} \\ &= -(v^*f)^*(\bar{y}^*) - (\bar{z}^*g)^*_S(-\bar{y}^*),\end{aligned}$$

which yields $(v^*, \bar{y}^*, \bar{z}^*, v) \in \mathcal{B}^{C_{FL}}$ and, consequently, $v \in h^{C_{FL}}(\mathcal{B}^{C_{FL}})$. □

Remark 4.3.9. (*a*) If X is a Fréchet space, $S \cap \operatorname{dom} g$ is closed and f is star K-lower semicontinuous, then the regularity condition in Theorem 4.3.13 can be replaced with the assumption that $0 \in \operatorname{sqri}(\operatorname{dom} f - (S \cap \operatorname{dom} g))$. If the linear subspace $\operatorname{lin}(\operatorname{dom} f - (S \cap \operatorname{dom} g))$ is finite dimensional, then one can assume that $0 \in \operatorname{ri}(\operatorname{dom} f - (S \cap \operatorname{dom} g))$.

(*b*) If X and Z are Fréchet spaces, S is closed and g is C-epi closed, then the regularity condition in Theorem 4.3.14 can be replaced with the assumption that $0 \in \operatorname{sqri}(g(S \cap \operatorname{dom} g) + C)$. If the linear subspace $\operatorname{lin}(g(S \cap \operatorname{dom} g) + C)$ is finite dimensional, then one can assume that $0 \in \operatorname{ri}(g(S \cap \operatorname{dom} g) + C)$. Assuming additionally that f is star K-lower semicontinuous, then one can ask instead of $(RCV^{C_{FL}})$ in Theorem 4.3.15 that $0 \in \operatorname{sqri}(\operatorname{dom} f \times C - \operatorname{epi}_{(-C)}(-g) \cap (S \times Z))$. If the linear subspace spanned by $\operatorname{dom} f \times C - \operatorname{epi}_{(-C)}(-g) \cap (S \times Z)$ has a finite dimension, then for guaranteeing the conclusion in Theorem 4.3.15 one needs only to assume that 0 belongs to the relative interior of this set.

(*c*) Consider $X = \mathbb{R}^n$, $Z = \mathbb{R}^m$, $C = \mathbb{R}^m_+$, $S \subseteq \mathbb{R}^n$ a nonempty convex set, $g : \mathbb{R}^n \to \mathbb{R}^m$ a given $\mathbb{R}^m_+$-convex function and $f_i : \mathbb{R}^n \to \overline{\mathbb{R}}$, $i = 1, ..., k$, proper and convex functions such that $\bigcap_{i=1}^k \operatorname{dom} f_i \cap S \cap g^{-1}(-\mathbb{R}^m_+) \neq \emptyset$. Let also be $V = \mathbb{R}^k$, $K = \mathbb{R}^k_+$, $\overline{V} = \overline{\mathbb{R}^k} = \mathbb{R}^k \cup \{\pm\infty_{\mathbb{R}^k_+}\}$ and $f : \mathbb{R}^n \to \overline{\mathbb{R}^k}$, defined by

$$f(x) = \begin{cases} (f_1(x), ..., f_k(x))^T, & \text{if } x \in \bigcap\limits_{i=1}^{k} \operatorname{dom} f_i, \\ +\infty_{\mathbb{R}^k_+}, & \text{otherwise.} \end{cases}$$

In this setting we introduce the following primal vector optimization problem

$$(\widetilde{PV}^C) \quad \underset{x \in \mathcal{A}}{\operatorname{Min}} f(x).$$
$$\mathcal{A} = \{x \in S : g(x) \leqq 0\}$$

As in the general case it is possible to construct for $(\widetilde{PV}^C)$ different vector dual problems in analogy to the ones introduced in this section. By Theorem 3.2.14 follows that, in order to have for these duals the equality in (4.19) fulfilled, it is enough to impose a sufficient condition which in this particular case looks like

$$(\widetilde{RCV}^{C_{FL}}) \left| \begin{array}{l} \exists x' \in \operatorname{ri}\left(\bigcap\limits_{i=1}^{k} \operatorname{dom} f_i\right) \cap \operatorname{ri}(S) \text{ such that} \\ g_i(x') \leq 0, i \in L, \text{ and } g_i(x') < 0, i \in N, \end{array}\right.$$

where L and N are the sets of indices defined in the end of section 3.2.

In the remaining part of the section we investigate the relations between the classical vector duality concepts due to Jahn and Nakayama for the vector optimization problem with geometric and cone constraints. We start by proving the following general result.

Proposition 4.3.16. *It holds* $h^{C_L}(\mathcal{B}^{C_L}) \subseteq h^{C_N}(\mathcal{B}^{C_N})$.

Proof. Let be an arbitrary $v \in h^{C_L}(\mathcal{B}^{C_L})$. Then there exists $(v^*, z^*) \in K^{*0} \times C^*$ such that $(v^*, z^*, v) \in \mathcal{B}^{C_L}$ or, equivalently, $\langle v^*, v\rangle \leq \inf_{x\in S}\{(v^*f)(x) + (z^*g)(x)\}$. Like in the proof of Theorem 4.2.2 one can provide an $U \in \mathcal{L}(Z, V)$ fulfilling $U^*v^* = z^*$ and $U(C) \subseteq K$. We prove that $(U, v) \in \mathcal{B}^{C_N}$ and to this end we assume the contrary, namely that there exists $x \in S \cap \operatorname{dom} g$ such that $v \geq_K f(x) + U(g(x))$. Thus

$$\langle v^*, v\rangle > \langle v^*, f(x)\rangle + \langle U^*v^*, g(x)\rangle = (v^*f)(x) + (z^*g)(x),$$

which leads to a contradiction. In conclusion, $v = h^{C_N}(U, v) \in h^{C_N}(\mathcal{B}^{C_N})$. □

Example 4.3.1. (cf. [37]) Let be $X = V = \mathbb{R}^2$, $Z = \mathbb{R}$, $C = \mathbb{R}_+$, $K = \mathbb{R}^2_+$,

$$S = \{x = (x_1, x_2)^T \in \mathbb{R}^2 : x_1 \geq 0, x_2 \geq 0 \text{ such that } x_2 > 0 \text{ if } x_1 \in [0, 1)\},$$

$g : \mathbb{R}^2 \to \mathbb{R}$, $g \equiv 0$ and $f_1, f_2 : \mathbb{R}^2 \to \overline{\mathbb{R}}$ defined by

$$f_1(x) = \begin{cases} x_1, & \text{if } x = (x_1, x_2)^T \in S, \\ +\infty, & \text{otherwise,} \end{cases}$$

and

$$f_2(x) = \begin{cases} x_2, & \text{if } x = (x_1, x_2)^T \in S, \\ +\infty, & \text{otherwise.} \end{cases}$$

For the dual problems of the primal vector optimization problem

$$\begin{array}{l} \operatorname*{Min}_{x \in \mathcal{A}} \begin{pmatrix} f_1(x) \\ f_2(x) \end{pmatrix} \\ \mathcal{A} = \{x \in S : g(x) \leq 0\} \end{array}$$

it holds $h^{C_{FL}}(\mathcal{B}^{C_{FL}}) = h^{C_F}(\mathcal{B}^{C_F}) = h^{C_L}(\mathcal{B}^{C_L}) = h^{C_P}(\mathcal{B}^{C_P})$. This is the case because of the fact that the regularity condition stated for (PV^C) in the particular formulation from Remark 4.3.9(c) is fulfilled (for $x' = (1,1)^T \in \operatorname{ri}(S)$ it holds $g(x') \leq 0$). In this particular case Nakayama's vector dual problem looks like

$$(DV^{C_N}) \quad \operatorname*{Max}_{(q,v) \in \mathcal{B}^{C_N}} h^{C_N}(q, v),$$

where

$$\begin{aligned} \mathcal{B}^{C_N} = \big\{(q,v) \in \mathbb{R}^2_+ \times \mathbb{R}^2 : q = (q_1, q_2)^T \text{ and } \\ \nexists x \in S \text{ such that } v \geq f(x) + (q_1 g(x), q_2 g(x))^T\big\}, \end{aligned}$$

$$= \{(q,v) \in \mathbb{R}^2_+ \times \mathbb{R}^2 : \nexists x \in S \text{ such that } v \geq x\}$$

and

$$h^{C_N}(q, v) = v.$$

Thus $h^{C_N}(\mathcal{B}^{C_N}) = (\mathbb{R}^2 \setminus S) \cup \{(1,0)^T\}$ and $\operatorname{Max}(h^{C_N}(\mathcal{B}^{C_N}), \mathbb{R}^2_+) = \{(1,0)^T\}$. Assuming that $(1,0)^T \in h^{C_L}(\mathcal{B}^{C_L}) = h^{C_P}(\mathcal{B}^{C_P})$ it follows that there exists $v^* = (v_1^*, v_2^*)^T \in \operatorname{int}(\mathbb{R}^2_+)$ fulfilling $v_1^* \leq \inf_{x \in \mathcal{A}}\{v_1^* f_1(x) + v_2^* f_2(x)\} = \inf_{x \in S}\{v_1^* x_1 + v_2^* x_2\}$. Since for all $l \geq 1$, $(1/l, 1/l) \in \mathcal{A}$, it must hold $v_1^* \leq 0$ and so we come to a contradiction. Thus $(1,0)^T \notin h^{C_P}(\mathcal{B}^{C_P}) = h^{C_L}(\mathcal{B}^{C_L})$, which means that even having $(\widetilde{RCV}^{C_{FL}})$ fulfilled, for $h^{C_N}(\mathcal{B}^{C_N})$ and $h^{C_L}(\mathcal{B}^{C_L})$ one can have in general a strict inclusion. More than that, $\operatorname{Max}(h^{C_N}(\mathcal{B}^{C_N}), \mathbb{R}^2_+) \not\subseteq \operatorname{Max}(h^{C_L}(\mathcal{B}^{C_L}), \mathbb{R}^2_+)$.

On the other hand, for $v^* = (1,1)^T \in \operatorname{int}(\mathbb{R}^2_+)$ it holds $(v^*, (0,0)^T) \in \mathcal{B}^{C_P}$ and so $(0,0)^T \in h^{C_P}(\mathcal{B}^{C_P})$. Moreover, one can easily see that $(0,0)^T \in \operatorname{Max}(h^{C_P}(\mathcal{B}^{C_P}), \mathbb{R}^2_+) = \operatorname{Max}(h^{C_L}(\mathcal{B}^{C_L}), \mathbb{R}^2_+)$. As $(0,0)^T$ is not a maximal element in $h^{C_N}(\mathcal{B}^{C_N})$, one can conclude that in general $\operatorname{Max}(h^{C_L}(\mathcal{B}^{C_L}), \mathbb{R}^2_+) \not\subseteq \operatorname{Max}(h^{C_N}(\mathcal{B}^{C_N}), \mathbb{R}^2_+)$. Sufficient conditions for having equality between the maximal sets of $h^{C_L}(\mathcal{B}^{C_L})$ and $h^{C_N}(\mathcal{B}^{C_N})$ are given in the following theorem.

Theorem 4.3.17. *Assume that the regularity condition (RCV^{C_L}) is fulfilled and that the set $f(\operatorname{dom} f \cap \mathcal{A}) + K$ is closed. Then the following statements are true*

(a) $\operatorname{Max}(h^{C_L}(\mathcal{B}^{C_L}), K) \subseteq \operatorname{Max}(h^{C_N}(\mathcal{B}^{C_N}), K)$*;*
(b) if, additionally, $\operatorname{qi}(K) \neq \emptyset$ *and every efficient solution to* (PV^C) *is also properly efficient, then* $\operatorname{Max}(h^{C_L}(\mathcal{B}^{C_L}), K) = \operatorname{Max}(h^{C_N}(\mathcal{B}^{C_N}), K)$.

Proof. The statement (a) follows from Theorem 4.3.5(c) and Theorem 4.2.2, while statement (b) follows from Theorem 4.2.5 and Theorem 4.3.5(b). □

Combining the assertions in the theorem above the following corollary can be stated.

Corollary 4.3.18. *Assume that* $\operatorname{qi}(K) \neq \emptyset$, *the regularity condition* (RCV^{C_L}) *is fulfilled, the set* $f(\operatorname{dom} f \cap \mathcal{A}) + K$ *is closed and every efficient solution to* (PV^C) *is also properly efficient. Then it holds* $\operatorname{Min}(f(\operatorname{dom} f \cap \mathcal{A}), K) = \operatorname{Max}(h^{C_N}(\mathcal{B}^{C_N}), K) = \operatorname{Max}(h^{C_L}(\mathcal{B}^{C_L}), K)$.

As one will see in section 5.5, all the assumptions of Corollary 4.3.18 are for instance fulfilled when one deals with linear vector optimization problems in finite dimensional spaces.

Remark 4.3.10. (a) If X and Z are Fréchet spaces, S is closed, f is star K-lower semicontinuous and g is C-epi closed, then one can replace the regularity condition (RCV^{C_L}) in Theorem 4.3.17 and Corollary 4.3.18 with the assumption that $0 \in \operatorname{sqri}(g(\operatorname{dom} f \cap S \cap \operatorname{dom} g) + C)$. If the linear subspace $\operatorname{lin}(g(\operatorname{dom} f \cap S \cap \operatorname{dom} g) + C)$ is finite dimensional, then one can alternatively assume that $0 \in \operatorname{ri}(g(\operatorname{dom} f \cap S \cap \operatorname{dom} g) + C)$.

(b) Corollary 4.3.18 improves some similar statements given by Nakayama in [142–144] in finite dimensional spaces and under stronger assumptions (see also Remark 4.2.5(b)).

(c) Replacing in the formulation of Corollary 4.3.18 (RCV^{C_L}) by the regularity condition $(RCV^{C_{FL}})$, by Theorem 4.3.15 one can conclude that

$$\operatorname{Min}(f(\operatorname{dom} f \cap \mathcal{A}), K) = \operatorname{Max}(h^{C_N}(\mathcal{B}^{C_N}), K) = \operatorname{Max}(h^{C_P}(\mathcal{B}^{C_P}), K)$$

$$= \operatorname{Max}(h^{C_L}(\mathcal{B}^{C_L}), K) = \operatorname{Max}(h^{C_F}(\mathcal{B}^{C_F}), K) = \operatorname{Max}(h^{C_{FL}}(\mathcal{B}^{C_{FL}}), K).$$

4.3.4 Duality with respect to weakly efficient solutions

In the following we suppose that $\operatorname{int}(K) \neq \emptyset$ and give a general approach for constructing vector dual problems with respect to weakly efficient solutions. In particular we construct different vector dual problems to (PV^C_w) and study the relations between them. In these investigations the set of weakly maximal elements of $h^{C_N}_w(\mathcal{B}^{C_N}_w)$ will be also involved.

For $F : X \to \overline{V} = V \cup \{\pm\infty_K\}$ a proper and K-convex functions we consider the general vector optimization problem

$$(PVG_w) \quad \underset{x \in X}{\operatorname{WMin}}\, F(x).$$

We say that $\bar{x} \in X$ is a weakly efficient solution to (PVG_w) if $\bar{x} \in \operatorname{dom} F$ and $F(\bar{x}) \in \operatorname{WMin}(F(\operatorname{dom} F), K)$. We consider as vector perturbation function $\Phi : X \times Y \to \overline{V}$ and define the following vector dual problem to (PVG_w) with respect to the weakly efficient solutions as being

$$(DVG_w) \quad \underset{(v^*,y^*,v)\in\mathcal{B}_w^G}{\operatorname{WMax}} h_w^G(v^*,y^*,v),$$

where

$$\mathcal{B}_w^G = \{(v^*,y^*,v) \in (K^* \setminus \{0\}) \times Y^* \times V : \langle v^*, v\rangle \leq -(v^*\Phi)^*(0,-y^*)\}$$

and

$$h_w^G(v^*,y^*,v) = v.$$

As follows from the following two results, for (PVG_w) and (DVG_w) weak and strong duality hold.

Theorem 4.3.19. *There is no $x \in X$ and no $(v^*,y^*,v) \in \mathcal{B}_w^G$ such that $F(x) <_K h_w^G(v^*,y^*,v)$.*

Proof. We assume the contrary, namely that there exist $x \in X$ and $(v^*,y^*,v) \in \mathcal{B}_w^G$ such that $F(x) <_K h_w^G(v^*,y^*,v) = v$. Then it holds $x \in \operatorname{dom} F$ and $\langle v^*, v\rangle > \langle v^*, F(x)\rangle$.

On the other hand, $\langle v^*, v\rangle \leq -(v^*\Phi)^*(0,-y^*) \leq \langle v^*, F(x)\rangle$, and this leads to a contradiction. □

Theorem 4.3.20. *Assume that the regularity condition (RCV^Φ) is fulfilled. If $\bar{x} \in X$ is a weakly efficient solution to (PVG_w), then there exists $(\bar{v}^*,\bar{y}^*,\bar{v})$, a weakly efficient solution to (DVG_w), such that $F(\bar{x}) = h_w^G(\bar{v}^*,\bar{y}^*,\bar{v}) = \bar{v}$.*

Proof. Since $\bar{x} \in X$ is a weakly efficient solution to (PVG_w), then $\bar{x} \in \operatorname{dom} F$ and $F(\bar{x})$ is a weakly minimal element of the set $F(\operatorname{dom} F) \subseteq V$. By Corollary 2.4.26 (see also Remark 2.4.11), there exists $v^* \in K^* \setminus \{0\}$ which satisfies

$$\langle \bar{v}^*, F(\bar{x})\rangle = \inf_{x\in X} \langle \bar{v}^*, F(x)\rangle = \inf_{x\in X} (\bar{v}^*\Phi)(x,0).$$

Applying now Theorem 3.2.1 one gets that there exists $\bar{y}^* \in Y^*$ such that

$$\langle \bar{v}^*, F(\bar{x})\rangle = \inf_{x\in X} (\bar{v}^*\Phi)(x,0) = \sup_{y^*\in Y^*} \{-(\bar{v}^*\Phi)^*(0,-y^*)\} = -(\bar{v}^*\Phi)^*(0,-\bar{y}^*).$$

For $\bar{v} = F(\bar{x})$ one has $(\bar{v}^*,\bar{y}^*,\bar{v}) \in \mathcal{B}_w^G$. That $(\bar{v}^*,\bar{y}^*,\bar{v})$ is a weakly efficient solution to (DVG_w) follows by Theorem 4.3.19. □

Before stating the converse duality theorem we prove the following preliminary result.

Theorem 4.3.21. *Assume that the regularity condition (RCV^Φ) is fulfilled. Then*

$$V \setminus \operatorname{cl}(F(\operatorname{dom} F) + K) \subseteq \operatorname{core}(h_w^G(\mathcal{B}_w^G)).$$

Proof. Consider $\bar{v}$ be an arbitrary element in $V \setminus \mathrm{cl}\,(F(\mathrm{dom}\, F) + K)$. Since the set $\mathrm{cl}\,(F(\mathrm{dom}\, F) + K) \subseteq V$ is convex and closed, by Theorem 2.1.5 there exists $\bar{v}^* \in K^* \setminus \{0\}$ and $\alpha \in \mathbb{R}$ such that

$$\langle \bar{v}^*, \bar{v} \rangle < \alpha < \langle \bar{v}^*, v \rangle \ \forall v \in \mathrm{cl}\,(F(\mathrm{dom}\, F) + K).$$

Thus

$$\langle \bar{v}^*, \bar{v} \rangle < \alpha \leq \inf_{x \in X} (\bar{v}^* F)(x) = \inf_{x \in X} (\bar{v}^* \Phi)(x, 0)$$

and, consequently, there exists $\bar{y}^* \in Y^*$ such that

$$\langle \bar{v}^*, \bar{v} \rangle < -(\bar{v}^* \Phi)(0, -\bar{y}^*).$$

Obviously, $\bar{v} \in h_w^G(\mathcal{B}_w^G)$. Like in the proof of Theorem 4.3.3 it can be proven that in fact we have more, namely that $\bar{v} \in \mathrm{core}(h_w^G(\mathcal{B}_w^G))$. □

The converse duality theorem is a direct consequence of the previous result.

Theorem 4.3.22. *Assume that the regularity condition (RCV^{Φ}) is fulfilled and that the set $F(\mathrm{dom}\, F) + K$ is closed. Then for every weakly efficient solution $(\bar{v}^*, \bar{y}^*, \bar{v})$ to (DVG_w) one has that $\bar{v}$ is a weakly minimal element of the set $F(\mathrm{dom}\, F) + K$.*

Proof. Assuming that $\bar{v} \notin F(\mathrm{dom}\, F) + K$, by Theorem 4.3.21 follows that $\bar{v} \in \mathrm{core}(h_w^G(\mathcal{B}_w^G))$. Considering an element $k \in \mathrm{int}(K)$ there exists $\lambda > 0$ such that $v_\lambda := \bar{v} + \lambda k >_K \bar{v}$ and $v_\lambda \in h_w^G(\mathcal{B}_w^G)$. This contradicts the fact that $(\bar{v}^*, \bar{y}^*, \bar{v})$ is a weakly efficient solution to (DVG_w). Supposing that $\bar{v}$ is not a weakly minimal element of $F(\mathrm{dom}\, F) + K$, it follows that there exist $\bar{x} \in \mathrm{dom}\, F$ and $\bar{k} \in K$ satisfying $\bar{v} >_K F(\bar{x}) + \bar{k} \geqq_K F(\bar{x})$, but this contradicts Theorem 4.3.19. □

Remark 4.3.11. (*a*) The observations pointed out in Remark 4.3.1 and Remark 4.3.2 apply also for the strong and converse duality theorems, respectively, given for the primal-dual vector pair $(PVG_w) - (DVG_w)$.

(*b*) The duality approach developed in subsection 4.1.2 for (PV_w^A) follows as a particular case of this general scheme by considering $\Phi : X \times Y \to \overline{V}$, $\Phi(x, y) = f(x) + g(Ax + y)$.

(*c*) In case $V = \mathbb{R}$ and $K = \mathbb{R}_+$ the vector dual problem (DVG_w) has the following formulation

$$(DG) \quad \sup_{v^* > 0, y^* \in Y^*} \left\{ -\Phi^* \left(0, -\frac{1}{v^*} y^* \right) \right\}$$

or, equivalently,

$$(DG) \quad \sup_{y^* \in Y^*} \left\{ -\Phi^* \left(0, -y^* \right) \right\},$$

and this is again (see the remark in the end of subsection 4.3.1) the general conjugate dual problem to (PG).

Similarly to the approach in subsection 4.3.2 one can construct by using the general scheme different vector dual problems to (PV_w^C) with respect to weakly efficient solutions. Even if their formulations are close to the formulations of the vector duals in subsection 4.3.2, we introduce them here for the sake of completeness along with the theorems which provide the weak, strong and converse duality.

We start with the *Lagrange type vector dual problem to* (PV_w^C) *with respect to the weakly efficient solutions* (see also [104])

$$(DV_w^{C_L}) \qquad \underset{(v^*,z^*,v)\in\mathcal{B}_w^{C_L}}{\text{WMax}} \; h_w^{C_L}(v^*,z^*,v),$$

where

$$\mathcal{B}_w^{C_L} = \left\{(v^*,z^*,v) \in (K^*\backslash\{0\})\times C^*\times V : \langle v^*,v\rangle \leq \inf_{x\in S}\{(v^*f)(x)+(z^*g)(x)\}\right\}$$

and

$$h^{C_L}(v^*,z^*,v) = v.$$

The weak, strong and converse duality results given above lead to the following statement.

Theorem 4.3.23. *(a) There is no $x \in \mathcal{A}$ and no $(v^*,z^*,v) \in \mathcal{B}_w^{C_L}$ such that $f(x) <_K h_w^{C_L}(v^*,z^*,v)$.*

(b) If (RCV^{C_L}) is fulfilled and $\bar{x} \in \mathcal{A}$ is a weakly efficient solution to (PV_w^C), then there exists $(\bar{v}^*,\bar{z}^*,\bar{v})$, a weakly efficient solution to $(DV_w^{C_L})$, such that $f(\bar{x}) = h_w^{C_L}(\bar{v}^*,\bar{z}^*,\bar{v}) = \bar{v}$.

(c) If (RCV^{C_L}) is fulfilled, $f(\text{dom}\, f \cap \mathcal{A}) + K$ is closed and $(\bar{v}^*,\bar{z}^*,\bar{v})$ is a weakly efficient solution to $(DV_w^{C_L})$, then $\bar{v}$ is a weakly minimal element of the set $f(\text{dom}\, f \cap \mathcal{A}) + K$.

The *Fenchel type vector dual problem to* (PV_w^C) *with respect to the weakly efficient solutions* has the following formulation

$$(DV_w^{C_F}) \qquad \underset{(v^*,y^*,v)\in\mathcal{B}_w^{C_F}}{\text{WMax}} \; h_w^{C_F}(v^*,y^*,v),$$

where

$$\mathcal{B}_w^{C_F} = \left\{(v^*,y^*,v) \in (K^*\setminus\{0\})\times X^*\times V : \langle v^*,v\rangle \leq -(v^*f)^*(y^*) - \sigma_{\mathcal{A}}(-y^*)\right\}$$

and

$$h_w^{C_F}(v^*,z^*,v) = v.$$

Theorem 4.3.24. *(a) There is no $x \in \mathcal{A}$ and no $(v^*,y^*,v) \in \mathcal{B}_w^{C_F}$ such that $f(x) <_K h_w^{C_F}(v^*,y^*,v)$.*

(b) If (RCV^{C_F}) is fulfilled and $\bar{x} \in \mathcal{A}$ is a weakly efficient solution to (PV_w^C), then there exists $(\bar{v}^*,\bar{y}^*,\bar{v})$, a weakly efficient solution to $(DV_w^{C_F})$, such that $f(\bar{x}) = h_w^{C_F}(\bar{v}^*,\bar{y}^*,\bar{v}) = \bar{v}$.

(c) If (RCV^{C_F}) is fulfilled, $f(\operatorname{dom} f \cap \mathcal{A}) + K$ is closed and $(\bar{v}^*, \bar{y}^*, \bar{v})$ is a weakly efficient solution to $(DV_w^{C_F})$, then $\bar{v}$ is a weakly minimal element of the set $f(\operatorname{dom} f \cap \mathcal{A}) + K$.

The next vector dual problem we introduce is the *Fenchel-Lagrange type vector dual problem to (PV_w^C) with respect to the weakly efficient solutions*

$$(DV_w^{C_{FL}}) \qquad \operatorname*{WMax}_{(v^*,y^*,z^*,v) \in \mathcal{B}_w^{C_{FL}}} h_w^{C_{FL}}(v^*, y^*, z^*, v),$$

where

$$\mathcal{B}_w^{C_{FL}} = \big\{(v^*, y^*, z^*, v) \in (K^* \setminus \{0\}) \times X^* \times C^* \times V : \\ \langle v^*, v \rangle \leq -(v^* f)^*(y^*) - (z^* g)_S^*(-y^*)\big\}$$

and

$$h_w^{C_{FL}}(v^*, y^*, z^*, v) = v.$$

Theorem 4.3.25. *(a) There is no $x \in \mathcal{A}$ and no $(v^*, y^*, z^*, v) \in \mathcal{B}_w^{C_{FL}}$ such that $f(x) <_K h_w^{C_{FL}}(v^*, y^*, z^*, v)$.*

(b) If $(RCV^{C_{FL}})$ is fulfilled and $\bar{x} \in \mathcal{A}$ is a weakly efficient solution to (PV_w^C), then there exists $(\bar{v}^*, \bar{y}^*, \bar{z}^*, \bar{v})$, a weakly efficient solution to $(DV_w^{C_{FL}})$, such that $f(\bar{x}) = h_w^{C_{FL}}(\bar{v}^*, \bar{y}^*, \bar{z}^*, \bar{v}) = \bar{v}$.

(c) If $(RCV^{C_{FL}})$ is fulfilled, $f(\operatorname{dom} f \cap \mathcal{A}) + K$ is closed and $(\bar{v}^*, \bar{y}^*, \bar{z}^*, \bar{v})$ is a weakly efficient solution to $(DV_w^{C_{FL}})$, then $\bar{v}$ is a weakly minimal element of the set $f(\operatorname{dom} f \cap \mathcal{A}) + K$.

In analogy to (DV^{C_P}) we define the following dual vector problem to (PC_w^V) with respect to the weakly efficient solutions

$$(DV_w^{C_P}) \qquad \operatorname*{WMax}_{(v^*,v) \in \mathcal{B}_w^{C_P}} h_w^{C_P}(v^*, v),$$

where

$$\mathcal{B}_w^{C_P} = \left\{(v^*, v) \in (K^* \setminus \{0\}) \times V : \langle v^*, v \rangle \leq \inf_{x \in \mathcal{A}} \langle v^*, f(x) \rangle \right\}$$

and

$$h_w^{C_P}(v^*, v) = v.$$

Theorem 4.3.26. *(a) There is no $x \in \mathcal{A}$ and no $(v^*, v) \in \mathcal{B}_w^{C_P}$ such that $f(x) <_K h_w^{C_P}(v^*, v)$.*

(b) If $\bar{x} \in \mathcal{A}$ is a weakly efficient solution to (PV_w^C), then there exists $(\bar{v}^*, \bar{v})$, a weakly efficient solution to $(DV_w^{C_P})$, such that $f(\bar{x}) = h_w^{C_P}(\bar{v}^*, \bar{v}) = \bar{v}$.

(c) If $f(\operatorname{dom} f \cap \mathcal{A}) + K$ is closed and $(\bar{v}^*, \bar{v})$ is a weakly efficient solution to $(DV_w^{C_P})$, then $\bar{v}$ is a weakly minimal element of the set $f(\operatorname{dom} f \cap \mathcal{A}) + K$.

Remark 4.3.12. The observations pointed out in Remark 4.3.5$(b)-(c)$, Remark 4.3.6, Remark 4.3.7 and Remark 4.3.8 apply also for the vector duals $(DV_w^{C_L})$, $(DV_w^{C_F})$, $(DV_w^{C_{FL}})$ and $(DV_w^{C_P})$, respectively.

Following the proofs of Proposition 4.3.11 and Proposition 4.3.12 one can easily show that the following relations of inclusion hold

$$h_w^{C_{FL}}(\mathcal{B}_w^{C_{FL}}) \subseteq \begin{matrix} h_w^{C_L}(\mathcal{B}_w^{C_L}) \\ h_w^{C_F}(\mathcal{B}_w^{C_F}) \end{matrix} \subseteq h_w^{C_P}(\mathcal{B}_w^{C_P}). \tag{4.20}$$

In analogy to Theorem 4.3.13, Theorem 4.3.14 and Theorem 4.3.15 the following results can be shown.

Theorem 4.3.27. *Assume that there exists $x' \in \operatorname{dom} f \cap S \cap \operatorname{dom} g$ such that f is continuous at x'. Then $h_w^{C_{FL}}(\mathcal{B}_w^{C_{FL}}) = h_w^{C_L}(\mathcal{B}_w^{C_L})$.*

Theorem 4.3.28. *Assume that there exists $x' \in \mathcal{A}$ such that $g(x') \in -\operatorname{int}(C)$. Then $h_w^{C_{FL}}(\mathcal{B}_w^{C_{FL}}) = h_w^{C_F}(\mathcal{B}_w^{C_F})$.*

Theorem 4.3.29. *Assume that the regularity condition $(RCV^{C_{FL}})$ is fulfilled. Then*

$$h_w^{C_{FL}}(\mathcal{B}_w^{C_{FL}}) = h_w^{C_F}(\mathcal{B}_w^{C_F}) = h_w^{C_L}(\mathcal{B}_w^{C_L}) = h_w^{C_P}(\mathcal{B}_w^{C_P}). \tag{4.21}$$

Consequently, under this hypothesis the maximal sets of the image sets of the feasible set through the objective functions of the vector dual problems $(DV_w^{C_{FL}}), (DV_w^{C_F}), (DV_w^{C_L})$ and $(DV_w^{C_P})$ are identical.

Remark 4.3.13. The theorems above remain valid even if one replaces the regularity conditions supposed in their hypotheses with different generalized interior point conditions. To this end one has to consider the observation in Remark 4.3.9 which applies also here.

Next we investigate the connections between the duality concepts introduced by Nakayama and Jahn, this time with respect to the weakly efficient solutions.

Theorem 4.3.30. *It holds $h_w^{C_L}(\mathcal{B}_w^{C_L}) = h_w^{C_N}(\mathcal{B}_w^{C_N})$.*

Proof. That $h_w^{C_L}(\mathcal{B}_w^{C_L}) \subseteq h_w^{C_N}(\mathcal{B}_w^{C_N})$ can be proven like in Proposition 4.3.16, by taking into consideration that the interior of K is nonempty. We prove here the opposite inclusion and to this end we consider an arbitrary element $v \in h_w^{C_N}(\mathcal{B}_w^{C_N})$. Thus there exists $U \in \mathcal{L}_+(Z, V)$ with the property that there is no $x \in S \cap \operatorname{dom} g$ fulfilling $v >_K f(x) + U(g(x))$ or, equivalently,

$$(v - \operatorname{int}(K)) \cap (f + U \circ g)(\operatorname{dom} f \cap S \cap \operatorname{dom} g) = \emptyset$$

$$\Leftrightarrow \operatorname{int}(v - K) \cap ((f + U \circ g)(\operatorname{dom} f \cap S \cap \operatorname{dom} g) + K) = \emptyset.$$

The sets $v - K$ and $(f + U \circ g)(\operatorname{dom} f \cap S \cap \operatorname{dom} g) + K$ are convex subsets of V. By Theorem 2.1.2, there exist $v^* \in V^* \setminus \{0\}$ and $\alpha \in \mathbb{R}$ which satisfy

$$\langle v^*, v - k_1 \rangle \leq \alpha \leq \langle v^*, (f + U \circ g)(x) \rangle + \langle v^*, k_2 \rangle \tag{4.22}$$

for all $x \in \operatorname{dom} f \cap S \cap \operatorname{dom} g$ and all $k_1, k_2 \in K$. One can easily show that $v^* \in K^* \setminus \{0\}$. Taking in (4.22) $k_1 = k_2 = 0$ it yields

$$\langle v^*, v \rangle \leq \langle v^*, f(x) \rangle + \langle U^* v^*, g(x) \rangle \ \forall x \in \operatorname{dom} f \cap S \cap \operatorname{dom} g.$$

Denote $z^* := U^* v^*$. Since $U(C) \subseteq K$, for all $c \in C$ it holds $\langle U^* v^*, c \rangle = \langle v^*, Uc \rangle \geq 0$, which implies that $z^* = U^* v^* \in C^*$. Consequently, for the element $(v^*, z^*, v) \in (K^* \setminus \{0\}) \times C^* \times V$ one has

$$\langle v^*, v \rangle \leq \inf_{x \in S} \{(v^* f)(x) + (z^* g)(x)\},$$

which means that $(v^*, z^*, v) \in \mathcal{B}_w^{C_L}$ and so $v \in h_w^{C_L}(\mathcal{B}_w^{C_L})$. □

Remark 4.3.14. If $(RCV^{C_{FL}})$ is fulfilled, then Theorem 4.3.29 and Theorem 4.3.30 yield

$$h_w^{C_{FL}}(\mathcal{B}_w^{C_{FL}}) = h_w^{C_F}(\mathcal{B}_w^{C_F}) = h_w^{C_L}(\mathcal{B}_w^{C_L}) = h_w^{C_P}(\mathcal{B}_w^{C_P}) = h_w^{C_N}(\mathcal{B}_w^{C_N}).$$

Thus, under this hypothesis, the weakly maximal elements of these sets coincide. If, additionally, $f(\operatorname{dom} f \cap \mathcal{A}) + K$ is closed, then one has

$$\begin{aligned} \operatorname{WMin}(f(\operatorname{dom} f \cap \mathcal{A}), K) &\subseteq \operatorname{WMax}(h_w^{C_{FL}}(\mathcal{B}_w^{C_{FL}}), K) = \operatorname{WMax}(h_w^{C_F}(\mathcal{B}_w^{C_F}), K) \\ &= \operatorname{WMax}(h_w^{C_L}(\mathcal{B}_w^{C_L}), K) = \operatorname{WMax}(h_w^{C_P}(\mathcal{B}_w^{C_P}), K) \\ &= \operatorname{WMax}(h_w^{C_N}(\mathcal{B}_w^{C_N}), K) \subseteq \operatorname{WMin}(f(\operatorname{dom} f \cap \mathcal{A}) + K, K). \end{aligned}$$

4.4 Vector duality via a general scalarization

In this section we first develop a general duality scheme for the vector optimization problem with geometric and cone constraints

$$(PV^C) \quad \operatorname*{Min}_{x \in \mathcal{A}} f(x), \\ \mathcal{A} = \{x \in S : g(x) \in -C\}$$

in the same setting as considered in section 4.2 with respect to a general class of efficient solutions introduced via a general scalarization function. Throughout this section we assume that $\operatorname{int}(K) \neq \emptyset$. In this framework we prove weak and strong duality theorems and derive necessary and sufficient optimality conditions. As particular instances we consider some particular scalarization functions widely used in the literature on vector optimization.

4.4.1 A general duality scheme with respect to a general scalarization

Let $\mathcal{S}$ be an arbitrary set of proper and convex functions $s : V \cup \{+\infty_K\} \to \overline{\mathbb{R}}$ fulfilling $s(+\infty_K) = +\infty$, $f(\operatorname{dom} f \cap \mathcal{A}) + K \subseteq \operatorname{dom} s$ and such that s is K-strongly increasing on the set $f(\operatorname{dom} f \cap \mathcal{A}) + K$. Let us recall that s is called K-strongly increasing on $f(\operatorname{dom} f \cap \mathcal{A}) + K$ if for $x, y \in f(\operatorname{dom} f \cap \mathcal{A}) + K$ such that $x - y \in K$ and $x \neq y$ it holds $s(x) > s(y)$. The elements of the set $\mathcal{S}$ are called *scalarization functions*. Following the ideas in [72, 74, 79] we consider the following notion.

Definition 4.4.1. *An element $\bar{x} \in \mathcal{A}$ is said to be an $\mathcal{S}$-properly efficient solution to (PV^C) if $\bar{x} \in \operatorname{dom} f$ and there exists an $s \in \mathcal{S}$ such that $s(f(\bar{x})) \leq s(f(x))$ for all $x \in \mathcal{A}$.*

If $\bar{x} \in \mathcal{A}$ is an $\mathcal{S}$-properly efficient solution to (PV^C), then $\bar{x}$ is also an efficient solution to (PV^C). On the other hand, $\bar{x}$ is an optimal solution to the scalar optimization problem

$$(P^{C_s}) \quad \inf_{x \in \mathcal{A}} s(f(x)).$$

Next we construct a scalar conjugate dual problem to (P^{C_s}) which will be the starting point for defining a vector dual to (PV^C) with respect to the set of scalarization functions $\mathcal{S}$. The duality schemes for a convex composed optimization problem developed in section 3.4 cannot be used at this point, as they are applicable only in case $s \in \mathcal{S}$ is K-increasing on $V \cup \{+\infty_K\}$. But this assumption is too strong, being not fulfilled by the majority of the scalarization functions one can meet in the literature. Nevertheless, even if we assume less, a conjugate dual problem to (P^{C_s}) can be constructed.

We start by rewriting (P^{C_s}) in the following equivalent form

$$(P^{C_s}) \quad \inf_{\substack{x \in \mathcal{A}, y \in V, \\ f(x) - y \leqq_K 0}} s(y),$$

which is in fact nothing else than

$$(P^{C_s}) \quad \inf_{\substack{(x,y) \in S \times V, g(x) \in -C, \\ f(x) - y \leqq_K 0}} s(y).$$

The Lagrange dual problem to (P^{C_s}) is

$$\sup_{k^* \in K^*, z^* \in C^*} \inf_{x \in S, y \in V} \{s(y) + \langle z^*, g(x) \rangle + \langle k^*, f(x) - y \rangle\}$$

$$= \sup_{k^* \in K^*, z^* \in C^*} \left\{ \inf_{x \in S} \{(k^* f)(x) + (z^* g)(x)\} + \inf_{y \in V} \{\langle -k^*, y \rangle + s(y)\} \right\}$$

$$= \sup_{k^* \in K^*, z^* \in C^*} \left\{ -s^*(k^*) + \inf_{x \in X} \{(k^* f)(x) + ((z^* g) + \delta_S)(x)\} \right\}.$$

Replacing the infimum problem in the objective function of the optimization problem from above with its Fenchel dual we finally get the following conjugate dual problem to (P^{C_S})

$$(D^{C_S}) \qquad \sup_{\substack{y^* \in X^*, k^* \in K^*, \\ z^* \in C^*}} \{-s^*(k^*) - (k^* f)^*(y^*) - (z^* g)_S^*(-y^*)\}.$$

That $v(P^{C_S}) \geq v(D^{C_S})$ follows automatically from the construction of the dual. Next we investigate the existence of strong duality in case the regularity condition $(RCV^{C_{FL}})$ is fulfilled.

Theorem 4.4.1. *If the regularity condition $(RCV^{C_{FL}})$ is fulfilled, then it holds $v(P^{C_S}) = v(D^{C_S})$ and the dual has an optimal solution.*

Proof. The regularity condition assumed along with the hypothesis that $\operatorname{int}(K) \neq \emptyset$ guarantee that there exists $y' \in Y$ such that $y' \in f(x') + \operatorname{int}(K) \subseteq f(\operatorname{dom} f \cap \mathcal{A}) + K \subseteq \operatorname{dom} s$. Thus there exists $(x', y') \in (\operatorname{dom} f \cap S) \times \operatorname{dom} s$ with the property that $g(x') \in -\operatorname{int}(C)$ and $f(x') - y' \in -\operatorname{int}(K)$. By Theorem 3.2.12 follows that there exist $\bar{k}^* \in K^*$ and $\bar{z}^* \in C^*$ such that

$$v(P^{C_S}) = \sup_{k^* \in K^*, z^* \in C^*} \left\{ -s^*(k^*) + \inf_{x \in S} \{(k^* f)(x) + (z^* g)(x)\} \right\} = -s^*(\bar{k}^*)$$

$$+ \inf_{x \in S} \{(\bar{k}^* f)(x) + (\bar{z}^* g)(x)\} = -s^*(\bar{k}^*) + \inf_{x \in X} \{(\bar{k}^* f)(x) + ((\bar{z}^* g) + \delta_S)(x)\}.$$

Taking again into consideration $(RCV^{C_{FL}})$, as $x' \in \operatorname{dom} f \cap S \cap \operatorname{dom} g = \operatorname{dom}(\bar{k}^* f) \cap \operatorname{dom}((\bar{z}^* g) + \delta_S)$ and $(\bar{k}^* f)$ is continuous at x', by Theorem 3.2.6 there exists $\bar{y}^* \in X^*$ which satisfies

$$\inf_{x \in X} \{(\bar{k}^* f)(x) + ((\bar{z}^* g) + \delta_S)(x)\} = \sup_{y^* \in X^*} \{-(\bar{k}^* f)^*(y^*) - (\bar{z}^* g)_S^*(-y^*)\}$$

$$= -(\bar{k}^* f)^*(\bar{y}^*) - (\bar{z}^* g)_S^*(-\bar{y}^*).$$

In conclusion,

$$v(P^{C_S}) = v(D^{C_S}) = -s^*(\bar{k}^*) - (\bar{k}^* f)^*(\bar{y}^*) - (\bar{z}^* g)_S^*(-\bar{y}^*)$$

and $(\bar{y}^*, \bar{k}^*, \bar{z}^*)$ is an optimal solution to (D^{C_S}). □

Remark 4.4.1. (a) The dual problem (D^{C_S}) can be seen as a Fenchel-Lagrange type dual problem to (P^{C_S}), since when constructing it we consider first the Lagrange dual to (P^{C_S}) and after that the Fenchel dual to the inner infimum optimization problem which appears in the objective function of the Lagrange dual.

(b) By defining in an appropriate way a perturbation function for (P^{C_S}) one can obtain the conjugate dual (D^{C_S}) by means of the general duality approach developed in section 3.1. The general regularity conditions (RC_i^{Φ}), $i \in \{2, 2', 2'', 3, 4\}$, would provide regularity conditions for the primal-dual pair $(P^{C_S}) - (D^{C_S})$ even if the interiors of K and C are empty. We leave this to the reader, as in this section our main purpose is not necessary to refine the regularity condition $(RCV^{C_{FL}})$, but to develop a general duality scheme for (PV^C) with respect to different general scalarization functions.

The dual vector optimization problem to (PV^C) we investigate in this section is the following (cf. [29])

$$(DV^{C_S}) \qquad \underset{(s,y^*,k^*,z^*,v)\in\mathcal{B}^{C_S}}{\mathrm{Max}} h^{C_S}(s, y^*, k^*, z^*, v),$$

where

$$\mathcal{B}^{C_S} = \{(s, y^*, k^*, z^*, v) \in \mathcal{S} \times X^* \times K^* \times C^* \times V : \\ s(v) \leq -s^*(k^*) - (k^* f)^*(y^*) - (z^* g)_S^*(-y^*)\}$$

and

$$h^{C_S}(s, y^*, k^*, z^*, v) = v.$$

The weak and strong duality statements follow.

Theorem 4.4.2. *There is no $x \in \mathcal{A}$ and no $(s, y^*, k^*, z^*, v) \in \mathcal{B}^{C_S}$ such that $f(x) \leq_K h^{C_S}(s, y^*, k^*, z^*, v)$.*

Proof. We assume that there exist $x \in \mathcal{A}$ and $(s, y^*, k^*, z^*, v) \in \mathcal{B}^{C_S}$ such that $f(x) \leq_K h^{C_S}(s, y^*, k^*, z^*, v) = v$. It is obvious that $x \in \operatorname{dom} f$ and so $x \in \operatorname{dom} f \cap \mathcal{A}$. Thus $v \in f(\operatorname{dom} f \cap \mathcal{A}) + K$ and, as s is K-strongly increasing on $f(\operatorname{dom} f \cap \mathcal{A}) + K$, it follows that $s(f(x)) < s(v)$. On the other hand,

$$s(v) \leq -s^*(k^*) - (k^* f)^*(y^*) - (z^* g)_S^*(-y^*) \leq \inf_{x\in\mathcal{A}} s(f(x)),$$

and this leads to a contradiction. □

Theorem 4.4.3. *Assume that the regularity condition $(RCV^{C_{FL}})$ is fulfilled. If $\bar{x} \in \mathcal{A}$ is an $\mathcal{S}$-properly efficient solution to (PV^C), then there exists $(\bar{s}, \bar{y}^*, \bar{k}^*, \bar{z}^*, \bar{v})$, an efficient solution to (DV^{C_S}), such that $f(\bar{x}) = h^{C_S}(\bar{s}, \bar{y}^*, \bar{k}^*, \bar{z}^*, \bar{v}) = \bar{v}$.*

Proof. According to Definition 4.4.1 there exists $\bar{s} \in \mathcal{S}$ such that $\bar{x}$ is an optimal solution to the optimization problem

$$\inf_{x\in\mathcal{A}} \bar{s}(f(x)).$$

As $(RCV^{C_{FL}})$ is assumed, by Theorem 4.4.1 there exists $(\bar{y}^*, \bar{k}^*, \bar{z}^*) \in X^* \times K^* \times C^*$ such that

$$\bar{s}(f(\bar{x})) = -\bar{s}^*(\bar{k}^*) - (\bar{k}^* f)^*(\bar{y}^*) - (\bar{z}^* g)_S^*(-\bar{y}^*).$$

For $\bar{v} := f(\bar{x})$ one obviously has $(\bar{s}, \bar{y}^*, \bar{k}^*, \bar{z}^*, \bar{v}) \in \mathcal{B}^{C_S}$. That $(\bar{s}, \bar{y}^*, \bar{k}^*, \bar{z}^*, \bar{v})$ is an efficient solution to (DV^{C_S}) follows by Theorem 4.4.2. □

The next result provides necessary and sufficient optimality conditions for the $\mathcal{S}$-properly efficient solutions to (PV^C). As we will see in the next subsection the linear scalarization can be seen as a particular instance of the general approach, thus one can derive from Theorem 4.4.4 necessary and sufficient optimality conditions for the primal-dual pairs investigated in the previous sections of this chapter.

Theorem 4.4.4. *(a) Let $\bar{x} \in \mathcal{A}$ be an $\mathcal{S}$-properly efficient solution to (PV^C) and the regularity condition $(RC^{C_{FL}})$ be fulfilled. Then there exists $(\bar{s}, \bar{y}^*, \bar{k}^*, \bar{z}^*, \bar{v}) \in \mathcal{B}^{C_S}$, an efficient solution to (DV^{C_S}), such that*
(i) $f(\bar{x}) = \bar{v}$;
(ii) $s^*(\bar{k}^*) + \bar{s}(f(\bar{x})) = \langle \bar{k}^*, f(\bar{x}) \rangle$;
(iii) $(\bar{k}^* f)^*(\bar{y}^*) + (\bar{k}^* f)(\bar{x}) = \langle \bar{y}^*, \bar{x} \rangle$;
(iv) $(\bar{z}^* g)_S^*(-\bar{y}^*) = -\langle \bar{y}^*, \bar{x} \rangle$;
(v) $(\bar{z}^* g)(\bar{x}) = 0$.
(b) Assume that $\bar{x} \in \mathcal{A}$ and $(\bar{s}, \bar{y}^*, \bar{k}^*, \bar{z}^*, \bar{v}) \in \mathcal{B}^{C_S}$ fulfill the relations $(i)-(v)$. Then $\bar{x}$ is an $\mathcal{S}$-properly efficient solution to (PV^C) and $(\bar{s}, \bar{y}^*, \bar{k}^*, \bar{z}^*, \bar{v})$ is an efficient solution to the dual problem (DV^{C_S}).

Proof. The previous theorem yields the existence of an efficient solution $(\bar{s}, \bar{y}^*, \bar{k}^*, \bar{z}^*, \bar{v})$ to the dual problem (DV^{C_S}) such that $\bar{v} = f(\bar{x})$ and

$$\bar{s}(f(\bar{x})) + \bar{s}^*(\bar{k}^*) + (\bar{k}^* f)^*(\bar{y}^*) + (\bar{z}^* g)_S^*(-\bar{y}^*) = 0. \tag{4.23}$$

By the Young-Fenchel inequality one has

$$s^*(\bar{k}^*) + \bar{s}(f(x)) - \langle \bar{k}^*, f(x) \rangle \geq 0,$$

$$(\bar{k}^* f)^*(\bar{y}^*) + (\bar{k}^* f)(x) - \langle \bar{y}^*, \bar{x} \rangle \geq 0,$$

$$(\bar{z}^* g)_S^*(-\bar{y}^*) + (\bar{z}^* g)(\bar{x}) - \langle -\bar{y}^*, \bar{x} \rangle \geq 0,$$

while since $\bar{z}^* \in C^*$ and $g(\bar{x}) \in -C$ it follows

$$-(\bar{z}^* g)(\bar{x}) \geq 0.$$

The sum of the terms in the left hand side of the inequalities above is equal to 0 (cf. (4.23)) and this means that they all must be equal to 0. This proves that assertion in (a) is true.

The assertion in (b) follows immediately even without the fulfillment of $(RC^{C_{FL}})$, because summing up the equalities in $(ii)-(v)$ yields (4.23), which, along with (i), implies that $\bar{x}$ is an $\mathcal{S}$-properly efficient solution to (PV^C). Relation (4.23) and the weak duality property imply that $(\bar{s}, \bar{y}^*, \bar{k}^*, \bar{z}^*, \bar{v})$ is an efficient solution to (DV^{C_S}). □

Remark 4.4.2. The optimality conditions $(i)-(v)$ in Theorem 4.4.4 can be equivalently written as

$$\bar{v} = f(\bar{x}), \bar{k}^* \in \partial f(\bar{x}), \bar{y}^* \in \partial k(\bar{x}) \cap (-\partial((\bar{z}^* g) + \delta_S)(\bar{x})) \text{ and } (\bar{z}^* g)(\bar{x}) = 0.$$

In the remaining part of the subsection we turn our attention to the weakly efficient solutions to (PV^C). We notice first that $\bar{x}$ is a weakly efficient solution to (PV^C) if and only if it is an efficient solution to (PV^C) with respect to the nontrivial pointed convex cone $\widehat{K} = \text{int}(K) \cup \{0\}$. One can easily prove that $\text{int}(\widehat{K}) = \text{int}(K)$.

Let $\mathcal{T}$ be an arbitrary set of proper and convex functions $s : V \cup \{+\infty_K\} \to \overline{\mathbb{R}}$ such that $s(+\infty_K) = +\infty$, $f(\text{dom} f \cap \mathcal{A}) + K \subseteq \text{dom} s$ and s is K-strictly increasing on the set $f(\text{dom} f \cap \mathcal{A}) + K$. Let us recall that s is called K-strictly increasing on $f(\text{dom} f \cap \mathcal{A}) + K$ if s is K-increasing on $f(\text{dom} f \cap \mathcal{A}) + K$ and for $x, y \in f(\text{dom} f \cap \mathcal{A}) + K$ such that $x - y \in \text{int}(K)$ it holds $s(x) > s(y)$.

A vector dual problem to (PV^C) with respect to the set of scalarization functions $\mathcal{T}$ will be now introduced by replacing $\mathcal{S}$ with $\mathcal{T}$ in the feasible set of the dual $(DV^{C_\mathcal{S}})$. This new dual will be denoted by $(DV^{C_\mathcal{T}})$ and its feasible set and objective function will be denoted by $\mathcal{B}^{C_\mathcal{T}}$ and $h^{C_\mathcal{T}}$, respectively. We come now to the weak and strong duality theorems.

Theorem 4.4.5. *There is no $x \in \mathcal{A}$ and no $(s, y^*, k^*, z^*, v) \in \mathcal{B}^{C_\mathcal{T}}$ such that $f(x) <_K h^{C_\mathcal{T}}(s, y^*, k^*, z^*, v)$.*

Proof. Let be $x \in \mathcal{A}$ and $(s, y^*, k^*, z^*, v) \in \mathcal{B}^{C_\mathcal{T}}$ such that $f(x) <_K h^{C_\mathcal{T}}(s, y^*, k^*, z^*, v)$ or, equivalently, $f(x) \leq_{\widehat{K}} h^{C_\mathcal{T}}(s, y^*, k^*, z^*, v)$. The fact that s is K-strictly increasing implies that s is $\widehat{K}$-strongly increasing. The conclusion follows now by taking into consideration Theorem 4.4.2. □

Theorem 4.4.6. *Assume that the regularity condition $(RCV^{C_{FL}})$ is fulfilled. If $\bar{x} \in \mathcal{A}$ is a $\mathcal{T}$-properly efficient solution to (PV^C), then there exists $(\bar{s}, \bar{y}^*, \bar{k}^*, \bar{z}^*, \bar{v})$, a weakly efficient solution to $(DV^{C_\mathcal{T}})$, such that $f(\bar{x}) = h^{C_\mathcal{T}}(\bar{s}, \bar{y}^*, \bar{k}^*, \bar{z}^*, \bar{v}) = \bar{v}$.*

Proof. The result follows from Theorem 4.4.3 by using that every K-strictly increasing function is also $\widehat{K}$-strongly increasing and that every efficient solution with respect to $\widehat{K}$ is a weakly efficient solution with respect to K. □

By a similar argument like in the proof of the theorem above and using Theorem 4.4.4 we obtain the following characterization for $\mathcal{T}$-properly efficient solutions to (PV^C).

Theorem 4.4.7. *(a) Let $\bar{x} \in \mathcal{A}$ be a $\mathcal{T}$-properly efficient solution to (PV^C) and the regularity conditions $(RC^{C_{FL}})$ be fulfilled. Then there exists $(\bar{s}, \bar{y}^*, \bar{k}^*, \bar{z}^*, \bar{v}) \in \mathcal{B}^{C_\mathcal{T}}$, a weakly efficient solution to $(DV^{C_\mathcal{T}})$, such that the relations $(i)-(v)$ in Theorem 4.4.4 are fulfilled.*

(b) Assume that $\bar{x} \in \mathcal{A}$ *and* $(\bar{s}, \bar{y}^*, \bar{k}^*, \bar{z}^*, \bar{v}) \in \mathcal{B}^{C_\mathcal{T}}$ *fulfill the relations* $(i)-(v)$ *in Theorem 4.4.4. Then* $\bar{x}$ *is a* $\mathcal{T}$*-properly efficient solution to* (PV^C) *and* $(\bar{s}, \bar{y}^*, \bar{k}^*, \bar{z}^*, \bar{v})$ *is a weakly efficient solution to the dual problem* $(DV^{C_\mathcal{T}})$.

In the remaining part of the section we consider different particular classes of scalarization functions given in the literature and adapt the vector dual problem and the duality results introduced above to these scalarization concepts. The scalarizations we consider in the following are the linear scalarization, the maximum(-linear) scalarization, the set scalarization and the (semi)norm scalarization.

4.4.2 Linear scalarization

The *linear scalarization* is the most famous and widely used scalarization in vector optimization and assumes that the scalarization functions are strongly increasing linear continuous functions. We consider the following set of scalarization functions

$$\mathcal{S}_l = \{s_{v^*} : V \cup \{+\infty_K\} \to \overline{\mathbb{R}} : v^* \in K^{*0},\\ s_{v^*}(v) = \langle v^*, v\rangle \ \forall v \in V \cup \{+\infty_K\}\}.$$

By the conventions we made, for $s_{v^*} \in \mathcal{S}_l$ it holds $s_{v^*}(+\infty_K) = +\infty$. The $\mathcal{S}_l$-properly efficient solutions to (PV^C) are the classical properly efficient solutions to (PV^C) in the sense of linear scalarization used in the sections 4.2 and 4.3. Obviously, for all $v^* \in K^{*0}$, $f(\operatorname{dom} f \cap \mathcal{A}) + K \subseteq V = \operatorname{dom} s_{v^*}$ and s_{v^*} is K-strongly increasing, linear and continuous. Noticing that for all $k^* \in K^*$ one has $s_{v^*}^*(k^*) = \delta_{v^*}(k^*)$, the dual vector problem $(DV^{C_\mathcal{S}})$ becomes

$$(DV^{C_{\mathcal{S}_l}}) \qquad \operatorname*{Max}_{(v^*,y^*,z^*,v) \in \mathcal{B}^{C_{\mathcal{S}_l}}} h^{C_{\mathcal{S}_l}}(v^*, y^*, z^*, v),$$

where

$$\mathcal{B}^{C_{\mathcal{S}_l}} = \{(v^*, y^*, z^*, v) \in K^{*0} \times X^* \times C^* \times V :\\ \langle v^*, v\rangle \leq -(v^* f)^*(y^*) - (z^* g)^*_S(-y^*)\}$$

and

$$h^{C_{\mathcal{S}_l}}(v^*, y^*, z^*, v) = v.$$

It is easy to observe that $(DV^{C_{\mathcal{S}_l}})$ is exactly the Fenchel-Lagrange type dual problem $(DV^{C_{FL}})$ investigated in the section 4.3. The weak duality result Theorem 4.4.2 and the strong duality result Theorem 4.4.3 become the statements (a) and (b) in Theorem 4.3.7, respectively.

Now considering as set of scalarization functions

$$\mathcal{T}_l = \{s_{v^*} : V \cup \{+\infty_K\} \to \overline{\mathbb{R}} : v^* \in K^* \setminus \{0\},\\ s_{v^*}(v) = \langle v^*, v\rangle \ \forall v \in V \cup \{+\infty_K\}\},$$

it yields that every scalarization function $s_{v^*} \in \mathcal{T}_l$ is K-strictly increasing, linear and continuous, while its domain contains $f(\operatorname{dom} f \cap \mathcal{A}) + K$. Because of Corollary 2.4.26 (see also Remark 2.4.11) one can easily see that $\bar{x} \in \mathcal{A}$ is a $\mathcal{T}_l$-properly efficient solution to (PV^C) if and only if $\bar{x}$ is a weakly efficient solution to (PV^C). The dual which we introduce with respect to the set of scalarization functions $\mathcal{T}_l$ is

$$(DV^{C\mathcal{T}_l}) \qquad \underset{(v^*,y^*,z^*,v)\in\mathcal{B}^{C\mathcal{T}_l}}{\operatorname{Max}} h^{C\mathcal{T}_l}(v^*,y^*,z^*,v),$$

where

$$\mathcal{B}^{C\mathcal{T}_l} = \{(v^*,y^*,z^*,v) \in (K^* \setminus \{0\}) \times X^* \times C^* \times V : \\ \langle v^*, v\rangle \leq -(v^*f)^*(y^*) - (z^*g)^*_S(-y^*)\}$$

and

$$h^{C\mathcal{T}_l}(v^*,y^*,z^*,v) = v$$

and is nothing else than $(DV_w^{C_{FL}})$. By particularizing the weak duality result Theorem 4.4.5 and the strong duality result in Theorem 4.4.6 for the set of scalarization functions $\mathcal{T}_l$ we rediscover the statements (a) and (b) in Theorem 4.3.25, respectively.

4.4.3 Maximum(-linear) scalarization

In case V is a finite dimensional space one of the scalarizations one can meet especially in the applications of vector optimization is the so-called *Tchebyshev (or, maximum) scalarization.* We deal here with a more general scalarization function defined by combining a weighted maximum scalarization function (cf. [104,179]) with a linear function. This so-called *maximum-linear scalarization* function was also investigated by Mitani and Nakayama in [133].

Assume that $V = \mathbb{R}^k$, $K = \mathbb{R}^k_+$, $V \cup \{+\infty_K\} = \mathbb{R}^k \cup \{+\infty_{\mathbb{R}^k_+}\}$ and that $f_i : \mathbb{R}^n \to \overline{\mathbb{R}}$, $i = 1,...,k$, are proper and convex functions such that $\bigcap_{i=1}^k \operatorname{dom} f_i \cap S \cap g^{-1}(-C) \neq \emptyset$. Further we define $f : X \to \mathbb{R}^k \cup \{+\infty_{\mathbb{R}^k_+}\}$ as being (see also Remark 4.3.9(c))

$$f(x) = \begin{cases} (f_1(x),...,f_k(x))^T, & \text{if } x \in \bigcap\limits_{i=1}^k \operatorname{dom} f_i, \\ +\infty_{\mathbb{R}^k_+}, & \text{otherwise.} \end{cases}$$

Let also be $\eta \geq 0$. For $w = (w_1,...,w_k)^T \in \operatorname{int}(\mathbb{R}^k_+)$ and $a = (a_1,...,a_k)^T \in \mathbb{R}^k$ we consider the scalarization function $s_{w,a} : \mathbb{R}^k \cup \{+\infty_{\mathbb{R}^k_+}\} \to \overline{\mathbb{R}}$, defined by

$$s_{w,a}(y) = \max_{j=1,\dots,k} \{w_j(y_j - a_j)\} + \eta \sum_{j=1}^k w_j y_j, \; y = (y_1,...,y_k)^T \in \mathbb{R}^k,$$

whereby $s_{w,a}(+\infty_{\mathbb{R}^k_+}) = +\infty$. For all $w \in \operatorname{int}(\mathbb{R}^k_+)$ and $a \in \mathbb{R}^k$, $s_{w,a}$ is convex and $\mathbb{R}^k_+$-strictly increasing and fulfills $f\left(\cap_{i=1}^k \operatorname{dom} f_i \cap \mathcal{A}\right) + \mathbb{R}^k_+ \subseteq \mathbb{R}^k = \operatorname{dom} s$. We introduce the following set of scalarization functions

$$\mathcal{T}_{ml} = \{s_{w,a} : \mathbb{R}^k \cup \{+\infty_{\mathbb{R}^k_+}\} \to \overline{\mathbb{R}} : (w,a) \in \operatorname{int}(\mathbb{R}^k_+) \times \mathbb{R}^k\}.$$

An element $\bar{x} \in \mathcal{A}$ is a $\mathcal{T}_{ml}$-*properly efficient solution* to (PV^C) if there exist $w \in \operatorname{int}(\mathbb{R}^k_+)$ and $a \in \mathbb{R}^k$ such that $\max\limits_{j=1,\dots,k}\{w_j(f_j(\bar{x}) - a_j)\} + \eta \sum\limits_{j=1}^k w_j f_j(\bar{x}) \leq \max\limits_{j=1,\dots,k}\{w_j(f_j(x) - a_j)\} + \eta \sum\limits_{j=1}^k w_j f_j(x)$ for all $x \in \mathcal{A}$.

Let be $w = (w_1, \dots, w_k)^T \in \operatorname{int}(\mathbb{R}^k_+)$ and $a = (a_1, \dots, a_k)^T \in \mathbb{R}^k$ fixed. The conjugate function of $s_{w,a} \in \mathcal{T}_{ml}$ has, for $\beta \in \mathbb{R}^k$ the following formulation

$$\begin{aligned}
s^*_{w,a}(\beta) &= \sup_{y \in \mathbb{R}^k} \left\{ \beta^T y - \max_{j=1,\dots,k}\{w_j(y_j - a_j)\} - \eta \sum_{j=1}^k w_j y_j \right\} \\
&= \sup_{y \in \mathbb{R}^k} \left\{ (\beta - \eta w)^T y - \max_{j=1,\dots,k}\{w_j(y_j - a_j)\} \right\} \\
&= \sup_{u \in \mathbb{R}^k} \left\{ (\beta - \eta w)^T (u + a) - \max_{j=1,\dots,k}\{w_j u_j\} \right\} \\
&= \begin{cases} (\beta - \eta w)^T a, & \text{if } \eta w \leqq \beta \text{ and } \sum\limits_{j=1}^k \frac{\beta_j}{w_j} = k\eta + 1, \\ +\infty, & \text{otherwise.} \end{cases}
\end{aligned}$$

By identifying the scalarization function $s_{w,a} \in \mathcal{T}_{ml}$ with the pair (w,a) for $w \in \operatorname{int}(\mathbb{R}^k_+)$ and $a \in \mathbb{R}^k$, the dual vector problem to (PV^C) with respect to the set of scalarization functions $\mathcal{T}_{ml}$ is

$$(DV^{C\mathcal{T}_{ml}}) \qquad \operatorname*{WMax}_{(w,a,y^*,\beta,z^*,v) \in \mathcal{B}^{C\mathcal{T}_{ml}}} h^{C\mathcal{T}_{ml}}(w,a,y^*,\beta,z^*,v),$$

where

$$\begin{aligned}
\mathcal{B}^{C\mathcal{T}_{ml}} = \Bigg\{ & (w,a,y^*,\beta,z^*,v) \in \operatorname{int}(\mathbb{R}^k_+) \times \mathbb{R}^k \times X^* \times \mathbb{R}^k_+ \times C^* \times \mathbb{R}^k : \\
& \eta w \leqq \beta,\ \sum_{j=1}^k \frac{\beta_j}{w_j} = k\eta + 1,\ \max_{j=1,\dots,k}\{w_j(v_j - a_j)\} + \eta \sum_{j=1}^k w_j v_j \\
& \leq (\beta - \eta w)^T a - \left(\sum_{j=1}^k \beta_j f_j\right)^*(y^*) - (z^* g)^*_S(-y^*) \Bigg\}
\end{aligned}$$

and

$$h^{C\mathcal{T}_{ml}}(w,a,y^*,\beta,z^*,v) = v.$$

Theorem 4.4.5 and Theorem 4.4.6 lead to the following results, respectively.

Theorem 4.4.8. *There is no $x \in \mathcal{A}$ and no $(w, a, y^*, \beta, z^*, v) \in \mathcal{B}^{C_{\mathcal{T}_{ml}}}$ such that $f_i(x) < h_i^{C_{\mathcal{T}_{ml}}}(w, a, y^*, \beta, z^*, v) = v_i, i = 1, ..., k$.*

Theorem 4.4.9. *Assume that the regularity condition $(RCV^{C_{FL}})$ is fulfilled. If $\bar{x} \in \mathcal{A}$ is a $\mathcal{T}_{ml}$-properly efficient solution to (PV^C), then there exists $(\bar{w}, \bar{a}, \bar{y}^*, \bar{\beta}, \bar{z}^*, \bar{v})$, a weakly efficient solution to $(DV^{C_{\mathcal{T}_{ml}}})$, such that $f_i(\bar{x}) = h_i^{C_{\mathcal{T}_{ml}}}(\bar{w}, \bar{a}, \bar{y}^*, \bar{\beta}, \bar{z}^*, \bar{v}) = \bar{v}_i, i = 1, ..., k$.*

In case $\eta = 0$ the maximum-linear scalarization becomes the weighted Tchebyshev scalarization. Moreover, when the scalarization function is the maximum function, i.e. $w_j = 1$ and $a_j = 0$ for all $j = 1, ..., k$, then the set of scalarization functions has only one element, namely

$$\mathcal{T}_m = \left\{ s : \mathbb{R}^k \cup \{+\infty_{\mathbb{R}^k_+}\} \to \overline{\mathbb{R}} : s(y) = \max_{j=1,\ldots,k} y_j \ \forall y \in \mathbb{R}^k, s(\infty_{\mathbb{R}^k_+}) = +\infty \right\}$$

and an element $\bar{x} \in \mathcal{A}$ is said to be a *$\mathcal{T}_m$-properly efficient solution* to (PV^C) if $\max_{j=1,\ldots,k} f_j(\bar{x}) \leq \max_{j=1,\ldots,k} f_j(x)$ for all $x \in \mathcal{A}$. The dual vector problem to (PV^C) with respect to the set of scalarization functions $\mathcal{T}_m$ is in this particular case

$$(DV^{C_{\mathcal{T}_m}}) \qquad \underset{(y^*, \beta, z^*, v) \in \mathcal{B}^{C_{\mathcal{T}_m}}}{\text{WMax}} \quad h^{C_{\mathcal{T}_m}}(y^*, \beta, z^*, v),$$

where

$$\mathcal{B}^{C_{\mathcal{T}_m}} = \left\{ (y^*, \beta, z^*, v) \in X^* \times \mathbb{R}^k_+ \times C^* \times \mathbb{R}^k : \sum_{j=1}^k \beta_j = 1, \right.$$
$$\left. \max_{j=1,\ldots,k} \{v_j\} \leq -\left(\sum_{j=1}^k \beta_j f_j\right)^* (y^*) - (z^* g)^*_S(-y^*) \right\}$$

and

$$h^{C_{\mathcal{T}_m}}(y^*, \beta, z^*, v) = v.$$

The weak and strong duality theorems for the vector primal-dual pair (PV^C)–$(DV^{C_{\mathcal{T}_m}})$ follow as particular instances of Theorem 4.4.8 and 4.4.9, respectively.

4.4.4 Set scalarization

Under the name *set scalarization* we include those scalarization approaches for which the scalarization functions are defined by means of some given sets. We consider here a quite general scalarization function in connection to the one due to Gerth and Weidner (cf. [75]). This scalarization function was investigated also in [173, 175, 188].

Consider the nonempty convex set $E \subseteq V$ which satisfies $\text{cl}(E) + \text{int}(K) \subseteq \text{int}(E)$. For all $\mu \in \text{int}(K)$ we define $s_\mu : V \cup \{+\infty_K\} \to \overline{\mathbb{R}}$ by

$$s_\mu(v) = \inf\left\{t \in \mathbb{R} : v \in t\mu - \operatorname{cl}(E)\right\}.$$

Notice that $s_\mu(+\infty_K) = +\infty$. According to [75, 188], for $\mu \in \operatorname{int}(K)$ the function s_μ is convex, K-strictly increasing and takes only real values, thus $f(\operatorname{dom} f \cap \mathcal{A}) + K \subseteq V = \operatorname{dom} s_\mu$. Further, let be

$$\mathcal{T}_s = \{s_\mu : V \cup \{+\infty_K\} \to \overline{\mathbb{R}} : \mu \in \operatorname{int}(K)\}.$$

An element $\bar{x} \in \mathcal{A}$ is said to be a $\mathcal{T}_s$*-properly efficient solution* to (PV^C) if there exist $\mu \in \operatorname{int}(K)$ such that $s_\mu(f(\bar{x})) \leq s_\mu(f(x))$ for all $x \in \mathcal{A}$.

In order to formulate the vector dual problem to (PV^C) that arises in this case we need the conjugate function of s_μ, when $\mu \in \operatorname{int}(K)$ is fixed. It is $s_\mu^* : V^* \to \overline{\mathbb{R}}$,

$$s_\mu^*(k^*) = \sup_{v \in V}\left\{\langle k^*, v\rangle - \inf_{\substack{t \in \mathbb{R},\\ v \in t\mu - \operatorname{cl}(E)}} t\right\} = \sup_{\substack{v \in V, t \in \mathbb{R},\\ v \in t\mu - \operatorname{cl}(E)}} \{\langle k^*, v\rangle - t\}$$

$$= \sup_{t \in \mathbb{R}}\left\{-t + \sup_{u \in -\operatorname{cl}(E)} \langle k^*, u + t\mu\rangle\right\} = \sup_{t \in \mathbb{R}}\left\{t(\langle k^*, \mu\rangle - 1) + \sup_{u \in -\operatorname{cl}(E)} \langle k^*, u\rangle\right\}$$

$$= \begin{cases} \sigma_{-\operatorname{cl}(E)}(k^*), & \text{if } \langle k^*, \mu\rangle = 1,\\ +\infty, & \text{otherwise.}\end{cases}$$

Now we are able to formulate the vector dual problem attached to (PV^C) via the set scalarization. It is

$$(DV^{C\mathcal{T}_s}) \qquad \operatorname*{WMax}_{(\mu, y^*, k^*, z^*, v) \in \mathcal{B}^{C\mathcal{T}_s}} h^{C\mathcal{T}_s}(\mu, y^*, k^*, z^*, v),$$

where

$$\mathcal{B}^{C\mathcal{T}_s} = \{(\mu, y^*, k^*, z^*, v) \in \operatorname{int}(K) \times X^* \times K^* \times C^* \times V : \langle k^*, \mu\rangle = 1,$$
$$s_\mu(v) \leq -\sigma_{-\operatorname{cl}(E)}(k^*) - (k^*f)^*(y^*) - (z^*g)^*_S(-y^*)\}$$

and

$$h^{C\mathcal{T}_s}(\mu, y^*, k^*, z^*, v) = v.$$

The weak and strong duality theorems are particularizations of Theorem 4.4.5 and Theorem 4.4.6, respectively.

Theorem 4.4.10. *There is no $x \in \mathcal{A}$ and no $(\mu, y^*, k^*, z^*, v) \in \mathcal{B}^{C\mathcal{T}_s}$ such that $f(x) <_K h^{C\mathcal{T}_s}(\mu, y^*, k^*, z^*, v)$.*

Theorem 4.4.11. *Assume that the regularity condition $(RCV^{C_{FL}})$ is fulfilled. If $\bar{x} \in \mathcal{A}$ is a $\mathcal{T}_s$-properly efficient solution to (PV^C), then there exists $(\bar{\mu}, \bar{y}^*, \bar{k}^*, \bar{z}^*, \bar{v})$, a weakly efficient solution to $(DV^{C\mathcal{T}_s})$, such that $f(\bar{x}) = h^{C\mathcal{T}_s}(\bar{\mu}, \bar{y}^*, \bar{k}^*, \bar{z}^*, \bar{v}) = \bar{v}$.*

Let us come now to a special case of the set scalarization, the so-called *set scalarization with a conical set*. To this end we take $E = K$ and notice that, since K is a convex cone, the condition $\operatorname{cl}(E) + \operatorname{int}(K) \subseteq \operatorname{int}(K)$ is automatically as equality fulfilled. For all $\nu \in \operatorname{int}(K)$ we define $s_\nu : V \cup \{+\infty_K\} \to \overline{\mathbb{R}}$ by

$$s_\nu(v) = \inf \big\{ t \in \mathbb{R} : v \in t\nu - \operatorname{cl}(K) \big\}.$$

We notice that therefore $s_\nu(+\infty_K) = +\infty$. Let be

$$\mathcal{T}_{sc} = \{ s_\nu : V \cup \{+\infty_K\} \to \overline{\mathbb{R}} : \nu \in \operatorname{int}(K) \}.$$

An element $\bar{x} \in \mathcal{A}$ is said to be a *$\mathcal{T}_{sc}$-properly efficient solution* to (PV^C) if there exists $\nu \in \operatorname{int}(K)$ such that $s_\nu(f(\bar{x})) \leq s_\nu(f(x))$ for all $x \in \mathcal{A}$. Among the authors who have used this scalarization approach we cite here Kaliszewski (cf. [107]), Rubinov and Gasimov (cf. [162]) and Tammer (cf. [172]). Since $\sigma_{-\operatorname{cl}(K)} = \delta_{K^*}$, for all $\nu \in \operatorname{int}(K)$ the conjugate function of s_ν has the following formulation

$$s_\nu^*(k^*) = \begin{cases} 0, & \text{if } k^* \in K^*, \langle k^*, \nu \rangle = 1, \\ +\infty, & \text{otherwise}, \end{cases}$$

which leads to the following dual vector problem

$$(DV^{C\mathcal{T}_{sc}}) \qquad \underset{(\nu, y^*, k^*, z^*, v) \in \mathcal{B}^{C\mathcal{T}_{sc}}}{\operatorname{WMax}} \quad h^{C\mathcal{T}_{sc}}(\nu, y^*, k^*, z^*, v),$$

where

$$\mathcal{B}^{C\mathcal{T}_{sc}} = \{ (\nu, y^*, k^*, z^*, v) \in \operatorname{int}(K) \times X^* \times K^* \times C^* \times V : \\ \langle k^*, \nu \rangle = 1, s_\nu(v) \leq -(k^* f)^*(y^*) - (z^* g)_S^*(-y^*) \}$$

and

$$h^{C\mathcal{T}_{sc}}(\nu, y^*, k^*, z^*, v) = v.$$

The weak and strong duality theorems have similar formulations to the ones given for the vector primal-dual pair $(PV^C) - (DV^{C\mathcal{T}_s})$, therefore we omit them.

In the framework provided by the set scalarization can be brought also the scalarization approach introduced in [189] which involves polyhedral sets in finite dimensional spaces as well as the scalarization approach treated in [175] which involves sets generated by norms. The reader is referred to [188] for a deeper analysis of an approach for embedding classical scalarization functions into the set scalarization concept.

4.4.5 (Semi)Norm scalarization

The investigations we make in this subsection have as starting point the fact that in some circumstances *(semi)norms* on V turn out to be K-strongly increasing functions. This has been noticed by many authors; we cite here only the works [104,165,201]. The scalarization functions we investigate in the following are based on K-strongly increasing gauges. This kind of scalarization functions has been used in [187] for location problems and in [45] for goal programming.

Assume first that there exists $b \in V$ such that $f(\operatorname{dom} f \cap \mathcal{A}) \subseteq b + K$. We consider $E \subseteq V$ a convex set such that $0 \in \operatorname{int}(E)$ and its (Minkowski) gauge γ_E is K-strongly increasing on K. Since $0 \in \operatorname{int}(E)$ it yields that $\gamma_E(v) \in \mathbb{R}$ for all $v \in V$.

Remark 4.4.3. If V is a Hilbert space, then the norm of V is K-strongly increasing on K if and only if $K \subseteq K^*$ (cf. [104]). This is the case if, for instance, $V = \mathbb{R}^k$ and K is the non-negative orthant in $\mathbb{R}^k$. Not only the Euclidean norm is $\mathbb{R}^k_+$-strongly increasing on $\mathbb{R}^k_+$, but also the oblique norms (cf. [165, 175]) are $\mathbb{R}^k_+$-strongly increasing on $\mathbb{R}^k_+$.

For all $a \in b - K$ define $s_a : V \cup \{+\infty_K\} \to \overline{\mathbb{R}}$ by

$$s_a(v) = \begin{cases} \gamma_E(v-a), & \text{if } v \in b + K, \\ +\infty, & \text{otherwise}, \end{cases}$$

while at $+\infty_K$ we take $s_a(+\infty_K) = +\infty$. Let be $a \in b - K$ fixed. Obviously, s_a is convex and it holds $f(\operatorname{dom} f \cap \mathcal{A}) \subseteq b + K = \operatorname{dom} s_a$. For all $v, w \in b + K$ such that $v \leq_K w$ it holds $v - a \leq_K w - a$. As $a \in b - K \Leftrightarrow -a \in -b + K$, $v - a$ and $w - a$ belong to K and so $\gamma_E(v-a) < \gamma_E(w-a) \Leftrightarrow s_a(v) < s_a(w)$. This means that s_a is K-strongly increasing on $b + K$. Consequently, since $f(\operatorname{dom} f \cap \mathcal{A}) \subseteq b + K$, s_a is K-strongly increasing on $f(\operatorname{dom} f \cap \mathcal{A})$.

Considering the following family of scalarization functions

$$\mathcal{S}_g = \{s_a : V \cup \{+\infty_K\} \to \overline{\mathbb{R}} : a \in b - K\},$$

we say that an element $\bar{x} \in \mathcal{A}$ is a *$\mathcal{S}_g$-properly efficient solution* to (PV^C) if there exists $a \in b - K$ such that $s_a(f(\bar{x})) \leq s_a(f(x))$ for all $x \in \mathcal{A}$.

We calculate next the conjugate function of s_a for $a \in b - K$ fixed. Let be $k^* \in V^*$. Due to Proposition 2.2.18(*b*) the gauge function γ_E is continuous. By using Theorem 3.5.6(*a*), we have that

$$(s_a)^*(k^*) = (\gamma_E(\cdot - a) + \delta_{b+K})^*(k^*)$$

$$= \min_{w^* \in V^*} \left\{ (\gamma_E(\cdot - a))^*(k^* - w^*) + \delta^*_{b+K}(w^*) \right\}.$$

Further,

$$(\gamma_E(\cdot - a))^*(k^* - w^*) = \sup_{v \in V} \{\langle k^* - w^*, v\rangle - \gamma_E(v - a)\}$$

$$= \sup_{u \in V} \{\langle k^* - w^*, u + a\rangle - \gamma_E(u)\} = \langle k^* - w^*, a\rangle + \gamma_E^*(k^* - w^*).$$

For the conjugate of a gauge we have (see Example 2.3.4(*a*))

$$\gamma_E^*(k^* - w^*) = \begin{cases} 0, & \text{if } \sigma_E(k^* - w^*) \leq 1, \\ +\infty, & \text{otherwise,} \end{cases}$$

and, on the other hand, it holds,

$$\delta_{b+K}^*(w^*) = \langle w^*, b \rangle + \delta_K^*(w^*).$$

Consequently, we get for the conjugate of s_a at k^* the following formula

$$\begin{aligned} (s_a)^*(k^*) &= \min_{\substack{w^* \in -K^*, \\ \sigma_E(k^* - w^*) \leq 1}} \{\langle k^* - w^*, a \rangle + \langle w^*, b \rangle\} \\ &= \langle k^*, a \rangle + \min_{\substack{w^* \in -K^*, \\ \sigma_E(k^* - w^*) \leq 1}} \langle w^*, b - a \rangle. \end{aligned}$$

The vector dual problem to (PV^C) with respect to gauge scalarization is

$$(DV^{C_{S_g}}) \qquad \underset{(a,y^*,k^*,z^*,w^*,v) \in \mathcal{B}^{C_{S_g}}}{\text{Max}} \quad h^{C_{S_g}}(a, y^*, k^*, z^*, w^*, v),$$

where

$$\mathcal{B}^{C_{S_g}} = \Big\{(a, y^*, k^*, z^*, w^*, v) \in (b - K) \times X^* \times K^* \times C^* \times (-K^*) \times (b + K) : \\ \sigma_E(k^* - w^*) \leq 1, \gamma_E(v - a) \leq \langle w^*, a - b \rangle \\ -\langle k^*, a \rangle - (k^* f)^*(y^*) - (z^* g)_S^*(-y^*)\Big\}$$

and

$$h^{C_{S_g}}(a, y^*, k^*, z^*, w^*, v) = v.$$

Remark 4.4.4. We emphasize that σ_E defines the so-called *dual gauge* to γ_E and if γ_E is a norm it turns out to be the dual norm.

The following results follow from Theorem 4.4.2 and Theorem 4.4.3, respectively.

Theorem 4.4.12. *There is no* $x \in \mathcal{A}$ *and no* $(a, y^*, k^*, z^*, w^*, v) \in \mathcal{B}^{C_{S_g}}$ *such that* $f(x) \leq_K h^{C_{S_g}}(a, y^*, k^*, z^*, w^*, v)$.

Theorem 4.4.13. *Assume that the regularity condition* $(RCV^{C_{FL}})$ *is fulfilled. If* $\bar{x} \in \mathcal{A}$ *is a* $\mathcal{S}_g$*-properly efficient solution to* (PV^C)*, then there exists* $(\bar{a}, \bar{y}^*, \bar{k}^*, \bar{z}^*, \bar{w}^*, \bar{v})$*, an efficient solution to* $(DV^{C_{S_g}})$*, such that* $f(\bar{x}) = h^{C_{S_g}}(\bar{a}, \bar{y}^*, \bar{k}^*, \bar{z}^*, \bar{w}^*, \bar{v}) = \bar{v}$.

Remark 4.4.5. The duality approach described in this section can be considered also in the particular case when γ_E is a norm with the unit ball E. Conditions which ensure that a norm is K-strongly increasing on a given set have been investigated in [102, 104, 201].

4.5 Linear vector duality

The duality theory for linear vector optimization problems has its starting point in the paper of Gale, Kuhn and Tucker [70]. Their work was continued by Kornbluth (cf. [118]), Rödder (cf. [161]) and Isermann (cf. [95–97]), who developed different duality concepts for the linear case. All these investigations have been done for problems stated in finite dimensional spaces, a framework in which one can provide duality results in analogy to the ones existing in the scalar linear duality theory. We intensively deal with linear vector optimization problems in finite dimensional spaces in the following chapter. What we do in this section is giving some preliminary results which hold for linear vector problems in general spaces.

Let X, Z and V be Hausdorff locally convex spaces and assume that Z is partially ordered by the convex cone $C \subseteq Z$, while V is partially ordered by the nontrivial pointed convex cone $K \subseteq V$. Further, assume that $S \subseteq X$ is a convex cone, $L \in \mathcal{L}(X,V)$, $A \in \mathcal{L}(X,Z)$ and $b \in Z$ is a given element such that $A(S) \cap (b+C) \neq \emptyset$. The primal vector optimization problem we investigate in this section is

$$(PV^{\mathcal{L}}) \quad \operatorname*{Min}_{x \in \mathcal{A}^{\mathcal{L}}} Lx.$$
$$\mathcal{A}^{\mathcal{L}} = \{x \in S : Ax - b \in C\}$$

4.5.1 The duals introduced via linear scalarization

We investigate here vector dual optimization problems to $(PV^{\mathcal{L}})$ with respect to properly efficient solutions as particular instances of the vector duals treated in section 4.3. To this end it is enough to notice that $(PV^{\mathcal{L}})$ is a particular formulation of (PV^C) in case $f : X \to V$, $f(x) = Lx$ and $g : X \to Z$, $g(x) = b - Ax$. The functions f and g satisfy the assumptions imposed when introducing the problem (PV^C).

As for $(v^*, z^*, v) \in K^{*0} \times C^* \times V$ it holds

$$\langle v^*, v\rangle \leq \inf_{x\in S}\{(v^*f)(x) + (z^*g)(x)\} \Leftrightarrow \langle v^*, v\rangle \leq \inf_{x\in S}\{\langle v^*, Lx\rangle + \langle z^*, b - Ax\rangle\}$$
$$\Leftrightarrow \langle v^*, v\rangle \leq \langle z^*, b\rangle + \inf_{x\in S}\{\langle L^*v^* - A^*z^*, x\rangle\}$$
$$\Leftrightarrow \langle v^*, v\rangle \leq \langle z^*, b\rangle \text{ and } L^*v^* - A^*z^* \in S^*,$$

the dual (DV^{C_L}) becomes in this special case (see also [101, 104])

$$(DV^{\mathcal{L}_L}) \quad \operatorname*{Max}_{(v^*,z^*,v)\in\mathcal{B}^{\mathcal{L}_L}} h^{\mathcal{L}_L}(v^*, z^*, v),$$

where

$$\mathcal{B}^{\mathcal{L}_L} = \big\{(v^*, z^*, v) \in K^{*0} \times C^* \times V : \langle v^*, v\rangle \leq \langle z^*, b\rangle \text{ and } L^*v^* - A^*z^* \in S^*\big\}$$

and

$$h^{\mathcal{L}_L}(v^*, z^*, v) = v.$$

Coming now to the Fenchel type dual (DV^{C_F}), since for all $(v^*, y^*, v) \in K^{*0} \times X^* \times V$ one has

$$\langle v^*, v \rangle \leq -(v^* f)^*(y^*) - \sigma_{\mathcal{A}}(-y^*) \Leftrightarrow \langle v^*, v \rangle \leq - \sup_{x \in X} \langle y^* - L^* v^*, x \rangle - \sigma_{\mathcal{A}}(-y^*)$$

$$\Leftrightarrow y^* = L^* v^* \text{ and } \langle v^*, v \rangle \leq \inf_{x \in \mathcal{A}} \langle y^*, x \rangle,$$

this dual is nothing else than the vector dual which follows by particularizing (DV^{C_P}) in the linear case (this is the reason why we denote it by $(DV^{\mathcal{L}_P})$)

$$(DV^{\mathcal{L}_P}) \qquad \operatorname*{Max}_{(v^*, v) \in \mathcal{B}^{\mathcal{L}_P}} h^{\mathcal{L}_P}(v^*, v),$$

where

$$\mathcal{B}^{\mathcal{L}_P} = \left\{ (v^*, v) \in K^{*0} \times V : \langle v^*, v \rangle \leq \inf_{x \in \mathcal{A}} \langle L^* v^*, x \rangle \right\}$$

and

$$h^{\mathcal{L}_P}(v^*, v) = v.$$

Now consider an element $(v^*, y^*, z^*, v) \in K^{*0} \times X^* \times C^* \times V$. The following equivalences hold

$$\langle v^*, v \rangle \leq -(v^* f)^*(y^*) - (z^* g)^*_S(-y^*) \Leftrightarrow y^* = L^* v^* \text{ and } \langle v^*, v \rangle \leq \langle z^*, b \rangle$$

$$+ \inf_{x \in S} \{ \langle L^* v^* - A^* z^*, x \rangle \} \Leftrightarrow y^* = L^* v^*, \langle v^*, v \rangle \leq \langle z^*, b \rangle \text{ and } L^* v^* - A^* z^* \in S^*.$$

Consequently, the vector dual problem $(DV^{C_{FL}})$ has in this case the same formulation like $(DV^{\mathcal{L}_L})$. This means that the vector dual problems investigated in subsection 4.3.2 can be resumed in this particular case to only two different vector duals. The weak, strong and converse duality theorems for the primal-dual vector pairs $(PV^{\mathcal{L}})-(DV^{\mathcal{L}_L})$ and $(PV^{\mathcal{L}})-(DV^{\mathcal{L}_P})$, respectively, follow as particular instances of the corresponding results given in subsection 4.3.2.

Remark 4.5.1. A sufficient condition for having strong duality for $(PV^{\mathcal{L}})$ and $(DV^{\mathcal{L}_L})$ which follows as a particularization of (RCV^{C_L}) is

$$(RCV^{\mathcal{L}_L}) \mid \exists x' \in S \text{ such that } Ax' - b \in \operatorname{int}(C).$$

In case $\operatorname{int}(C) = \emptyset$ and X and Z are Fréchet spaces, S is closed and C is closed (this guarantees that g is C-epi closed) one can assume instead, that $b \in \operatorname{sqri}(A(S) - C)$. If the linear subspace $\operatorname{lin}(A(S) - C)$ has a finite dimension, then one can assume that $b \in \operatorname{ri}(A(S) - C)$ in order to achieve strong duality.

With respect to the duals $(DV^{\mathcal{L}_L})$ and $(DV^{\mathcal{L}_P})$ one has always that (cf. (4.18)) $h^{\mathcal{L}_L}(\mathcal{B}^{\mathcal{L}_L}) \subseteq h^{\mathcal{L}_P}(\mathcal{B}^{\mathcal{L}_P})$, while under any of the regularity conditions mentioned in Remark 4.5.1 these sets become equal.

In what follows we put the dual $(DV^{\mathcal{L}_L})$ in connection to another dual vector problem to $(PV^{\mathcal{L}})$, which generalizes in a direct way the dual in scalar linear programming. These investigations are based on ideas due to Jahn published in [101, 104]. Let this vector dual, named in [104] *abstract linear optimization problem*, be

$$(DV^{\mathcal{L}_J}) \qquad \operatorname*{Max}_{(v^*,U)\in\mathcal{B}^{\mathcal{L}_J}} h^{\mathcal{L}_J}(v^*,U),$$

where

$$\mathcal{B}^{\mathcal{L}_J} = \big\{(v^*,U) \in K^{*0} \times \mathcal{L}(Z,V) : U^*v^* \in C^* \text{ and } (L - U \circ A)^*v^* \in S^*\big\}$$

and

$$h^{\mathcal{L}_J}(v^*,U) = Ub.$$

We prove first the following preliminary result.

Proposition 4.5.1. *It holds* $h^{\mathcal{L}_J}(\mathcal{B}^{\mathcal{L}_J}) \subseteq h^{\mathcal{L}_L}(\mathcal{B}^{\mathcal{L}_L})$.

Proof. Let be $(v^*,U) \in K^{*0} \times \mathcal{L}(Z,V)$ such that $U^*v^* \in C^*$ and $(L - U \circ A)^*v^* \in S^*$. Take $v = Ub$ and $z^* = U^*v^*$. Thus it holds $\langle v^*,v\rangle = \langle v^*,Ub\rangle = \langle U^*v^*,b\rangle = \langle z^*,b\rangle$ and $(L-U\circ A)^*v^* = L^*v^* - A^*(U^*v^*) = L^*v^* - A^*z^* \in S^*$. This means that $(v^*,z^*,v) \in \mathcal{B}^{\mathcal{L}_L}$ and $Ub = v \in h^{\mathcal{L}_L}(\mathcal{B}^{\mathcal{L}_L})$. □

The next statement (cf. [104, Theorem 8.13]) gives a sufficient condition which ensures the coincidence of the maximal elements of $h^{\mathcal{L}_J}(\mathcal{B}^{\mathcal{L}_J})$ and $h^{\mathcal{L}_L}(\mathcal{B}^{\mathcal{L}_L})$ with respect to the partial ordering induced by the cone K.

Theorem 4.5.2. *The following statements are fulfilled*

(a) $\operatorname{Max}(h^{\mathcal{L}_J}(\mathcal{B}^{\mathcal{L}_J}),K) \subseteq \operatorname{Max}(h^{\mathcal{L}_L}(\mathcal{B}^{\mathcal{L}_L}),K)$;
(b) If $b \neq 0$, *then* $\operatorname{Max}(h^{\mathcal{L}_J}(\mathcal{B}^{\mathcal{L}_J}),K) = \operatorname{Max}(h^{\mathcal{L}_L}(\mathcal{B}^{\mathcal{L}_L}),K)$.

Proof. (*a*) If the set $\operatorname{Max}(h^{\mathcal{L}_J}(\mathcal{B}^{\mathcal{L}_J}),K)$ is empty, then the conclusion follows.

Let be $\operatorname{Max}(h^{\mathcal{L}_J}(\mathcal{B}^{\mathcal{L}_J}),K) \neq \emptyset$. Suppose that $b \neq 0$ and consider $(v^*,U) \in \mathcal{B}^{\mathcal{L}_J}$ such that Ub is a maximal element of $h^{\mathcal{L}_J}(\mathcal{B}^{\mathcal{L}_J})$. By Proposition 4.5.1 one has that $Ub \in h^{\mathcal{L}_L}(\mathcal{B}^{\mathcal{L}_L})$. Assume that $Ub \notin \operatorname{Max}(h^{\mathcal{L}_L}(\mathcal{B}^{\mathcal{L}_L}),K)$. Thus there exists $(\bar{v}^*,\bar{z}^*,\bar{v}) \in \mathcal{B}^{\mathcal{L}_L}$ such that $Ub \leq_K \bar{v}$.

We can assume that $\langle \bar{v}^*,\bar{v}\rangle = \langle \bar{z}^*,b\rangle$. In case $\langle \bar{v}^*,\bar{v}\rangle < \langle \bar{z}^*,b\rangle$, it is easy to find an element $v \in K\setminus\{0\}$ such that $(\bar{v}^*,\bar{z}^*,\bar{v}+v) \in \mathcal{B}^{\mathcal{L}_L}$, $Ub \leq_K \bar{v}+v$ and $\langle \bar{v}^*,\bar{v}+v\rangle = \langle \bar{z}^*,b\rangle$. Therefore the element $\bar{v}+v$ would satisfy the requirements. Having this fulfilled, we prove that there exists $\overline{U} \in \mathcal{L}(Z,V)$ such that $\overline{U}b = \bar{v}$ and $\overline{U}^*\bar{v}^* = \bar{z}^*$.

First we suppose that $\langle \bar{z}^*,b\rangle \neq 0$ and define $\overline{U} : Z \to V$ by $\overline{U}z = (\langle \bar{z}^*,z\rangle/\langle \bar{z}^*,b\rangle)\bar{v}$. Obviously, $\overline{U} \in \mathcal{L}(Z,V)$ and it fulfills $\overline{U}b = \bar{v}$ and $\overline{U}^*\bar{v}^* = \bar{z}^*$.

In case $\langle \bar{z}^*, b\rangle = 0$, as $b \neq 0$ and $\bar{v}^* \in K^{*0}$, there exist $\tilde{z}^* \in Z^*$ and $\tilde{v} \in V$ such that $\langle \tilde{z}^*, b\rangle = \langle \bar{v}^*, \tilde{v}\rangle = 1$. We define in this case $\overline{U} : Z \to V$ as being $\overline{U}z = \langle \bar{z}^*, z\rangle \tilde{v} + \langle \tilde{z}^*, z\rangle \bar{v}$. It is evident that $\overline{U} \in \mathcal{L}(Z,V)$ and $\overline{U}b = \bar{v}$. More than that, for all $z \in Z$ it holds

$$\langle \overline{U}^* \bar{v}^*, z\rangle = \langle \bar{v}^*, \overline{U}z\rangle = \langle \bar{z}^*, z\rangle + \langle \tilde{z}^*, z\rangle\langle \bar{v}^*, \bar{v}\rangle = \langle \bar{z}^*, z\rangle$$

and so $\overline{U}^* \bar{v}^* = \bar{z}^*$. Since $L^*\bar{v}^* - A^*\bar{z}^* = (L - U \circ A)^*\bar{v}^* \in S^*$, one has that $(\bar{v}^*, \overline{U}) \in \mathcal{B}^{\mathcal{L}_J}$. But $Ub \leq_K \bar{v} = \overline{U}b$ leads to a contradiction and, consequently, $Ub \in \operatorname{Max}(h^{\mathcal{L}_L}(\mathcal{B}^{\mathcal{L}_L}), K)$.

In case $b = 0$ we have that $\operatorname{Max}(h^{\mathcal{L}_J}(\mathcal{B}^{\mathcal{L}_J}), K) = \{0\}$ and, by Proposition 4.5.1, $0 \in h^{\mathcal{L}_L}(\mathcal{B}^{\mathcal{L}_L})$. Assume that $0 \notin \operatorname{Max}(h^{\mathcal{L}_L}(\mathcal{B}^{\mathcal{L}_L}), K)$. Thus there exists $(\bar{v}^*, \bar{z}^*, \bar{v}) \in \mathcal{B}^{\mathcal{L}_L}$ such that $0 \leq_K \bar{v}$. But this means that $0 < \langle \bar{v}^*, \bar{v}\rangle$, which contradicts the relation $\langle \bar{v}^*, \bar{v}\rangle \leq \langle \bar{z}^*, b\rangle = 0$. In this way we get the desired result.

(*b*) Assume that $b \neq 0$ and consider an arbitrary $\bar{v} \in \operatorname{Max}(h^{\mathcal{L}_L}(\mathcal{B}^{\mathcal{L}_L}), K)$. Thus there exist $\bar{v}^* \in K^{*0}$ and $\bar{z}^* \in C^*$ such that $\langle \bar{v}^*, \bar{v}\rangle \leq \langle \bar{z}^*, b\rangle$ and $L^*\bar{v}^* - A^*\bar{z}^* \in S^*$. If $\langle \bar{v}^*, \bar{v}\rangle < \langle \bar{z}^*, b\rangle$, then there exists $v \in K \setminus \{0\}$ such that $(\bar{v}^*, \bar{z}^*, \bar{v} + v) \in \mathcal{B}^{\mathcal{L}_L}$, which contradicts the maximality of $\bar{v}$ in $h^{\mathcal{L}_L}(\mathcal{B}^{\mathcal{L}_L})$. Thus $\langle \bar{v}^*, \bar{v}\rangle = \langle \bar{z}^*, b\rangle$. As in the proof of statement (*a*) one can construct $\overline{U} \in \mathcal{L}(Z,V)$ such that $\overline{U}b = \bar{v}$ and $\overline{U}^* \bar{v}^* = \bar{z}^*$. This yields that $(\bar{v}^*, \overline{U}) \in \mathcal{B}^{\mathcal{L}_J}$. Proposition 4.5.1 ensures that $\bar{v} = \overline{U}b \in \operatorname{Max}(h^{\mathcal{L}_J}(\mathcal{B}^{\mathcal{L}_J}), K)$. □

Remark 4.5.2. The previous result states that in case $b \neq 0$ one can consider as dual problem to $(PV^{\mathcal{L}})$ the following vector optimization problem

$$(DV^{\mathcal{L}_J}) \qquad \operatorname*{Max}_{\substack{v^* \in K^{*0}, U \in \mathcal{L}(Z,V),\\ U^*v^* \in C^*, (L - U\circ A)^* v^* \in S^*}} Ub,$$

which has a very close formulation to the one of the dual problem in scalar linear programming. An elaborated discussion on the connections between the different formulations for the dual vector linear problems in finite dimensional spaces existing in the literature will be done in the next chapter.

4.5.2 Linear vector duality with respect to weakly efficient solutions

When $\operatorname{int}(K) \neq \emptyset$, similar to the investigations above, one can introduce two vector dual problems to

$$(PV_w^{\mathcal{L}}) \quad \operatorname*{WMin}_{x \in \mathcal{A}} Lx.$$
$$\mathcal{A}^{\mathcal{L}} = \{x \in S : Ax - b \in C\}$$

with respect to the weakly efficient solutions by particularizing the duals introduced in subsection 4.3.4. The vector duals $(DV_w^{C_L})$ and $(DV_w^{C_{FL}})$ turn out to be

$$(DV_w^{\mathcal{L}_L}) \qquad \underset{(v^*,z^*,v)\in\mathcal{B}_w^{\mathcal{L}_L}}{\text{WMax}} h_w^{\mathcal{L}_L}(v^*,z^*,v),$$

where

$$\mathcal{B}_w^{\mathcal{L}_L} = \big\{(v^*,z^*,v) \in (K^* \setminus \{0\}) \times C^* \times V : \\ \langle v^*, v\rangle \leq \langle z^*, b\rangle \text{ and } L^*v^* - A^*z^* \in S^*\big\}$$

and

$$h_w^{\mathcal{L}_L}(v^*,z^*,v) = v,$$

while the vector duals $(DV_w^{C_F})$ and $(DV_w^{C_P})$ become

$$(DV_w^{\mathcal{L}_P}) \qquad \underset{(v^*,v)\in\mathcal{B}_w^{\mathcal{L}_P}}{\text{WMax}} h_w^{\mathcal{L}_P}(v^*,v),$$

where

$$\mathcal{B}_w^{\mathcal{L}_P} = \left\{(v^*,v) \in (K^* \setminus \{0\}) \times V : \langle v^*, v\rangle \leq \inf_{x\in\mathcal{A}} \langle L^*v^*, x\rangle\right\}$$

and

$$h_w^{\mathcal{L}_L}(v^*,v) = v.$$

We always have that $h_w^{\mathcal{L}_L}(\mathcal{B}_w^{\mathcal{L}_L}) \subseteq h_w^{\mathcal{L}_P}(\mathcal{B}_w^{\mathcal{L}_P})$ and under the regularity conditions considered in Remark 4.5.1 these sets coincide.

In analogy to $(DV^{\mathcal{L}_J})$ one can introduce the following dual problem to $(PV_w^{\mathcal{L}})$ with respect to the weakly efficient solutions (cf. [104])

$$(DV_w^{\mathcal{L}_J}) \qquad \underset{(v^*,U)\in\mathcal{B}_w^{\mathcal{L}_J}}{\text{WMax}} h_w^{\mathcal{L}_J}(v^*,U),$$

where

$$\mathcal{B}_w^{\mathcal{L}_J} = \big\{(v^*,U) \in (K^*\setminus\{0\}) \times \mathcal{L}(Z,V) : U^*v^* \in C^* \text{ and } (L - U\circ A)^*v^* \in S^*\big\}$$

and

$$h_w^{\mathcal{L}_J}(v^*,U) = Ub.$$

The proof of the following results can be done in the lines of the proofs of Proposition 4.5.1 and Theorem 4.5.2, respectively.

Proposition 4.5.3. *It holds* $h_w^{\mathcal{L}_J}(\mathcal{B}_w^{\mathcal{L}_J}) \subseteq h_w^{\mathcal{L}_L}(\mathcal{B}_w^{\mathcal{L}_L})$.

Theorem 4.5.4. *The following statements are fulfilled*

(a) $\text{WMax}(h_w^{\mathcal{L}_J}(\mathcal{B}_w^{\mathcal{L}_J}), K) \subseteq \text{WMax}(h_w^{\mathcal{L}_L}(\mathcal{B}_w^{\mathcal{L}_L}), K)$;
(b) If $b \neq 0$, *then* $\text{WMax}(h_w^{\mathcal{L}_J}(\mathcal{B}_w^{\mathcal{L}_J}), K) = \text{WMax}(h_w^{\mathcal{L}_L}(\mathcal{B}_w^{\mathcal{L}_L}), K)$.

Remark 4.5.3. In case $b \neq 0$ one can consider as dual problem to $(PV_w^{\mathcal{L}})$ with respect to the weakly efficient solutions the following vector optimization problem

$$(DV_w^{\mathcal{L}_J}) \qquad \underset{\substack{v^*\in K^*\setminus\{0\}, U\in\mathcal{L}(Z,V),\\ U^*v^*\in C^*, (L-U\circ A)^*v^*\in S^*}}{\text{WMax}} \quad Ub.$$

4.5.3 Nakayama's geometric dual in the linear case

The next dual vector problem to $(PV^{\mathcal{L}})$ we investigate with respect to properly efficient solutions is Nakayama's geometric dual. The problem (DV^{C_N}) looks in this particular case like

$$(DV^{\mathcal{L}_N}) \quad \underset{(U,v)\in\mathcal{B}^{\mathcal{L}_N}}{\text{Max}} \; h^{\mathcal{L}_N}(U,v),$$

where

$$\mathcal{B}^{\mathcal{L}_N} = \{(U,v) \in \mathcal{L}_+(Z,V) \times V : \\ \nexists x \in S \text{ such that } v - Ub \geqq_K (L - U \circ A)x\}$$

and

$$h^{\mathcal{L}_N}(U,v) = v.$$

Proposition 4.3.16 and Proposition 4.5.1 yield

$$h^{\mathcal{L}_J}(\mathcal{B}^{\mathcal{L}_J}) \subseteq h^{\mathcal{L}_L}(\mathcal{B}^{\mathcal{L}_L}) \subseteq h^{\mathcal{L}_N}(\mathcal{B}^{\mathcal{L}_N}).$$

If $(RCV^{\mathcal{L}})$ is fulfilled and $L\left(S \cap A^{-1}(b+C)\right)+K$ is closed, Theorem 4.3.17(a) and Theorem 4.5.2(a) imply

$$\text{Max}(h^{\mathcal{L}_J}(\mathcal{B}^{\mathcal{L}_J}), K) \subseteq \text{Max}(h^{\mathcal{L}_L}(\mathcal{B}^{\mathcal{L}_L}), K) \subseteq \text{Max}(h^{\mathcal{L}_N}(\mathcal{B}^{\mathcal{L}_N}), K).$$

If, additionally, $b \neq 0$, $\text{qi}(K) \neq \emptyset$ and every efficient solution to $(PV^{\mathcal{L}})$ is properly efficient, then, by Theorem 4.3.17(b) and Theorem 4.5.2(b), follows that

$$\text{Max}(h^{\mathcal{L}_J}(\mathcal{B}^{\mathcal{L}_J}), K) = \text{Max}(h^{\mathcal{L}_L}(\mathcal{B}^{\mathcal{L}_L}), K) = \text{Max}(h^{\mathcal{L}_N}(\mathcal{B}^{\mathcal{L}_N}), K). \quad (4.24)$$

In the next chapter we show that for the linear vector optimization problems stated in finite dimensional spaces relation (4.24) is automatically fulfilled. Obviously, in the general case, in the statements above the assumption that $(RCV^{\mathcal{L}})$ is fulfilled can be replaced with some weaker regularity conditions (cf. Remark 4.5.1).

Assuming next that $\text{int}(K) \neq \emptyset$, via $(DV_w^{C_N})$ we get the following vector dual problem, this time to $(PV_w^{\mathcal{L}})$ with respect to the weakly efficient solutions

$$(DV_w^{\mathcal{L}_N}) \quad \underset{(U,v)\in\mathcal{B}_w^{\mathcal{L}_N}}{\text{WMax}} \; h_w^{\mathcal{L}_N}(U,v),$$

where

$$\mathcal{B}_w^{\mathcal{L}_N} = \{(U,v) \in \mathcal{L}_+(Z,V) \times V : \\ \nexists x \in S \text{such that } v - Ub >_K (L - U \circ A)x\}$$

and

$$h_w^{\mathcal{L}_N}(U,v) = v.$$

Theorem 4.3.30 and Proposition 4.5.3 yield

$$h_w^{\mathcal{L}_J}(\mathcal{B}_w^{\mathcal{L}_J}) \subseteq h_w^{\mathcal{L}_L}(\mathcal{B}_w^{\mathcal{L}_L}) = h_w^{\mathcal{L}_N}(\mathcal{B}_w^{\mathcal{L}_N}),$$

while from Theorem 4.5.4(a) follows that

$$\mathrm{WMax}(h_w^{\mathcal{L}_J}(\mathcal{B}_w^{\mathcal{L}_J}), K) \subseteq \mathrm{WMax}(h_w^{\mathcal{L}_L}(\mathcal{B}_w^{\mathcal{L}_L}), K) = \mathrm{WMax}(h_w^{\mathcal{L}_N}(\mathcal{B}_w^{\mathcal{L}_N}), K).$$

If $b \neq 0$, then the sets in the relation above are all equal, namely it holds

$$\mathrm{WMax}(h_w^{\mathcal{L}_J}(\mathcal{B}_w^{\mathcal{L}_J}), K) = \mathrm{WMax}(h_w^{\mathcal{L}_L}(\mathcal{B}_w^{\mathcal{L}_L}), K) = \mathrm{WMax}(h_w^{\mathcal{L}_N}(\mathcal{B}_w^{\mathcal{L}_N}), K). \tag{4.25}$$

Bibliographical notes

The use of the scalar duality theory when introducing different vector duality approaches is very fruitful and this fact is underlined by the large number of publications one can find in the literature which deals with this issue. In the majority of these works to the primal vector problem a scalar optimization problem is attached to which a dual problem is established. The vector dual is then defined by involving in the formulation of its feasible set this scalar dual. This idea was first used for defining a Fenchel type vector dual by Breckner and Kolumbán in [42, 43] and also by Gerstewitz and Göpfert in [73]. Deeper investigations concerning this problem have been done in finite dimensional spaces in [28].

Richer is the literature on vector duality for the vector optimization problem with geometric and cone constraints. A so-called geometric duality approach for this kind of vector problems was introduced by Nakayama in [142–144]. A more general approach based on linear scalarization was suggested by Göpfert in [78], while Jahn investigated in [101] (see also [103,104]) a vector dual problem, the definition of which is inspired by the scalar Lagrange duality theory. In [36, 37] (see also [24]), for a primal vector problem with geometric and cone constraints stated in finite dimensional spaces, Boţ and Wanka extended these investigations by considering also other duality concepts. An overview on the relations between the sets of maximal elements of the image sets of the feasible set through the objective functions of these vector dual problems have been done in [24, 37]. In these investigations Nakayama's geometric dual problem is also included.

The idea of replacing the linear scalarization by different general scalarization functions in connection to the definition of different classes of solutions was used for the first time in [72,74,79]. The scalar duality theory for composed convex optimization problems has been used for the first time in [29] for defining a general vector dual in finite dimensional spaces. This approach, which is considered here for problems in Hausdorff locally convex spaces, turns out to be very fruitful since it can be employed for different particular scalarization functions, like those in [75, 107, 133, 172, 175, 179, 187, 188, 201], etc.

The literature on duality for linear vector optimization problems is very large. We mention here only the works of Gale, Kuhn and Tucker (cf. [70]), Kornbluth (cf. [118]), Rödder (cf. [161]), Isermann (cf. [95–97]) and Nakayama (cf. [144]). Very important references for the infinite dimensional case are the works of Jahn [101, 104], where also connections to earlier results due to Isermann are established.

5

Conjugate duality for vector optimization problems with finite dimensional image spaces

In this chapter we introduce new conjugate vector dual problems to the primal problems treated in the previous chapter in case their objective functions have finite dimensional image spaces. Weak, strong and converse duality assertions are proven and these duals are compared with the ones introduced in chapter 4. Note that the properly efficient solutions considered in this chapter are in the sense of linear scalarization.

5.1 Another Fenchel type vector dual problem

Throughout this section we consider two Hausdorff locally convex spaces X and Y, the proper functions $f_i : X \to \overline{\mathbb{R}}$ and $g_i : Y \to \overline{\mathbb{R}}$, $i = 1, \dots, k$, and $A \in \mathcal{L}(X, Y)$ such that $\cap_{i=1}^k (\operatorname{dom} f_i \cap A^{-1}(\operatorname{dom} g_i)) \neq \emptyset$. Further, we assume that the image space $V = \mathbb{R}^k$ is partially ordered by the cone $K = \mathbb{R}^k_+$ and denote according to the notations in section 2.1 $\overline{\mathbb{R}^k} = \mathbb{R}^k \cup \{\pm\infty_{\mathbb{R}^k_+}\}$. For

$$f : X \to \overline{\mathbb{R}^k}, f(x) = \begin{cases} (f_1(x), \dots, f_k(x))^T, & \text{if } x \in \bigcap\limits_{i=1}^k \operatorname{dom} f_i, \\ +\infty_{\mathbb{R}^k_+}, & \text{otherwise,} \end{cases}$$

and

$$g : Y \to \overline{\mathbb{R}^k}, g(y) = \begin{cases} (g_1(y), \dots, g_k(y))^T, & \text{if } y \in \bigcap\limits_{i=1}^k \operatorname{dom} g_i, \\ +\infty_{\mathbb{R}^k_+}, & \text{otherwise,} \end{cases}$$

we introduce the following primal vector optimization problem

$$(PVF^A) \quad \operatorname*{Min}_{x \in X} \{f(x) + g(Ax)\},$$

which will constitute the object of our investigations in this section. The primal problem (PVF^A) can be explicitly written as

R.I. Boţ et al., *Duality in Vector Optimization*, Vector Optimization,
DOI 10.1007/978-3-642-02886-1_5,

$$(PVF^A) \quad \operatorname*{Min}_{x \in X} \begin{pmatrix} f_1(x) + g_1(Ax) \\ \vdots \\ f_k(x) + g_k(Ax) \end{pmatrix}.$$

The aim of our investigation is to introduce a new vector dual problem to (PVF^A) and to establish weak, strong and converse duality assertions with respect to both properly efficient solutions in the sense of linear scalarization (called simply properly efficient solutions) and weakly efficient solutions to (PVF^A).

5.1.1 Duality with respect to properly efficient solutions

In this subsection we provide a vector dual problem to (PVF^A) with respect to its properly efficient solutions. According to Definition 2.5.1 we say that $\bar{x} \in X$ is a properly efficient solution to (PVF^A) in the sense of linear scalarization if $\bar{x} \in \cap_{i=1}^k (\operatorname{dom} f_i \cap A^{-1}(\operatorname{dom} g_i))$ and $(f + g \circ A)(\bar{x}) \in \operatorname{PMin}_{LSc}\big((f + g \circ A)(\cap_{i=1}^k(\operatorname{dom} f_i \cap A^{-1}(\operatorname{dom} g_i))), \mathbb{R}^k_+\big)$. Since $K^{*0} = \operatorname{int}(\mathbb{R}^k_+)$ this is the case if there exists $\lambda = (\lambda_1, \ldots, \lambda_k)^T \in \operatorname{int}(\mathbb{R}^k_+)$ such that (cf. Definition 2.4.12)

$$\sum_{i=1}^k \lambda_i(f_i(\bar{x}) + g_i(A\bar{x})) \leq \sum_{i=1}^k \lambda_i(f_i(x) + g_i(Ax)) \ \forall x \in X.$$

This is the reason why we first investigate, for a fixed $\lambda = (\lambda_1, \ldots, \lambda_k)^T \in \operatorname{int}(\mathbb{R}^k_+)$, the following scalar optimization problem

$$(PF_\lambda^A) \quad \inf_{x \in X} \left\{ \sum_{i=1}^k \lambda_i(f_i(x) + g_i(Ax)) \right\}.$$

The vector dual problem to (PVF^A) that we introduce in this section will have its origins in the conjugate scalar dual to (PF_λ^A). Via the investigations done in subsection 3.1.2 one can associate to (PF_λ^A) (see the primal-dual pair $(P^A) - (D^A)$) the dual problem

$$\sup_{y^* \in Y^*} \left\{ -\left(\sum_{i=1}^k \lambda_i f_i\right)^*(A^* y^*) - \left(\sum_{i=1}^k \lambda_i g_i\right)^*(-y^*) \right\},$$

and, on the other hand (see the primal-dual pair $(P^\Sigma) - (D^\Sigma)$ in subsection 3.1.2 and Proposition 2.3.2(e)), the dual problem

$$\sup_{\substack{x^{i*} \in X^*, i=1,\ldots,k, \\ \sum_{i=1}^k \lambda_i x^{i*} = 0}} \left\{ -\sum_{i=1}^k \lambda_i (f_i + g_i \circ A)^*(x^{i*}) \right\}.$$

But a valuable scalar dual problem to (PF_λ^A), which should constitute the starting point for the formulation of the vector dual needs to have separated the functions f_i and g_i, $i = 1, \dots, k$, in its formula. In order to construct such a problem we employ the general approach from section 3.1. To this aim, let us consider the perturbation function $\Phi_\lambda^A : X \times X^k \times Y^k \to \overline{\mathbb{R}}$,

$$\Phi_\lambda^A(x, x^1, \dots, x^k, y^1, \dots, y^k) = \sum_{i=1}^k \lambda_i(f_i(x + x^i) + g_i(Ax + y^i)),$$

with $(x^1, \dots, x^k, y^1, \dots, y^k) \in X^k \times Y^k$ as perturbation variables. The conjugate of Φ_λ^A, $(\Phi_\lambda^A)^* : X^* \times (X^*)^k \times (Y^*)^k \to \overline{\mathbb{R}}$, is given by the following formula

$$(\Phi_\lambda^A)^*(x^*, x^{1*}, \dots, x^{k*}, y^{1*}, \dots, y^{k*}) = \sup_{\substack{x, x^i \in X, y^i \in Y, \\ i=1,\dots,k}} \left\{ \langle x^*, x\rangle + \sum_{i=1}^k \langle x^{i*}, x^i\rangle \right.$$

$$\left. + \sum_{i=1}^k \langle y^{i*}, y^i\rangle - \sum_{i=1}^k (\lambda_i f_i)(x + x^i) - \sum_{i=1}^k (\lambda_i g_i)(Ax + y^i) \right\}$$

$$= \begin{cases} \sum\limits_{i=1}^k \left((\lambda_i f_i)^*(x^{i*}) + (\lambda_i g_i)^*(y^{i*})\right), & \text{if } \sum\limits_{i=1}^k (x^{i*} + A^* y^{i*}) = x^*, \\ +\infty, & \text{otherwise.} \end{cases}$$

This provides the following conjugate dual to (PF_λ^A)

$$(DF_\lambda^A) \quad \sup_{\substack{x^{i*} \in X^*, y^{i*} \in Y_i^*, i=1,\dots,k, \\ \sum_{i=1}^k (x^{i*} + A^* y^{i*}) = 0}} \left\{ -\sum_{i=1}^k \Big((\lambda_i f_i)^*(x^{i*}) + (\lambda_i g_i)^*(y^{i*}) \Big) \right\},$$

which, via Proposition 2.3.2(*e*), can be equivalently written as

$$(DF_\lambda^A) \quad \sup_{\substack{x^{i*} \in X^*, y^{i*} \in Y_i^*, i=1,\dots,k, \\ \sum_{i=1}^k \lambda_i (x^{i*} + A^* y^{i*}) = 0}} \left\{ -\sum_{i=1}^k \lambda_i f_i^*(x^{i*}) - \sum_{i=1}^k \lambda_i g_i^*(y^{i*}) \right\}.$$

In what follows we provide regularity conditions for the primal-dual pair $(PF_\lambda^A) - (DF_\lambda^A)$ which are deduced from the general ones given in section 3.2. Let us notice that we consider here only generalized interior point conditions, since they do not depend on the choice of λ, which would not be the case for the closedness type ones.

The reason why we proceed in this way is given by the fact that these regularity conditions will be employed in guaranteeing strong duality for the vector optimization problem (PVF^A) and its dual that is introduced below.

Guaranteeing the scalar strong duality for $(PF_\lambda^A) - (DF_\lambda^A)$ will be an intermediate step in this approach and we have to ensure that this is the case for all $\lambda \in \operatorname{int}(\mathbb{R}_+^k)$. This fact motivates the need to use regularity conditions that are independent of λ.

One can notice that whenever f_i and g_i, $i = 1, \ldots, k$, are convex, then Φ_λ^A is convex, too. For this primal-dual pair one can consider the following regularity condition (see the discussion made in subsection 3.2.2 in connection to the primal-dual pair $(P^\Sigma) - (D^\Sigma)$)

$$(RCF_1^A) \left| \begin{array}{l} \exists x' \in \bigcap_{i=1}^k (\operatorname{dom} f_i \cap A^{-1}(\operatorname{dom} g_i)) \text{ such that } k-1 \text{ of the} \\ \text{functions } f_i,\ i = 1, \ldots, k, \text{ are continuous at } x' \text{ and} \\ g_i \text{ is continuous at } Ax',\ i = 1, \ldots, k. \end{array} \right.$$

Before stating further regularity conditions, we also note that if f_i and g_i, $i = 1, \ldots, k$, are lower semicontinuous, then Φ_λ^A is lower semicontinuous, too. Further, it holds $(x^1, \ldots, x^k, y^1, \ldots, y^k) \in \operatorname{Pr}_{X^k \times Y^k}(\operatorname{dom} \Phi_\lambda^A)$ if and only if there exists an $x \in X$ such that $x^i \in \operatorname{dom} f_i - x$ and $y^i \in \operatorname{dom} g_i - Ax$ for $i = 1, \ldots, k$. This is further equivalent to $(x^1, \ldots, x^k, y^1, \ldots, y^k) \in \prod_{i=1}^k \operatorname{dom} f_i \times \prod_{i=1}^k \operatorname{dom} g_i - \Delta_{X^k, A}$, where $\Delta_{X^k,A} = \{(x, \ldots, x, Ax, \ldots, Ax) : x \in X\} \subseteq X^k \times Y^k$. This leads to the following regularity condition (obtained via (RC_2^Φ))

$$(RCF_2^A) \left| \begin{array}{l} X \text{ and } Y \text{ are Fréchet spaces, } f_i \text{ and } g_i \text{ are lower semicontinuous,} \\ i = 1, \ldots, k, \text{ and } 0 \in \operatorname{sqri}\left(\prod_{i=1}^k \operatorname{dom} f_i \times \prod_{i=1}^k \operatorname{dom} g_i - \Delta_{X^k,A} \right), \end{array} \right.$$

along with its stronger versions

$$(RCF_{2'}^A) \left| \begin{array}{l} X \text{ and } Y \text{ are Fréchet spaces, } f_i \text{ and } g_i \text{ are lower semicontinuous,} \\ i = 1, \ldots, k, \text{ and } 0 \in \operatorname{core}\left(\prod_{i=1}^k \operatorname{dom} f_i \times \prod_{i=1}^k \operatorname{dom} g_i - \Delta_{X^k,A} \right) \end{array} \right.$$

and

$$(RCF_{2''}^A) \left| \begin{array}{l} X \text{ and } Y \text{ are Fréchet spaces, } f_i \text{ and } g_i \text{ are lower semicontinuous,} \\ i = 1, \ldots, k, \text{ and } 0 \in \operatorname{int}\left(\prod_{i=1}^k \operatorname{dom} f_i \times \prod_{i=1}^k \operatorname{dom} g_i - \Delta_{X^k,A} \right), \end{array} \right.$$

which are in fact equivalent. In the finite dimensional case one has from (RC_3^Φ)

$$(RCF_3^A) \left| \begin{array}{l} \dim\left(\operatorname{lin}\left(\prod_{i=1}^k \operatorname{dom} f_i \times \prod_{i=1}^k \operatorname{dom} g_i - \Delta_{X^k,A} \right)\right) < +\infty \\ \text{and } 0 \in \operatorname{ri}\left(\prod_{i=1}^k \operatorname{dom} f_i \times \prod_{i=1}^k \operatorname{dom} g_i - \Delta_{X^k,A} \right). \end{array} \right.$$

We can state now the strong duality theorem for the scalar primal-dual pair $(PF_\lambda^A) - (DF_\lambda^A)$.

Theorem 5.1.1. *Let $f_i : X \to \overline{\mathbb{R}}$ and $g_i : Y \to \overline{\mathbb{R}}$, $i = 1, \ldots, k$, be proper and convex functions, $A \in \mathcal{L}(X, Y)$ such that $\cap_{i=1}^k (\operatorname{dom} f_i \cap A^{-1}(\operatorname{dom} g_i)) \neq \emptyset$ and*

$\lambda \in \operatorname{int}(\mathbb{R}^k_+)$ be arbitrarily chosen. If one of the regularity conditions (RCF_i^A), $i \in \{1,2,3\}$, is fulfilled, then $v(PF_\lambda^A) = v(DF_\lambda^A)$ and the dual has an optimal solution.

Remark 5.1.1. In order to deliver strong duality statements for (PF_λ^A) and (DF_λ^A) one can also combine the regularity conditions given in subsection 3.2.2. We exemplify this here by the ones expressed via the strong quasi-relative interior. Thus, assuming

$$\left| \begin{array}{l} X \text{ and } Y \text{ are Fréchet spaces, } f_i \text{ and } g_i \text{ are lower semicontinuous,} \\ i=1,\dots,k,\ 0 \in \operatorname{sqri}\left(\bigcap\limits_{i=1}^k \operatorname{dom} g_i - A\left(\bigcap\limits_{i=1}^k \operatorname{dom} f_i \right)\right), \\ 0 \in \operatorname{sqri}\left(\prod\limits_{i=1}^k \operatorname{dom} f_i - \Delta_{X^k} \right) \text{ and } 0 \in \operatorname{sqri}\left(\prod\limits_{i=1}^k \operatorname{dom} g_i - \Delta_{Y^k} \right), \end{array} \right.$$

or

$$\left| \begin{array}{l} X \text{ and } Y \text{ are Fréchet spaces, } f_i \text{ and } g_i \text{ are lower semicontinuous,} \\ i=1,\dots,k,\ 0 \in \operatorname{sqri}\left(\prod\limits_{i=1}^k \big(\operatorname{dom} f_i \cap A^{-1}(\operatorname{dom} g_i)\big) - \Delta_{X^k} \right) \\ \text{and } 0 \in \operatorname{sqri}\big(\operatorname{dom} g_i - A(\operatorname{dom} f_i)\big),\ i=1,\dots,k, \end{array} \right.$$

guarantees the existence of strong duality for the primal-dual pair $(PF_\lambda^A) - (DF_\lambda^A)$ for all $\lambda \in \operatorname{int}(\mathbb{R}^k_+)$.

Let us come now to the formulation of the necessary and sufficient optimality conditions for the primal-dual pair $(PF_\lambda^A) - (DF_\lambda^A)$.

Theorem 5.1.2. *(a) Let $f_i : X \to \overline{\mathbb{R}}$ and $g_i : Y \to \overline{\mathbb{R}}$, $i = 1,\dots,k$, be proper and convex functions, $A \in \mathcal{L}(X,Y)$ such that $\cap_{i=1}^k(\operatorname{dom} f_i \cap A^{-1}(\operatorname{dom} g_i)) \neq \emptyset$ and $\lambda \in \operatorname{int}(\mathbb{R}^k_+)$ be arbitrarily chosen. If $\bar{x} \in X$ is an optimal solution to (PF_λ^A) and one of the regularity conditions (RCF_i^A), $i \in \{1,2,3\}$, is fulfilled, then there exists $(\bar{x}^{1*},\dots,\bar{x}^{k*},\bar{y}^{1*},\dots,\bar{y}^{k*}) \in (X^*)^k \times (Y^*)^k$, an optimal solution to the dual problem (DF_λ^A), such that*

(i) $\sum_{i=1}^k \lambda_i(\bar{x}^{i*} + A^*\bar{y}^{i*}) = 0$;

(ii) $f_i(\bar{x}) + f_i^*(\bar{x}^{i*}) = \langle \bar{x}^{i*}, \bar{x}\rangle$, $i = 1,\dots,k$;

(iii) $g_i(A\bar{x}) + g_i^*(\bar{y}^{i*}) = \langle \bar{y}^{i*}, A\bar{x}\rangle$, $i = 1,\dots,k$.

(b) For a given $\lambda \in \operatorname{int}(\mathbb{R}^k_+)$ assume that $\bar{x} \in X$ and $(\bar{x}^{1*},\dots,\bar{x}^{k*},\bar{y}^{1*},\dots,\bar{y}^{k*}) \in (X^*)^k \times (Y^*)^k$ fulfill the relations $(i)-(iii)$. Then $\bar{x}$ is an optimal solution to (PF_λ^A), $(\bar{x}^{1*},\dots,\bar{x}^{k*},\bar{y}^{1*},\dots,\bar{y}^{k*})$ is an optimal solution to (DF_λ^A) and $v(PF_\lambda^A) = v(DF_\lambda^A)$.

Proof. The proof follows in the lines of the ones given for Theorem 3.3.4 and Theorem 3.3.13. □

Remark 5.1.2. The optimality conditions $(i)-(iii)$ in Theorem 5.1.2 can be equivalently written as

$$\sum_{i=1}^{k} \lambda_i(\bar{x}^{i*} + A^*\bar{y}^{i*}) = 0, \bar{x}^{i*} \in \partial f_i(\bar{x}) \text{ and } \bar{y}_i^* \in \partial g_i(A\bar{x}), i = 1, ..., m.$$

We can introduce now the following multiobjective dual problem to (PVF^A), which can also be seen as a *Fenchel type vector dual* (see also section 4.1),

$$(DVF^A) \qquad \underset{(\lambda,x^*,y^*,t)\in\mathcal{B}^{FA}}{\text{Max}} h^{FA}(\lambda, x^*, y^*, t),$$

where

$$\mathcal{B}^{FA} = \Big\{(\lambda, x^*, y^*, t) \in \text{int}(\mathbb{R}^k_+) \times \prod_{i=1}^{k} \text{dom} f_i^* \times \prod_{i=1}^{k} \text{dom} g_i^* \times \mathbb{R}^k : \\ \lambda = (\lambda_1, \dots, \lambda_k)^T, x^* = (x^{1*}, \dots, x^{k*}), \\ y^* = (y^{1*}, \dots, y^{k*}), t = (t_1, \dots, t_k)^T, \\ \sum_{i=1}^{k} \lambda_i(x^{i*} + A^*y^{i*}) = 0, \sum_{i=1}^{k} \lambda_i t_i = 0\Big\}$$

and

$$h^{FA}(\lambda, x^*, y^*, t) = \begin{pmatrix} -f_1^*(x^{1*}) - g_1^*(y^{1*}) + t_1 \\ \vdots \\ -f_k^*(x^{k*}) - g_k^*(y^{k*}) + t_k \end{pmatrix}.$$

The properness of the functions f_i and g_i, $i = 1, \dots, k$, and Lemma 2.3.1(a) ensure that $h^{FA}(\mathcal{B}^{FA}) \subseteq \mathbb{R}^k$. According to Definition 2.5.1, an element $(\bar{\lambda}, \bar{x}^*, \bar{y}^*, \bar{t})$ is said to be efficient to the problem (DVF^A) if $h^{FA}(\bar{\lambda}, \bar{x}^*, \bar{y}^*, \bar{t}) \in \text{Max}(h^{FA}(\mathcal{B}^{FA}), \mathbb{R}^k_+)$. Moreover, an element $(\bar{\lambda}, \bar{x}^*, \bar{y}^*, \bar{t})$ is weakly efficient to (DVF^A) if $h^{FA}(\bar{\lambda}, \bar{x}^*, \bar{y}^*, \bar{t}) \in \text{WMax}(h^{FA}(\mathcal{B}^{FA}), \mathbb{R}^k_+)$. The weak duality property for (PVF^A) and (DVF^A) follows.

Theorem 5.1.3. *There is no $x \in X$ and no $(\lambda, x^*, y^*, t) \in \mathcal{B}^{FA}$ such that $f_i(x) + g_i(Ax) \leq h_i^{FA}(\lambda, x^*, y^*, t)$, $i = 1, \dots, k$, and $f_j(x) + g_j(Ax) < h_j^{FA}(\lambda, x^*, y^*, t)$ for at least one $j \in \{1, \dots, k\}$.*

Proof. Let us suppose the contrary, namely that there exist $x \in X$ and $(x^*, y^*, \lambda, t) \in \mathcal{B}^{FA}$ fulfilling $f_i(x)+g_i(Ax) \leq h_i^{FA}(\lambda, x^*, y^*, t)$, for $i = 1, \dots, k$, and $f_j(x) + g_j(Ax) < h_j^{FA}(\lambda, x^*, y^*, t)$ for at least one $j \in \{1, \dots, k\}$. As $\lambda_i > 0$ for $i = 1, \dots, k$, we get $\sum_{i=1}^k \lambda_i(f_i(x)+g_i(Ax)) < \sum_{i=1}^k \lambda_i(-f_i^*(x^{i*}) - g_i^*(y^{i*})+t_i) = -\sum_{i=1}^k \lambda_i(f_i^*(x^{i*})+g_i^*(y^{i*}))$. Thus we acquire $\sum_{i=1}^k \lambda_i(f_i(x)+g_i(Ax)) < -\sum_{i=1}^k \lambda_i(f_i^*(x^{i*}) + g_i^*(y^{i*}))$, which is impossible because of the weak duality for the problems (PF_λ^A) and (DF_λ^A), which secures the reverse inequality. □

We come now to the vector strong duality statement for (PVF^A) and (DVF^A). In what follows we assume that the functions f_i and g_i, $i = 1, \dots, k$, are convex. In order to maintain the symmetry to the investigations made in the previous chapter, we assume also here for the strong vector duality results the fulfillment of a regularity condition in general Hausdorff locally convex spaces and then remark that these remain valid also when some other regularity conditions are verified. At first we work with (RCF_1^A), renamed as

$$(RCVF^A) \left| \begin{array}{l} \exists x' \in \bigcap_{i=1}^{k} (\operatorname{dom} f_i \cap A^{-1}(\operatorname{dom} g_i)) \text{ such that } k-1 \text{ of the} \\ \text{functions } f_i,\ i=1,\dots,k, \text{ are continuous at } x' \text{ and} \\ g_i \text{ is continuous at } Ax',\ i=1,\dots,k. \end{array} \right.$$

Theorem 5.1.4. *Assume that the regularity condition $(RCVF^A)$ is fulfilled. If $\bar{x} \in X$ is a properly efficient solution to (PVF^A) then there exists $(\bar{\lambda}, \bar{x}^*, \bar{y}^*, \bar{t})$, an efficient solution to (DVF^A), such that $f_i(\bar{x}) + g_i(A\bar{x}) = h_i^{FA}(\bar{\lambda}, \bar{x}^*, \bar{y}^*, \bar{t})$ for $i = 1, \dots, k$.*

Proof. As $\bar{x}$ is a properly efficient solution to (PVF^A), there exists a vector $\bar{\lambda} \in \operatorname{int}(\mathbb{R}^k_+)$ such that $\bar{x}$ is an optimal solution to the scalar problem $(PF^A_{\bar{\lambda}})$. Even more, according to Theorem 5.1.2 its dual problem $(DF^A_{\bar{\lambda}})$ admits an optimal solution $(\bar{x}^{1*}, \dots, \bar{x}^{k*}, \bar{y}^{1*}, \dots, \bar{y}^{k*})$ such that the optimality conditions $(i) - (iii)$ of Theorem 5.1.2 are fulfilled. Further let be $\bar{x}^* := (\bar{x}^{1*}, \dots, \bar{x}^{k*})$, $\bar{y}^* := (\bar{y}^{1*}, \dots, \bar{y}^{k*})$ and $\bar{t} := (\bar{t}_1, \dots, \bar{t}_k)^T$ with $\bar{t}_i = \langle \bar{x}^{i*} + A^*\bar{y}^{i*}, \bar{x} \rangle$, for $i = 1, \dots, k$. Thus $\bar{x}^* \in \prod_{i=1}^k \operatorname{dom} f_i^*$, $\bar{y}^* \in \prod_{i=1}^k \operatorname{dom} g_i^*$, $\sum_{i=1}^k \bar{\lambda}_i \bar{t}_i = 0$ and $(\bar{\lambda}, \bar{x}^*, \bar{y}^*, \bar{t}) \in \mathcal{B}^{FA}$. Using the assertions $(ii) - (iii)$ of Theorem 5.1.2, we get for $i = 1, \dots, k$, $h_i(\bar{\lambda}, \bar{x}^*, \bar{y}^*, \bar{t}) = -f_i^*(\bar{x}^{i*}) - g_i^*(\bar{y}^{i*}) + \bar{t}_i = f_i(\bar{x}) - \langle \bar{x}^{i*}, \bar{x} \rangle + g_i(A\bar{x}) - \langle \bar{y}^{i*}, A\bar{x} \rangle + \langle \bar{x}^{i*} + A^*\bar{y}^{i*}, \bar{x} \rangle = f_i(\bar{x}) + g_i(A\bar{x})$. The efficiency of $(\bar{\lambda}, \bar{x}^*, \bar{y}^*, \bar{t})$ is a direct consequence of Theorem 5.1.3. □

To be able to give a converse duality assertion for the vector problems (PVF^A) and (DVF^A) we need the following statement.

Theorem 5.1.5. *Assume that $\mathcal{B}^{FA}$ is nonempty and that the regularity condition $(RCVF^A)$ is fulfilled. Then*

$$\mathbb{R}^k \backslash \operatorname{cl}((f + g \circ A)(\cap_{i=1}^k (\operatorname{dom} f_i \cap A^{-1}(\operatorname{dom} g_i))) + \mathbb{R}^k_+) \subseteq h^{FA}(\mathcal{B}^{FA}) - \operatorname{int}(\mathbb{R}^k_+).$$

Proof. Let be $\bar{v} \in \mathbb{R}^k \backslash \operatorname{cl}((f + g \circ A)(\cap_{i=1}^k (\operatorname{dom} f_i \cap A^{-1}(\operatorname{dom} g_i))) + \mathbb{R}^k_+)$. Similarly to the proof of Theorem 4.1.3 (see also Theorem 4.3.3) one can prove that there exists $\bar{\lambda} \in \operatorname{int}(\mathbb{R}^k_+)$ such that

$$\sum_{i=1}^k \bar{\lambda}_i \bar{v}_i < \inf_{x \in X} \left\{ \sum_{i=1}^k \bar{\lambda}_i \big(f_i(x) + g_i(Ax)\big) \right\}.$$

According to Theorem 5.1.1, there exist $\bar{x}^* = (\bar{x}^{1*}, \dots, \bar{x}^{k*}) \in \prod_{i=1}^k \operatorname{dom} f_i^*$ and $\bar{y}^* = (\bar{y}^{1*}, \dots, \bar{y}^{k*}) \in \prod_{i=1}^k \operatorname{dom} g_i^*$ such that $\sum_{i=1}^k \bar{\lambda}_i (\bar{x}^{i*} + A^*\bar{y}^{i*}) = 0$ and

$$\inf_{x \in X} \left\{ \sum_{i=1}^{k} \bar{\lambda}_i (f_i(x) + g_i(Ax)) \right\} = - \sum_{i=1}^{k} \bar{\lambda}_i (f_i^*(\bar{x}^{i*}) + g_i^*(\bar{y}^{i*})).$$

Thus $(\bar{\lambda}, \bar{x}^*, \bar{y}^*, 0) \in \mathcal{B}^{FA}$ and it holds

$$\sum_{i=1}^{k} \bar{\lambda}_i \bar{v}_i < \sum_{i=1}^{k} \bar{\lambda}_i h_i^{FA}(\bar{\lambda}, \bar{x}^*, \bar{y}^*, 0). \tag{5.1}$$

Consider the hyperplane with the normal vector $\bar{\lambda}$

$$H = \left\{ h^{FA}(\bar{\lambda}, \bar{x}^*, \bar{y}^*, 0) + t : t \in \mathbb{R}^k, \sum_{i=1}^{k} \bar{\lambda}_i t_i = 0 \right\}$$

$$= \left\{ v \in \mathbb{R}^k : \sum_{i=1}^{k} \bar{\lambda}_i v_i = \sum_{i=1}^{k} \bar{\lambda}_i h_i^{FA}(\bar{\lambda}, \bar{x}^*, \bar{y}^*, 0) \right\}.$$

One can easily see that $H \subseteq h^{FA}(\mathcal{B}^{FA})$, while from (5.1) we deduce that $\bar{v}$ is an element of the open halfspace $H^- = \{v \in \mathbb{R}^k : \sum_{i=1}^{k} \bar{\lambda}_i v_i < \sum_{i=1}^{k} \bar{\lambda}_i h_i^{FA}(\bar{\lambda}, \bar{x}^*, \bar{y}^*, 0)\}$. The orthogonal projection of $\bar{v}$ on H is an element $\tilde{v} := \bar{v} + \delta\bar{\lambda} \in H \subseteq h^{FA}(\mathcal{B}^{FA})$ with $\delta > 0$ (we refer for instance to [58] for an explicit formula for δ). Since $\bar{\lambda} \in \operatorname{int}(\mathbb{R}^k_+)$, it follows that $\bar{v} \in h^{FA}(\mathcal{B}^{FA}) - \operatorname{int}(\mathbb{R}^k_+)$. □

Remark 5.1.3. We refer the reader to a comparison of the result in Theorem 5.1.5 with the one given in Theorem 4.3.3 for the primal-dual pair $(PVG) - (DVG)$. A direct consequence of the latter result is that, when (RCV^{Φ}) is fulfilled and $(\bar{\lambda}, \bar{y}^*, \bar{v})$ is an efficient solution to (DVG), then $h^G(\bar{\lambda}, \bar{y}^*, \bar{v}) \in \operatorname{cl}(F(\operatorname{dom} F) + K)$. As proven below, a similar result can be given for the primal-dual pair $(PVF^A) - (DVF^A)$, but this does not result as directly as for $(PVG) - (DVG)$. This result ensures the fact that the duality gap may be excluded in the sense that the objective value of every dual weakly efficient solution and implicitly of every dual efficient solution is the limit of a sequence of elements from the image set of the primal problem.

Theorem 5.1.6. *Assume that the regularity condition $(RCVF^A)$ is fulfilled and that $(\bar{\lambda}, \bar{x}^*, \bar{y}^*, \bar{t}) \in \mathcal{B}^{FA}$ is a weakly efficient solution to (DVF^A). Then $h^{FA}(\bar{\lambda}, \bar{x}^*, \bar{y}^*, \bar{t}) \in \operatorname{cl}((f + g \circ A)(\cap_{i=1}^{k}(\operatorname{dom} f_i \cap A^{-1}(\operatorname{dom} g_i))))$.*

Proof. Since $(\bar{\lambda}, \bar{x}^*, \bar{y}^*, \bar{t}) \in \mathcal{B}^{FA}$ is a weakly efficient solution to (DVF^A), by Theorem 5.1.5 follows that

$$h^{FA}(\bar{\lambda}, \bar{x}^*, \bar{y}^*, \bar{t}) \in \operatorname{cl}\left((f + g \circ A)\left(\bigcap_{i=1}^{k} (\operatorname{dom} f_i \cap A^{-1}(\operatorname{dom} g_i)) \right) + \mathbb{R}^k_+ \right).$$

Assuming the contrary implies the existence of $\bar{v} \in h^{FA}(\mathcal{B}^{FA})$ such that $h^{FA}(\bar{\lambda}, \bar{x}^*, \bar{y}^*, \bar{t}) < \bar{v}$. But this contradicts the weak efficiency of $(\bar{\lambda}, \bar{x}^*, \bar{y}^*, \bar{t})$ to the vector dual problem.

Then there exist $\{v^l\} \subseteq (f + g \circ A)(\cap_{i=1}^k(\operatorname{dom} f_i \cap A^{-1}(\operatorname{dom} g_i)))$ and $r^l \in \mathbb{R}^k_+$ such that $v^l + r^l \to h^{FA}(\bar{\lambda}, \bar{x}^*, \bar{y}^*, \bar{t})$ when $l \to +\infty$. For all $l \geq 1$ let be $x^l \in \cap_{i=1}^k(\operatorname{dom} f_i \cap A^{-1}(\operatorname{dom} g_i))$ with $v^l = (f + g \circ A)(x^l)$. The weak duality theorem for $(PF_{\bar{\lambda}}^A) - (DF_{\bar{\lambda}}^A)$ yields for all $l \geq 1$

$$\sum_{i=1}^k \bar{\lambda}_i h_i^{FA}(\bar{\lambda}, \bar{x}^*, \bar{y}^*, \bar{t}) = -\sum_{i=1}^k \bar{\lambda}_i (f_i^*(\bar{x}^{i*}) + g_i^*(\bar{y}^{i*}))$$

$$\leq \sum_{i=1}^k \bar{\lambda}_i (f_i(x^l) + g_i(Ax^l)) = \sum_{i=1}^k \bar{\lambda}_i v_i^l \leq \sum_{i=1}^k \bar{\lambda}_i (v_i^l + r_i^l).$$

We have that $\lim_{l\to+\infty} \sum_{i=1}^k \bar{\lambda}_i (v_i^l + r_i^l) = \sum_{i=1}^k \bar{\lambda}_i h_i^{FA}(\bar{\lambda}, \bar{x}^*, \bar{y}^*, \bar{t})$ and this implies that

$$\lim_{l\to+\infty} \sum_{i=1}^k \bar{\lambda}_i v_i^l = \sum_{i=1}^k \bar{\lambda}_i h_i^{FA}(\bar{\lambda}, \bar{x}^*, \bar{y}^*, \bar{t}).$$

Consequently, $\lim_{l\to+\infty} \sum_{i=1}^k \bar{\lambda}_i r_i^l = 0$. Since $r^l \in \mathbb{R}^k_+$ for all $l \geq 1$, this can be the case only if $\lim_{l\to+\infty} r^l = 0$. In conclusion, $\lim_{l\to+\infty} v^l = h^{FA}(\bar{\lambda}, \bar{x}^*, \bar{y}^*, \bar{t})$ and so $h^{FA}(\bar{\lambda}, \bar{x}^*, \bar{y}^*, \bar{t}) \in \operatorname{cl}((f + g \circ A)(\cap_{i=1}^k(\operatorname{dom} f_i \cap A^{-1}(\operatorname{dom} g_i))))$. □

A direct consequence of Theorem 5.1.6 is the following converse duality statement, which we state for dual weakly efficient solutions, its validity for dual efficient solutions being an immediate consequence.

Theorem 5.1.7. *Assume that the regularity condition $(RCVF^A)$ is fulfilled and that the set $(f + g \circ A)(\cap_{i=1}^k(\operatorname{dom} f_i \cap A^{-1}(\operatorname{dom} g_i))) + \mathbb{R}^k_+$ is closed. Then for every weakly efficient solution $(\bar{\lambda}, \bar{x}^*, \bar{y}^*, \bar{t})$ to (DVF^A) there exists a properly efficient solution $\bar{x} \in X$ to (PVF^A) such that $f_i(\bar{x}) + g_i(A\bar{x}) = h_i^{FA}(\bar{\lambda}, \bar{x}^*, \bar{y}^*, \bar{t})$ for $i = 1, \ldots, k$.*

Proof. As seen in the proof of Theorem 5.1.6, the weak efficiency of $(\bar{\lambda}, \bar{x}^*, \bar{y}^*, \bar{t})$ to the dual problem ensures that $h^{FA}(\bar{\lambda}, \bar{x}^*, \bar{y}^*, \bar{t}) \in \operatorname{cl}((f+g\circ A)(\cap_{i=1}^k(\operatorname{dom} f_i \cap A^{-1}(\operatorname{dom} g_i))) + \mathbb{R}^k_+) = (f + g \circ A)(\cap_{i=1}^k(\operatorname{dom} f_i \cap A^{-1}(\operatorname{dom} g_i))) + \mathbb{R}^k_+$. This means that $h^{FA}(\bar{\lambda}, \bar{x}^*, \bar{y}^*, \bar{t}) = (f + g \circ A)(\bar{x}) + \bar{r}$ for some $\bar{x} \in \cap_{i=1}^k(\operatorname{dom} f_i \cap A^{-1}(\operatorname{dom} g_i))$ and $\bar{r} \in \mathbb{R}^k_+$. But $\bar{r} = 0$, otherwise the weak duality (see Theorem 5.1.3) would be violated. It remains to prove that $\bar{x}$ is properly efficient to (PVF^A). Since $h^{FA}(\bar{\lambda}, \bar{x}^*, \bar{y}^*, \bar{t}) = (f + g \circ A)(\bar{x})$ it holds $\sum_{i=1}^k \bar{\lambda}_i (f_i + g_i \circ A)(\bar{x}) = \sum_{i=1}^k \bar{\lambda}_i h_i^{FA}(\bar{\lambda}, \bar{x}^*, \bar{y}^*, \bar{t}) = \sum_{i=1}^k \bar{\lambda}_i (-f_i^*(\bar{x}^{i*}) - g_i^*(\bar{y}^{i*})) \leq \inf_{x\in X} \left\{ \sum_{i=1}^k \bar{\lambda}_i (f_i + g_i \circ A)(x) \right\}$. Therefore $\bar{x}$ is an optimal solution to $(P_{\bar{\lambda}}^{FA})$ and this proves that $\bar{x}$ is properly efficient to (PVF^A). □

Remark 5.1.4. (a) One can easily notice that all the results given above remain valid if the vector dual (DVF^A) is slightly modified by replacing in the formulation of the feasible set $\mathcal{B}^{FA}$ the restriction $\sum_{i=1}^k \lambda_i t_i = 0$ with $\sum_{i=1}^k \lambda_i t_i \leq 0$.

(*b*) Since the proof of the theorem given above uses as a main tool the scalar strong duality result for (PF_λ^A) and (DF_λ^A) when $\lambda \in \text{int}(\mathbb{R}_+^k)$, it is clear that the regularity condition $(RCVF^A)$ can be replaced by any of the regularity conditions (RCF_i^A), $i \in \{2, 2', 2'', 3\}$. In theorems 5.1.5-5.1.7, the regularity condition can be replaced by the assumption that for all $\lambda \in \text{int}(\mathbb{R}_+^k)$ the optimization problem (PF_λ^A) is normal with respect to its conjugate dual problem (DF_λ^A).

(*c*) Theorem 5.1.7 remains valid even if instead of asking that $(f + g \circ A)(\cap_{i=1}^k (\text{dom}\, f_i \cap A^{-1}(\text{dom}\, g_i))) + \mathbb{R}_+^k$ is closed one assumes that $(f + g \circ A)(\cap_{i=1}^k (\text{dom}\, f_i \cap A^{-1}(\text{dom}\, g_i)))$ is closed.

(*d*) Combining the assertions of Theorem 5.1.4 and Theorem 5.1.7 makes it immediately clear that under the hypotheses of the latter any weakly efficient solution to (DVF^A) is also efficient to it.

(*e*) In case $k = 1$, i.e. if $V = \mathbb{R}$ and $K = \mathbb{R}_+$, the problems (PVF^A) and (DVF^A) become (see also subsection 3.1.2)

$$(P^A) \quad \inf_{x \in X} \{f(x) + g(Ax)\}$$

and, respectively,

$$(D^A) \quad \sup_{(\lambda, x^*, y^*, t) \in \mathcal{B}^{FA}} \{-f^*(x^*) - g^*(y^*) + t\},$$

where

$$\mathcal{B}^{FA} = \{(\lambda, x^*, y^*, t) \in \text{int}(\mathbb{R}_+) \times X^* \times Y^* \times \mathbb{R} : \lambda(x^* + A^* y^*) = 0, \lambda t = 0\}.$$

Obviously the dual problem can be equivalently written as

$$\sup_{y^* \in Y^*} \{-f^*(-A^* y^*) - g^*(y^*)\},$$

which is, indeed, the classical Fenchel dual problem to (P^A) (cf. subsection 3.1.2). This motivates giving the name Fenchel type vector dual problem to (DVF^A).

Remark 5.1.5. (*a*) In case $X = Y$, $A = \text{id}_X$ and $f_i, g_i : X \to \overline{\mathbb{R}}$, $i = 1, \dots, k$, are given proper functions, (PVF^A) becomes

$$(PVF^{\text{id}}) \quad \underset{x \in X}{\text{Min}} \begin{pmatrix} f_1(x) + g_1(x) \\ \vdots \\ f_k(x) + g_k(x) \end{pmatrix}.$$

With (DVF^A) one gets the following Fenchel type vector dual to (PVF^{id})

$$(DVF^{\mathrm{id}}) \quad \underset{(\lambda,x^*,y^*,t)\in\mathcal{B}^{F\,\mathrm{id}}}{\mathrm{Max}} h^{F\,\mathrm{id}}(\lambda,x^*,y^*,t),$$

where

$$\mathcal{B}^{F\,\mathrm{id}} = \Bigg\{(\lambda,x^*,y^*,t) \in \mathrm{int}(\mathbb{R}^k_+) \times \prod_{i=1}^k \mathrm{dom}\, f_i^* \times \prod_{i=1}^k \mathrm{dom}\, g_i^* \times \mathbb{R}^k : \\ \lambda = (\lambda_1,\ldots,\lambda_k)^T, x^* = (x^{1*},\ldots,x^{k*}), \\ y^* = (y^{1*},\ldots,y^{k*}), t = (t_1,\ldots,t_k)^T, \\ \sum_{i=1}^k \lambda_i(x^{i*}+y^{i*}) = 0, \sum_{i=1}^k \lambda_i t_i = 0\Bigg\}$$

and

$$h^{F\,\mathrm{id}}(\lambda,x^*,y^*,t) = \begin{pmatrix} -f_1^*(x^{1*}) - g_1^*(y^{1*}) + t_1 \\ \vdots \\ -f_k^*(x^{k*}) - g_k^*(y^{k*}) + t_k \end{pmatrix}.$$

The weak, strong and converse duality for $(PVF^{F\,\mathrm{id}}) - (DVF^{F\,\mathrm{id}})$ follow as particular instances of the corresponding statements given in this subsection.

(*b*) If one takes in the above setting that $g_i \equiv 0$ for $i = 1,\ldots,k$, the primal vector problem becomes

$$(PVF^f) \ \underset{x\in X}{\mathrm{Min}} \begin{pmatrix} f_1(x) \\ \vdots \\ f_k(x) \end{pmatrix},$$

while its Fenchel type vector dual turns into

$$(DVF^f) \quad \underset{(\lambda,x^*,t)\in\mathcal{B}^f}{\mathrm{Max}} h^f(\lambda,x^*,t),$$

where

$$\mathcal{B}^f = \Bigg\{(\lambda,x^*,t) \in \mathrm{int}(\mathbb{R}^k_+) \times \prod_{i=1}^k \mathrm{dom}\, f_i^* \times \mathbb{R}^k : \lambda = (\lambda_1,\ldots,\lambda_k)^T, \\ x^* = (x^{1*},\ldots,x^{k*}), t = (t_1,\ldots,t_k)^T, \\ \sum_{i=1}^k \lambda_i x^{i*} = 0, \sum_{i=1}^k \lambda_i t_i = 0\Bigg\}$$

and

$$h^f(\lambda,x^*,t) = \begin{pmatrix} -f_1^*(x^{1*}) + t_1 \\ \vdots \\ -f_k^*(x^{k*}) + t_k \end{pmatrix}.$$

By particularizing the corresponding statements given in this subsection we get weak, strong and converse duality for the primal-dual pair $(PVF^f) - (DVF^f)$.

5.1.2 Comparisons to (DV^A) and (DV^A_{BK})

Working in the same setting as above, in this subsection we investigate the relations between the image sets of the feasible sets through their objective functions for (DVF^A) and the other two Fenchel type vector duals introduced in section 4.1, (DV^A) and (DV^A_{BK}), when $V = \mathbb{R}^k$. In this special instance the duals considered in the above mentioned section look like

$$(DV^A) \qquad \operatorname*{Max}_{(\lambda,y^*,v)\in\mathcal{B}^A} h^A(\lambda, y^*, v),$$

where

$$\mathcal{B}^A = \Bigg\{(\lambda, y^*, v) \in \operatorname{int}(\mathbb{R}^k_+) \times Y^* \times \mathbb{R}^k : \\ \sum_{i=1}^k \lambda_i v_i \leq -\Big(\sum_{i=1}^k \lambda_i f_i\Big)^*(-A^*y^*) - \Big(\sum_{i=1}^k \lambda_i g_i\Big)^*(y^*)\Bigg\}$$

and

$$h^A(\lambda, y^*, v) = v,$$

and, respectively,

$$(DV^A_{BK}) \qquad \operatorname*{Max}_{(\lambda,y^*,v)\in\mathcal{B}^A_{BK}} h^A_{BK}(\lambda, y^*, v),$$

where

$$\mathcal{B}^A_{BK} = \Bigg\{(\lambda, y^*, v) \in \operatorname{int}(\mathbb{R}^k_+) \times Y^* \times \mathbb{R}^k : \\ \sum_{i=1}^k \lambda_i v_i = -\Big(\sum_{i=1}^k \lambda_i f_i\Big)^*(-A^*y^*) - \Big(\sum_{i=1}^k \lambda_i g_i\Big)^*(y^*)\Bigg\}$$

and

$$h^A_{BK}(\lambda, y^*, v) = v.$$

We noticed in subsection 4.1.1 that $h^A_{BK}(\mathcal{B}^A_{BK}) \subseteq h^A(\mathcal{B}^A)$. In the following we prove that whenever a regularity condition is fulfilled it holds $h^A_{BK}(\mathcal{B}^A_{BK}) \subseteq h^{FA}(\mathcal{B}^{FA}) \subseteq h^A(\mathcal{B}^A)$. We begin by proving a general result.

Proposition 5.1.8. *It holds $h^{FA}(\mathcal{B}^{FA}) \subseteq h^A(\mathcal{B}^A)$.*

Proof. Let be $(\lambda, x^*, y^*, t) \in \mathcal{B}^{FA}$ arbitrarily chosen and set $z^* := \sum_{i=1}^k \lambda_i y^{i*} \in Y^*$. It holds $h^{FA}(\lambda, x^*, y^*, t) \in \mathbb{R}^k$, $\sum_{i=1}^k \lambda_i x^{i*} = -A^*(\sum_{i=1}^k \lambda_i^{i*}) = -A^*z^*$ and, consequently,

$$\sum_{i=1}^k \lambda_i h_i^{FA}(\lambda, x^*, y^*, t) = \sum_{i=1}^k \lambda_i\big(-f_i^*(x^{i*}) - g_i^*(y^{i*}) + t_i\big) = -\sum_{i=1}^k \big((\lambda_i f_i)^*(\lambda_i x^{i*})$$

$$+(\lambda_i g_i)^*(\lambda_i y^{i*})) \leq -\left(\sum_{i=1}^k \lambda_i f_i\right)^*(-A^*z^*) - \left(\sum_{i=1}^k \lambda_i g_i\right)^*(z^*).$$

Thus for $v := h^{FA}(\lambda, x^*, y^*, t)$ it holds $(\lambda, z^*, v) \in \mathcal{B}^A$ and, consequently, $h^{FA}(\lambda, x^*, y^*, t) \in h^A(\mathcal{B}^A)$. □

Next we investigate the relation between $h^A_{BK}(\mathcal{B}^A_{BK})$ and $h^{FA}(\mathcal{B}^{FA})$.

Proposition 5.1.9. *Assume that one of the regularity conditions* (RC_i^{Σ}), $i \in \{1,2,3\}$, *stated for* $\{f_1, \ldots, f_k\}$ *and, respectively,* $\{g_1, \ldots, g_k\}$ *is fulfilled. Then it holds* $h^A_{BK}(\mathcal{B}^A_{BK}) \subseteq h^{FA}(\mathcal{B}^{FA})$.

Proof. Let be $v \in h^A_{BK}(\mathcal{B}^A_{BK})$. Then there exist $\lambda \in \operatorname{int}(\mathbb{R}^k_+)$ and $z^* \in Y^*$ such that $(\lambda, z^*, v) \in \mathcal{B}^A_{BK}$. Furthermore, $\sum_{i=1}^k \lambda_i v_i = -\left(\sum_{i=1}^k \lambda_i f_i\right)^*(-A^*z^*) - \left(\sum_{i=1}^k \lambda_i g_i\right)^*(z^*)$. By Theorem 3.5.8(a), there exist $x^* = (x^{1*}, \ldots, x^{k*}) \in (X^*)^k$ and $y^* = (y^{1*}, \ldots, y^{k*}) \in (Y^*)^k$ such that $\sum_{i=1}^k \lambda_i x^{i*} = -A^*z^*$, $\sum_{i=1}^k \lambda_i y^{i*} = z^*$, $\left(\sum_{i=1}^k \lambda_i f_i\right)^*(-A^*z^*) = \sum_{i=1}^k \lambda_i f_i^*(x^{i*})$ and $\left(\sum_{i=1}^k \lambda_i g_i\right)^*(z^*) = \sum_{i=1}^k \lambda_i g_i^*(y^{i*})$. Therefore, $\sum_{i=1}^k \lambda_i(x^{i*} + A^*y^{i*}) = 0$ and $\sum_{i=1}^k \lambda_i v_i = -\sum_{i=1}^k \lambda_i f_i^*(x^{i*}) - \sum_{i=1}^k \lambda_i g_i^*(y^{i*})$. Taking $t_i := v_i + f_i^*(x^{i*}) + g_i^*(y^{i*}) \in \mathbb{R}$, for $i = 1, \ldots, k$, we have that $\sum_{i=1}^k \lambda_i t_i = 0$, and so $v = h^{FA}(\lambda, x^*, y^*, t) \in h^{FA}(\mathcal{B}^{FA})$. □

Under the hypotheses of Proposition 5.1.9, it holds

$$h^A_{BK}(\mathcal{B}^A_{BK}) \subseteq h^{FA}(\mathcal{B}^{FA}) \subseteq h^A(\mathcal{B}^A).$$

Obviously, $(RCVF^A)$ is a sufficient condition which guarantees these inclusions. Next we discuss two examples which prove that the inclusions of these image sets are in general strict, i.e.

$$h^A_{BK}(\mathcal{B}^A_{BK}) \subsetneq h^{FA}(\mathcal{B}^{FA}) \subsetneq h^A(\mathcal{B}^A).$$

Example 5.1.1. (a) Take $V = \mathbb{R}^2$, $K = \mathbb{R}^2_+$, $f, g : \mathbb{R} \to \mathbb{R}^2$ given by $f(x) = (x-1, -x-1)^T$ and $g(x) = (x, -x)^T$ for $x \in \mathbb{R}$, and $A = \mathrm{id}_{\mathbb{R}}$. We show that in this situation $h^{FA}(\mathcal{B}^{FA}) \subsetneq h^A(\mathcal{B}^A)$.

For $\lambda = (1,1)^T$, $z^* = 0$ and $v = (-2,-2)^T$, there is $(\lambda, z^*, v) \in \mathcal{B}^A$ and $v \in h^A(\mathcal{B}^A)$, since $\lambda_1 v_1 + \lambda_2 v_2 = -4 < -2 = -(f_1 + f_2)^*(-z^*) - (g_1 + g_2)^*(z^*)$. We show that $v \notin h^{FA}(\mathcal{B}^{FA})$. Let us suppose by contradiction that there exist $(\bar{\lambda}, \bar{x}^*, \bar{y}^*, \bar{t}) \in \mathcal{B}^{FA}$ such that $h^{FA}(\bar{\lambda}, \bar{x}^*, \bar{y}^*, \bar{t}) = v$. This means that $h_i^{FA}(\bar{\lambda}, \bar{x}^*, \bar{y}^*, \bar{t}) = -f_i^*(\bar{x}^{i*}) - g_i^*(\bar{y}^{i*}) + \bar{t}_i = -2$, for $i = 1,2$ and one must necessarily have that $\bar{x}^* = (1,-1)^T$ and $\bar{y}^* = (1,-1)^T$. Moreover, $\sum_{i=1}^2 \bar{\lambda}_i(\bar{x}^{i*} + \bar{y}^{i*}) = 0$, which means that $\bar{\lambda}_1 - \bar{\lambda}_2 = 0$. We obtain $-f_i^*(\bar{x}^{i*}) - g_i^*(\bar{y}^{i*}) + \bar{t}_i = -1 + \bar{t}_i = -2$, for $i = 1,2$, meaning that $\bar{t}_1 = \bar{t}_2 = -1$. Since we have supposed that $(\bar{\lambda}, \bar{x}^*, \bar{y}^*, \bar{t}) \in \mathcal{B}^{FA}$, the equality $\sum_{i=1}^2 \bar{\lambda}_i \bar{t}_i = -2\bar{\lambda}_1$ must hold. But this is a contradiction to $\lambda \in \operatorname{int}(\mathbb{R}^2_+)$.

Consequently, $v = (-2,-2)^T \in h^A(\mathcal{B}^A)$, there exists no $(\bar{\lambda}, \bar{x}^*, \bar{y}^*, \bar{t}) \in \mathcal{B}^{FA}$ such that $h^{FA}(\bar{\lambda}, \bar{x}^*, \bar{y}^*, \bar{t}) = v$, which shows that $h^{FA}(\mathcal{B}^{FA}) \subsetneq h^A(\mathcal{B}^A)$.

(*b*) Consider again $V = \mathbb{R}^2$, $K = \mathbb{R}^2_+$, while $f, g : \mathbb{R} \to \mathbb{R}^2$ are given by $f(x) = (2x^2 - 1, x^2)^T$ and $g(x) = (-2x, -x+1)^T$ for $x \in \mathbb{R}$, and $A = \mathrm{id}_{\mathbb{R}}$. We prove that $h^A_{BK}(\mathcal{B}^A_{BK}) \subsetneq h^{FA}(\mathcal{B}^{FA})$.

For $\lambda = (1,1)^T$, $x^* = (3,0)^T \in \operatorname{dom} f_1^* \times \operatorname{dom} f_2^*$, $y^* = (-2,-1)^T \in \operatorname{dom} g_1^* \times \operatorname{dom} g_2^*$ and $t = (3/8, -3/8)^T$, we have both relations $\sum_{i=1}^2 \lambda_i(x^{i*} + y^{i*}) = 0$ and $\sum_{i=1}^2 \lambda_i t_i = 0$ fulfilled. Thus, $(\lambda, x^*, y^*, t) \in \mathcal{B}^{FA}$ and $h^{FA}(\lambda, x^*, y^*, t) = (-14/8, 5/8)^T \in h^{FA}(\mathcal{B}^{FA})$. Suppose that there exist $(\bar{\lambda}, \bar{z}^*, \bar{v}) \in \mathcal{B}^A_{BK}$ such that $\bar{v} = h^{FA}(\lambda, x^*, y^*, t)$. Then

$$\bar{\lambda}_1\bar{v}_1 + \bar{\lambda}_2\bar{v}_2 = \inf_{x\in\mathbb{R}} \left\{z^* x + x^2(2\bar{\lambda}_1 + \bar{\lambda}_2) - \bar{\lambda}_1\right\} + \inf_{x\in\mathbb{R}} \left\{x(-z^* - 2\bar{\lambda}_1 - \bar{\lambda}_2) + \bar{\lambda}_2\right\}.$$

This means that $(-14/8)\bar{\lambda}_1 + (5/8)\bar{\lambda}_2 = -(2\bar{\lambda}_1 + \bar{\lambda}_2)/4 - \bar{\lambda}_1 + \bar{\lambda}_2$, which is equivalent to $2\bar{\lambda}_1 + \bar{\lambda}_2 = 0$, obviously a contradiction to $\lambda \in \operatorname{int}(\mathbb{R}^2_+)$. Therefore there is no $(\bar{\lambda}, \bar{z}^*, \bar{v}) \in \mathcal{B}^A_{BK}$ such that $\bar{v} = h^{FA}(\lambda, x^*, y^*, t)$. Then $h^A_{BK}(\mathcal{B}^A_{BK}) \subsetneq h^{FA}(\mathcal{B}^{FA})$.

We close the subsection by the following result.

Combining Proposition 5.1.8, Proposition 5.1.9 and Theorem 4.1.5, one obtains the following statement.

Theorem 5.1.10. *Assume that one of the regularity conditions* (RC_i^{Σ}), $i \in \{1,2,3\}$, *stated for* $\{f_1, \ldots, f_k\}$ *and, respectively,* $\{g_1, \ldots, g_k\}$ *is fulfilled. Then it holds*

$$\operatorname{Max}\left(h^A_{BK}(\mathcal{B}^A_{BK}), \mathbb{R}^k_+\right) = \operatorname{Max}\left(h^{FA}(\mathcal{B}^{FA}), \mathbb{R}^k_+\right) = \operatorname{Max}\left(h^A(\mathcal{B}^A), \mathbb{R}^k_+\right).$$

Proof. By Theorem 4.1.5 follows that $\operatorname{Max}\left(h^A_{BK}(\mathcal{B}^A_{BK}), \mathbb{R}^k_+\right) = \operatorname{Max}\left(h^A(\mathcal{B}^A), \mathbb{R}^k_+\right)$. Take now $d \in \operatorname{Max}\left(h^{FA}(\mathcal{B}^{FA}), \mathbb{R}^k_+\right)$. Then $d \in h^{FA}(\mathcal{B}^{FA})$ and, by Proposition 5.1.8, $d \in h^A(\mathcal{B}^A)$. Assuming $d \notin \operatorname{Max}\left(h^A(\mathcal{B}^A), \mathbb{R}^k_+\right)$ implies $d' \in h^A(\mathcal{B}^A)$ with $d \leq d'$. Thus one can easily construct an element $d'' \in h^A_{BK}(\mathcal{B}^A_{BK})$ with $d' \leqq d''$ and so $d \leq d''$. But, by Proposition 5.1.9, d'' must belong to $h^{FA}(\mathcal{B}^{FA})$ and this leads to a contradiction. Consequently, $\operatorname{Max}\left(h^{FA}(\mathcal{B}^{FA}), \mathbb{R}^k_+\right) \subseteq \operatorname{Max}\left(h^A(\mathcal{B}^A), \mathbb{R}^k_+\right)$. In order to prove the opposite inclusion we consider an arbitrary element $d \in \operatorname{Max}\left(h^A(\mathcal{B}^A), \mathbb{R}^k_+\right)$. Then $d \in \operatorname{Max}\left(h^A_{BK}(\mathcal{B}^A_{BK}), \mathbb{R}^k_+\right) \subseteq h^A_{BK}(\mathcal{B}^A_{BK}) \subseteq h^{FA}(\mathcal{B}^{FA})$. Were d not a maximal element of $h^{FA}(\mathcal{B}^{FA})$, we would again obtain a contradiction. This completes the proof. □

5.1.3 Duality with respect to weakly efficient solutions

In the last part of this section we discuss a duality concept similar to the one introduced above for (PVF^A), but this time with respect to the weakly efficient solutions. To this aim, we assume in the following that X and Y

are Hausdorff locally convex spaces, the convex functions $f_i : X \to \mathbb{R}$ and $g_i : Y \to \mathbb{R}$, $i = 1, \ldots, k$, have their effective domains equal to the whole space (see Remark 5.1.6 for a comment concerning this choice) and $A \in \mathcal{L}(X, Y)$. The vector dual problem with respect to weakly efficient solutions that we introduce to the primal vector optimization problem

$$(PVF_w^A) \quad \operatorname*{WMin}_{x \in X} \begin{pmatrix} f_1(x) + g_1(Ax) \\ \vdots \\ f_k(x) + g_k(Ax) \end{pmatrix}$$

is

$$(DVF_w^A) \quad \operatorname*{WMax}_{(\lambda, x^*, y^*, t) \in \mathcal{B}_w^{FA}} h_w^{FA}(\lambda, x^*, y^*, t),$$

where

$$\mathcal{B}_w^{FA} = \Big\{ (\lambda, x^*, y^*, t) \in (\mathbb{R}_+^k \backslash \{0\}) \times \prod_{i=1}^k \operatorname{dom} f_i^* \times \prod_{i=1}^k \operatorname{dom} g_i^* \times \mathbb{R}^k : \\ \lambda = (\lambda_1, \ldots, \lambda_k)^T, x^* = (x^{1*}, \ldots, x^{k*}), \\ y^* = (y^{1*}, \ldots, y^{k*}), t = (t_1, \ldots, t_k)^T, \\ \sum_{i=1}^k \lambda_i (x^{i*} + A^* y^{i*}) = 0, \sum_{i=1}^k \lambda_i t_i = 0 \Big\}$$

and

$$h_w^{FA}(\lambda, x^*, y^*, t) = \begin{pmatrix} -f_1^*(x^{1*}) - g_1^*(y^{1*}) + t_1 \\ \vdots \\ -f_k^*(x^{k*}) - g_k^*(y^{k*}) + t_k \end{pmatrix}.$$

According to Definition 2.5.1, an element $\bar{x} \in X$ is a weakly efficient solution to (PVF_w^A) if $(f + g \circ A)(\bar{x}) \in \operatorname{WMin}((f + g \circ A)(X), \mathbb{R}_+^k)$, where $f = (f_1, \ldots, f_k)^T$ and $g = (g_1, \ldots, g_k)^T$. One can note that $h_w^{FA}(\lambda, x^*, y^*, t) \subseteq \mathbb{R}^k$ and, according to the same statement, $(\bar{\lambda}, \bar{x}^*, \bar{y}^*, \bar{t})$ is said to be a weakly efficient solution to (DVF_w^A) when $h_w^{FA}(\bar{\lambda}, \bar{x}^*, \bar{y}^*, \bar{t}) \in \operatorname{WMax}(h_w^{FA}(\mathcal{B}_W^{FA}), \mathbb{R}_+^k)$. The following result presents the weak duality statement for (PVF_w^A) and (DVF_w^A). We omit its proof since it follows analogous to the proof of Theorem 5.1.3.

Theorem 5.1.11. *There is no $x \in X$ and no $(\lambda, x^*, y^*, t) \in \mathcal{B}_w^{FA}$ such that $f_i(x) + g_i(Ax) < h_{wi}^{FA}(\lambda, x^*, y^*, t)$ for $i = 1, \ldots, k$.*

Before stating the strong duality result, we consider the following regularity condition, the choice of which being discussed in Remark 5.1.7

$(RCVF_w^A)$ | $k - 1$ of the functions f_i, $i = 1, ..., k$, and g_i, $i = 1, \ldots, k$, are continuous.

Theorem 5.1.12. *Assume that the regularity condition* $(RCVF_w^A)$ *is fulfilled. If* $\bar{x} \in X$ *is a weakly efficient solution to* (PVF_w^A) *then there exists* $(\bar{\lambda}, \bar{x}^*, \bar{y}^*, \bar{t})$*, a weakly efficient solution to* (DVF_w^A)*, such that* $f_i(\bar{x})+g_i(A\bar{x}) = h_{wi}^{FA}(\bar{\lambda}, \bar{x}^*, \bar{y}^*, \bar{t})$ *for* $i = 1, \ldots, k$.

Proof. Let $\bar{x}$ be a weakly efficient solution to (PVF_w^A). Then $(f + g \circ A)(\bar{x}) \in \operatorname{WMin}((f + g \circ A)(X), \mathbb{R}_+^k)$. Obviously, $(f + g \circ A)(X) + \mathbb{R}_+^k$ is a convex set, thus by Corollary 2.4.26 there exists $\bar{\lambda} \in \mathbb{R}_+^k \backslash \{0\}$ such that $\bar{x}$ is an optimal solution to the scalar problem

$$(PF_{w\bar{\lambda}}^A) \quad \inf_{x \in X} \left\{ \sum_{i \in I(\bar{\lambda})} \bar{\lambda}_i (f_i(x) + g_i(Ax)) \right\},$$

where $I(\bar{\lambda}) = \{i \in \{1, \ldots, k\} : \bar{\lambda}_i > 0\}$ is a nonempty set. Applying the same scheme as for $(PF_{\bar{\lambda}}^A)$ and $(DF_{\bar{\lambda}}^A)$, we obtain as dual to $(PF_{\bar{\lambda}}^A)$ the following optimization problem

$$(DF_{w\bar{\lambda}}^A) \quad \sup_{\substack{x^{i*} \in X^*, y^{i*} \in Y_i^*, i \in I(\bar{\lambda}), \\ \sum\limits_{i \in I(\bar{\lambda})} \bar{\lambda}_i (x^{i*} + A^* y^{i*}) = 0}} \left\{ - \sum_{i \in I(\bar{\lambda})} \bar{\lambda}_i f_i^*(x^{i*}) - \sum_{i \in I(\bar{\lambda})} \bar{\lambda}_i g_i^*(y^{i*}) \right\}.$$

By Theorem 5.1.1 follows that for the primal-dual pair $(PF_{w\bar{\lambda}}^A) - (DF_{w\bar{\lambda}}^A)$ strong duality holds, while Theorem 5.1.2 ensures that there exist $\bar{x}^{i*} \in \operatorname{dom} f_i^*$ and $\bar{y}^{i*} \in \operatorname{dom} g_i^*$, $i \in I(\bar{\lambda})$, such that

(i) $\sum_{i \in I(\bar{\lambda})} \bar{\lambda}_i (\bar{x}^{i*} + A\bar{y}^{i*}) = 0$;
(ii) $f_i(\bar{x}) + f_i^*(\bar{x}^{i*}) = \langle \bar{x}^{i*}, \bar{x} \rangle$, $i \in I(\bar{\lambda})$;
(iii) $g_i(A\bar{x}) + g_i^*(\bar{y}^{i*}) = \langle \bar{y}^{i*}, A\bar{x} \rangle$, $i \in I(\bar{\lambda})$.

As $(RCVF^A)$ holds, one can choose some $\bar{x}^{i*} \in \operatorname{dom} f_i^*$ and $\bar{y}^{i*} \in \operatorname{dom} g_i^*$ for $i \notin I(\bar{\lambda})$. Taking $\bar{x}^* = (\bar{x}^{1*}, \ldots, \bar{x}^{k*}) \in \prod_{i=1}^k \operatorname{dom} f_i^*$ and $\bar{y}^* = (\bar{y}^{1*}, \ldots, \bar{y}^{k*}) \in \prod_{i=1}^k \operatorname{dom} g_i^*$, one has, via (i) that $\sum_{i=1}^k \bar{\lambda}_i (\bar{x}^{i*} + A\bar{y}^{i*}) = 0$. On the other hand, let us take, for each $i \in I(\bar{\lambda})$, $\bar{t}_i := \langle \bar{x}^{i*} + A^* \bar{y}^{i*}, \bar{x} \rangle \in \mathbb{R}$ and, for each $i \notin I(\bar{\lambda})$, $\bar{t}_i := f_i(\bar{x}) + g_i(A\bar{x}) + f_i^*(\bar{x}^{i*}) + g_i^*(\bar{y}^{i*})$. Therefore, $\sum_{i=1}^k \bar{\lambda}_i \bar{t}_i = 0$ and $(\bar{\lambda}, \bar{x}^*, \bar{y}^*, \bar{t}) \in \mathcal{B}_w^{FA}$. Moreover, for $i = 1, \ldots, k$, it holds $f_i(\bar{x}) + g_i(A\bar{x}) = -f_i^*(\bar{x}^{i*}) - g_i^*(\bar{y}^{i*}) + \bar{t}_i$. The fact that $(\bar{\lambda}, \bar{x}^*, \bar{y}^*, \bar{t})$ is weakly efficient to (DVF_w^A) is a direct consequence of Theorem 5.1.11. □

Remark 5.1.6. A closer look into the proof of Theorem 5.1.12 makes clear why it is necessary to consider when dealing with weakly efficient solutions that the functions f_i and g_i, $i = 1, \ldots, k$, have full domains. Otherwise, if $\bar{x}$ is a weakly efficient solution to (PVF_w^A), there must exist $\bar{\lambda} = (\bar{\lambda}_1, \ldots, \bar{\lambda}_k)^T \in \mathbb{R}_+^k \backslash \{0\}$ such that $\bar{x}$ is an optimal solution to the scalar optimization problem

$$\inf_{x \in X} \left\{ \sum_{i \in I(\bar{\lambda})} \bar{\lambda}_i (f_i(x) + g(Ax)) + \sum_{i \notin I(\bar{\lambda})} \left(\delta_{\operatorname{dom} f_i}(x) + \delta_{\operatorname{dom} g_i}(Ax) \right) \right\}.$$

This is because $0(+\infty) = (+\infty)$. By means of the duality scheme from subsection 5.1.1, one can assign as dual problem to it

$$\sup_{\substack{x^{i*} \in X^*, y^{i*} \in Y_i^*, i=1,\dots,k, \\ \sum\limits_{i \in I(\bar{\lambda})} \bar{\lambda}_i (x^{i*} + A^* y^{i*}) + \sum\limits_{i \notin I(\bar{\lambda})} (x^{i*} + A^* y^{i*}) = 0}} \Bigg\{ - \sum_{i \in I(\bar{\lambda})} \bar{\lambda}_i \big(f_i^*(x^{i*}) - g_i^*(y^{i*}) \big) - \sum_{i \notin I(\bar{\lambda})} \big(\sigma_{\operatorname{dom} f_i}(x^{i*}) + \sigma_{\operatorname{dom} g_i}(y^{i*}) \big) \Bigg\}.$$

One can prove for this primal-dual pair the existence of strong duality under some sufficient regularity conditions and also provide necessary and sufficient optimality conditions. Nevertheless, the optimality conditions corresponding to $(i) - (iii)$ from the proof of Theorem 5.1.12 do not lead to vector strong duality for (PVF_w^A) and (DVF_w^A). This was the reason why we work when treating the vector duality with respect to weakly efficient solutions in the setting $\operatorname{dom} f_i = X$ and $\operatorname{dom} g_i = Y$, for $i = 1, \dots, k$, which is in fact the usual one in the literature when dealing with the same topic.

Remark 5.1.7. Assuming that f_i and g_i, $i = 1, \dots, k$, are convex, let us take a look at the regularity condition $(RCVF_w^A)$, which seems to be very strong. Even if $(RCVF^A)$, which is a renaming for (RCF_1^A) in section 5.1, seems at the first look to be weaker, stating that in case there exists $x' \in X$ such that $k-1$ of the functions f_i, $i = 1, ..., k$, are continuous at x' and g_i is continuous at Ax', for $i = 1, \dots, k$, then by Theorem 2.2.17 follows that those $k-1$ functions f_i, $i = 1, \dots, k$, as well as the functions g_i, $i = 1, \dots, k$, must be continuous on the whole space X. Considering the condition (RCF_i^A), $i \in \{2, 2', 2''\}$, let us notice that even if the spaces X and Y are assumed to be Fréchet, consequently barreled, and f_i and g_i, $i = 1, \dots, k$, are lower semicontinuous, all of them must be continuous on X (see for instance [207, Theorem 2.2.20]). So in this case any of (RCF_i^A), $i \in \{2, 2', 2''\}$, is nothing else than $(RCVF_w^A)$. Finally, we can see that condition (RCF_3^A) asks nothing else than X and Y to be finite dimensional. But in this context the continuity of f_i and g_i, $i = 1, \dots, k$, is obviously guaranteed, so (RCF_3^A) implies $(RCVF_w^A)$.

By employing the techniques from the proofs of Theorem 4.1.8 and Theorem 5.1.5 one can show the following result.

Theorem 5.1.13. *Assume that the regularity condition $(RCVF_w^A)$ is fulfilled. Then*

$$\mathbb{R}^k \setminus \operatorname{cl}((f + g \circ A)(X) + \mathbb{R}_+^k) \subseteq h_w^{FA}(\mathcal{B}_w^{FA}) - (\mathbb{R}_+^k \setminus \{0\}).$$

Nevertheless, different to the investigations in subsection 5.1.1, one cannot directly obtain from here the converse duality theorem for $(PVF_w^A)-(DVF_w^A)$. To this aim we have to consider first the vector duals (DV_w^A) and (DV_{BKw}^A) introduced in subsection 4.1.2. It can be always proven that $h_w^{FA}(\mathcal{B}_w^{FA}) \subseteq$

$h_w^A(\mathcal{B}_w^A)$ and, under $(RCVF_w^A)$, that $h_{BKw}^A(\mathcal{B}_{BKw}^A) \subseteq h_w^{FA}(\mathcal{B}_w^{FA})$. Since we always have that $\mathrm{WMax}(h_{BKw}^A(\mathcal{B}_{BKw}^A), \mathbb{R}_+^k) = \mathrm{WMax}(h_w^A(\mathcal{B}_w^A), \mathbb{R}_+^k)$, when the condition $(RCVF_w^A)$ is fulfilled, it holds

$$\mathrm{WMax}(h_{BKw}^A(\mathcal{B}_{BKw}^A), \mathbb{R}_+^k) = \mathrm{WMax}(h_w^{FA}(\mathcal{B}_w^{FA}), \mathbb{R}_+^k) = \mathrm{WMax}(h_w^A(\mathcal{B}_w^A), \mathbb{R}_+^k).$$

Thus the converse duality result for the primal-dual pair $(PVF_w^A) - (DVF_w^A)$ follows as consequence of Theorem 4.1.9.

Theorem 5.1.14. *Assume that the regularity condition $(RCVF_w^A)$ is fulfilled and that the set $(f + g \circ A)(X) + \mathbb{R}_+^k$ is closed. Then for every weakly efficient solution $(\bar{\lambda}, \bar{x}^*, \bar{y}^*, \bar{t})$ to (DVF_w^A) one has that $h_w^{FA}(\bar{\lambda}, \bar{x}^*, \bar{y}^*, \bar{t})$ is a weakly minimal element of the set $(f + g \circ A)(X) + \mathbb{R}_+^k$.*

Remark 5.1.8. (a) The results given above remain valid if we slightly modify the vector dual (DVF_w^A) by replacing in the formulation of the feasible set $\sum_{i=1}^k \lambda_i t_i = 0$ by $\sum_{i=1}^k \lambda_i t_i \leq 0$.

(b) In case $k = 1$, i.e. $V = \mathbb{R}$ and $K = \mathbb{R}_+$, the vector dual problem (DVF_w^A) becomes the classical scalar Fenchel dual problem introduced in chapter 3.

(c) In case $X = Y$, $A = \mathrm{id}_X$ and $f_i, g_i : X \to \mathbb{R}$, $i = 1, \ldots, k$, are given functions, one can formulate a dual for

$$(PVF_w^{\mathrm{id}}) \quad \underset{x \in X}{\mathrm{WMin}} \begin{pmatrix} f_1(x) + g_1(x) \\ \vdots \\ f_k(x) + g_k(x) \end{pmatrix},$$

with respect to weakly efficient solutions by slightly modifying (DVF^{id}) by replacing $\lambda \in \mathrm{int}(\mathbb{R}_+^k)$ with $\lambda \in \mathbb{R}_+^k \backslash \{0\}$. The same applies for the problem

$$(PVF_w^f) \quad \underset{x \in X}{\mathrm{WMin}} \begin{pmatrix} f_1(x) \\ \vdots \\ f_k(x) \end{pmatrix},$$

whose dual being obtained by slightly modifying in the same way (DVF^f). For these primal-dual pairs of vector problems weak, strong and converse duality statements follow from the ones given in the general case for (PVF_w^A) and (DVF_w^A).

5.2 A family of Fenchel-Lagrange type vector duals

Different duality approaches were proposed for the cone constrained vector optimization problems with finite dimensional image sets of the objective vector functions. In the following we present one which is based on the Fenchel-Lagrange dual problems attached to the family obtained by linearly scalarizing the primal vector optimization problem, introduced in [182,183] and generalized and refined in [24,36,37,184,185]. In these papers all the spaces involved were taken finite dimensional. Here we work in a more general framework.

Let X and Z be Hausdorff locally convex spaces, with the latter partially ordered by the convex cone $C \subseteq Z$. Further, let $S \subseteq X$, $f_i : X \to \overline{\mathbb{R}}$, $i = 1,\dots,k$, be proper functions and $g : X \to \overline{Z}$ a proper function such that $\cap_{i=1}^k \operatorname{dom} f_i \cap S \cap g^{-1}(-C) \neq \emptyset$. Further, assume that the image space $V = \mathbb{R}^k$ is partially ordered by the cone $K = \mathbb{R}_+^k$ and consider the vector function

$$f : X \to \overline{\mathbb{R}^k},\, f(x) = \begin{cases} (f_1(x),\dots,f_k(x))^T, & \text{if } x \in \bigcap\limits_{i=1}^k \operatorname{dom} f_i, \\ +\infty_{\mathbb{R}_+^k}, & \text{otherwise.} \end{cases}$$

Due to the hypotheses on the functions f_i, $i = 1,\dots,k$, f is proper. When f_i, $i = 1,\dots,k$, are convex, f is also $\mathbb{R}_+^k$-convex. The primal vector optimization problem with geometric and cone constraints we work in this section with is

$$(PVF^C) \quad \operatorname*{Min}_{x\in\mathcal{A}} f(x).$$
$$\mathcal{A} = \{x \in S : g(x) \in -C\}$$

5.2.1 Duality with respect to properly efficient solutions

Similar to the previous section, we say that $\bar{x} \in \mathcal{A}$ is a properly efficient solution to (PVF^C) in the sense of linear scalarization if $\bar{x} \in \cap_{i=1}^k \operatorname{dom} f_i$ and $f(\bar{x}) \in \operatorname{PMin}_{LSc}\left(f(\cap_{i=1}^k \operatorname{dom} f_i \cap \mathcal{A}), \mathbb{R}_+^k\right)$. This means that there exists $\lambda \in \operatorname{int}(\mathbb{R}_+^k)$ such that $\sum_{i=1}^k \lambda_i f_i(\bar{x}) \leq \sum_{i=1}^k \lambda_i f_i(x)$ for all $x \in \mathcal{A}$ (cf. section 2.4). This is the reason why we first investigate, for a fixed $\lambda = (\lambda_1,\dots,\lambda_k)^T \in \operatorname{int}(\mathbb{R}_+^k)$, the following scalar optimization problem

$$(PF_\lambda^C) \quad \inf_{x\in\mathcal{A}} \left\{ \sum_{i=1}^k \lambda_i f_i(x) \right\}.$$

The vector dual problem to (PVF^C) which we introduce in this section will have its origins in the conjugate scalar dual to (PF_λ^C). Via the investigations done in subsection 3.1.3 one can associate to (PF_λ^C) as there (see the primal-dual pair $(P^C) - (D^{C_{FL}})$) the dual problem

$$\sup_{y^*\in X^*, z^*\in C^*} \left\{ -\left(\sum_{i=1}^k \lambda_i f_i\right)^*(y^*) - (z^*g)_S^*(-y^*) \right\},$$

which is not satisfactory for our purposes, since for the formulation of the vector dual we need to have the functions f_i, $i = 1,\dots,k$, separated in the formula of the scalar dual to (PF_λ^C). In order to construct such a problem, we employ the general approach investigated in section 3.1. To this aim, let us consider the following perturbation function $\Phi_\lambda^C : X \times X^k \times Z \to \overline{\mathbb{R}}$,

$$\Phi_\lambda^C(x, x^1, \ldots, x^k, z) = \begin{cases} \sum\limits_{i=1}^k \lambda_i f_i(x + x^i), & \text{if } x \in S, g(x) - z \in -C, \\ +\infty, & \text{otherwise}, \end{cases}$$

with $(x^1, \ldots, x^k, z) \in X^k \times Z$ as perturbation variables. The conjugate of Φ_λ^C, $(\Phi_\lambda^C)^* : X^* \times (X^*)^k \times Z^* \to \overline{\mathbb{R}}$, is given by the following formula

$$(\Phi_\lambda^C)^*(x^*, y^{1*}, \ldots, y^{k*}, z^*) = \sup_{\substack{x \in S, x^i \in X, i=1,\ldots,k, \\ z \in Z, g(x) - z \in -C}} \Bigg\{ \langle x^*, x \rangle + \sum_{i=1}^k \langle y^{i*}, x^i \rangle + \langle z^*, z \rangle$$

$$- \sum_{i=1}^k (\lambda_i f_i)(x + x^i) \Bigg\} = \sum_{i=1}^k (\lambda_i f_i)^*(y^{i*}) + (z^* g)_S^* \left(x^* - \sum_{i=1}^k y^{i*} \right) + \delta_{C^*}(z^*).$$

This provides the following conjugate dual to (PF_λ^C)

$$(DF_\lambda^C) \quad \sup_{\substack{y^{i*} \in X^*, i=1,\ldots,k, \\ z^* \in C^*}} \left\{ - \sum_{i=1}^k (\lambda_i f_i)^*(y^{i*}) - (z^* g)_S^* \left(- \sum_{i=1}^k y^{i*} \right) \right\},$$

which, via Proposition 2.3.2(*e*), can be equivalently written as

$$(DF_\lambda^C) \quad \sup_{\substack{y^{i*} \in X^*, i=1,\ldots,k, \\ z^* \in C^*}} \left\{ - \sum_{i=1}^k \lambda_i f_i^*(y^{i*}) - (z^* g)_S^* \left(- \sum_{i=1}^k \lambda_i y^{i*} \right) \right\}.$$

In what follows we provide regularity conditions for the primal-dual pair $(PF_\lambda^C) - (DF_\lambda^C)$ that are independent of λ, which we deduce from the general ones given in section 3.2.

One can notice that in case the set S is convex, the functions f_i, $i = 1, \ldots, k$, are convex and the function g is C-convex, Φ_λ^C is convex, too. The regularity condition (RC_1^Φ) (cf. section 3.2) becomes in this case

$$(RCF_1^C) \left| \begin{array}{l} \exists x' \in \bigcap\limits_{i=1}^k \operatorname{dom} f_i \cap S \text{ such that } f_i \text{ is continuous at } x', \\ i = 1, \ldots, k, \text{ and } g(x') \in -\operatorname{int}(C). \end{array} \right.$$

Before stating further regularity conditions, we also note that if S is closed, f_i is lower semicontinuous, $i = 1, \ldots, k$, and g is C-epi closed, then Φ_λ^C is lower semicontinuous, too. Further, it holds $(x^1, \ldots, x^k, z) \in \Pr_{X^k \times Z}(\operatorname{dom} \Phi_\lambda^C)$ if and only if there exists an $x \in S \cap \operatorname{dom} g$ such that $x^i \in \operatorname{dom} f_i - x$ for $i = 1, \ldots, k$ and $z \in g(x) + C$. This is further equivalent to the existence of an $x \in S \cap \operatorname{dom} g$ such that $(x^1, \ldots, x^k, z) \in \prod_{i=1}^k \operatorname{dom} f_i \times C - (x, \ldots, x, -g(x))$, which can also be written as $(x^1, \ldots, x^k, z) \in \prod_{i=1}^k \operatorname{dom} f_i \times C - \Delta_{S^k, g}$, where $\Delta_{S^k, g} = \{(x, \ldots, x, -g(x)) : x \in S \cap \operatorname{dom} g\} \subseteq X^k \times Z$. This leads to the following regularity condition (obtained via (RC_2^Φ))

$$(RCF_2^C) \left| \begin{array}{l} X \text{ and } Z \text{ are Fréchet spaces, } S \text{ is closed, } f_i \text{ is lower} \\ \text{semicontinuous, } i = 1,\dots,k, \ g \text{ is } C\text{-epi closed and} \\ 0 \in \operatorname{sqri}\left(\prod_{i=1}^{k} \operatorname{dom} f_i \times C - \varDelta_{S^k,g} \right), \end{array} \right.$$

along with its stronger versions

$$(RCF_{2'}^C) \left| \begin{array}{l} X \text{ and } Z \text{ are Fréchet spaces, } S \text{ is closed, } f_i \text{ is lower} \\ \text{semicontinuous, } i = 1,\dots,k, \ g \text{ is } C\text{-epi closed and} \\ 0 \in \operatorname{core}\left(\prod_{i=1}^{k} \operatorname{dom} f_i \times C - \varDelta_{S^k,g} \right) \end{array} \right.$$

and

$$(RCF_{2''}^C) \left| \begin{array}{l} X \text{ and } Z \text{ are Fréchet spaces, } S \text{ is closed, } f_i \text{ is lower} \\ \text{semicontinuous, } i = 1,\dots,k, \ g \text{ is } C\text{-epi closed and} \\ 0 \in \operatorname{int}\left(\prod_{i=1}^{k} \operatorname{dom} f_i \times C - \varDelta_{S^k,g} \right), \end{array} \right.$$

which are in fact equivalent. In the finite dimensional case one has from (RC_3^Φ)

$$(RCF_3^C) \left| \begin{array}{l} \dim\left(\operatorname{lin}\left(\prod_{i=1}^{k} \operatorname{dom} f_i \times C - \varDelta_{S^k,g} \right) \right) < +\infty \\ \text{and } 0 \in \operatorname{ri}\left(\prod_{i=1}^{k} \operatorname{dom} f_i \times C - \varDelta_{S^k,g} \right). \end{array} \right.$$

We can state now the strong duality theorem for the scalar primal-dual pair $(PF_\lambda^C) - (DF_\lambda^C)$.

Theorem 5.2.1. *Let $S \subseteq X$ be a nonempty convex set, $f_i : X \to \overline{\mathbb{R}}$, $i = 1,\dots,k$, be proper and convex functions, $g : X \to \overline{Z}$ a proper and C-convex function such that $\cap_{i=1}^k \operatorname{dom} f_i \cap S \cap g^{-1}(-C) \neq \emptyset$ and $\lambda \in \operatorname{int}(\mathbb{R}_+^k)$ be arbitrarily chosen. If one of the regularity conditions (RCF_i^C), $i \in \{1,2,3\}$, is fulfilled, then $v(PF_\lambda^C) = v(DF_\lambda^C)$ and the dual has an optimal solution.*

Remark 5.2.1. In order to deliver strong duality statements for (PF_λ^C) and (DF_λ^C) one can also combine the regularity conditions given in subsection 3.2.2. We exemplify this here by the ones expressed via the strong quasi-relative interior. Thus, assuming

$$\left| \begin{array}{l} X \text{ and } Z \text{ are Fréchet spaces, } S \text{ is closed,} \\ f_i \text{ is lower semicontinuous, } i = 1,\dots,k, \ g \text{ is } C\text{-epi closed,} \\ 0 \in \operatorname{sqri}\left(\bigcap_{i=1}^{k} \operatorname{dom} f_i \times C - \operatorname{epi}_{-C}(-g) \cap (S \times Z) \right), \\ \text{and } 0 \in \operatorname{sqri}\left(\prod_{i=1}^{k} \operatorname{dom} f_i - \varDelta_{X^k} \right), \end{array} \right.$$

guarantees the existence of strong duality for the primal-dual pair $(PF_\lambda^C) - (DF_\lambda^C)$ for all $\lambda \in \operatorname{int}(\mathbb{R}_+^k)$.

Let us come now to the formulation of the necessary and sufficient optimality conditions for the primal-dual pair $(PF_\lambda^C) - (DF_\lambda^C)$.

Theorem 5.2.2. *(a) Let $S \subseteq X$ be a nonempty convex set, $f_i : X \to \overline{\mathbb{R}}$, $i = 1, \ldots, k$, be proper and convex functions, $g : X \to \overline{Z}$ a proper and C-convex function such that $\cap_{i=1}^k \operatorname{dom} f_i \cap S \cap g^{-1}(-C) \neq \emptyset$ and $\lambda \in \operatorname{int}(\mathbb{R}_+^k)$ be arbitrarily chosen. If $\bar{x} \in X$ is an optimal solution to (PF_λ^C) and one of the regularity conditions (RCF_i^C), $i \in \{1, 2, 3\}$, is fulfilled, then there exists $(\bar{y}^{1*}, \ldots, \bar{y}^{k*}, \bar{z}^*) \in (X^*)^k \times C^*$, an optimal solution to the dual problem (DF_λ^C), such that*

(i) $(\bar{z}^* g)_S^*\big(-\sum_{i=1}^k \lambda_i \bar{y}^{i*}\big) = \langle -\sum_{i=1}^k \lambda_i \bar{y}^{i*}, \bar{x} \rangle$;

(ii) $(\bar{z}^* g)(\bar{x}) = 0$;

(iii) $f_i(\bar{x}) + f_i^*(\bar{y}^{i*}) = \langle \bar{y}^{i*}, \bar{x} \rangle$, $i = 1, \ldots, k$.

(b) For a given $\lambda \in \operatorname{int}(\mathbb{R}_+^k)$ assume that $\bar{x} \in X$ and $(\bar{y}^{1*}, \ldots, \bar{y}^{k*}, \bar{z}^*) \in (X^*)^k \times C^*$ fulfill the relations $(i) - (iii)$. Then $\bar{x}$ is an optimal solution to (PF_λ^C), $(\bar{y}^{1*}, \ldots, \bar{y}^{k*}, \bar{z}^*)$ is an optimal solution to (DF_λ^C) and $v(PF_\lambda^C) = v(DF_\lambda^C)$.

Proof. The proof follows in the lines of the ones given for Theorem 3.3.13 and Theorem 3.3.22. □

Remark 5.2.2. The optimality conditions $(i) - (iii)$ in Theorem 5.2.2 can be equivalently written as

$$-\sum_{i=1}^k \lambda_i \bar{y}^{i*} \in \partial((\bar{z}^* g) + \delta_S)(\bar{x}), (\bar{z}^* g)(\bar{x}) = 0 \text{ and } \bar{y}^{i*} \in \partial f_i(\bar{x}), i = 1, ..., k.$$

We introduce in the following not only one vector dual problem to (PVF^C), but a family of such problems, which are of *Fenchel-Lagrange type* (see also section 4.3). For this, the following set is required,

$$\mathcal{F} = \Big\{ \alpha = (\alpha_1, \ldots, \alpha_k)^T : \operatorname{int}(\mathbb{R}_+^k) \to \operatorname{int}(\mathbb{R}_+^k) : \\ \sum_{i=1}^k \lambda_i \alpha_i(\lambda) = 1 \ \forall \lambda = (\lambda_1, \ldots, \lambda_k)^T \in \operatorname{int}(\mathbb{R}_+^k) \Big\}.$$

For each $\alpha \in \mathcal{F}$ we attach to (PVF^C) the following dual vector optimization problem

$$(DVF^{C_\alpha}) \qquad \underset{(\lambda, y^*, z^*, t) \in \mathcal{B}^{C_\alpha}}{\operatorname{Max}} h^{C_\alpha}(\lambda, y^*, z^*, t),$$

where

$$\mathcal{B}^{C_\alpha} = \Big\{ (\lambda, y^*, z^*, t) \in \operatorname{int}(\mathbb{R}^k_+) \times \prod_{i=1}^k \operatorname{dom} f_i^* \times (Z^*)^k \times \mathbb{R}^k : \lambda = (\lambda_1, \ldots, \lambda_k)^T,$$
$$y^* = (y^{1*}, \ldots, y^{k*}), z^* = (z^{1*}, \ldots, z^{k*}), t = (t_1, \ldots, t_k)^T,$$
$$-\alpha_i(\lambda) \sum_{j=1}^k \lambda_j y^{j*} \in \operatorname{dom}(z^{i*}g)^*_S, i = 1, \ldots, k,$$
$$\sum_{i=1}^k \lambda_i z^{i*} \in C^*, \sum_{i=1}^k \lambda_i t_i = 0 \Big\}$$

and

$$h^{C_\alpha}(\lambda, y^*, z^*, t) = \begin{pmatrix} -f_1^*(y^{1*}) - (z^{1*}g)^*_S\Big(-\alpha_1(\lambda) \sum\limits_{j=1}^k \lambda_j y^{j*} \Big) + t_1 \\ \vdots \\ -f_k^*(y^{k*}) - (z^{k*}g)^*_S\Big(-\alpha_k(\lambda) \sum\limits_{j=1}^k \lambda_j y^{j*} \Big) + t_k \end{pmatrix}.$$

Whenever $\alpha \in \mathcal{F}$, the properness of the functions f_i, $i = 1, \ldots, k$, and g along with Lemma 2.3.1(a) ensure that $h^{C_\alpha}(\mathcal{B}^{C_\alpha}) \subseteq \mathbb{R}^k$. According to Definition 2.5.1, an element $(\bar{\lambda}, \bar{y}^*, \bar{z}^*, \bar{t}) \in \mathcal{B}^{C_\alpha}$ is said to be efficient to (DVF^{C_α}) if $h^{C_\alpha}(\bar{\lambda}, \bar{y}^*, \bar{z}^*, \bar{t}) \in \operatorname{Max}(h^{C_\alpha}(\mathcal{B}^{C_\alpha}), \mathbb{R}^k_+)$, while if if $h^{C_\alpha}(\bar{\lambda}, \bar{y}^*, \bar{z}^*, \bar{t}) \in \operatorname{WMax}(h^{C_\alpha}(\mathcal{B}^{C_\alpha}), \mathbb{R}^k_+)$ we call $(\bar{\lambda}, \bar{y}^*, \bar{z}^*, \bar{t}) \in \mathcal{B}^{C_\alpha}$ weakly efficient to (DVF^{C_α}). Next we prove that for each of the vector duals we just introduced there is weak duality.

Theorem 5.2.3. *Let $\alpha \in \mathcal{F}$ be fixed. Then there is no $x \in \mathcal{A}$ and no $(\lambda, y^*, z^*, t) \in \mathcal{B}^{C_\alpha}$ such that $f_i(x) \leq h_i^{C_\alpha}(\lambda, y^*, z^*, t)$ for $i = 1, \ldots, k$, and $f_j(x) < h_j^{C_\alpha}(\lambda, y^*, z^*, t)$ for at least one $j \in \{1, \ldots, k\}$.*

Proof. Assume the contrary, namely that there are some $x \in X$ and $(\lambda, y^*, z^*, t) \in \mathcal{B}^{C_\alpha}$ such that $f_i(x) \leq h_i^{C_\alpha}(\lambda, y^*, z^*, t)$ for each $i = 1, \ldots, k$, and $f_j(x) < h_j^{C_\alpha}(\lambda, y^*, z^*, t)$ for at least one $j \in \{1, \ldots, k\}$. Consequently, $\sum_{i=1}^k \lambda_i f_i(x) < \sum_{i=1}^k \lambda_i h_i^{C_\alpha}(\lambda, y^*, z^*, t) = -\sum_{i=1}^k \lambda_i \big(f_i^*(y^{i*}) + (z^{i*}g)^*_S\big(-\alpha_i(\lambda) \sum_{j=1}^k \lambda_j y^{j*}\big)\big) + \sum_{i=1}^k \lambda_i t_i = -\sum_{i=1}^k \lambda_i \big(f_i^*(y^{i*}) + (z^{i*}g)^*_S\big(-\alpha_i(\lambda) \sum_{j=1}^k \lambda_j y^{j*}\big)\big)$, which, by the Young-Fenchel inequality, is less than or equal to $\sum_{i=1}^k \lambda_i f_i(x)$ since $\langle (\sum_{i=1}^k \lambda_i \alpha_i(\lambda)) \sum_{j=1}^k \lambda_j y^{j*}, x \rangle = \langle \sum_{i=1}^k \lambda_i y^{i*}, x \rangle$ and $\big(\sum_{i=1}^k \lambda_i z^{i*} g\big)(x) \leq 0$ and we reached a contradiction. □

We come now to the vector strong duality statement for (PVF^C) and (DVF^{C_α}). In what follows we assume that the nonempty set S is convex, the functions f_i, $i = 1, \ldots, k$, are convex and the function g is C-convex. In order to maintain the analogy to the investigations made in the previous chapter, we assume also here for the strong vector duality results the fulfillment of a regularity condition and then remark that these remain valid when others are verified, too. In this case we work with (RCF_1^C), renamed as

$$(RCVF^C) \left| \begin{array}{l} \exists x' \in \bigcap\limits_{i=1}^{k} \operatorname{dom} f_i \cap S \text{ such that } f_i \text{ is continuous at } x', \\ i = 1, \ldots, k, \text{ and } g(x') \in -\operatorname{int}(C). \end{array} \right.$$

Theorem 5.2.4. *Let $\alpha \in \mathcal{F}$ be fixed and assume that the regularity condition $(RCVF^C)$ is fulfilled. If $\bar{x} \in \mathcal{A}$ is a properly efficient solution to (PVF^C) then there exists $(\bar{\lambda}, \bar{y}^*, \bar{z}^*, \bar{t})$, an efficient solution to (DVF^{C_α}), such that $f_i(\bar{x}) = h_i^{C_\alpha}(\bar{\lambda}, \bar{y}^*, \bar{z}^*, \bar{t})$ for $i = 1, \ldots, k$.*

Proof. Given the proper efficiency of $\bar{x}$ to (PVF^C), there is some $\bar{\lambda} \in \operatorname{int}(\mathbb{R}_+^k)$ such that $\bar{x}$ is an optimal solution to $(PF_{\bar{\lambda}}^C)$. Because of Theorem 5.2.1, there is strong duality for this scalar optimization problem and its Fenchel-Lagrange type dual $(DF_{\bar{\lambda}}^C)$, which has an optimal solution, say $(\bar{y}^{1*}, \ldots, \bar{y}^{k*}, \bar{w}^*)$ such that the optimality conditions $(i) - (iii)$ in Theorem 5.2.2 hold. Take $\bar{z}^{i*} := \alpha_i(\bar{\lambda})\bar{w}^*$ and $\bar{t}_i := \langle \bar{y}^{i*}, \bar{x} \rangle + (\bar{z}^{i*} g)_S^*\big(-\alpha_i(\bar{\lambda}) \sum_{j=1}^k \bar{\lambda}_j \bar{y}^{j*}\big)$, $i = 1, \ldots, k$, and let $\bar{z}^* := (\bar{z}^{1*}, \ldots, \bar{z}^{k*})$ and $\bar{t} := (\bar{t}_1, \ldots, \bar{t}_k)^T$. Then $\sum_{i=1}^k \bar{\lambda}_i \bar{z}^{i*} = \bar{w}^* \in C^*$. Note that whenever $i \in \{1, \ldots, k\}$ one has $(\bar{z}^{i*} g)_S^*\big(-\alpha_i(\bar{\lambda}) \sum_{j=1}^k \bar{\lambda}_j \bar{y}^{j*}\big) = \alpha_i(\bar{\lambda})(\bar{w}^* g)_S^*\big(-\sum_{j=1}^k \bar{\lambda}_j \bar{y}^{j*}\big)$ and thus $-\alpha_i(\bar{\lambda}) \sum_{j=1}^k \bar{\lambda}_j \bar{y}^{j*} \in \operatorname{dom}(\bar{z}^{i*} g)_S^*$ for $i = 1, \ldots, k$. Then, using the above mentioned optimality conditions, we get

$$\sum_{i=1}^k \bar{\lambda}_i \bar{t}_i = \left\langle \sum_{j=1}^k \bar{\lambda}_j \bar{y}^{j*}, \bar{x} \right\rangle + \sum_{i=1}^k \bar{\lambda}_i \alpha_i(\bar{\lambda})(\bar{w}^* g)_S^*\Bigg(-\sum_{j=1}^k \bar{\lambda}_j \bar{y}^{j*}\Bigg)$$

$$= \left\langle \sum_{j=1}^k \bar{\lambda}_j \bar{y}^{j*}, \bar{x} \right\rangle + (\bar{w}^* g)_S^*\Bigg(-\sum_{j=1}^k \bar{\lambda}_j \bar{y}^{j*}\Bigg) = 0,$$

which yields that $(\bar{\lambda}, \bar{y}^*, \bar{z}^*, \bar{t})$ is feasible to (DVF^{C_α}), where $\bar{y}^* = (\bar{y}^{1*}, \ldots, \bar{y}^{k*})$. Moreover, whenever $i \in \{1, \ldots, k\}$ $h_i^{C_\alpha}(\bar{\lambda}, \bar{y}^*, \bar{z}^*, \bar{t}) = -f_i^*(\bar{y}^{i*}) + \langle \bar{y}^{i*}, \bar{x} \rangle = f_i(\bar{x})$, via the optimality condition (iii) of Theorem 5.2.2. The efficiency of $(\bar{\lambda}, \bar{y}^*, \bar{z}^*, \bar{t})$ to (DVF^{C_α}) follows by applying Theorem 5.2.3. Because $\alpha \in \mathcal{F}$ was arbitrarily chosen, the conclusion follows. □

To give the converse duality statement for the vector problems (PVF^C) and (DVF^{C_α}), $\alpha \in \mathcal{F}$, two preliminary statements are needed.

Theorem 5.2.5. *Let $\alpha \in \mathcal{F}$ be fixed and assume that $\mathcal{B}^{C_\alpha}$ is nonempty and that the regularity condition $(RCVF^C)$ is fulfilled. Then*

$$\mathbb{R}^k \setminus \operatorname{cl}(f(\cap_{i=1}^k \operatorname{dom} f_i \cap \mathcal{A}) + \mathbb{R}_+^k) \subseteq h^{C_\alpha}(\mathcal{B}^{C_\alpha}) - \operatorname{int}(\mathbb{R}_+^k).$$

Proof. Let $\bar{v} \in \mathbb{R}^k \setminus \operatorname{cl}(f(\cap_{i=1}^k \operatorname{dom} f_i \cap \mathcal{A}) + \mathbb{R}_+^k)$. Similarly to the proof of Theorem 4.3.3 one can prove that there exists $\bar{\lambda} \in \operatorname{int}(\mathbb{R}_+^k)$ such that

$$\sum_{i=1}^{k} \bar{\lambda}_i \bar{v}_i < \inf_{x \in \mathcal{A}} \left\{ \sum_{i=1}^{k} \bar{\lambda}_i f_i(x) \right\}.$$

According to Theorem 5.2.1, there exist $\bar{y}^* = (\bar{y}^{1*}, \ldots, \bar{y}^{k*}) \in \prod_{i=1}^k \operatorname{dom} f_i^*$ and $\bar{w}^* \in C^*$ such that

$$\inf_{x \in X} \left\{ \sum_{i=1}^{k} \bar{\lambda}_i f_i(x) \right\} = -\sum_{i=1}^{k} \bar{\lambda}_i f_i^*(\bar{y}^{i*}) - (\bar{w}^* g)_S^* \left(-\sum_{i=1}^{k} \bar{\lambda}_i \bar{y}^{i*} \right).$$

Let $\bar{z}^{i*} := \alpha_i(\bar{\lambda})\bar{w}^*$, $i = 1, \ldots, k$, and $\bar{z}^* := (\bar{z}^{1*}, \ldots, \bar{z}^{k*})$. Then $\sum_{i=1}^k \bar{\lambda}_i \bar{z}^{i*} = \bar{w}^* \in C^*$. Thus $(\bar{\lambda}, \bar{y}^*, \bar{z}^*, 0) \in \mathcal{B}^{C_\alpha}$ and it holds

$$\sum_{i=1}^{k} \bar{\lambda}_i \bar{v}_i < \sum_{i=1}^{k} \bar{\lambda}_i h_i^{C_\alpha}(\bar{\lambda}, \bar{y}^*, \bar{z}^*, 0). \tag{5.2}$$

Consider the hyperplane with the normal vector $\bar{\lambda}$

$$H = \left\{ h^{C_\alpha}(\bar{\lambda}, \bar{y}^*, \bar{z}^*, 0) + t : t \in \mathbb{R}^k, \sum_{i=1}^{k} \bar{\lambda}_i t_i = 0 \right\}$$

$$= \left\{ v \in \mathbb{R}^k : \sum_{i=1}^{k} \bar{\lambda}_i v_i = \sum_{i=1}^{k} \bar{\lambda}_i h_i^{C_\alpha}(\bar{\lambda}, \bar{y}^*, \bar{z}^*, 0) \right\}.$$

One can easily see that $H \subseteq h^{C_\alpha}(\mathcal{B}^{C_\alpha})$, while from (5.2) we deduce that $\bar{v}$ is an element of the open halfspace $H^- = \{v \in \mathbb{R}^k : \sum_{i=1}^k \bar{\lambda}_i v_i < \sum_{i=1}^k \bar{\lambda}_i h_i^{C_\alpha}(\bar{\lambda}, \bar{y}^*, \bar{z}^*, 0)\}$. The orthogonal projection of $\bar{v}$ on H is an element $\tilde{v} = \bar{v} + \delta\bar{\lambda} \in H \subseteq h^{C_\alpha}(\mathcal{B}^{C_\alpha})$ with $\delta > 0$. Since $\bar{\lambda} \in \operatorname{int}(\mathbb{R}_+^k)$, it follows that $\bar{v} \in h^{C_\alpha}(\mathcal{B}^{C_\alpha}) - \operatorname{int}(\mathbb{R}_+^k)$. □

Remark 5.2.3. We refer the reader to a comparison of the result in Theorem 5.2.5 with the one given in Theorem 4.3.3 for the primal-dual pair $(PVG) - (DVG)$. A direct consequence of this latter result is that, when (RCV^{Φ}) is fulfilled and $(\bar{\lambda}, \bar{y}^*, \bar{v})$ is an efficient solution to (DVG), then $h^G(\bar{\lambda}, \bar{y}^*, \bar{v}) \in \operatorname{cl}(F(\operatorname{dom} F) + K)$. As proven below, a similar result can be given, when $\alpha \in \mathcal{F}$, for the primal-dual pair $(PVF^C) - (DVF^{C_\alpha})$, but this does not result directly as for $(PVG) - (DVG)$.

Theorem 5.2.6. *Let $\alpha \in \mathcal{F}$ be fixed and assume the regularity condition $(RCVF^C)$ fulfilled. If $(\bar{\lambda}, \bar{y}^*, \bar{z}^*, \bar{t}) \in \mathcal{B}^{C_\alpha}$ is a weakly efficient solution to (DVF^{C_α}), then $h^{C_\alpha}(\bar{\lambda}, \bar{y}^*, \bar{z}^*, \bar{t}) \in \operatorname{cl}(f(\cap_{i=1}^k \operatorname{dom} f_i \cap \mathcal{A}))$.*

Proof. Since $(\bar{\lambda}, \bar{y}^*, \bar{z}^*, \bar{t}) \in \mathcal{B}^{C_\alpha}$ is a weakly efficient solution to (DVF^{C_α}), by Theorem 5.2.5 follows that

$$h^{C_\alpha}(\bar{\lambda}, \bar{y}^*, \bar{y}^*, \bar{t}) \in \operatorname{cl}\left(f\left(\bigcap_{i=1}^{k} \operatorname{dom} f_i \cap \mathcal{A} \right) + \mathbb{R}_+^k \right).$$

Assuming the contrary implies the existence of $\bar{v} \in h^{C_\alpha}(\mathcal{B}^{C_\alpha})$ such that $h^{C_\alpha}(\bar{\lambda}, \bar{y}^*, \bar{z}^*, \bar{t}) < \bar{v}$. But this contradicts the weak efficiency of $(\bar{\lambda}, \bar{y}^*, \bar{z}^*, \bar{t})$ to the vector dual problem.

Then there exist $\{v^l\} \subseteq f(\cap_{i=1}^k \operatorname{dom} f_i \cap \mathcal{A})$ and $r^l \in \mathbb{R}^k_+$ such that $v^l + r^l \to h^{C_\alpha}(\bar{\lambda}, \bar{y}^*, \bar{z}^*, \bar{t})$ when $l \to +\infty$. For all $l \geq 1$ let be $x^l \in \cap_{i=1}^k \operatorname{dom} f_i \cap \mathcal{A}$ with $v^l = f(x^l)$. But the weak duality theorem for $(PF_\lambda^C) - (DF_\lambda^C)$ yields for all $l \geq 1$

$$\sum_{i=1}^k \bar{\lambda}_i h_i^{C_\alpha}(\bar{\lambda}, \bar{y}^*, \bar{z}^*, \bar{t}) = \sum_{i=1}^k \bar{\lambda}_i \Bigg(- f_i^*(\bar{y}^{i*}) - (\bar{z}^{i*} g)_S^* \Bigg(- \alpha_i(\bar{\lambda}_i) \sum_{j=1}^k \bar{\lambda}_j \bar{y}^{j*} \Bigg) \Bigg)$$

$$+ \sum_{i=1}^k \bar{\lambda}_i \bar{t}_i \leq \sum_{i=1}^k \bar{\lambda}_i f_i(x^l) - \left\langle \sum_{i=1}^k \bar{\lambda}_i \bar{y}^{i*}, x^l \right\rangle + \sum_{i=1}^k \bar{\lambda}_i (\bar{z}^{i*} g + \delta_S)(x^l)$$

$$+ \left\langle \sum_{i=1}^k \bar{\lambda}_i \bar{y}^{i*}, x^l \right\rangle \leq \sum_{i=1}^k \bar{\lambda}_i f_i(x^l) = \sum_{i=1}^k \bar{\lambda}_i v_i^l \leq \sum_{i=1}^k \bar{\lambda}_i (v_i^l + r_i^l).$$

We have $\lim_{l \to +\infty} \sum_{i=1}^k \bar{\lambda}_i (v_i^l + r_i^l) = \sum_{i=1}^k \bar{\lambda}_i h_i^{C_\alpha}(\bar{\lambda}, \bar{y}^*, \bar{z}^*, \bar{t})$ and this implies that

$$\lim_{l \to +\infty} \sum_{i=1}^k \bar{\lambda}_i v_i^l = \sum_{i=1}^k \bar{\lambda}_i h_i^{C_\alpha}(\bar{\lambda}, \bar{y}^*, \bar{z}^*, \bar{t}).$$

Consequently, $\lim_{l \to +\infty} \sum_{i=1}^k \bar{\lambda}_i r_i^l = 0$. Since $r^l \in \mathbb{R}^k_+$ for all $l \geq 1$, this can be the case only if $\lim_{l \to +\infty} r^l = 0$. In conclusion, $\lim_{l \to +\infty} v^l = h^{C_\alpha}(\bar{\lambda}, \bar{y}^*, \bar{z}^*, \bar{t})$ and so $h^{C_\alpha}(\bar{\lambda}, \bar{y}^*, \bar{z}^*, \bar{t}) \in \operatorname{cl}(f(\cap_{i=1}^k \operatorname{dom} f_i \cap \mathcal{A}))$. □

A direct consequence of Theorem 5.2.6 is the following converse duality statement.

Theorem 5.2.7. *Let $\alpha \in \mathcal{F}$ be fixed and assume that the regularity condition $(RCVF^C)$ is fulfilled and that the set $f(\cap_{i=1}^k \operatorname{dom} f_i \cap \mathcal{A}) + \mathbb{R}^k_+$ is closed. Then for every weakly efficient solution $(\bar{\lambda}, \bar{y}^*, \bar{z}^*, \bar{t})$ to (DVF^{C_α}), there exists $\bar{x} \in \mathcal{A}$, a properly efficient solution to (PVF^C), such that $f_i(\bar{x}) = h_i^{C_\alpha}(\bar{\lambda}, \bar{y}^*, \bar{z}^*, \bar{t})$, $i = 1, \ldots, k$.*

Proof. As seen in the proof of Theorem 5.2.6, the weak efficiency of $(\bar{\lambda}, \bar{y}^*, \bar{z}^*, \bar{t})$ to the dual ensures that $h^{C_\alpha}(\bar{\lambda}, \bar{y}^*, \bar{z}^*, \bar{t}) \in \operatorname{cl}(f(\cap_{i=1}^k \operatorname{dom} f_i \cap \mathcal{A}) + \mathbb{R}^k_+) = f(\cap_{i=1}^k \operatorname{dom} f_i \cap \mathcal{A}) + \mathbb{R}^k_+$.

This means that $h^{C_\alpha}(\bar{\lambda}, \bar{y}^*, \bar{z}^*, \bar{t}) = f(\bar{x}) + \bar{r}$ for some $\bar{x} \in \cap_{i=1}^k \operatorname{dom} f_i \cap \mathcal{A}$ and $\bar{r} \in \mathbb{R}^k_+$. But $\bar{r} = 0$, otherwise the weak duality (see Theorem 5.2.3) would be violated. Even more, the weak duality proves also the efficiency of $\bar{x}$ to (PVF^C). It remains to prove that $\bar{x}$ is a properly efficient to (PVF^C). Since $h^{C_\alpha}(\bar{\lambda}, \bar{y}^*, \bar{z}^*, \bar{t}) = f(\bar{x})$ it holds

$$\sum_{i=1}^k \bar{\lambda}_i f_i(\bar{x}) = \sum_{i=1}^k \bar{\lambda}_i h_i^{C_\alpha}(\bar{\lambda}, \bar{y}^*, \bar{z}^*, \bar{t}) = - \sum_{i=1}^k \bar{\lambda}_i \Bigg(f_i^*(\bar{y}^{i*})$$

$$+(\bar{z}^{i*}g)^*_S\Big(-\alpha_i(\bar{\lambda})\sum_{j=1}^k \bar{\lambda}_j\bar{y}^{j*}\Big)+\bar{t}_i\Big) \leq \inf_{x\in\mathcal{A}}\Big\{\sum_{i=1}^k \bar{\lambda}_i f_i(x)\Big\}.$$

Therefore $\bar{x}$ is an optimal solution to $(PF^C_{\bar{\lambda}})$ and this proves that $\bar{x}$ is a properly efficient solution to (PVF^C). □

Remark 5.2.4. Note also that an alternative converse duality statement can be obtained following the method used in [184].

Remark 5.2.5. (*a*) One can easily notice that all the results given above remain valid whenever $\alpha \in \mathcal{F}$ if the vector dual (DVF^{C_α}) is slightly modified by replacing in the formulation of the feasible set $\mathcal{B}^{C_\alpha}$ the restriction $\sum_{i=1}^k \lambda_i t_i = 0$ by $\sum_{i=1}^k \lambda_i t_i \leq 0$.

(*b*) Since the proof of the theorem given above uses as a main tool the scalar strong duality result for (PF^C_λ) and (DF^C_λ) when $\lambda \in \operatorname{int}(\mathbb{R}^k_+)$, it is always true that the regularity condition $(RCVF^C)$ can be replaced by any of the regularity conditions (RCF^C_i), $i \in \{2, 2', 2'', 3\}$. In theorems 5.2.5-5.2.7, the regularity condition can be replaced by the assumption that for all $\lambda \in \operatorname{int}(\mathbb{R}^k_+)$ the optimization problem (PF^C_λ) is normal with respect to its conjugate dual problem (DF^C_λ).

(*c*) Theorem 5.2.7 remains valid even if instead of asking that the set $f(\cap_{i=1}^k \operatorname{dom} f_i \cap \mathcal{A}) + \mathbb{R}^k_+$ is closed one assumes that $f(\cap_{i=1}^k \operatorname{dom} f_i \cap \mathcal{A})$ is closed.

(*d*) Combining the assertions of Theorem 5.2.4 and Theorem 5.2.7 makes it immediately clear that under the hypotheses of the latter any weakly efficient solution to (DVF^{C_α}) is also efficient to it for any $\alpha \in \mathcal{F}$.

(*e*) In case $k = 1$, i.e. if $V = \mathbb{R}$ and $K = \mathbb{R}_+$, the problems (PVF^C) and (DVF^{C_α}), where $\alpha \in \mathcal{F}$, become (see also subsection 3.2.2)

$$(P^C) \quad \inf_{x\in\mathcal{A}} f(x)$$

and, respectively,

$$(D^{C_{FL}}) \qquad \sup_{(\lambda,y^*,z^*,t)\in\mathcal{B}^{C_\alpha}} \{-f^*(y^*) - (z^*g)^*_S(-y^*) + t\},$$

where

$$\mathcal{B}^{C_\alpha} = \{(\lambda, y^*, z^*, t) \in \operatorname{int}(\mathbb{R}_+) \times \operatorname{dom} f^* \times C^* \times \mathbb{R} : \lambda t = 0\}.$$

Obviously the dual problem can be equivalently written as

$$\sup_{y^*\in X^*, z^*\in C^*} \{-f^*(y^*) - (z^*g)^*_S(-y^*)\},$$

which is, indeed, the Fenchel-Lagrange dual problem to (P^C) (cf. subsection 3.1.3). This motivates giving the name Fenchel-Lagrange type vector dual problem to (DVF^{C_α}).

Remark 5.2.6. For particular choices of the function $\alpha \in \mathcal{F}$ one obtains different vector duals to (PVF^C). For instance, when

$$\alpha(\lambda) = \left(\frac{1}{k\lambda_1}, \ldots, \frac{1}{k\lambda_k}\right)^T \quad \forall \lambda = (\lambda_1, \ldots, \lambda_k)^T \in \operatorname{int}(\mathbb{R}^k_+),$$

(DVF^{C_α}) turns out to be the dual problem introduced in [184] for the case $S = X = \mathbb{R}^n$, which generalizes the one considered in [182, 183]. Another interesting special case occurs when we take

$$\alpha(\lambda) = \left(\frac{1}{\sum\limits_{i=1}^k \lambda_i}, \ldots, \frac{1}{\sum\limits_{i=1}^k \lambda_i}\right)^T \quad \forall \lambda = (\lambda_1, \ldots, \lambda_k)^T \in \operatorname{int}(\mathbb{R}^k_+).$$

Considering in the feasible set of (DVF^{C_α}) formulated in this case that all z^{i*}, $i = 1, \ldots, k$, are equal, one obtains a vector dual problem introduced to (PVF^C) in [36, 37] for the case $S = X = \mathbb{R}^n$, namely

$$(DVF^{C_M}) \qquad \operatorname*{Max}_{(\lambda, y^*, z^*, t) \in \mathcal{B}^{C_M}} h^{C_M}(\lambda, y^*, z^*, t),$$

where

$$\begin{aligned}\mathcal{B}^{C_M} = \Bigg\{(\lambda, y^*, z^*, t) \in \operatorname{int}(\mathbb{R}^k_+) \times \prod_{i=1}^k \operatorname{dom} f_i^* \times C^* \times \mathbb{R}^k : \lambda = (\lambda_1, \ldots, \lambda_k)^T,\\ y^* = (y^{1*}, \ldots, y^{k*}), t = (t_1, \ldots, t_k)^T,\\ \frac{-1}{\sum\limits_{j=1}^k \lambda_j} \sum_{j=1}^k \lambda_j y^{j*} \in \operatorname{dom}(z^*g)^*_S, \sum_{i=1}^k \lambda_i t_i = 0\Bigg\}\end{aligned}$$

and

$$h^{C_M}(\lambda, y^*, z^*, t) = \begin{pmatrix} -f_1^*(y^{1*}) - (z^*g)^*_S\left(-\frac{1}{\sum_{j=1}^k \lambda_j} \sum_{j=1}^k \lambda_j y^{j*}\right) + t_1 \\ \vdots \\ -f_k^*(y^{k*}) - (z^*g)^*_S\left(-\frac{1}{\sum_{j=1}^k \lambda_j} \sum_{j=1}^k \lambda_j y^{j*}\right) + t_k \end{pmatrix}.$$

The vector duality statements given in Theorem 5.2.3, Theorem 5.2.4 and Theorem 5.2.7 hold when replacing (DVF^{C_α}) with (DVF^{C_M}), too. In section 5.3 we prove that $h^{C_M}(\mathcal{B}^{C_M})$ is a subset of $h^{C_\alpha}(\mathcal{B}^{C_\alpha})$ whenever $\alpha \in \mathcal{F}$.

5.2.2 Duality with respect to weakly efficient solutions

In the second part of this section we discuss a duality concept similar to the one introduced above for (PVF^C), but this time with respect to the weakly efficient solutions. To this aim, we assume in the following that X and Z are Hausdorff locally convex spaces, with the latter partially ordered by the convex cone $C \subseteq Z$. Further, let the convex set $S \subseteq X$, the convex functions $f_i : X \to \mathbb{R}$, $i = 1,\dots,k$, having their effective domains equal to the whole space X (see Remark 5.2.7 for a comment concerning this choice) and the proper and C-convex function $g : X \to \overline{Z}$ such that $S \cap g^{-1}(-C) \neq \emptyset$. Further, assume that the image space $V = \mathbb{R}^k$ is partially ordered by the cone $K = \mathbb{R}^k_+$ and consider the vector function $f = (f_1, ..., f_k)^T$. The primal vector optimization problem with geometric and cone constraints with respect to weakly efficient solutions we work in this section with is

$$(PVF^C_w) \quad \underset{x \in \mathcal{A}}{\operatorname{WMin}} f(x).$$
$$\mathcal{A} = \{x \in S : g(x) \in -C\}$$

Considering the set

$$\mathcal{F}_w = \Big\{ \beta = (\beta_1, \dots, \beta_k)^T : (\mathbb{R}^k_+ \backslash \{0\}) \to \operatorname{int}(\mathbb{R}^k_+) : \sum_{i=1}^k \lambda_i \beta_i(\lambda) = 1 \ \forall \lambda = (\lambda_1, \dots, \lambda_k)^T \in \mathbb{R}^k_+ \backslash \{0\} \Big\},$$

for each $\beta \in \mathcal{F}_w$ we attach to (PVF^C_w) the following dual vector problem with respect to weakly efficient solutions

$$(DVF^{C_\beta}_w) \qquad \underset{(\lambda, y^*, z^*, t) \in \mathcal{B}^{C_\beta}_w}{\operatorname{WMax}} h^{C_\beta}_w(\lambda, y^*, z^*, t),$$

where

$$\mathcal{B}^{C_\beta}_w = \Big\{ (\lambda, y^*, z^*, t) \in (\mathbb{R}^k_+ \backslash \{0\}) \times \prod_{i=1}^k \operatorname{dom} f_i^* \times (Z^*)^k \times \mathbb{R}^k : \lambda = (\lambda_1, \dots, \lambda_k)^T,$$
$$y^* = (y^{1*}, \dots, y^{k*}), z^* = (z^{1*}, \dots, z^{k*}), t = (t_1, \dots, t_k)^T,$$
$$-\beta_i(\lambda) \sum_{j=1}^k \lambda_j y^{j*} \in \operatorname{dom}(z^{i*} g)^*_S, i = 1, \dots, k,$$
$$\sum_{i=1}^k \lambda_i z^{i*} \in C^*, \sum_{i=1}^k \lambda_i t_i = 0 \Big\}$$

and

$$h_w^{C_\beta}(\lambda, y^*, z^*, t) = \begin{pmatrix} -f_1^*(y^{1*}) - (z^{1*}g)_S^*\left(-\beta_1(\lambda) \sum\limits_{j=1}^k \lambda_j y^{j*}\right) + t_1 \\ \vdots \\ -f_k^*(y^{k*}) - (z^{k*}g)_S^*\left(-\beta_k(\lambda) \sum\limits_{j=1}^k \lambda_j y^{j*}\right) + t_k \end{pmatrix}.$$

Whenever $\beta \in \mathcal{F}_w$ one has that $h_w^{C_\beta}(\mathcal{B}_w^{C_\beta}) \subseteq \mathbb{R}^k$. According to Definition 2.5.1, while an element $\bar{x} \in \mathcal{A}$ is a weakly efficient solution to (PVF_w^C) if $f(\bar{x}) \in \operatorname{WMin}(f(\mathcal{A}), \mathbb{R}_+^k)$, an element $(\bar{\lambda}, \bar{y}^*, \bar{z}^*, \bar{t}) \in \mathcal{B}_w^{C_\beta}$ is said to be weakly efficient to $(DVF_w^{C_\beta})$ if $h_w^{C_\beta}(\bar{\lambda}, \bar{y}^*, \bar{z}^*, \bar{t}) \in \operatorname{WMax}(h_w^{C_\beta}(\mathcal{B}_w^{C_\beta}), \mathbb{R}_+^k)$.

The next statement gives a weak duality result for (PVF_w^C) and $(DVF_w^{C_\beta})$, $\beta \in \mathcal{F}_w$, which can be proven analogously to Theorem 5.2.3.

Theorem 5.2.8. *Let $\beta \in \mathcal{F}_w$ be fixed. Then there is no $x \in X$ and no $(\lambda, y^*, z^*, t) \in \mathcal{B}_w^{C_\beta}$ such that $f_i(x) < h_{wi}^{C_\beta}(\lambda, y^*, z^*, t)$ for $i = 1, \dots, k$.*

Before stating the strong duality result, we consider the following regularity condition, the choice of which can be sustained by a discussion similar to the one in Remark 5.1.7,

$$(RCVF_w^C) \mid f_i \text{ is continuous, } i = 1, \dots, k, \text{ and } 0 \in g(S \cap \operatorname{dom} g) + \operatorname{int}(C).$$

Theorem 5.2.9. *Let $\beta \in \mathcal{F}_w$ be fixed and assume that the regularity condition $(RCVF_w^C)$ is fulfilled. If $\bar{x} \in \mathcal{A}$ is a weakly efficient solution to (PVF_w^C) then there exists $(\bar{\lambda}, \bar{y}^*, \bar{z}^*, \bar{t})$, a weakly efficient solution to $(DVF_w^{C_\beta})$, such that $f_i(\bar{x}) = h_{wi}^{C_\beta}(\bar{\lambda}, \bar{y}^*, \bar{z}^*, \bar{t})$, $i = 1, \dots, k$.*

Proof. Let $\bar{x}$ be a weakly efficient solution to (PVF_w^C). This means that $f(\bar{x}) \in \operatorname{WMin}(f(\mathcal{A}), \mathbb{R}_+^k)$. Obviously, $f(\mathcal{A}) + \mathbb{R}_+^k$ is a convex set, thus by Corollary 2.4.26 there exists $\bar{\lambda} \in \mathbb{R}_+^k \backslash \{0\}$ such that $\bar{x}$ is an optimal solution to the scalar problem

$$(PF_{w\bar{\lambda}}^C) \quad \inf_{x \in X} \left\{ \sum_{i \in I(\bar{\lambda})} \bar{\lambda}_i f_i(x) \right\},$$

where $I(\bar{\lambda}) = \left\{ i \in \{1, \dots, k\} : \bar{\lambda}_i > 0 \right\}$ is a nonempty set. Applying the same scheme as for $(PF_{\bar{\lambda}}^C)$ and $(DF_{\bar{\lambda}}^C)$, we obtain as a scalar dual to $(PF_{\bar{\lambda}}^C)$ the following optimization problem

$$(DF_{w\bar{\lambda}}^C) \quad \sup_{\substack{y^{i*} \in X^*, i \in I(\bar{\lambda}), \\ w^* \in C^*}} \left\{ -\sum_{i \in I(\bar{\lambda})} \bar{\lambda}_i f_i^*(y^{i*}) - (w^* g)_S^*\left(-\sum_{i \in I(\bar{\lambda})} \bar{\lambda}_i y^{i*} \right) \right\}.$$

By Theorem 5.2.1 follows that for the primal-dual pair $(PF_{w\bar{\lambda}}^C) - (DF_{w\bar{\lambda}}^C)$ strong duality holds, while Theorem 5.2.2 ensures that there exist $\bar{x}^{i*} \in \operatorname{dom} f_i^*$, $i \in I(\bar{\lambda})$, and $\bar{w}^* \in C^*$ such that

(i) $(\bar{w}^*g)^*_S\big(-\sum_{i\in I(\bar{\lambda})}\lambda_i\bar{y}^{i*}\big) = \langle -\sum_{i\in I(\bar{\lambda})}\lambda_i\bar{y}^{i*}, \bar{x}\rangle$;
(ii) $(\bar{w}^*g)(\bar{x}) = 0$;
(iii) $f_i(\bar{x}) + f_i^*(\bar{y}^{i*}) = \langle \bar{y}^{i*}, \bar{x}\rangle$, $i \in I(\bar{\lambda})$.

As $(RCVF_w^C)$ holds, one can choose some $\bar{y}^{i*} \in \operatorname{dom} f_i^*$ for $i \notin I(\bar{\lambda})$. Taking $\bar{y}^* := (\bar{y}^{1*}, \ldots, \bar{y}^{k*}) \in \prod_{i=1}^k \operatorname{dom} f_i^*$, one has, via (i) that $(\bar{w}^*g)^*_S\big(-\sum_{i=1}^k \bar{\lambda}_i\bar{y}^{i*}\big) = \langle -\sum_{i=1}^k \bar{\lambda}_i\bar{y}^{i*}, \bar{x}\rangle$. Let us take, for each $i = 1, \ldots, k$, $\bar{z}^{i*} := \beta_i(\bar{\lambda})\bar{w}^*$ and denote $\bar{z}^* = (\bar{z}^{1*}, \ldots, \bar{z}^{k*})$. Then $\sum_{i=1}^k \bar{\lambda}_i\bar{z}^{i*} = \bar{w}^* \in C^*$. Note that whenever $i \in \{1, \ldots, k\}$ one has, as explained in the proof of Theorem 5.2.4, $(\bar{z}^{i*}g)^*_S\big(-\beta_i(\bar{\lambda})\sum_{j=1}^k \bar{\lambda}_j\bar{y}^{j*}\big) = \beta_i(\bar{\lambda})(\bar{w}^*g)^*_S\big(-\sum_{j=1}^k \bar{\lambda}_j\bar{y}^{j*}\big)$ and so $-\beta_i(\bar{\lambda})\sum_{j=1}^k \bar{\lambda}_j\bar{y}^{j*} \in \operatorname{dom}(\bar{z}^{i*}g)^*_S$. Take also

$$\bar{t}_i := \langle \bar{y}^{i*}, \bar{x}\rangle + (\bar{z}^{i*}g)^*_S\left(-\beta_i(\bar{\lambda})\sum_{j=1}^k \bar{\lambda}_j\bar{y}^{j*}\right)$$

when $i \in I(\bar{\lambda})$,

$$\bar{t}_i := f_i(\bar{x}) + f_i^*(\bar{y}^{i*}) + (\bar{z}^{i*}g)^*_S\left(-\beta_i(\bar{\lambda})\sum_{j=1}^k \bar{\lambda}_j\bar{y}^{j*}\right)$$

when $i \notin I(\bar{\lambda})$ and denote $\bar{t} := (\bar{t}_1, \ldots, \bar{t}_k)^T$. It can be verified that $\sum_{i=1}^k \bar{\lambda}_i\bar{t}_i = 0$ and $(\bar{\lambda}, \bar{y}^*, \bar{z}^*, \bar{t}) \in \mathcal{B}^{C_\beta}$. Moreover, for $i = 1, \ldots, k$, it holds $f_i(\bar{x}) = -f_i^*(\bar{y}^{i*}) - (\bar{z}^{i*}g)^*_S\big(-\beta_i(\bar{\lambda})\sum_{j=1}^k \bar{\lambda}_j\bar{y}^{j*}\big) + \bar{t}_i$. The fact that $(\bar{\lambda}, \bar{y}^*, \bar{z}^*, \bar{t})$ is weakly efficient to $(DVF_w^{C_\beta})$ is a direct consequence of Theorem 5.2.8. □

Remark 5.2.7. A discussion similar to the one in Remark 5.1.6 makes clear why it is necessary to consider when dealing with weakly efficient solutions that the functions f_i, $i = 1, \ldots, k$, have full domains.

Remark 5.2.8. Analogously to Remark 5.1.7, note that assuming continuity for f_i, $i = 1, ..., k$, in $(RCVF_w^C)$ is not too strong. Even if $(RCVF^C)$ seems at the first look to be weaker, by Theorem 2.2.17 follows that in the setting considered in this subsection it turns out to be nothing but $(RCVF_w^C)$. Nevertheless, this condition can be replaced in the results gathered in this subsection with $(RCF_i^C), i \in \{2, 2', 2'', 3\}$, adapted to the situation where $\operatorname{dom} f_i = X$ for $i = 1, ..., k$.

The proof of the following theorem can be done by combining the ideas from the proofs of Theorem 4.3.21 and Theorem 5.2.5.

Theorem 5.2.10. *Let $\beta \in \mathcal{F}_w$ be fixed and assume that the regularity condition $(RCVF_w^C)$ is fulfilled. Then*

$$\mathbb{R}^k \setminus \operatorname{cl}(f(\mathcal{A}) + \mathbb{R}^k_+) \subseteq h_w^{C_\beta}(\mathcal{B}_w^{C_\beta}) - (\mathbb{R}^k_+ \setminus \{0\}).$$

The converse duality theorem for the primal-dual pair treated here does not follow as a direct consequence of this result, as it was the case for $(PVF^C) - (DVF^C)$. Nevertheless, we are able to state such a result with the mention that it turns out to be an implication of Theorem 4.3.25(c) and Theorem 5.3.9, which we state in the forthcoming section.

Theorem 5.2.11. *Let $\beta \in \mathcal{F}_w$ be fixed and assume that the regularity condition $(RCVF_w^C)$ is fulfilled and that the set $f(\mathcal{A})+\mathbb{R}^k_+$ is closed. Then for every weakly efficient solution $(\bar{\lambda}, \bar{y}^*, \bar{z}^*, \bar{t})$ to $(DVF_w^{C_\beta})$ one has that $h_w^{C_\beta}(\bar{\lambda}, \bar{y}^*, \bar{z}^*, \bar{t})$ is a weakly minimal element of the set $f(\mathcal{A}) + \mathbb{R}^k_+$.*

Remark 5.2.9. The observations pointed our in Remark 5.2.1 and Remark 5.2.8 are valid for Theorem 5.2.9 and Theorem 5.2.11, too, while the ones from Remark 5.2.6, except for the first particular case, can be applied to vector duality with respect to weakly efficient solutions, too. Note that the analogous vector dual to (DVF^{C_M}) from the latter remark is

$$(DVF_w^{C_M}) \qquad \underset{(\lambda,y^*,z^*,t)\in\mathcal{B}_w^{C_M}}{\text{WMax}} h_w^{C_M}(\lambda, y^*, z^*, t),$$

where

$$\mathcal{B}_w^{C_M} = \Bigg\{(\lambda, y^*, z^*, t) \in (\mathbb{R}^k_+\backslash\{0\}) \times \prod_{i=1}^k \operatorname{dom} f_i^* \times C^* \times \mathbb{R}^k : \lambda = (\lambda_1, \dots, \lambda_k)^T,$$
$$y^* = (y^{1*}, \dots, y^{k*}), t = (t_1, \dots, t_k)^T, \sum_{i=1}^k \lambda_i t_i = 0,$$
$$\frac{-1}{\sum\limits_{j=1}^k \lambda_j} \sum_{j=1}^k \lambda_j y^{j*} \in \operatorname{dom}(z^*g)_S^*\Bigg\}$$

and

$$h_w^{C_M}(\lambda, y^*, z^*, t) = \begin{pmatrix} -f_1^*(y^{1*}) - (z^*g)_S^*\left(-\frac{1}{\sum\limits_{j=1}^k \lambda_j} \sum\limits_{j=1}^k \lambda_j y^{j*}\right) + t_1 \\ \vdots \\ -f_k^*(y^{k*}) - (z^*g)_S^*\left(-\frac{1}{\sum\limits_{j=1}^k \lambda_j} \sum\limits_{j=1}^k \lambda_j y^{j*}\right) + t_k \end{pmatrix}.$$

5.2.3 Duality for linearly constrained vector optimization problems

In the following we see what happens to the duals introduced in the previous two subsections when working with the linearly constrained vector minimization primal problem

$$(PVF^{\mathcal{L}}) \quad \underset{x\in\mathcal{A}^{\mathcal{L}}}{\operatorname{Min}} f(x),$$
$$\mathcal{A}^{\mathcal{L}} = \{x \in S : Ax - b \in C\}$$

where X and Z are Hausdorff locally convex spaces, with the latter partially ordered by the convex cone $C \subseteq Z$, $S \subseteq X$ is a convex cone, $f_i : X \to \overline{\mathbb{R}}$, $i = 1,\dots,k$, are proper functions, $A \in \mathcal{L}(X,Z)$ and $b \in Z$ fulfill $A(\cap_{i=1}^k \operatorname{dom} f_i \cap S) \cap (b + C) \neq \emptyset$. Assume that the image space $\mathbb{R}^k$ is partially ordered by the cone $\mathbb{R}^k_+$. Consider the vector function $f : X \to \overline{\mathbb{R}^k}$ defined as in the beginning of the section. One can immediately notice that $(PVF^{\mathcal{L}})$ is a special case of (PVF^C), for $g(x) = b - Ax$ for $x \in X$ and S being moreover a cone. Like in subsection 5.2.1, for each $\alpha \in \mathcal{F}$ we attach to $(PVF^{\mathcal{L}})$ a dual problem obtained by particularizing (DVF^{C_α}). Noting that

$$(z^*g)^*_S(x^*) = \sup_{x\in S}\{\langle x^*, x\rangle + \langle z^*, Ax - b\rangle\} = -\langle z^*, b\rangle + \sup_{x\in S}\langle x^* + A^*z^*, x\rangle$$

$$= -\langle z^*, b\rangle + \delta_{\{A^*z^*+x^*\in -S^*\}}(x^*)$$

for any $x^* \in X^*$ and $z^* \in Z^*$, the dual in discussion turns into

$$(DVF^{\mathcal{L}_\alpha}) \qquad \underset{(\lambda,y^*,z^*,t)\in\mathcal{B}^{\mathcal{L}_\alpha}}{\operatorname{Max}} h^{\mathcal{L}_\alpha}(\lambda, y^*, z^*, t),$$

where

$$\mathcal{B}^{\mathcal{L}_\alpha} = \Bigg\{(\lambda, y^*, z^*, t) \in \operatorname{int}(\mathbb{R}^k_+) \times \prod_{i=1}^k \operatorname{dom} f_i^* \times (Z^*)^k \times \mathbb{R}^k : \lambda = (\lambda_1, \dots, \lambda_k)^T,$$
$$y^* = (y^{1*}, \dots, y^{k*}), z^* = (z^{1*}, \dots, z^{k*}), t = (t_1, \dots, t_k)^T,$$
$$\sum_{i=1}^k \lambda_i z^{i*} \in C^*, \sum_{i=1}^k \lambda_i t_i = 0,$$
$$\alpha_i(\lambda)\sum_{j=1}^k \lambda_j y^{j*} - A^*z^{i*} \in S^*,\ i = 1,\dots,k\Bigg\}$$

and

$$h^{\mathcal{L}_\alpha}(\lambda, y^*, z^*, t) = \begin{pmatrix} \langle z^{1*}, b\rangle + t_1 - f_1^*(y^{1*}) \\ \vdots \\ \langle z^{k*}, b\rangle + t_k - f_k^*(y^{k*}) \end{pmatrix}.$$

One can notice that at $(DVF^{\mathcal{L}_\alpha})$ the parameter α is present only in the feasible set. However, we show that all $(DVF^{\mathcal{L}_\alpha})$, $\alpha \in \mathcal{F}$, are equivalent to each other and α plays no role in this case.

Let $\alpha \in \mathcal{F}$. First we demonstrate that taking the set

$$\mathcal{B}^{\mathcal{L}}=\Bigg\{(\lambda, y^*, w^*, v) \in \operatorname{int}(\mathbb{R}^k_+) \times \prod_{i=1}^k \operatorname{dom} f_i^* \times (Z^*)^k \times \mathbb{R}^k : \lambda = (\lambda_1, \ldots, \lambda_k)^T,$$
$$y^* = (y^{1*}, \ldots, y^{k*}), w^* = (w^{1*}, \ldots, w^{k*}), v = (v_1, \ldots, v_k)^T,$$
$$\sum_{i=1}^k \lambda_i w^{i*} \in C^*, \sum_{i=1}^k \lambda_i v_i = 0, \sum_{i=1}^k \lambda_i (y^{i*} - A^* w^{i*}) \in S^*\Bigg\},$$

one has $h^{\mathcal{L}_\alpha}(\mathcal{B}^{\mathcal{L}_\alpha}) = h^{\mathcal{L}_\alpha}(\mathcal{B}^{\mathcal{L}})$.

The inclusion "$\subseteq$" follows immediately, since whenever $(\lambda, y^*, z^*, t) \in \mathcal{B}^{\mathcal{L}_\alpha}$ one can directly verify that $(\lambda, y^*, z^*, t) \in \mathcal{B}^{\mathcal{L}}$. To prove the opposite inclusion, when $(\lambda, y^*, w^*, v) \in \mathcal{B}^{\mathcal{L}}$ take $z^{i*} := \alpha_i(\lambda) \sum_{j=1}^k \lambda_j w^{j*}$, $t_i := v_i + \langle w^{i*}, b\rangle - \langle \alpha_i(\lambda) \sum_{j=1}^k \lambda_j w^{j*}, b\rangle$, $i = 1, \ldots, k$, and $t = (t_1, ..., t_k)^T$. Therefore it follows that $(\lambda, y^*, z^*, t) \in \mathcal{B}^{\mathcal{L}_\alpha}$. Moreover, for $i = 1, \ldots, k$, $h_i^{\mathcal{L}_\alpha}(\lambda, y^*, z^*, t) = \langle z^{i*}, b\rangle + t_i - f_i^*(y^{i*}) = \langle \alpha_i(\lambda) \sum_{j=1}^k \lambda_j w^{j*}, b\rangle + v_i + \langle w^{i*}, b\rangle - \langle \alpha_i(\lambda) \sum_{j=1}^k \lambda_j w^{j*}, b\rangle - f_i^*(y^{i*}) = \langle w^{i*}, b\rangle + v_i - f_i^*(y^{i*}) = h_i^{\mathcal{L}_\alpha}(\lambda, y^*, w^*, v)$.

Therefore for all $\alpha \in \mathcal{F}$ the duals $(DVF^{\mathcal{L}_\alpha})$ collapse into

$$(DVF^{\mathcal{L}}) \qquad \operatorname*{Max}_{(\lambda, y^*, w^*, v) \in \mathcal{B}^{\mathcal{L}}} h^{\mathcal{L}}(\lambda, y^*, w^*, v),$$

where

$$h^{\mathcal{L}}(\lambda, y^*, w^*, v) = \begin{pmatrix} \langle w^{1*}, b\rangle + v_1 - f_1^*(y^{1*}) \\ \vdots \\ \langle w^{k*}, b\rangle + v_k - f_k^*(y^{k*}) \end{pmatrix}.$$

The formulation of this dual problem can be further simplified, depending on the way b is taken.

When $b = 0$, $(DVF^{\mathcal{L}})$ turns into

$$(DVF^{\mathcal{L}}) \qquad \operatorname*{Max}_{(\lambda, y^*, v) \in \mathcal{B}^{\mathcal{L}}} h^{\mathcal{L}}(\lambda, y^*, v),$$

where

$$\mathcal{B}^{\mathcal{L}} = \Bigg\{(\lambda, y^*, v) \in \operatorname{int}(\mathbb{R}^k_+) \times \prod_{i=1}^k \operatorname{dom} f_i^* \times \mathbb{R}^k, \lambda = (\lambda_1, \ldots, \lambda_k)^T,$$
$$y^* = (y^{1*}, \ldots, y^{k*}), v = (v_1, \ldots, v_k)^T, \sum_{i=1}^k \lambda_i v_i = 0,$$
$$\sum_{i=1}^k \lambda_i y^{i*} \in A^*(C^*) + S^*\Bigg\}$$

and

$$h^{\mathcal{L}}(\lambda, y^*, v) = \begin{pmatrix} v_1 - f_1^*(y^{1*}) \\ \vdots \\ v_k - f_k^*(y^{k*}) \end{pmatrix}.$$

On the other hand, when $b \neq 0$ the vector dual problem $(DVF^{\mathcal{L}})$ can be equivalently rewritten as

$$(DVF^{\mathcal{L}}) \qquad \underset{(\lambda,y^*,q^*)\in\mathcal{B}^{\mathcal{L}}}{\operatorname{Max}} h^{\mathcal{L}}(\lambda, y^*, q^*),$$

where

$$\mathcal{B}^{\mathcal{L}} = \Big\{(\lambda, y^*, q^*) \in \operatorname{int}(\mathbb{R}^k_+) \times \prod_{i=1}^k \operatorname{dom} f_i^* \times (Z^*)^k : \lambda = (\lambda_1, \ldots, \lambda_k)^T,$$
$$y^* = (y^{1*}, \ldots, y^{k*}), q^* = (q^{1*}, \ldots, q^{k*}),$$
$$\sum_{i=1}^k \lambda_i q^{i*} \in C^*, \sum_{i=1}^k \lambda_i (y^{i*} - A^* q_i^*) \in S^*\Big\}$$

and

$$h^{\mathcal{L}}(\lambda, y^*, q^*) = \begin{pmatrix} \langle q^{1*}, b\rangle - f_1^*(y^{1*}) \\ \vdots \\ \langle q^{k*}, b\rangle - f_k^*(y^{k*}) \end{pmatrix}.$$

To prove this, take $w^{i*} := q^{i*}$ and $v_i := 0$, $i = 1, \ldots, k$, while for the inverse inclusion, let $q^{i*} := w^{i*} + v_i\zeta$, $i = 1, \ldots, k$, where $\zeta \in Z^*$ is taken such that $\langle \zeta, b\rangle = 1$.

The vector dual problem (DVF^{C_M}) introduced in Remark 5.2.6 turns in the special case treated in this subsection into

$$(DVF^{\mathcal{L}_M}) \qquad \underset{(\lambda,y^*,z^*,v)\in\mathcal{B}^{\mathcal{L}_M}}{\operatorname{Max}} h^{\mathcal{L}_M}(\lambda, y^*, z^*, v),$$

where

$$\mathcal{B}^{\mathcal{L}_M} = \Big\{(\lambda, y^*, z^*, v) \in \operatorname{int}(\mathbb{R}^k_+) \times \prod_{i=1}^k \operatorname{dom} f_i^* \times C^* \times \mathbb{R}^k : \lambda = (\lambda_1, \ldots, \lambda_k)^T,$$
$$y^* = (y^{1*}, \ldots, y^{k*}), v = (v_1, \ldots, v_k)^T, \sum_{i=1}^k \lambda_i v_i = 0,$$
$$\sum_{i=1}^k \lambda_i (y^{i*} - A^* z^*) \in S^*\Big\}$$

and

$$h^{\mathcal{L}_M}(\lambda, y^*, z^*, v) = \begin{pmatrix} \langle z^*, b\rangle + v_1 - f_1^*(y^{1*}) \\ \vdots \\ \langle z^*, b\rangle + v_k - f_k^*(y^{k*}) \end{pmatrix}.$$

Note that when $b = 0$ this turns into $(DVF^{\mathcal{L}})$, too.

Remark 5.2.10. Weak duality for the primal problem $(PVF^{\mathcal{L}})$ and its vector dual $(DVF^{\mathcal{L}})$ holds from the general case. To obtain strong and converse duality for this pair of problems, one can consider the regularity condition

$$(RCVF^{\mathcal{L}}) \left| \begin{array}{l} \exists x' \in \bigcap\limits_{i=1}^{k} \operatorname{dom} f_i \cap S \text{ such that } f_i \text{ is continuous at } x', \\ i = 1, \ldots, k, \text{ and } Ax' - b \in \operatorname{int}(C), \end{array} \right.$$

or one of the regularity conditions mentioned in Remark 5.2.5(b) adapted to this particular situation.

Remark 5.2.11. The results obtained within this subsection extend the ones in [182–184], where all the spaces involved were taken finite dimensional and those in [174] where the entries of the objective vector function were sums of norms and linear functions.

With the only change in the framework consisting in taking the functions $f_i : X \to \mathbb{R}$, $i = 1, \ldots, k$, like in subsection 5.2.2, namely having full domains, similar considerations can be made when we are dealing with the weakly efficient solutions to the primal problem

$$(PVF_w^{\mathcal{L}}) \quad \operatorname*{WMin}_{x \in \mathcal{A}^{\mathcal{L}}} f(x).$$
$$\mathcal{A}^{\mathcal{L}} = \{x \in S : Ax - b \in C\}$$

Like in subsection 5.2.2, for each $\beta \in \mathcal{F}_w$ we can attach to $(PVF_w^{\mathcal{L}})$ a dual problem obtained by particularizing $(DVF_w^{C_\beta})$, but all these duals coincide, being actually

$$(DVF_w^{\mathcal{L}}) \quad \operatorname*{WMax}_{(\lambda, y^*, w^*, v) \in \mathcal{B}_w^{\mathcal{L}}} h_w^{\mathcal{L}}(\lambda, y^*, w^*, v),$$

where

$$\mathcal{B}_w^{\mathcal{L}} = \Bigg\{ (\lambda, y^*, w^*, v) \in (\mathbb{R}_+^k \backslash \{0\}) \times \prod_{i=1}^{k} \operatorname{dom} f_i^* \times (Z^*)^k \times \mathbb{R}^k : \lambda = (\lambda_1, \ldots, \lambda_k)^T,$$
$$y^* = (y^{1*}, \ldots, y^{k*}), w^* = (w^{1*}, \ldots, w^{k*}), v = (v_1, \ldots, v_k)^T,$$
$$\sum_{i=1}^{k} \lambda_i w^{i*} \in C^*, \sum_{i=1}^{k} \lambda_i v_i = 0, \sum_{i=1}^{k} \lambda_i (y^{i*} - A^* w^{i*}) \in S^* \Bigg\}$$

and

$$h_w^{\mathcal{L}}(\lambda, y^*, w^*, v) = \begin{pmatrix} \langle w^{1*}, b \rangle + v_1 - f_1^*(y^{1*}) \\ \vdots \\ \langle w^{k*}, b \rangle + v_k - f_k^*(y^{k*}) \end{pmatrix}.$$

The formulation of this dual problem can be further simplified.
When $b = 0$ $(DVF_w^{\mathcal{L}})$ turns into

$$(DVF_w^{\mathcal{L}}) \quad \underset{(\lambda,y^*,v)\in\mathcal{B}_w^{\mathcal{L}}}{\mathrm{WMax}} h_w^{\mathcal{L}}(\lambda,y^*,v),$$

where

$$\mathcal{B}_w^{\mathcal{L}} = \Big\{(\lambda,y^*,v) \in (\mathbb{R}_+^k\backslash\{0\}) \times \prod_{i=1}^k \mathrm{dom}\, f_i^* \times \mathbb{R}^k : \lambda = (\lambda_1,\ldots,\lambda_k)^T, \\ y^* = (y^{1*},\ldots,y^{k*}), v = (v_1,\ldots,v_k)^T, \sum_{i=1}^k \lambda_i v_i = 0, \\ \sum_{i=1}^k \lambda_i y^{i*} \in A^*(C^*) + S^*\Big\}$$

and

$$h_w^{\mathcal{L}}(\lambda,y^*,v) = \begin{pmatrix} v_1 - f_1^*(y^{1*}) \\ \vdots \\ v_k - f_k^*(y^{k*}) \end{pmatrix}.$$

On the other hand, when $b \neq 0$, $(DVF_w^{\mathcal{L}})$ can be equivalently rewritten as

$$(DVF_w^{\mathcal{L}}) \quad \underset{(\lambda,y^*,q^*)\in\mathcal{B}_w^{\mathcal{L}}}{\mathrm{WMax}} h_w^{\mathcal{L}}(\lambda,y^*,q^*),$$

where

$$\mathcal{B}_w^{\mathcal{L}} = \Big\{(\lambda,y^*,q^*) \in (\mathbb{R}_+^k\backslash\{0\}) \times \prod_{i=1}^k \mathrm{dom}\, f_i^* \times (Z^*)^k : \lambda = (\lambda_1,\ldots,\lambda_k)^T, \\ y^* = (y^{1*},\ldots,y^{k*}), q^* = (q^{1*},\ldots,q^{k*}), \\ \sum_{i=1}^k \lambda_i q^{i*} \in C^*, \sum_{i=1}^k \lambda_i (y^{i*} - A^* q^{i*}) \in S^*\Big\}$$

and

$$h_w^{\mathcal{L}}(\lambda,y^*,q^*) = \begin{pmatrix} \langle q^{1*}, b\rangle - f_1^*(y^{1*}) \\ \vdots \\ \langle q^{k*}, b\rangle - f_k^*(y^{k*}) \end{pmatrix}.$$

The vector dual problem $(DVF_w^{C_M})$ turns in the special case treated in this subsection into

$$(DVF_w^{\mathcal{L}_M}) \quad \underset{(\lambda,y^*,z^*,v)\in\mathcal{B}_w^{\mathcal{L}_M}}{\mathrm{WMax}} h_w^{\mathcal{L}_M}(\lambda,y^*,z^*,v),$$

where

$$\mathcal{B}_w^{\mathcal{L}_M} = \Bigg\{(\lambda, y^*, z^*, v) \in (\mathbb{R}_+^k \backslash \{0\}) \times \prod_{i=1}^k \operatorname{dom} f_i^* \times C^* \times \mathbb{R}^k : \lambda = (\lambda_1, \ldots, \lambda_k)^T,$$
$$y^* = (y^{1*}, \ldots, y^{k*}), v = (v_1, \ldots, v_k)^T,$$
$$\sum_{i=1}^k \lambda_i v_i = 0, \sum_{i=1}^k \lambda_i (y^{i*} - A^* z^*) \in S^* \Bigg\}$$

and

$$h_w^{\mathcal{L}_M}(\lambda, y^*, z^*, v) = \begin{pmatrix} \langle z^*, b\rangle + v_1 - f_1^*(y^{1*}) \\ \vdots \\ \langle z^*, b\rangle + v_k - f_k^*(y^{k*}) \end{pmatrix}$$

and when $b = 0$ it coincides with $(DVF_w^{\mathcal{L}})$.

Remark 5.2.12. Weak duality for the primal problem $(PVF_w^{\mathcal{L}})$ and its vector dual $(DVF_w^{\mathcal{L}})$ holds from the general case. To obtain strong and converse duality for this pair of problems, one can consider the regularity condition

$$(RCVF_w^{\mathcal{L}}) \mid f_i \text{ is continuous, } i = 1, \ldots, k, \text{ and } b \in A(S) - \operatorname{int}(C),$$

or one of the regularity conditions mentioned in Remark 5.2.8 adapted to this particular context.

5.3 Comparisons between different duals to (PVF^C)

In the following we investigate, like in subsection 4.3.3, inclusion relations between the image sets of the feasible sets through the objective functions of different vector duals to (PVF^C). Besides (DVF^{C_α}) and (DVF^{C_M}), given in section 5.2, in chapter 4 we introduced several vector duals to (PVF^C) from which we recall $(DV^{C_{FL}})$, formulated in the framework of this chapter as

$$(DVF^{C_{FL}}) \qquad \operatorname*{Max}_{(\lambda, y^*, z^*, v) \in \mathcal{B}^{C_{FL}}} h^{C_{FL}}(\lambda, y^*, z^*, v),$$

where

$$\mathcal{B}^{C_{FL}} = \Bigg\{(\lambda, y^*, z^*, v) \in \operatorname{int}(\mathbb{R}_+^k) \times X^* \times C^* \times \mathbb{R}^k : \lambda = (\lambda_1, \ldots, \lambda_k)^T,$$
$$v = (v_1, \ldots, v_k)^T,$$
$$\sum_{i=1}^k \lambda_i v_i \leq -\left(\sum_{i=1}^k \lambda_i f_i\right)^*(y^*) - (z^* g)_S^*(-y^*)\Bigg\}$$

and

$$h^{C_{FL}}(\lambda, y^*, z^*, v) = v.$$

By slightly modifying the formulation of $(DVF^{C_{FL}})$ we consider another vector dual to (PVF^C) that looks like

$$(DVF^{C_{\widetilde{FL}}}) \qquad \underset{(\lambda,y^*,z^*,v)\in\mathcal{B}^{C_{\widetilde{FL}}}}{\text{Max}} \quad h^{C_{\widetilde{FL}}}(\lambda,y^*,z^*,v),$$

where

$$\mathcal{B}^{C_{\widetilde{FL}}} = \bigg\{(\lambda,y^*,z^*,v) \in \operatorname{int}(\mathbb{R}^k_+) \times (X^*)^k \times C^* \times \mathbb{R}^k : \lambda = (\lambda_1,\dots,\lambda_k)^T,$$
$$y^* = (y^{1*},\dots,y^{k*}), v = (v_1,\dots,v_k)^T,$$
$$\sum_{i=1}^k \lambda_i v_i \leq -\sum_{i=1}^k \lambda_i f_i^*(y^{i*}) - (z^*g)_S^*\Big(-\sum_{j=1}^k \lambda_j y^{j*}\Big)\bigg\}$$

and

$$h^{C_{\widetilde{FL}}}(\lambda,y^*,z^*,v) = v.$$

We begin by comparing the image sets of the feasible sets through the objective functions of the vector duals to (PVF^C) introduced in this chapter, (DVF^{C_α}), for $\alpha \in \mathcal{F}$, and (DVF^{C_M}).

Theorem 5.3.1. *Let $\alpha \in \mathcal{F}$ be fixed. Then $h^{C_M}(\mathcal{B}^{C_M}) \subseteq h^{C_\alpha}(\mathcal{B}^{C_\alpha})$.*

Proof. Take an arbitrary element $d = (d_1,\dots,d_k)^T \in h^{C_M}(\mathcal{B}^{C_M})$. Then there is a feasible point $(\lambda,y^*,z^*,t) \in \mathcal{B}^{C_M}$ such that

$$d_i = -f_i^*(y^{i*}) - (z^*g)_S^*\left(-\frac{1}{\sum\limits_{j=1}^k \lambda_j}\sum_{j=1}^k \lambda_j y^{j*}\right) + t_i, i = 1,\dots,k.$$

Take $\bar{z}^{i*} := \alpha_i(\lambda)\big(\sum_{j=1}^k \lambda_j\big)z^*$, $i = 1,\dots,k$, and denote $\bar{z}^* := (\bar{z}^{1*},\dots,\bar{z}^{k*})$. Then

$$\sum_{i=1}^k \lambda_i \bar{z}^{i*} = \sum_{i=1}^k \lambda_i\alpha_i(\lambda)\Big(\sum_{j=1}^k \lambda_j\Big)z^* = \Big(\sum_{i=1}^k \lambda_i\Big)z^* \in C^*$$

and, for $i = 1,\dots,k$,

$$(\bar{z}^{i*}g)_S^*\Big(-\alpha_i(\lambda)\sum_{j=1}^k \lambda_j y^{j*}\Big) = \Big(\alpha_i(\lambda)\sum_{j=1}^k \lambda_j\Big)(z^*g)_S^*\left(-\frac{1}{\sum\limits_{j=1}^k \lambda_j}\sum_{j=1}^k \lambda_j y^{j*}\right).$$

Consequently, $-\alpha_i(\lambda)\sum_{j=1}^k \lambda_j y^{j*} \in \operatorname{dom}(\bar{z}^{i*}g)_S^*$ for $i = 1,\dots,k$. For all $i = 1,\dots,k$, let

$$\bar{t}_i := t_i + (\bar{z}^{i*}g)^*_S\left(- \alpha_i(\lambda)\sum_{j=1}^k \lambda_j y^{j*}\right) - (z^*g)^*_S\left(- \frac{1}{\sum\limits_{j=1}^k \lambda_j}\sum_{j=1}^k \lambda_j y^{j*}\right).$$

We obtain immediately that $\bar{t} := (\bar{t}_1, \ldots, \bar{t}_k)^T \in \mathbb{R}^k$ and $\sum_{i=1}^k \lambda_i \bar{t}_i = \sum_{i=1}^k \lambda_i t_i = 0$, which yields $(\lambda, y^*, \bar{z}^*, \bar{t}) \in \mathcal{B}^{C_\alpha}$. Moreover,

$$d_i = -f_i^*(y^{i*}) - (\bar{z}^{i*}g)^*_S\left(- \alpha_i(\lambda)\sum_{j=1}^k \lambda_j y^{j*}\right) + \bar{t}_i \ \forall i = 1, \ldots, k,$$

i.e. $d = h^{C_\alpha}(\lambda, y^*, \bar{z}^*, \bar{t})$. Therefore, $h^{C_M}(\mathcal{B}^{C_M}) \subseteq h^{C_\alpha}(\mathcal{B}^{C_\alpha})$. □

With the following example (from [24, 36]) we show that the inclusion just proven can sometimes be strict.

Example 5.3.1. Let be $\alpha \in \mathcal{F}$ fixed, $k = 2$, $X = S = \mathbb{R}$, $Z = C = \mathbb{R}^2$, $f_1, f_2 : \mathbb{R} \to \mathbb{R}$, defined by $f_1 = f_2 \equiv 0$, and $g : \mathbb{R} \to \mathbb{R}^2$, $g(x) = (g_1(x), g_2(x))^T$, where $g_1, g_2 : \mathbb{R} \to \mathbb{R}$, defined by

$$g_1(x) = \begin{cases} 1, & \text{if } x < 0, \\ e^{-x}, & \text{if } x \geq 0, \end{cases} \text{ and } g_2(x) = \begin{cases} e^x, & \text{if } x < 0, \\ 1, & \text{if } x \geq 0. \end{cases}$$

For $z^* = (z_1^*, z_2^*)$, $z_1^* = (1, -1)^T, z_2^* = (-1, 1)^T$, we have

$$(z_1^*g)(x) = \begin{cases} 1 - e^x, & \text{if } x < 0, \\ e^{-x} - 1, & \text{if } x \geq 0, \end{cases} \text{ and } (z_2^*g)(x) = \begin{cases} -1 + e^x, & \text{if } x < 0, \\ 1 - e^{-x}, & \text{if } x \geq 0. \end{cases}$$

Taking $y^* = (y_1^*, y_2^*)^T = (0, 0)^T$, $\lambda = (1, 1)^T$ and $t = (1/2, -1/2)^T$, we have $\lambda_1 z_1^* + \lambda_2 z_2^* = (0, 0)^T \in C^* = \{(0, 0)^T\}$, $\lambda_1 t_1 + \lambda_2 t_2 = 0$ and $f_1^*(0) = f_2^*(0) = 0$. This means that

$$d = \begin{pmatrix} -\frac{1}{2} \\ \\ -\frac{3}{2} \end{pmatrix} = \begin{pmatrix} -(z_1^*g)^*(0) + t_1 \\ \\ -(z_2^*g)^*(0) + t_2 \end{pmatrix} = h^{C_\alpha}(\lambda, y^*, z^*, t),$$

i.e. $d \in h^{C_\alpha}(\mathcal{B}^{C_\alpha})$. Let us show now that $d \notin h^{C_M}(\mathcal{B}^{C_M})$. If this were not true, then there would exist an element $(\bar{\lambda}, \bar{y}^*, \bar{z}^*, \bar{t}) \in \mathcal{B}^{C_M}$ such that

$$\begin{pmatrix} -\frac{1}{2} \\ \\ -\frac{3}{2} \end{pmatrix} = \begin{pmatrix} -f_1^*(\bar{y}_1^*) - (\bar{z}^*g)^*\left(-\frac{\bar{\lambda}_1\bar{y}_1^* + \bar{\lambda}_2\bar{y}_2^*}{\bar{\lambda}1 + \bar{\lambda}_2}\right) + \bar{t}_1 \\ \\ -f_2^*(\bar{y}_2^*) - (\bar{z}^*g)^*\left(-\frac{\bar{\lambda}_1\bar{y}_1^* + \bar{\lambda}_2\bar{y}_2^*}{\bar{\lambda}1 + \bar{\lambda}_2}\right) + \bar{t}_2 \end{pmatrix}.$$

It follows that $f_1^*(\bar{y}_1^*), f_2^*(\bar{y}_2^*) \in \mathbb{R}$, but, in order to happen this, we must have $\bar{y}_1^* = \bar{y}_2^* = 0$, $f_1^*(\bar{y}_1^*) = f_2^*(\bar{y}_2^*) = 0$ and $\bar{z}^* = 0$, thus $(\bar{z}^*g)^*(0) = 0$. These yield $\bar{t}_1 = -1/2$ and $\bar{t}_2 = -3/2$. Consequently, $\bar{\lambda}_1\bar{t}_1 + \bar{\lambda}_2\bar{t}_2 = -(\bar{\lambda}_1 + 3\bar{\lambda}_2)/2 < 0$. This contradicts $\bar{\lambda}_1\bar{t}_1 + \bar{\lambda}_2\bar{t}_2 = 0$, therefore $d \notin h^{C_M}(\mathcal{B}^{C_M})$, i.e. the inclusion proven in Theorem 5.3.1 may be strict.

The next pair of vector duals to (PVF^C) we deal with consists of (DVF^{C_α}) and $(DVF^{C_{\widetilde{FL}}})$.

Theorem 5.3.2. *Let $\alpha \in \mathcal{F}$ be fixed. Then $h^{C_\alpha}(\mathcal{B}^{C_\alpha}) \subseteq h^{C_{\widetilde{FL}}}(\mathcal{B}^{C_{\widetilde{FL}}})$.*

Proof. Let $d = (d_1, \ldots, d_k)^T \in h^{C_\alpha}(\mathcal{B}^{C_\alpha})$. Then there exists $(\lambda, y^*, z^*, t) \in \mathcal{B}^{C_\alpha}$ such that $d = h^{C_\alpha}(\lambda, y^*, z^*, t)$. Denote $\bar{z}^* := \sum_{i=1}^k \lambda_i z^{i*}$. We have $\bar{z}^* \in C^*$ and

$$\sum_{i=1}^k \lambda_i d_i = -\sum_{i=1}^k \lambda_i f_i^*(y^{i*}) - \sum_{i=1}^k \lambda_i (z^{i*}g)_S^*\Big(- \alpha_i(\lambda) \sum_{j=1}^k \lambda_j y^{j*}\Big) + \sum_{j=1}^k \lambda_j t_j.$$

The Young-Fenchel inequality yields

$$-(z^{i*}g)_S^*\Big(- \alpha_i(\lambda) \sum_{j=1}^k \lambda_j y^{j*}\Big) \leq (z^{i*}g)(x) + \Big\langle \alpha_i(\lambda) \sum_{j=1}^k \lambda_j y^{j*}, x\Big\rangle$$

for $i = 1, \ldots, k$, and all $x \in S$. Consequently,

$$\sum_{i=1}^k \lambda_i d_i \leq -\sum_{i=1}^k \lambda_i f_i^*(y^{i*}) + \Big(\Big(\sum_{i=1}^k \lambda_i z^{i*}\Big) g\Big)(x) + \Big(\sum_{i=1}^k \lambda_i \alpha_i(\lambda)\Big)$$

$$\Big\langle \sum_{i=1}^k \lambda_i y^{i*}, x\Big\rangle = -\sum_{i=1}^k \lambda_i f_i^*(y^{i*}) + (\bar{z}^* g)(x) + \Big\langle \sum_{i=1}^k \lambda_i y^{i*}, x\Big\rangle \quad \forall x \in S.$$

This yields

$$\sum_{i=1}^k \lambda_i d_i \leq -\sum_{i=1}^k \lambda_i f_i^*(y^{i*}) + \inf_{x \in S} \Big\{(\bar{z}^* g)(x) + \Big\langle \sum_{i=1}^k \lambda_i y^{i*}, x\Big\rangle\Big\}$$
$$= -\sum_{i=1}^k \lambda_i f_i^*(y^{i*}) - (\bar{z}^* g)_S^*\Big(- \sum_{i=1}^k \lambda_i y^{i*}\Big),$$

i.e. $(\lambda, y^*, \bar{z}^*, d) \in \mathcal{B}^{C_{\widetilde{FL}}}$. Therefore $d \in h^{C_{\widetilde{FL}}}(\mathcal{B}^{C_{\widetilde{FL}}})$ and, since d was arbitrarily chosen, the desired inclusion is proven. □

With the following example we show that the inclusion just proven can be strict in general.

Example 5.3.2. Fix an $\alpha \in \mathcal{F}$ and take $k = 2$, $X = S = \mathbb{R}$, $Z = \mathbb{R}^2$, $C = \mathbb{R}^2_+$, $f_1, f_2 : \mathbb{R} \to \mathbb{R}$, defined by $f_1 = f_2 \equiv 0$ and $g : \mathbb{R} \to \mathbb{R}^2$, $g(x) = (x+1, -x)^T$ for $x \in \mathbb{R}$. For $y^* = (y_1^*, y_2^*)^T = (0,0)^T$, $z^* = (1,1)^T \in C^* = \mathbb{R}^2_+$, $\lambda = (\lambda_1, \lambda_2)^T$, $\lambda_1 = \lambda_2 = 1$ and $d = (d_1, d_2)^T$, $d_1 = d_2 = -1$, we have $f_1^*(y_1^*) = 0$, $f_2^*(y_2^*) = 0$ and $(z^*g)^*\big(- \lambda_1 y_1^* - \lambda_2 y_2^*\big) = -1$, so

$$\lambda_1 d_1 + \lambda_2 d_2 = -2 < 1 = -f_1^*(y_1^*) - f_2^*(y_2^*) - (z^*g)^*\big(- \lambda_1 y_1^* - \lambda_2 y_2^*\big),$$

which implies that $d = (-1,-1)^T \in h^{C_{\overline{FL}}}(\mathcal{B}^{C_{\overline{FL}}})$.

Assuming that $d \in h^{C_\alpha}(\mathcal{B}^{C_\alpha})$ yields $(\bar\lambda, \bar y^*, \bar z^*, \bar t) \in \mathcal{B}^{C_\alpha}$, with $\bar t = (\bar t_1, \bar t_2)^T \in \mathbb{R}^2$, $\bar y^* = (\bar y_1^*, \bar y_2^*) \in \mathbb{R}^2$ and $\bar z^* = (\bar z^{1*}, \bar z^{2*}), \bar z^{1*}, \bar z^{2*} \in \mathbb{R}^2$, such that

$$\begin{pmatrix} -1 \\ \\ -1 \end{pmatrix} = \begin{pmatrix} -f_1^*(\bar y_1^*) - (\bar z^{1*} g)^*(-\alpha_1(\bar\lambda)(\bar\lambda_1 \bar y_1^* + \bar\lambda_2 \bar y_2^*)) + \bar t_1 \\ \\ -f_2^*(\bar y_2^*) - (\bar z^{2*} g)^*(-\alpha_2(\bar\lambda)(\bar\lambda_1 \bar y_1^* + \bar\lambda_2 \bar y_2^*)) + \bar t_2 \end{pmatrix}.$$

Because $f_1^*(\bar y_1^*), f_2^*(\bar y_2^*) \in \mathbb{R}$, we must have $\bar y_1^* = \bar y_2^* = 0$ and $f_1^*(\bar y_1^*) = f_2^*(\bar y_2^*) = 0$. This yields $-1 = -(\bar z^{1*} g)^*(0) + \bar t_1 = -(\bar z^{2*} g)^*(0) + \bar t_2$. Denoting $\bar z^{i*} = (\bar z_1^{i*}, \bar z_2^{i*})^T$, $i = 1,2$, we obtain $(\bar z^{i*} g)^*(0) = -\bar z_1^{i*}$, $i = 1,2$. From $(\bar\lambda, \bar y^*, \bar z^*, \bar t) \in \mathcal{B}^{C_\alpha}$ one has $\bar\lambda_1 \bar t_1 + \bar\lambda_2 \bar t_2 = 0$ and $\bar\lambda_1 \bar z^{1*} + \bar\lambda_2 \bar z^{2*} \geqq 0$, the latter implying $\bar\lambda_1 \bar z_1^{1*} + \bar\lambda_2 \bar z_1^{2*} \geq 0$. We get

$$\begin{pmatrix} -1 \\ \\ -1 \end{pmatrix} = \begin{pmatrix} \bar z_1^{1*} + \bar t_1 \\ \\ \bar z_1^{2*} + \bar t_2 \end{pmatrix},$$

which yields $-\bar\lambda_1 - \bar\lambda_2 = \bar\lambda_1 \bar z_1^{1*} + \bar\lambda_2 \bar z_1^{2*}$. The sum in the left-hand side is negative, while in the right-hand side there is, as proven above, a nonnegative term. Therefore we reached a contradiction and, consequently, $d \notin h^{C_\alpha}(\mathcal{B}^{C_\alpha})$, thus the inclusion proven in Theorem 5.3.2 is strict in this example.

Using eventually Proposition 2.3.2(*e*), (*j*) one can easily prove the next inclusion. It is followed by an example which shows that it can be strict in general.

Theorem 5.3.3. *There is* $h^{C_{\overline{FL}}}(\mathcal{B}^{C_{\overline{FL}}}) \subseteq h^{C_{FL}}(\mathcal{B}^{C_{FL}})$.

Example 5.3.3. Take $X = S = Z = \mathbb{R}$, $C = \mathbb{R}_+$, $g : \mathbb{R} \to \mathbb{R}$, defined by $g(x) = x$ for $x \in \mathbb{R}$ and $f_1, f_2 : \mathbb{R} \to \overline{\mathbb{R}}$, defined by

$$f_1(x) = \begin{cases} x \ln x - x, & \text{if } x > 0, \\ 0, & \text{if } x = 0, \\ +\infty, & \text{otherwise}, \end{cases} \quad \text{and } f_2(x) = \begin{cases} \frac{x^2}{2}, & \text{if } x \leq 0, \\ \\ +\infty, & \text{otherwise}. \end{cases}$$

The conjugates of f_1 and f_2 are (see Example 2.3.1) $f_1^*, f_2^* : \mathbb{R} \to \mathbb{R}$, defined by $f_1^*(x^*) = e^{x^*}$, for $x^* \in \mathbb{R}$ and, respectively,

$$f_2^*(x^*) = \begin{cases} \frac{x^{*2}}{2}, & \text{if } x^* \leq 0, \\ 0, & \text{if } x^* > 0, \end{cases}$$

while $(\lambda_1 f_1 + \lambda_2 f_2)^*(x^*) = 0$ for $\lambda_1, \lambda_2 > 0$ and $x^* \in \mathbb{R}$.

For $\lambda = (\lambda_1, \lambda_2)^T$, $\lambda_1 = \lambda_2 = 1$, $v = (v_1, v_2)^T$, $v_1 = v_2 = 0$, $z^* = 1$ and $y^* = -1$, we have $\lambda_1 v_1 + \lambda_2 v_2 = 0 = -(\lambda_1 f_1 + \lambda_2 f_2)^*(-1) - (z^* g)^*(1)$, thus $v \in h^{C_{FL}}(\mathcal{B}^{C_{FL}})$.

Assuming that $v \in h^{C_{\widetilde{FL}}}(\mathcal{B}^{C_{\widetilde{FL}}})$, there exist some $\bar{\lambda} = (\bar{\lambda}_1, \bar{\lambda}_2)^T > 0$, $\bar{y}^* = (\bar{y}^{1*}, \bar{y}^{2*})^T \in \mathbb{R}^2$, $\bar{z}^* \in \mathbb{R}_+$ and $\bar{v} = (\bar{v}_1, \bar{v}_2)^T \in \mathbb{R}^2$ such that

$$\bar{\lambda}_1\bar{v}_1 + \bar{\lambda}_2\bar{v}_2 = 0 \leq -\bar{\lambda}_1 f_1^*(\bar{y}^{1*}) - \bar{\lambda}_2 f_2^*(\bar{y}^{2*}) - (\bar{z}^* g)^*\big(-\bar{\lambda}_1\bar{y}^{1*} - \bar{\lambda}_2\bar{y}^{2*}\big). \quad (5.3)$$

As $(\bar{z}^* g)^* = \delta_{\{\bar{z}^*\}}$, we get $\bar{\lambda}_1\bar{y}^{1*} + \bar{\lambda}_2\bar{y}^{2*} = -\bar{z}^*$ and $(\bar{z}^* g)^*\big(-\bar{\lambda}_1\bar{y}^{1*} - \bar{\lambda}_2\bar{y}^{2*}\big) = 0$. On the other hand, $\bar{\lambda}_1 f_1^*(\bar{y}^{1*}) = \bar{\lambda}_1 e^{\bar{y}^{1*}} > 0$ and $\bar{\lambda}_2 f_2^*(\bar{y}^{2*}) \geq 0$ for $\bar{y}^{2*} \in \mathbb{R}$, thus the term on the right-hand side of (5.3) is negative and we reached a contradiction. Consequently, $d \notin h^{C_{\widetilde{FL}}}(\mathcal{B}^{C_{\widetilde{FL}}})$.

Remark 5.3.1. The inclusions proven in Theorem 5.3.1, Theorem 5.3.2 and Theorem 5.3.3 hold in the most general case. Moreover, as can be seen in Example 5.3.1, Example 5.3.2 and Example 5.3.3 these inclusions can be strict. Consequently, for all $\alpha \in \mathcal{F}$ there is

$$h^{C_M}(\mathcal{B}^{C_M}) \subsetneqq h^{C_\alpha}(\mathcal{B}^{C_\alpha}) \subsetneqq h^{C_{\widetilde{FL}}}(\mathcal{B}^{C_{\widetilde{FL}}}) \subsetneqq h^{C_{FL}}(\mathcal{B}^{C_{FL}}).$$

Further we show for the duals dealt with in this subsection that even if the images of their feasible sets through their objective vector functions satisfy sometimes strict inclusions, the maximal elements of three of these sets coincide and, under a weak regularity condition, also the fourth set is equal to them.

Theorem 5.3.4. *Let $\alpha \in \mathcal{F}$ be fixed. Then there is*

$$\operatorname{Max}\big(h^{C_M}(\mathcal{B}^{C_M}), \mathbb{R}^k_+\big) = \operatorname{Max}\big(h^{C_\alpha}(\mathcal{B}^{C_\alpha}), \mathbb{R}^k_+\big) = \operatorname{Max}\big(h^{C_{\widetilde{FL}}}(\mathcal{B}^{C_{\widetilde{FL}}}), \mathbb{R}^k_+\big).$$

Proof. We show first that $\operatorname{Max}\big(h^{C_M}(\mathcal{B}^{C_M}), \mathbb{R}^k_+\big) \subseteq \operatorname{Max}\big(h^{C_{\widetilde{FL}}}(\mathcal{B}^{C_{\widetilde{FL}}}), \mathbb{R}^k_+\big)$. Let be $d \in \operatorname{Max}\big(h^{C_M}(\mathcal{B}^{C_M}), \mathbb{R}^k_+\big)$. Then $d \in h^{C_M}(\mathcal{B}^{C_M})$, thus, by Theorem 5.3.1 and Theorem 5.3.2, $d \in h^{C_{\widetilde{FL}}}(\mathcal{B}^{C_{\widetilde{FL}}})$, too. This means that there is an element $(\lambda, y^*, z^*, v) \in \mathcal{B}^{C_{\widetilde{FL}}}$ such that $v = h^{C_{\widetilde{FL}}}(\lambda, y^*, z^*, v) = d$. Suppose that this is not a maximal element in $h^{C_{\widetilde{FL}}}(\mathcal{B}^{C_{\widetilde{FL}}})$. Then there exists $(\bar{\lambda}, \bar{y}^*, \bar{z}^*, \bar{v}) \in \mathcal{B}^{C_{\widetilde{FL}}}$ such that $d \leq \bar{v}$. Denote $\bar{\lambda} = (\bar{\lambda}_1, \ldots, \bar{\lambda}_k)^T \in \operatorname{int}(\mathbb{R}^k_+)$ and $\bar{y}^* = (\bar{y}^{1*}, \ldots, \bar{y}^{k*}) \in \prod_{i=1}^k \operatorname{dom} f_i^*$. Then

$$\sum_{i=1}^k \bar{\lambda}_i d_i < \sum_{i=1}^k \bar{\lambda}_i \bar{v}_i \leq -\sum_{i=1}^k \bar{\lambda}_i f_i^*(\bar{y}^{i*}) - (\bar{z}^* g)_S^*\bigg(-\sum_{i=1}^k \bar{\lambda}_i \bar{y}^{i*}\bigg). \quad (5.4)$$

As this cannot happen if $(\bar{z}^* g)_S^*(-\sum_{i=1}^k \bar{\lambda}_i \bar{y}^{i*}) = +\infty$, it follows $-\sum_{i=1}^k \bar{\lambda}_i \bar{y}^{i*} \in \operatorname{dom}(\bar{z}^* g)_S^*$. Without losing the generality we can assume the second inequality in (5.4) fulfilled as equality.

Considering $\tilde{z}^* := \big(1/(\sum_{i=1}^k \bar{\lambda}_i)\big)\bar{z}^* \in C^*$ and, for $i = 1, \ldots, k$,

$$\bar{t}_i := f_i^*(\bar{y}^{i*}) + (\tilde{z}^* g)_S^*\left(-\frac{1}{\sum\limits_{j=1}^k \bar{\lambda}_j}\sum_{j=1}^k \bar{\lambda}_j \bar{y}^{j*}\right) + \bar{v}_i$$

$$= f_i^*(\bar{y}^{i*}) + \frac{1}{\sum\limits_{j=1}^{k} \bar{\lambda}_j}(\bar{z}^* g)_S^*\Big(-\sum_{j=1}^{k} \bar{\lambda}_j \bar{y}^{j*}\Big) + \bar{v}_i \in \mathbb{R},$$

we obtain an element $(\bar{\lambda}, \bar{y}^*, \tilde{z}^*, \bar{t})$ satisfying $\tilde{z}^* \in C^*$, $\bar{\lambda} \in \operatorname{int}(\mathbb{R}^k_+)$ and

$$\sum_{i=1}^{k} \bar{\lambda}_i \bar{t}_i = \sum_{i=1}^{k} \bar{\lambda}_i \bar{v}_i + \sum_{i=1}^{k} \bar{\lambda}_i f_i^*(\bar{y}^{i*}) + (\bar{z}^* g)_S^*\Big(-\sum_{j=1}^{k} \bar{\lambda}_j \bar{y}^{j*}\Big) = 0.$$

Therefore $(\bar{\lambda}, \bar{y}^*, \tilde{z}^*, \bar{t}) \in h^{C_M}(\mathcal{B}^{C_M})$ and $h^{C_M}(\bar{\lambda}, \bar{y}^*, \tilde{z}^*, \bar{t}) = \bar{v}$. As this contradicts the maximality of d in $h^{C_M}(\mathcal{B}^{C_M})$, our supposition is false, consequently the maximal elements of $h^{C_M}(\mathcal{B}^{C_M})$ are maximal in $h^{C_{\widetilde{FL}}}(\mathcal{B}^{C_{\widetilde{FL}}})$, too.

To prove that this holds conversely, too, let v be maximal in $h^{C_{\widetilde{FL}}}(\mathcal{B}^{C_{\widetilde{FL}}})$. Then there are some $\lambda \in \operatorname{int}(\mathbb{R}^k_+)$, $y^* = (y^{1*}, \dots, y^{k*}) \in (X^*)^k$ and $z^* \in C^*$ such that $(\lambda, y^*, z^*, v) \in \mathcal{B}^{C_{\widetilde{FL}}}$, i.e. $\sum_{i=1}^k \lambda_i v_i = -\sum_{i=1}^k \lambda_i f_i^*(y^{i*}) - (z^* g)_S^*(-\sum_{i=1}^k \lambda_i y^{i*})$. Thus $-\sum_{i=1}^k \lambda_i y^{i*} \in \operatorname{dom}(z^* g)_S^*$. Taking $\tilde{z}^* = (1/(\sum_{i=1}^k \lambda_i)) z^* \in C^*$ and $t = (t_1, \dots, t_k)^T$ where for each $i = 1, \dots, k$,

$$t_i := f_i^*(y^{i*}) + (\tilde{z}^* g)_S^* \left(-\frac{1}{\sum\limits_{j=1}^{k} \lambda_j} \sum_{j=1}^{k} \lambda_j y^{j*} \right) + v_i \in \mathbb{R},$$

we get $\sum_{i=1}^k \lambda_i t_i = 0$, which means that $(\lambda, y^*, \tilde{z}^*, t) \in \mathcal{B}^{C_M}$. Therefore $v = h^{C_M}(\lambda, y^*, \tilde{z}^*, t) \in h^{C_M}(\mathcal{B}^{C_M})$. Assuming that v is not maximal in $h^{C_M}(\mathcal{B}^{C_M})$, we obtain the existence of some $\bar{d} \in h^{C_M}(\mathcal{B}^{C_M})$ such that $v \leq \bar{d}$. Theorem 5.3.1 yields then $\bar{d} \in h^{C_{\widetilde{FL}}}(\mathcal{B}^{C_{\widetilde{FL}}})$, but, since $v \leq \bar{d}$, the maximality of v in $h^{C_{\widetilde{FL}}}(\mathcal{B}^{C_{\widetilde{FL}}})$ is contradicted. Therefore, the maximal elements of $h^{C_{\widetilde{FL}}}(\mathcal{B}^{C_{\widetilde{FL}}})$ are maximal in $h^{C_M}(\mathcal{B}^{C_M})$, too. Consequently,

$$\operatorname{Max}\big(h^{C_M}(\mathcal{B}^{C_M}), \mathbb{R}^k_+\big) = \operatorname{Max}\big(h^{C_{\widetilde{FL}}}(\mathcal{B}^{C_{\widetilde{FL}}}), \mathbb{R}^k_+\big).$$

Take now $d \in \operatorname{Max}\big(h^{C_\alpha}(\mathcal{B}^{C_\alpha}), \mathbb{R}^k_+\big)$. Then $d \in h^{C_\alpha}(\mathcal{B}^{C_\alpha})$ and by Theorem 5.3.2 it follows $d \in h^{C_{\widetilde{FL}}}(\mathcal{B}^{C_{\widetilde{FL}}})$. Suppose that d is not maximal in $h^{C_{\widetilde{FL}}}(\mathcal{B}^{C_{\widetilde{FL}}})$. Then there exists an element $(\bar{\lambda}, \bar{y}^*, \bar{z}^*, \bar{v}) \in \mathcal{B}^{C_{\widetilde{FL}}}$ such that $d \leq \bar{v}$ and

$$\sum_{i=1}^{k} \bar{\lambda}_i \bar{v}_i = -\sum_{i=1}^{k} \bar{\lambda}_i f^*(\bar{y}^{i*}) - (\bar{z}^* g)_S^*\Big(-\sum_{j=1}^{k} \bar{\lambda}_j \bar{y}^{j*}\Big),$$

where $\bar{y}^* = (\bar{y}^{1*}, \dots, \bar{y}^{k*})$, $\bar{v} = (\bar{v}_1, \dots, \bar{v}_k)^T$ and $\bar{\lambda} = (\bar{\lambda}_1, \dots, \bar{\lambda}_k)^T$. Consequently, $-\sum_{j=1}^k \bar{\lambda}_j \bar{y}^{j*} \in \operatorname{dom}(\bar{z}^* g)_S^*$.

Taking $\tilde{z}^* := (\tilde{z}^{1*}, \dots, \tilde{z}^{k*})$ where for each $i = 1, \dots, k$, $\tilde{z}^{i*} := \alpha_i(\bar{\lambda})\bar{z}^*$,

$$\bar{t}_i := f_i^*(\bar{y}^{i*}) + (\tilde{z}^{i*}g)_S^*\Big(- \alpha_i(\bar{\lambda}) \sum_{j=1}^k \bar{\lambda}_j \bar{y}^{j*}\Big) + \bar{v}_i$$

$$= f_i^*(\bar{y}^{i*}) + \alpha_i(\bar{\lambda})(\bar{z}^*g)_S^*\Big(- \sum_{j=1}^k \bar{\lambda}_j \bar{y}^{j*}\Big) + \bar{v}_i,$$

for $i = 1, ..., k$, and $\bar{t} := (\bar{t}_1, ..., \bar{t}_k)^T \in \mathbb{R}^k$, we obtain an element $(\bar{\lambda}, \bar{y}^*, \tilde{z}^*, \bar{t}) \in \mathcal{B}^{C_\alpha}$ for which $h^{C_\alpha}(\bar{\lambda}, \bar{y}^*, \tilde{z}^*, \bar{t}) = \bar{v} \geq d$. This contradicts the maximality of d in $h^{C_\alpha}(\mathcal{B}^{C_\alpha})$, therefore our supposition fails and consequently d is maximal in $h^{C_{\widetilde{FL}}}(\mathcal{B}^{C_{\widetilde{FL}}})$, too.

Take now v to be maximal in $h^{C_{\widetilde{FL}}}(\mathcal{B}^{C_{\widetilde{FL}}})$. Then it is maximal in $h^{C_M}(\mathcal{B}^{C_M})$, too. By Theorem 5.3.1 we obtain then $v \in h^{C_\alpha}(\mathcal{B}^{C_\alpha})$. Assuming it to be not maximal in the latter set, there should be a $d \in h^{C_\alpha}(\mathcal{B}^{C_\alpha})$ such that $v \leq d$. By Theorem 5.3.2 it follows $d \in h^{C_{\widetilde{FL}}}(\mathcal{B}^{C_{\widetilde{FL}}})$, which contradicts the maximality of v in this set. Consequently, the maximal elements of $h^{C_{\widetilde{FL}}}(\mathcal{B}^{C_{\widetilde{FL}}})$ are maximal in $h^{C_\alpha}(\mathcal{B}^{C_\alpha})$, too, and we are done. □

Theorem 5.3.5. *If one of the regularity conditions* (RC_i^Σ), $i \in \{1, 2, 3\}$, *is fulfilled, then* $\operatorname{Max}\big(h^{C_{\widetilde{FL}}}(\mathcal{B}^{C_{\widetilde{FL}}}), \mathbb{R}^k_+\big) = \operatorname{Max}\big(h^{C_{FL}}(\mathcal{B}^{C_{FL}}), \mathbb{R}^k_+\big)$.

Proof. Using Theorem 3.5.8(a) one can immediately show that under the fulfillment of any of the regularity conditions (RC_i^Σ), $i \in \{1, 2, 3\}$, the vector duals in discussion coincide. □

Combining Theorem 5.3.4 and Theorem 5.3.5, we see that under a weak regularity condition the maximal elements of the vector dual problems considered in this section coincide.

Theorem 5.3.6. *Let* $\alpha \in \mathcal{F}$ *be fixed. If one of the regularity conditions* (RC_i^Σ), $i \in \{1, 2, 3\}$, *is fulfilled, then*

$$\operatorname{Max}\big(h^{C_M}(\mathcal{B}^{C_M}), \mathbb{R}^k_+\big) = \operatorname{Max}\big(h^{C_\alpha}(\mathcal{B}^{C_\alpha}), \mathbb{R}^k_+\big) =$$

$$\operatorname{Max}\big(h^{C_{\widetilde{FL}}}(\mathcal{B}^{C_{\widetilde{FL}}}), \mathbb{R}^k_+\big) = \operatorname{Max}\big(h^{C_{FL}}(\mathcal{B}^{C_{FL}}), \mathbb{R}^k_+\big).$$

Remark 5.3.2. Recall that in chapter 4 besides $(DVF^{C_{FL}})$ several vector duals to (PV^C) were introduced, and in (4.18) and Proposition 4.3.16 other inclusions similar to the ones in Remark 5.3.1 were given. Combining Theorem 5.3.6 and Theorem 4.3.15 one gets that under $(RCVF^C)$ (see also Remark 4.3.9(b)) the maximal elements of all the sets mentioned in (4.18) and Remark 5.3.1 coincide.

Remark 5.3.3. The converse duality statement in Theorem 5.2.7 can be proven alternatively by making use of Theorem 5.3.6 and Theorem 4.3.7.

Similar considerations can be made when dealing with weakly efficient solutions, too. For the remainder of this section take the functions f_i, $i = 1,\dots,k$, with full domains. The following statements are given without proofs, since these are similar to the ones of the corresponding statements concerning efficient solutions. Recall the Fenchel-Lagrange type vector dual with respect to weakly efficient solutions introduced in chapter 4

$$(DVF_w^{C_{FL}}) \qquad \underset{(\lambda,y^*,z^*,v)\in\mathcal{B}_w^{C_{FL}}}{\operatorname{WMax}} h_w^{C_{FL}}(\lambda,y^*,z^*,v),$$

where

$$\mathcal{B}_w^{C_{FL}} = \left\{(\lambda,y^*,z^*,v) \in (\mathbb{R}_+^k\backslash\{0\}) \times X^* \times C^* \times \mathbb{R}^k : \lambda = (\lambda_1,\dots,\lambda_k)^T,\right.$$
$$v = (v_1,\dots,v_k)^T,$$
$$\left.\sum_{i=1}^k \lambda_i v_i \leq -\left(\sum_{i=1}^k \lambda_i f_i\right)^*(y^*) - (z^*g)_S^*(-y^*)\right\}$$

and

$$h_w^{C_{FL}}(\lambda,y^*,z^*,v) = v.$$

Analogously to the vector dual introduced in the beginning of the section, consider the following dual problem to (PVF_w^C) with respect to weakly efficient solutions

$$(DVF_w^{C_{\widetilde{FL}}}) \qquad \underset{(\lambda,y^*,z^*,v)\in\mathcal{B}_w^{C_{\widetilde{FL}}}}{\operatorname{WMax}} h_w^{C_{\widetilde{FL}}}(\lambda,y^*,z^*,v),$$

where

$$\mathcal{B}_w^{C_{\widetilde{FL}}} = \left\{(\lambda,y^*,z^*,v) \in (\mathbb{R}_+^k\backslash\{0\}) \times (X^*)^k \times C^* \times \mathbb{R}^k : \lambda = (\lambda_1,\dots,\lambda_k)^T,\right.$$
$$y^* = (y^{1*},\dots,y^{k*}), v = (v_1,\dots,v_k)^T,$$
$$\left.\sum_{i=1}^k \lambda_i v_i \leq -\sum_{i=1}^k \lambda_i f_i^*(y^{i*}) - (z^*g)_S^*\left(-\sum_{i=1}^k \lambda_i y^{i*}\right)\right\}$$

and

$$h_w^{C_{\widetilde{FL}}}(\lambda,y^*,z^*,v) = v.$$

First we compare the images of the feasible sets through their objective functions of the vector duals with respect to weakly efficient solutions dealt with so far.

Theorem 5.3.7. *Let $\beta \in \mathcal{F}_w$ be fixed. Then there is*

$$h_w^{C_M}(\mathcal{B}_w^{C_M}) \subseteq h_w^{C_\beta}(\mathcal{B}_w^{C_\beta}) \subseteq h_w^{C_{\widetilde{FL}}}(\mathcal{B}_w^{C_{\widetilde{FL}}}) \subseteq h_w^{C_{FL}}(\mathcal{B}_w^{C_{FL}}).$$

Next we obtain that the weakly maximal elements of the first three sets mentioned in Theorem 5.3.7 coincide.

Theorem 5.3.8. *Let $\beta \in \mathcal{F}_w$ be fixed. Then there is*

$$\operatorname{WMax}(h_w^{C_M}(\mathcal{B}_w^{C_M}), \mathbb{R}^k_+) = \operatorname{WMax}(h_w^{C_\beta}(\mathcal{B}_w^{C_\beta}), \mathbb{R}^k_+) = \operatorname{WMax}\left(h_w^{C_{\widetilde{FL}}}(\mathcal{B}_w^{C_{\widetilde{FL}}}), \mathbb{R}^k_+\right).$$

Under a weak regularity condition, the weakly maximal elements of these sets coincide with the weakly efficient solutions to $(DVF_w^{C_{FL}})$, too.

Theorem 5.3.9. *Let $\beta \in \mathcal{F}_w$ be fixed. If $k-1$ of the functions f_i, $i = 1, \ldots, k$, are continuous, one has*

$$\operatorname{WMax}\left(h_w^{C_M}(\mathcal{B}_w^{C_M}), \mathbb{R}^k_+\right) = \operatorname{WMax}\left(h_w^{C_\beta}(\mathcal{B}_w^{C_\beta}), \mathbb{R}^k_+\right) =$$

$$\operatorname{WMax}\left(h_w^{C_{\widetilde{FL}}}(\mathcal{B}_w^{C_{\widetilde{FL}}}), \mathbb{R}^k_+\right) = \operatorname{WMax}\left(h_w^{C_{FL}}(\mathcal{B}_w^{C_{FL}}), \mathbb{R}^k_+\right).$$

Remark 5.3.4. From Remark 4.3.14 and Theorem 5.3.9 we deduce that under the fulfillment of $(RCVF_w^C)$ (see also Remark 4.3.9(*b*)) the weakly maximal elements of all the sets mentioned in Theorem 5.3.7 and Remark 4.3.14 coincide.

5.4 Linear vector duality for problems with finite dimensional image spaces

In this section we deal with vector duality for linear vector optimization problems with objective functions mapping into finite dimensional spaces, continuing in this framework the work from section 4.5. There, some vector duals to a primal linear vector optimization problem $(PV^{\mathcal{L}})$ were obtained by particularizing the vector dual problems introduced in sections 4.2 and 4.3. In the following we see what happens to the duals introduced in section 5.2 in this particular instance and we compare all these mentioned duals.

5.4.1 Duality with respect to properly efficient solutions

In subsection 5.2.3 we have already considered a primal vector optimization problem with both geometric and linear cone inequality constraints. Maintaining the framework introduced in section 5.2, we go now further by taking the objective function to be linear and continuous, namely for $L = (L_1, \ldots, L_k)^T : X \to \mathbb{R}^k$ we consider the primal vector optimization problem

$$(PVF^{\mathcal{L}}) \quad \operatorname*{Min}_{x \in \mathcal{A}^{\mathcal{L}}} Lx.$$
$$\mathcal{A}^{\mathcal{L}} = \{x \in S : Ax - b \in C\}$$

First we assign to it the vector duals which are particular instances of $(DVF^{\mathcal{L}})$ and $(DVF^{\mathcal{L}_M})$, respectively. Noting that $(L_i)^* = \delta_{\{L_i\}}$ for $i = 1, \ldots, k$, we see that the variable $y^* = (y^{1*}, \ldots, y^{k*}) \in \prod_{i=1}^k \operatorname{dom} f_i^*$ can be eliminated and these vector duals are

$$(DVF^{\mathcal{L}}) \quad \operatorname*{Max}_{(\lambda,z^*,v)\in\mathcal{B}^{\mathcal{L}}} h^{\mathcal{L}}(\lambda,z^*,v),$$

where

$$\mathcal{B}^{\mathcal{L}} = \Big\{(\lambda,z^*,v) \in \operatorname{int}(\mathbb{R}^k_+) \times (Z^*)^k \times \mathbb{R}^k : \lambda = (\lambda_1,\ldots,\lambda_k)^T,$$
$$z^* = (z^{1*},\ldots,z^{k*}), v = (v_1,\ldots,v_k)^T, \sum_{i=1}^k \lambda_i z^{i*} \in C^*,$$
$$\sum_{i=1}^k \lambda_i v_i = 0, \sum_{i=1}^k \lambda_i (L_i - A^* z^{i*}) \in S^*\Big\}$$

and

$$h^{\mathcal{L}}(\lambda,z^*,v) = \begin{pmatrix} \langle z^{1*}, b\rangle + v_1 \\ \vdots \\ \langle z^{k*}, b\rangle + v_k \end{pmatrix}$$

and, respectively,

$$(DVF^{\mathcal{L}_M}) \quad \operatorname*{Max}_{(\lambda,z^*,v)\in\mathcal{B}^{\mathcal{L}_M}} h^{\mathcal{L}_M}(\lambda,z^*,v),$$

where

$$\mathcal{B}^{\mathcal{L}_M} = \Big\{(\lambda,z^*,v) \in \operatorname{int}(\mathbb{R}^k_+) \times C^* \times \mathbb{R}^k : \lambda = (\lambda_1,\ldots,\lambda_k)^T, v = (v_1,\ldots,v_k)^T,$$
$$\sum_{i=1}^k \lambda_i v_i = 0, \sum_{i=1}^k \lambda_i (L_i - A^* z^*) \in S^*\Big\}$$

and

$$h^{\mathcal{L}_M}(\lambda,z^*,v) = \begin{pmatrix} \langle z^*, b\rangle + v_1 \\ \vdots \\ \langle z^*, b\rangle + v_k \end{pmatrix}.$$

The latter can be rewritten also as

$$(DVF^{\mathcal{L}_M}) \quad \operatorname*{Max}_{(\lambda,z^*,v)\in\mathcal{B}^{\mathcal{L}_M}} h^{\mathcal{L}_M}(\lambda,z^*,v),$$

where

$$\mathcal{B}^{\mathcal{L}_M} = \Big\{(\lambda,z^*,v) \in \operatorname{int}(\mathbb{R}^k_+) \times C^* \times \mathbb{R}^k : \lambda = (\lambda_1,\ldots,\lambda_k)^T,$$
$$v = (v_1,\ldots,v_k)^T, \sum_{i=1}^k \lambda_i v_i = 0, \sum_{i=1}^k \lambda_i L_i - A^* z^* \in S^*\Big\}$$

and

$$h^{\mathcal{L}_M}(\lambda, z^*, v) = \begin{pmatrix} \frac{1}{\sum\limits_{i=1}^{k} \lambda_i} \langle z^*, b \rangle + v_1 \\ \vdots \\ \frac{1}{\sum\limits_{i=1}^{k} \lambda_i} \langle z^*, b \rangle + v_k \end{pmatrix}.$$

Like in subsection 5.2.3, to simplify the formulation of $(DVF^{\mathcal{L}})$ we consider two cases.

When $b = 0$ we have already noticed that $(DVF^{\mathcal{L}})$ and $(DVF^{\mathcal{L}_M})$ coincide, having the following formulation

$$(DVF^{\mathcal{L}}) \quad \operatorname*{Max}_{(\lambda,v)\in\mathcal{B}^{\mathcal{L}}} h^{\mathcal{L}}(\lambda, v),$$

where

$$\mathcal{B}^{\mathcal{L}} = \Big\{(\lambda, v) \in \operatorname{int}(\mathbb{R}^k_+) \times \mathbb{R}^k : \lambda = (\lambda_1, \dots, \lambda_k)^T, v = (v_1, \dots, v_k)^T,$$
$$\sum_{i=1}^{k} \lambda_i v_i = 0, \sum_{i=1}^{k} \lambda_i L_i \in A^*(C^*) + S^* \Big\}$$

and

$$h^{\mathcal{L}}(\lambda, v) = v.$$

In the other case, namely when $b \neq 0$, $(DVF^{\mathcal{L}})$ becomes

$$(DVF^{\mathcal{L}}) \quad \operatorname*{Max}_{(\lambda,q^*)\in\mathcal{B}^{\mathcal{L}}} h^{\mathcal{L}}(\lambda, q^*),$$

where

$$\mathcal{B}^{\mathcal{L}} = \Big\{(\lambda, q^*) \in \operatorname{int}(\mathbb{R}^k_+) \times (Z^*)^k : \lambda = (\lambda_1, \dots, \lambda_k)^T, q^* = (q^{1*}, \dots, q^{k*}),$$
$$\sum_{i=1}^{k} \lambda_i q^{i*} \in C^*, \sum_{i=1}^{k} \lambda_i (L_i - A^* q^{i*}) \in S^* \Big\}$$

and

$$h^{\mathcal{L}}(\lambda, q^*) = \begin{pmatrix} \langle q^{1*}, b \rangle \\ \vdots \\ \langle q^{k*}, b \rangle \end{pmatrix}.$$

Comparing $(DVF^{\mathcal{L}})$ in case $b \neq 0$ with the vector dual problem introduced in [101], denoted in section 4.5 by $(DVF^{\mathcal{L}_J})$, one can see that these two vector dual problems coincide.

The weak, strong and converse duality theorems for the primal-dual vector pairs $(PVF^{\mathcal{L}}) - (DVF^{\mathcal{L}_M})$ and $(PVF^{\mathcal{L}}) - (DVF^{\mathcal{L}})$, respectively, follow as particular instances of the corresponding results given in subsection 5.2.1 (see also Remark 5.2.10).

Now let us compare the images of the feasible sets through their objective functions of these duals and the maximal elements of these sets with the ones of the other duals assigned to $(PVF^{\mathcal{L}})$ in subsection 4.5.1. Some inclusion relations concerning these image sets were already proven in more general contexts. In the following we show other inclusions.

Proposition 5.4.1. *One has* $h^{\mathcal{L}_J}(\mathcal{B}^{\mathcal{L}_J}) \subseteq h^{\mathcal{L}_M}(\mathcal{B}^{\mathcal{L}_M})$.

Proof. Let $d \in h^{\mathcal{L}_J}(\mathcal{B}^{\mathcal{L}_J})$, i.e. there are some $\lambda = (\lambda_1, \ldots, \lambda_k)^T \in \operatorname{int}(\mathbb{R}^k_+)$ and $U = (U_1, \ldots, U_k)^T \in \mathcal{L}(Z, \mathbb{R}^k)$ such that $(\lambda, U) \in \mathcal{B}^{\mathcal{L}_J}$ and $d = (d_1, \ldots, d_k)^T = Ub$. Denote $z^* := \sum_{i=1}^k \lambda_i U_i$ and

$$v := Ub - \frac{1}{\sum\limits_{i=1}^k \lambda_i} \begin{pmatrix} \langle z^*, b \rangle \\ \vdots \\ \langle z^*, b \rangle \end{pmatrix}.$$

Thus $\sum_{i=1}^k \lambda_i L_i - A^* z^* = \sum_{i=1}^k \lambda_i (L_i - A^* U_i) \in S^*$ and $d_i = U_i b = (1/\sum_{j=1}^k \lambda_j)\langle z^*, b \rangle + v_i$ for $i = 1, \ldots, k$. Further have

$$\sum_{i=1}^k \lambda_i v_i = \sum_{i=1}^k \lambda_i \left(U_i b - \frac{\langle z^*, b \rangle}{\sum\limits_{j=1}^k \lambda_j} \right) = \langle z^*, b \rangle - \langle z^*, b \rangle = 0,$$

which yields $d \in h^{\mathcal{L}_M}(\mathcal{B}^{\mathcal{L}_M})$. □

Remark 5.4.1. Regarding the vector duals to $(PVF^{\mathcal{L}})$ just mentioned, using Theorem 5.3.1 and Proposition 5.4.1 one can see that in general it holds

$$h^{\mathcal{L}_J}(\mathcal{B}^{\mathcal{L}_J}) \subseteq h^{\mathcal{L}_M}(\mathcal{B}^{\mathcal{L}_M}) \subseteq h^{\mathcal{L}}(\mathcal{B}^{\mathcal{L}}).$$

In case $b \neq 0$, via the observation following the last formulation of $(DVF^{\mathcal{L}})$, we obtain that there is

$$h^{\mathcal{L}_J}(\mathcal{B}^{\mathcal{L}_J}) = h^{\mathcal{L}_M}(\mathcal{B}^{\mathcal{L}_M}) = h^{\mathcal{L}}(\mathcal{B}^{\mathcal{L}}),$$

while in case $b = 0$ one has

$$h^{\mathcal{L}_J}(\mathcal{B}^{\mathcal{L}_J}) \subseteq h^{\mathcal{L}_M}(\mathcal{B}^{\mathcal{L}_M}) = h^{\mathcal{L}}(\mathcal{B}^{\mathcal{L}}).$$

Remark 5.4.2. Given the results and considerations from above, we have the following scheme involving the images of the feasible sets through their objective functions of the linear vector duals with respect to properly efficient solutions assigned to $(PVF^{\mathcal{L}})$ in both chapter 4 and chapter 5

$$h^{\mathcal{L}_J}(\mathcal{B}^{\mathcal{L}_J}) \subseteq h^{\mathcal{L}_M}(\mathcal{B}^{\mathcal{L}_M}) = h^{\mathcal{L}}(\mathcal{B}^{\mathcal{L}}) \subseteq h^{\mathcal{L}_L}(\mathcal{B}^{\mathcal{L}_L}) \subseteq \begin{matrix} h^{\mathcal{L}_N}(\mathcal{B}^{\mathcal{L}_N}) \\ h^{\mathcal{L}_P}(\mathcal{B}^{\mathcal{L}_P}). \end{matrix}$$

In chapter 4 and chapter 5 there were introduced other vector duals, too, but $(DVF^{C_{FL}})$ and $(DVF^{C_{\widetilde{FL}}})$ coincide in this particular case with (DV^{C_L}) and (DV^{C_F}) turns out to be exactly (DV^{C_P}), thus they will be not mentioned further in this chapter, as the framework becomes more particular.

In more specialized settings some of the inclusions given in this scheme become equalities as follows.

(a) In case $b \neq 0$, the scheme from above becomes, via Remark 5.4.1,

$$h^{\mathcal{L}_J}(\mathcal{B}^{\mathcal{L}_J}) = h^{\mathcal{L}_M}(\mathcal{B}^{\mathcal{L}_M}) = h^{\mathcal{L}}(\mathcal{B}^{\mathcal{L}}) \subseteq h^{\mathcal{L}_L}(\mathcal{B}^{\mathcal{L}_L}) \subseteq \begin{matrix} h^{\mathcal{L}_N}(\mathcal{B}^{\mathcal{L}_N}) \\ h^{\mathcal{L}_P}(\mathcal{B}^{\mathcal{L}_P}). \end{matrix}$$

(b) Provided the fulfillment of any of the regularity conditions mentioned in Remark 4.5.1, the scheme turns into

$$h^{\mathcal{L}_J}(\mathcal{B}^{\mathcal{L}_J}) \subseteq h^{\mathcal{L}_M}(\mathcal{B}^{\mathcal{L}_M}) = h^{\mathcal{L}}(\mathcal{B}^{\mathcal{L}}) \subseteq h^{\mathcal{L}_L}(\mathcal{B}^{\mathcal{L}_L}) = h^{\mathcal{L}_P}(\mathcal{B}^{\mathcal{L}_P}) \subseteq h^{\mathcal{L}_N}(\mathcal{B}^{\mathcal{L}_N}).$$

Remark 5.4.3. Since the regularity conditions required in Theorem 5.3.6 are all automatically fulfilled in this case, it follows that the sets of maximal elements of $h^{\mathcal{L}_M}(\mathcal{B}^{\mathcal{L}_M})$ and $h^{\mathcal{L}}(\mathcal{B}^{\mathcal{L}})$ coincide with the ones of $h^{\mathcal{L}_L}(\mathcal{B}^{\mathcal{L}_L})$. Thus we have the following scheme concerning the maximal elements of the sets considered in Remark 5.4.2

$$\begin{aligned} \operatorname{Max}\left(h^{\mathcal{L}_J}(\mathcal{B}^{\mathcal{L}_J}), \mathbb{R}^k_+\right) &\subseteq \operatorname{Max}\left(h^{\mathcal{L}_M}(\mathcal{B}^{\mathcal{L}_M}), \mathbb{R}^k_+\right) = \operatorname{Max}\left(h^{\mathcal{L}}(\mathcal{B}^{\mathcal{L}}), \mathbb{R}^k_+\right) \\ &= \operatorname{Max}\left(h^{\mathcal{L}_L}(\mathcal{B}^{\mathcal{L}_L}), \mathbb{R}^k_+\right) \subseteq \operatorname{Max}\left(h^{\mathcal{L}_N}(\mathcal{B}^{\mathcal{L}_N}), \mathbb{R}^k_+\right). \end{aligned}$$

In more particular frameworks some of the inclusions given in this scheme become equalities, as follows.

(a) When $b \neq 0$ the scheme becomes (see also Theorem 4.5.2)

$$\begin{aligned} \operatorname{Max}\left(h^{\mathcal{L}_J}(\mathcal{B}^{\mathcal{L}_J}), \mathbb{R}^k_+\right) &= \operatorname{Max}\left(h^{\mathcal{L}_M}(\mathcal{B}^{\mathcal{L}_M}), \mathbb{R}^k_+\right) = \operatorname{Max}\left(h^{\mathcal{L}}(\mathcal{B}^{\mathcal{L}}), \mathbb{R}^k_+\right) \\ &= \operatorname{Max}\left(h^{\mathcal{L}_L}(\mathcal{B}^{\mathcal{L}_L}), \mathbb{R}^k_+\right) \subseteq \operatorname{Max}\left(h^{\mathcal{L}_N}(\mathcal{B}^{\mathcal{L}_N}), \mathbb{R}^k_+\right). \end{aligned}$$

(b) If every efficient solution to $(PVF^{\mathcal{L}})$ is also properly efficient the scheme becomes (cf. Theorem 4.3.17)

$$\begin{aligned} \operatorname{Max}\left(h^{\mathcal{L}_J}(\mathcal{B}^{\mathcal{L}_J}), \mathbb{R}^k_+\right) &\subseteq \operatorname{Max}\left(h^{\mathcal{L}_M}(\mathcal{B}^{\mathcal{L}_M}), \mathbb{R}^k_+\right) = \operatorname{Max}\left(h^{\mathcal{L}}(\mathcal{B}^{\mathcal{L}}), \mathbb{R}^k_+\right) \\ &= \operatorname{Max}\left(h^{\mathcal{L}_L}(\mathcal{B}^{\mathcal{L}_L}), \mathbb{R}^k_+\right) = \operatorname{Max}\left(h^{\mathcal{L}_N}(\mathcal{B}^{\mathcal{L}_N}), \mathbb{R}^k_+\right). \end{aligned}$$

(c) Provided the fulfillment of one of the regularity conditions mentioned in Remark 4.5.1 the scheme can be enriched as follows

$$\operatorname{Max}\left(h^{\mathcal{L}_J}(\mathcal{B}^{\mathcal{L}_J}), \mathbb{R}^k_+\right) \subseteq \operatorname{Max}\left(h^{\mathcal{L}_M}(\mathcal{B}^{\mathcal{L}_M}), \mathbb{R}^k_+\right) = \operatorname{Max}\left(h^{\mathcal{L}}(\mathcal{B}^{\mathcal{L}}), \mathbb{R}^k_+\right) =$$

$$\operatorname{Max}\left(h^{\mathcal{L}_L}(\mathcal{B}^{\mathcal{L}_L}), \mathbb{R}^k_+\right) = \operatorname{Max}\left(h^{\mathcal{L}_P}(\mathcal{B}^{\mathcal{L}_P}), \mathbb{R}^k_+\right) \subseteq \operatorname{Max}\left(h^{\mathcal{L}_N}(\mathcal{B}^{\mathcal{L}_N}), \mathbb{R}^k_+\right).$$

Note also that in general $\operatorname{Max}\left(h^{\mathcal{L}_L}(\mathcal{B}^{\mathcal{L}_L}), \mathbb{R}^k_+\right)$ is not a subset of $\operatorname{Max}\left(h^{\mathcal{L}_P}(\mathcal{B}^{\mathcal{L}_P}), \mathbb{R}^k_+\right)$ even if we have $h^{\mathcal{L}_L}(\mathcal{B}^{\mathcal{L}_L}) \subseteq h^{\mathcal{L}_P}(\mathcal{B}^{\mathcal{L}_P})$.

5.4.2 Duality with respect to weakly efficient solutions

Parallelly to subsection 5.4.1 where the vector optimization problems were considered with respect to efficient and properly efficient solutions, we work with weakly efficient ones, too. Consider the framework from the beginning of subsection 5.4.1 and the primal vector optimization problem with respect to weakly efficient solutions

$$(PVF_w^{\mathcal{L}}) \quad \underset{x\in\mathcal{A}^{\mathcal{L}}}{\operatorname{WMin}}\, Lx.$$
$$\mathcal{A}^{\mathcal{L}} = \{x \in S : Ax - b \in C\}$$

As L takes values in $\mathbb{R}^k$, we are in the framework in which the investigations involving weakly efficient solutions are carried out in this chapter. First we assign to $(PVF_w^{\mathcal{L}})$ the vector duals which are particular instances of $(DVF_w^{\mathcal{L}})$ and $(DVF_w^{\mathcal{L}_M})$, respectively. They are

$$(DVF_w^{\mathcal{L}}) \quad \underset{(\lambda,z^*,v)\in\mathcal{B}_w^{\mathcal{L}}}{\operatorname{WMax}} h_w^{\mathcal{L}}(\lambda, z^*, v),$$

where

$$\mathcal{B}_w^{\mathcal{L}} = \Big\{(\lambda, z^*, v) \in (\mathbb{R}_+^k\backslash\{0\}) \times (Z^*)^k \times \mathbb{R}^k : \lambda = (\lambda_1, \dots, \lambda_k)^T,$$
$$v = (v_1, \dots, v_k)^T, z^* = (z^{1*}, \dots, z^{k*}), \sum_{i=1}^k \lambda_i v_i = 0,$$
$$\sum_{i=1}^k \lambda_i z^{i*} \in C^*, \sum_{i=1}^k \lambda_i (L_i - A^* z^{i*}) \in S^*\Big\}$$

and

$$h_w^{\mathcal{L}}(\lambda, z^*, v) = \begin{pmatrix} \langle z^{1*}, b\rangle + v_1 \\ \vdots \\ \langle z^{k*}, b\rangle + v_k \end{pmatrix},$$

and, respectively (see the reformulation of $(DVF^{\mathcal{L}_M})$ from subsection 5.4.1),

$$(DVF_w^{\mathcal{L}_M}) \quad \underset{(\lambda,z^*,v)\in\mathcal{B}_w^{\mathcal{L}_M}}{\operatorname{WMax}} h_w^{\mathcal{L}_M}(\lambda, z^*, v),$$

where

$$\mathcal{B}_w^{\mathcal{L}_M} = \Big\{(\lambda, z^*, v) \in (\mathbb{R}_+^k\backslash\{0\}) \times C^* \times \mathbb{R}^k : \lambda = (\lambda_1, \dots, \lambda_k)^T,$$
$$v = (v_1, \dots, v_k)^T, \sum_{i=1}^k \lambda_i v_i = 0, \sum_{i=1}^k \lambda_i L_i - A^* z^* \in S^*\Big\}$$

and

$$h_w^{\mathcal{L}_M}(\lambda, z^*, v) = \begin{pmatrix} \frac{1}{\sum\limits_{i=1}^{k} \lambda_i} \langle z^*, b\rangle + v_1 \\ \vdots \\ \frac{1}{\sum\limits_{i=1}^{k} \lambda_i} \langle z^*, b\rangle + v_k \end{pmatrix}.$$

Note that we have $h^{\mathcal{L}}(\mathcal{B}^{\mathcal{L}}) \subseteq h_w^{\mathcal{L}}(\mathcal{B}_w^{\mathcal{L}})$ and $h^{\mathcal{L}_M}(\mathcal{B}^{\mathcal{L}_M}) \subseteq h_w^{\mathcal{L}_M}(\mathcal{B}_w^{\mathcal{L}_M})$.

Like in subsection 5.2.3, to simplify the formulation of $(DVF_w^{\mathcal{L}})$ we consider two cases. When $b = 0$ we get

$$(DVF_w^{\mathcal{L}}) \quad \underset{(\lambda,v)\in\mathcal{B}_w^{\mathcal{L}}}{\text{WMax}}\ h_w^{\mathcal{L}}(\lambda, v),$$

where

$$\mathcal{B}_w^{\mathcal{L}} = \Big\{(\lambda, v) \in (\mathbb{R}_+^k \backslash \{0\}) \times \mathbb{R}^k : \lambda = (\lambda_1, \dots, \lambda_k)^T, v = (v_1, \dots, v_k)^T,$$
$$\sum_{i=1}^{k} \lambda_i v_i = 0, \sum_{i=1}^{k} \lambda_i L_i \in A^*(C^*) + S^*\Big\}$$

and

$$h_w^{\mathcal{L}}(\lambda, v) = v,$$

and this coincides with $(DVF_w^{\mathcal{L}_M})$, while in case $b \neq 0$, $(DVF_w^{\mathcal{L}})$ becomes

$$(DVF_w^{\mathcal{L}}) \quad \underset{(\lambda,q^*)\in\mathcal{B}_w^{\mathcal{L}}}{\text{WMax}}\ h_w^{\mathcal{L}}(\lambda, q^*),$$

where

$$\mathcal{B}_w^{\mathcal{L}} = \Big\{(\lambda, q^*) \in (\mathbb{R}_+^k \backslash \{0\}) \times (Z^*)^k : \lambda = (\lambda_1, \dots, \lambda_k)^T, q^* = (q^{1*}, \dots, q^{k*}),$$
$$\sum_{i=1}^{k} \lambda_i q^{i*} \in C^*, \sum_{i=1}^{k} \lambda_i (L_i - A^* q^{i*}) \in S^*\Big\},$$

and

$$h_w^{\mathcal{L}}(\lambda, q^*) = \begin{pmatrix} \langle q^{1*}, b\rangle \\ \vdots \\ \langle q^{k*}, b\rangle \end{pmatrix}.$$

Analogously to the similar conjugate vector dual obtained in the considered framework with respect to properly efficient solutions, $(DVF_w^{\mathcal{L}})$ turns out to coincide, when $b \neq 0$, with the vector dual $(DV_w^{\mathcal{L}_J})$ from subsection 4.5.2.

The weak, strong and converse duality theorems for the primal-dual vector pairs $(PVF_w^{\mathcal{L}}) - (DVF_w^{\mathcal{L}_M})$ and $(PVF_w^{\mathcal{L}}) - (DVF_w^{\mathcal{L}})$, respectively, follow as particular instances of the corresponding results given in subsection 5.2.2 (see also Remark 5.2.12).

Now let us compare the images of the feasible sets through their objective functions of these duals and the weakly maximal elements of these sets with the ones of the other duals assigned to $(PVF_w^{\mathcal{L}})$ in sections 4.2 and 4.3.

Some inclusion relations concerning these image sets were already proven in more general contexts. In the following we show another inclusion, whose proof is omitted being analogous to the one of Proposition 5.4.1, followed by the general schemes involving the image sets of all the duals in discussion.

Proposition 5.4.2. *One has* $h_w^{\mathcal{L}_J}(\mathcal{B}_w^{\mathcal{L}_J}) \subseteq h_w^{\mathcal{L}_M}(\mathcal{B}_w^{\mathcal{L}_M})$.

Remark 5.4.4. Like in subsection 5.4.1, we give a scheme involving the images of the feasible sets through their objective functions of the vector duals with respect to weakly efficient solutions assigned so far to $(PVF_w^{\mathcal{L}})$

$$h_w^{\mathcal{L}_J}(\mathcal{B}_w^{\mathcal{L}_J}) \subseteq h_w^{\mathcal{L}_M}(\mathcal{B}_w^{\mathcal{L}_M}) = h_w^{\mathcal{L}}(\mathcal{B}_w^{\mathcal{L}}) \subseteq h_w^{\mathcal{L}_L}(\mathcal{B}_w^{\mathcal{L}_L}) = h_w^{\mathcal{L}_N}(\mathcal{B}_w^{\mathcal{L}_N}) \subseteq h_w^{\mathcal{L}_P}(\mathcal{B}_w^{\mathcal{L}_P}).$$

In chapter 4 and chapter 5 there were introduced other vector duals, too, like $(DVF_w^{C_{FL}})$ and $(DVF_w^{C_{\widetilde{FL}}})$, which coincide in this particular case with $(DV_w^{C_L})$, and $(DV_w^{C_F})$ which is nothing but $(DV_w^{C_P})$, thus they will be not mentioned further in this chapter, as the framework becomes more particular.

In more specialized settings some of the inclusions given in this scheme become equalities, as follows.

(a) In case $b \neq 0$ this scheme turns into

$$h_w^{\mathcal{L}_J}(\mathcal{B}_w^{\mathcal{L}_J}) = h_w^{\mathcal{L}_M}(\mathcal{B}_w^{\mathcal{L}_M}) = h_w^{\mathcal{L}}(\mathcal{B}_w^{\mathcal{L}}) \subseteq h_w^{\mathcal{L}_L}(\mathcal{B}_w^{\mathcal{L}_L}) = h_w^{\mathcal{L}_N}(\mathcal{B}_w^{\mathcal{L}_N}) \subseteq h_w^{\mathcal{L}_P}(\mathcal{B}_w^{\mathcal{L}_P}).$$

(b) Under one of the regularity conditions mentioned in Remark 4.5.1, the scheme becomes

$$h_w^{\mathcal{L}_J}(\mathcal{B}_w^{\mathcal{L}_J}) \subseteq h_w^{\mathcal{L}_M}(\mathcal{B}_w^{\mathcal{L}_M}) = h_w^{\mathcal{L}}(\mathcal{B}_w^{\mathcal{L}}) \subseteq h_w^{\mathcal{L}_L}(\mathcal{B}_w^{\mathcal{L}_L}) = h_w^{\mathcal{L}_N}(\mathcal{B}_w^{\mathcal{L}_N}) = h_w^{\mathcal{L}_P}(\mathcal{B}_w^{\mathcal{L}_P}).$$

Remark 5.4.5. Since the regularity condition required in Theorem 5.3.9 is automatically fulfilled in this particular case, it follows that concerning the weakly maximal elements of the sets mentioned in Remark 5.4.4 we have

$$\operatorname{WMax}\left(h_w^{\mathcal{L}_J}(\mathcal{B}_w^{\mathcal{L}_J}), \mathbb{R}_+^k\right) \subseteq \operatorname{WMax}\left(h_w^{\mathcal{L}_M}(\mathcal{B}_w^{\mathcal{L}_M}), \mathbb{R}_+^k\right) = \operatorname{WMax}\left(h_w^{\mathcal{L}}(\mathcal{B}_w^{\mathcal{L}}), \mathbb{R}_+^k\right)$$
$$= \operatorname{WMax}\left(h_w^{\mathcal{L}_L}(\mathcal{B}_w^{\mathcal{L}_L}), \mathbb{R}_+^k\right) = \operatorname{WMax}\left(h_w^{\mathcal{L}_N}(\mathcal{B}_w^{\mathcal{L}_N}), \mathbb{R}_+^k\right).$$

In more particular frameworks this scheme can be developed as follows.

(a) In case $b \neq 0$ this scheme turns into

$$\operatorname{WMax}\left(h_w^{\mathcal{L}_J}(\mathcal{B}_w^{\mathcal{L}_J}), \mathbb{R}_+^k\right) = \operatorname{WMax}\left(h_w^{\mathcal{L}_M}(\mathcal{B}_w^{\mathcal{L}_M}), \mathbb{R}_+^k\right) = \operatorname{WMax}\left(h_w^{\mathcal{L}}(\mathcal{B}_w^{\mathcal{L}}), \mathbb{R}_+^k\right)$$
$$= \operatorname{WMax}\left(h_w^{\mathcal{L}_L}(\mathcal{B}_w^{\mathcal{L}_L}), \mathbb{R}_+^k\right) = \operatorname{WMax}\left(h_w^{\mathcal{L}_N}(\mathcal{B}_w^{\mathcal{L}_N}), \mathbb{R}_+^k\right).$$

(b) Under one of the regularity conditions mentioned in Remark 4.5.1, the scheme becomes

$$\text{WMax}\left(h_w^{\mathcal{L}_J}(\mathcal{B}_w^{\mathcal{L}_J}), \mathbb{R}_+^k\right) \subseteq \text{WMax}\left(h_w^{\mathcal{L}_M}(\mathcal{B}_w^{\mathcal{L}_M}), \mathbb{R}_+^k\right) =$$

$$\text{WMax}\left(h_w^{\mathcal{L}}(\mathcal{B}_w^{\mathcal{L}}), \mathbb{R}_+^k\right) = \text{WMax}\left(h_w^{\mathcal{L}_L}(\mathcal{B}_w^{\mathcal{L}_L}), \mathbb{R}_+^k\right) =$$

$$\text{WMax}\left(h_w^{\mathcal{L}_N}(\mathcal{B}_w^{\mathcal{L}_N}), \mathbb{R}_+^k\right) = \text{WMax}\left(h_w^{\mathcal{L}_P}(\mathcal{B}_w^{\mathcal{L}_P}), \mathbb{R}_+^k\right).$$

Note also that in general $\text{WMax}\left(h_w^{\mathcal{L}_L}(\mathcal{B}_w^{\mathcal{L}_L}), \mathbb{R}_+^k\right)$ is not a subset of $\text{WMax}\left(h_w^{\mathcal{L}_P}(\mathcal{B}_w^{\mathcal{L}_P}), \mathbb{R}_+^k\right)$ even if we have $h_w^{\mathcal{L}_L}(\mathcal{B}_w^{\mathcal{L}_L}) \subseteq h_w^{\mathcal{L}_P}(\mathcal{B}_w^{\mathcal{L}_P})$.

5.5 Classical linear vector duality in finite dimensional spaces

The primal problem dealt with in the previous section generalizes the classical linear vector optimization problem considered in the literature and treated further. In the following we see what happens to the duals considered so far by us when the primal is the classical linear vector optimization problem in finite dimensional spaces and we recall some of the classical duals from the literature on linear vector duality. Comparing the image sets of the feasible sets through the objective functions of the vector duals and then their subsets of (weakly) maximal elements allows us to give an overview over the linear vector duality concepts considered so far in the literature.

5.5.1 Duality with respect to efficient solutions

To deal with the classical linear vector duality in finite dimensional spaces, let the matrices $A \in \mathbb{R}^{m\times n}$ and $L \in \mathbb{R}^{k\times n}$ and the vector $b \in \mathbb{R}^m$ be such that $b \in A(\mathbb{R}_+^n)$. We consider the primal linear vector optimization problem treated first by Isermann in [97]

$$(PV^{\mathcal{L}}) \quad \underset{x\in\mathcal{A}^{\mathcal{L}}}{\text{Min}}\, Lx.$$
$$\mathcal{A}^{\mathcal{L}} = \{x \in \mathbb{R}_+^n : Ax = b\}$$

This primal linear vector problem is actually a special case of $(PVF^{\mathcal{L}})$ considered in section 5.4 (or section 4.5) for $X = \mathbb{R}^n$, $C = \{0\}$ and $S = \mathbb{R}_+^n$ and, A, b and L as introduced above.

An important result regarding the solutions to $(PV^{\mathcal{L}})$ in the case treated in this subsection was given in [98] and in a slightly simplified presentation in [64]. Two preparatory lemmata precede it.

Lemma 5.5.1. *A point $\bar{x} \in \mathcal{A}^{\mathcal{L}}$ is efficient to $(PV^{\mathcal{L}})$ if and only if the linear optimization problem*

$$\sup\left\{\sum_{i=1}^{k} y_i : (x,y) \in \mathbb{R}^n_+ \times \mathbb{R}^k_+, y = (y_1,\dots,y_k)^T, Ax = b, Lx + Iy = L\bar{x}\right\},$$

has an optimal solution $(\hat{x}, 0)$.

Lemma 5.5.2. *A point* $\bar{x} \in \mathcal{A}^{\mathcal{L}}$ *is efficient to* $(PV^{\mathcal{L}})$ *if and only if the linear optimization problem*

$$\inf\left\{u^T b + v^T L\bar{x} : (u,v) \in \mathbb{R}^m \times \mathbb{R}^k, v = (v_1,\dots,v_k)^T, v_i \geq 1, i = 1,\dots,k, \right. \\ \left. u^T A + v^T L \geqq 0\right\}$$

has an optimal solution $(\hat{u}, \hat{v})$ *fulfilling* $\hat{u}^T b + \hat{v}^T L\bar{x} = 0$.

Remark 5.5.1. The linear scalar optimization problem considered in Lemma 5.5.2 is the dual of the one used in Lemma 5.5.1.

The following statement, proven in [98], characterizes the properly efficient solutions in the sense of linear scalarization to $(PV^{\mathcal{L}})$, by showing that they coincide with the efficient solutions to the same vector problem.

Theorem 5.5.3. *Every efficient solution to* $(PV^{\mathcal{L}})$ *is also properly efficient and vice versa.*

Proof. Since it is known by Proposition 2.4.12 that any properly efficient solution to $(PV^{\mathcal{L}})$ is efficient, too, it remains to show only the converse implication.

Let $\bar{x} \in \mathcal{A}^{\mathcal{L}}$ be efficient to $(PV^{\mathcal{L}})$. Then the optimal objective value of the linear minimization considered in Lemma 5.5.2 is 0 and it is attained at some feasible $(\hat{u}, \hat{v})$. Then $\hat{u}$ solves also the linear programming problem

$$\inf\left\{u^T b : u \in \mathbb{R}^m, u^T A \geqq -\hat{v}^T L\right\},$$

hence its dual

$$\sup\left\{-\hat{v}^T Lx : x \in \mathbb{R}^n_+, Ax = b\right\}$$

has as optimal objective value $\hat{u}^T b = -\hat{v}^T L\bar{x}$. From here it follows that $\hat{v}^T L\bar{x} \leq \hat{v}^T Lx$ for all $x \in \mathcal{A}^{\mathcal{L}}$, i.e. $\bar{x}$ is properly efficient to $(PV^{\mathcal{L}})$. □

Therefore in the framework considered in this subsection we deal only with efficient solutions to $(PV^{\mathcal{L}})$, as the study of the properly efficient ones reduces to them, too. Given this, it becomes clear why considering in this section the vector duals to $(PV^{\mathcal{L}})$ with respect to efficient solutions and not with respect to properly efficient solutions as in most of this book.

Before particularizing into this special setting the vector duals considered before to $(PVF^{\mathcal{L}})$, we introduce the celebrated vector dual to $(PV^{\mathcal{L}})$ due to Isermann (cf. [97])

$$(DV^{\mathcal{L}_I}) \quad \underset{U \in \mathcal{B}^{\mathcal{L}_I}}{\text{Max}}\, h^{\mathcal{L}_I}(U),$$

where

$$\mathcal{B}^{\mathcal{L}_I} = \left\{U \in \mathbb{R}^{k \times m} : \nexists x \in \mathbb{R}^n_+ \text{ such that } (L - UA)x \leq 0\right\}$$

and

$$h^{\mathcal{L}_I}(U) = Ub.$$

Next we see what happens to the vector duals to $(PVF^{\mathcal{L}})$ considered so far in the present framework, where the primal linear vector optimization problem is $(PV^{\mathcal{L}})$. We begin with the dual stated in [101, 104]

$$(DV^{\mathcal{L}_J}) \quad \operatorname*{Max}_{(\lambda,U)\in\mathcal{B}^{\mathcal{L}_J}} h^{\mathcal{L}_J}(\lambda, U),$$

where

$$\mathcal{B}^{\mathcal{L}_J} = \left\{(\lambda, U) \in \operatorname{int}(\mathbb{R}^k_+) \times \mathbb{R}^{k \times m} : (L - UA)^T \lambda \in \mathbb{R}^n_+\right\}$$

and

$$h^{\mathcal{L}_J}(\lambda, U) = Ub,$$

followed by the two vector duals we introduced in section 5.2

$$(DV^{\mathcal{L}_M}) \quad \operatorname*{Max}_{(\lambda,z,v)\in\mathcal{B}^{\mathcal{L}_M}} h^{\mathcal{L}_M}(\lambda, z, v),$$

where

$$\mathcal{B}^{\mathcal{L}_M} = \left\{(\lambda, z, v) \in \operatorname{int}(\mathbb{R}^k_+) \times \mathbb{R}^m \times \mathbb{R}^k : \lambda^T v = 0 \text{ and } L^T\lambda - A^T z \in \mathbb{R}^n_+\right\}$$

and

$$h^{\mathcal{L}_M}(\lambda, z, v) = \frac{1}{\sum\limits_{i=1}^{k} \lambda_i} z^T b + v$$

and, respectively,

$$(DV^{\mathcal{L}}) \quad \operatorname*{Max}_{(\lambda,U,v)\in\mathcal{B}^{\mathcal{L}}} h^{\mathcal{L}}(\lambda, U, v),$$

where

$$\mathcal{B}^{\mathcal{L}} = \left\{(\lambda, U, v) \in \operatorname{int}(\mathbb{R}^k_+) \times \mathbb{R}^{k \times m} \times \mathbb{R}^k : \lambda^T v = 0 \text{ and } (L - UA)^T\lambda \in \mathbb{R}^n_+\right\}$$

and

$$h^{\mathcal{L}}(\lambda, U, v) = Ub + v.$$

From section 4.3 we have other two vector duals to $(PV^{\mathcal{L}})$, namely the Lagrange type one, which, as noted before, is equivalent to the Fenchel-Lagrange type ones,

$$(DV^{\mathcal{L}_L}) \quad \operatorname*{Max}_{(\lambda,z,v)\in\mathcal{B}^{\mathcal{L}_L}} h^{\mathcal{L}_L}(\lambda, z, v),$$

where

$$\mathcal{B}^{\mathcal{L}_L} = \{(\lambda, z, v) \in \operatorname{int}(\mathbb{R}^k_+) \times \mathbb{R}^m \times \mathbb{R}^k : \lambda^T v - z^T b \leq 0 \text{ and } L^T\lambda - A^T z \in \mathbb{R}^n_+\}$$

and

$$h^{\mathcal{L}_L}(\lambda, z, v) = v,$$

and

$$(DV^{\mathcal{L}_P}) \quad \underset{(\lambda,v)\in\mathcal{B}^{\mathcal{L}_P}}{\operatorname{Max}} h^{\mathcal{L}_P}(\lambda, v),$$

where

$$\mathcal{B}^{\mathcal{L}_P} = \left\{(\lambda, v) \in \operatorname{int}(\mathbb{R}^k_+) \times \mathbb{R}^k : \lambda^T v \leq \inf_{x\in\mathcal{A}^{\mathcal{L}}} \lambda^T Lx\right\}$$

and

$$h^{\mathcal{L}_P}(\lambda, v) = v,$$

which is equivalent to the Fenchel type one, while in section 4.2 we considered Nakayama's vector dual

$$(DV^{\mathcal{L}_N}) \quad \underset{(U,v)\in\mathcal{B}^{\mathcal{L}_N}}{\operatorname{Max}} h^{\mathcal{L}_N}(U, v),$$

where

$$\mathcal{B}^{\mathcal{L}_N} = \{(U, v) \in \mathbb{R}^{k\times m}_+ \times \mathbb{R}^k : \nexists x \in \mathbb{R}^n_+ \text{ such that } v - Ub \geq (L - UA)x\}$$

and

$$h^{\mathcal{L}_N}(U, v) = v.$$

As noted in Remark 5.4.2 the images of the feasible sets through their objective functions of these vector duals satisfy the following scheme of inclusions

$$h^{\mathcal{L}_J}(\mathcal{B}^{\mathcal{L}_J}) \subseteq h^{\mathcal{L}_M}(\mathcal{B}^{\mathcal{L}_M}) = h^{\mathcal{L}}(\mathcal{B}^{\mathcal{L}}) \subseteq h^{\mathcal{L}_L}(\mathcal{B}^{\mathcal{L}_L}) = h^{\mathcal{L}_P}(\mathcal{B}^{\mathcal{L}_P}) \subseteq h^{\mathcal{L}_N}(\mathcal{B}^{\mathcal{L}_N}),$$

while via Remark 5.4.3 we know that their sets of maximal elements are in the following relations

$$\operatorname{Max}\left(h^{\mathcal{L}_J}(\mathcal{B}^{\mathcal{L}_J}), \mathbb{R}^k_+\right) \subseteq \operatorname{Max}\left(h^{\mathcal{L}_M}(\mathcal{B}^{\mathcal{L}_M}), \mathbb{R}^k_+\right) = \operatorname{Max}\left(h^{\mathcal{L}}(\mathcal{B}^{\mathcal{L}}), \mathbb{R}^k_+\right) =$$
$$\operatorname{Max}\left(h^{\mathcal{L}_L}(\mathcal{B}^{\mathcal{L}_L}), \mathbb{R}^k_+\right) = \operatorname{Max}\left(h^{\mathcal{L}_P}(\mathcal{B}^{\mathcal{L}_P}), \mathbb{R}^k_+\right) = \operatorname{Max}\left(h^{\mathcal{L}_N}(\mathcal{B}^{\mathcal{L}_N}), \mathbb{R}^k_+\right), \tag{5.5}$$

where the last equality follows by using Remark 5.4.3(b) and Theorem 5.5.3.

When $b \neq 0$ the first inclusion turns into equality in each of these schemes, while when $b = 0$, as noted in the proof of Theorem 4.5.2, either $\operatorname{Max}\left(h^{\mathcal{L}_J}(\mathcal{B}^{\mathcal{L}_J}), \mathbb{R}^k_+\right) = \{0\}$ or this set is empty. We refer to [101] for an example which shows that the first inclusion relation in (5.5) can be strict when $b = 0$.

Now let us see where can $(DV^{\mathcal{L}_I})$ be integrated into these chains of inclusions. Via [101] we have the following statement.

Proposition 5.5.4. *One has* $h^{\mathcal{L}_I}(\mathcal{B}^{\mathcal{L}_I}) = h^{\mathcal{L}_J}(\mathcal{B}^{\mathcal{L}_J})$.

Proof. Let $(\lambda, U) \in \mathcal{B}^{\mathcal{L}_J}$. Then $(L - UA)^T \lambda \in \mathbb{R}^n_+$. Assume that there is some $x \in \mathbb{R}^n_+$ such that $UAx \geq Lx$. Then $\lambda^T(L - UA)x < 0$. On the other hand, $\left((L - UA)^T\lambda\right)^T x = \lambda^T(L - UA)x \geq 0$, therefore we obtained a contradiction. Consequently, for all $x \in \mathbb{R}^n_+$ there is $UAx \not\geq Lx$, i.e. $U \in \mathcal{B}^{\mathcal{L}_I}$. As the objective function of both these problems is Ub, it follows $h^{\mathcal{L}_I}(\mathcal{B}^{\mathcal{L}_I}) \supseteq h^{\mathcal{L}_J}(\mathcal{B}^{\mathcal{L}_J})$.

To prove the opposite inclusion, let be $U \in \mathcal{B}^{\mathcal{L}_I}$. Then, for $i = 1, \ldots, k$, the system of inequalities

$$\begin{cases} x \geqq 0, (L - UA)x \leqq 0, \\ (L - UA)_i x < 0, \end{cases}$$

has no solution. Consequently, for $i = 1, \ldots, k$,

$$\inf\{(L - UA)_i x : x \in \mathbb{R}^n_+, (L - UA)x \leqq 0\} \geq 0.$$

Due to Theorem 3.2.14 (see also Remark 3.2.6), there are some $\nu^i \in \mathbb{R}^n_+$ and $\mu^i \in \mathbb{R}^k_+$ such that $\inf\{(L - UA)_i x - (\nu^i)^T x + (\mu^i)^T (L - UA)x : x \in \mathbb{R}^n\} \geq 0$ for $i = 1, \ldots, k$. This yields

$$\inf\{x^T((L - UA)_i - \nu^i + (\mu^i)^T (L - UA)) : x \in \mathbb{R}^n\} \geq 0,$$

which implies

$$(L - UA)_i - \nu^i + (\mu^i)^T (L - UA) = 0 \ \forall i = 1, \ldots, k.$$

Thus $(L - UA)^T(\mu^i + e_i) = \nu^i \in \mathbb{R}^n_+$ for $i = 1, \ldots, k$. This yields $(L - UA)^T(\sum_{i=1}^k \mu^i + e) = \sum_{i=1}^k \nu^i \in \mathbb{R}^n_+$. As $\sum_{i=1}^k \mu^i + e \in \operatorname{int}(\mathbb{R}^k_+)$ it follows $(\sum_{i=1}^k \mu^i + e, U) \in \mathcal{B}^{\mathcal{L}_J}$ and, consequently, $h^{\mathcal{L}_I}(\mathcal{B}^{\mathcal{L}_I}) = h^{\mathcal{L}_J}(\mathcal{B}^{\mathcal{L}_J})$. □

Combining (5.5), the comments following it and Proposition 5.5.4, we see that the sets of efficient solutions to all the vector duals to $(PV^{\mathcal{L}})$ considered in this section coincide when $b \neq 0$, while in case $b = 0$ the last five of them coincide, while ones of the duals formulated by Jahn in [101, 104] and by Isermann in [97] either contain only the element 0 or they are empty. Using Remark (5.5), it is clear that it is enough to give only a duality statement for all the vector duals with respect to efficient solutions we considered here, with the notable exception of $(DV^{\mathcal{L}_J})$ in case $b = 0$. As the weak duality follows directly from the general case, the regularity conditions that ensure strong and converse duality are automatically satisfied (see Theorem 3.2.14 and Remark 3.2.6), and the set $L(\mathcal{A}^{\mathcal{L}}) + \mathbb{R}^k_+$ is closed since $\mathcal{A}^{\mathcal{L}}$ is polyhedral and L is linear, the duality assertions given before for $(DVF^{\mathcal{L}})$ yield the following statement.

Theorem 5.5.5. *(a) There is no* $x \in \mathcal{A}^{\mathcal{L}}$ *and no* $(\lambda, U, v) \in \mathcal{B}^{\mathcal{L}}$ *such that* $Lx \leq Ub + v$.

(b) If $\bar{x} \in \mathcal{A}^{\mathcal{L}}$ *is an efficient solution to* $(PV^{\mathcal{L}})$*, then there exists* $(\bar{\lambda}, \overline{U}, \bar{v}) \in \mathcal{B}^{\mathcal{L}}$*, an efficient solution to* $(DV^{\mathcal{L}})$*, such that* $L\bar{x} = \overline{U}b + \bar{v}$.

(c) If $(\bar{\lambda}, \overline{U}, \bar{v}) \in \mathcal{B}^{\mathcal{L}}$ is an efficient solution to $(DV^{\mathcal{L}})$, then there exists $\bar{x} \in \mathcal{A}^{\mathcal{L}}$, an efficient solution to $(PV^{\mathcal{L}})$, such that $L\bar{x} = \overline{U}b + \bar{v}$.

Remark 5.5.2. (*a*) From a historical point of view it is interesting to notice that the first pair of primal-dual linear vector optimization problems was considered by Gale, Kuhn and Tucker in [70]. For the primal problem

$$(PV^{\mathcal{L}_{GKT}}) \qquad \operatorname*{Min}_{(D,x,y)\in\mathcal{A}^{\mathcal{L}}_{GKT}} D,$$

where

$$\mathcal{A}^{\mathcal{L}}_{GKT} = \{(D, x, y) \in \mathbb{R}^{k\times p} \times \mathbb{R}^n_+ \times \operatorname{int}(\mathbb{R}^p_+) : Ax = By \text{ and } Dy - Lx \in \mathbb{R}^k_+\},$$

they took as vector dual

$$(DV^{\mathcal{L}_{GKT}}) \qquad \operatorname*{Max}_{(V,z,v)\in\mathcal{B}^{\mathcal{L}}_{GKT}} V,$$

where

$$\mathcal{B}^{\mathcal{L}}_{GKT} = \{(V, z, v) \in \mathbb{R}^{k\times p} \times \mathbb{R}^m \times \operatorname{int}(\mathbb{R}^k_+) : L^T v - A^T z \in \mathbb{R}^n_+, V^T v - B^T z \leqq 0\}.$$

Here $B \in \mathbb{R}^{m\times p}$ is a given matrix and for matrices a componentwise partial ordering analogously to the one for vectors introduced via the nonnegative orthant is considered.

One of the classical particular instances of $(PV^{\mathcal{L}_{GKT}})$ is obtained by taking $p = 1$ and $y = 1$ (see also [96]). Thus one gets as primal problem

$$(PV^{\mathcal{L}_G}) \qquad \operatorname*{Min}_{(d,x)\in\mathcal{A}^{\mathcal{L}_G}} d,$$

where

$$\mathcal{A}^{\mathcal{L}_G} = \{(d, x) \in \mathbb{R}^k \times \mathbb{R}^n_+ : Ax = b \text{ and } d - Lx \in \mathbb{R}^k_+\},$$

which is actually equivalent to $(PV^{\mathcal{L}})$. Here it is worth noticing that the vector dual of this problem, attached to it via the duality scheme of Gale, Kuhn and Tucker,

$$(DV^{\mathcal{L}_G}) \qquad \operatorname*{Max}_{(\lambda,z,v)\in\mathcal{B}^{\mathcal{L}}_G} v,$$

where

$$\mathcal{B}^{\mathcal{L}}_G = \{(\lambda, z, v) \in \operatorname{int}(\mathbb{R}^k_+) \times \mathbb{R}^m \times \mathbb{R}^k : L^T\lambda - A^T z \in \mathbb{R}^n_+ \text{ and } \lambda^T v - z^T b \leq 0\}$$

is nothing else than $(DV^{\mathcal{L}_L})$, the particularization to the linear case of the vector dual from section 4.3.

(*b*) Another pair of primal-dual linear vector optimization problems important mainly from the historical point of view was considered by Kornbluth in [118], namely

$$(PV^{\mathcal{L}_K}) \qquad \operatorname*{Min}_{(x,y)\in\mathcal{A}^{\mathcal{L}_K}} Lx,$$
$$\mathcal{A}^{\mathcal{L}_K} = \{(x,y) \in \mathbb{R}^n_+ \times \operatorname{int}(\mathbb{R}^k_+) : Ax + By = 0\}$$

and its vector dual

$$(DV^{\mathcal{L}_K}) \qquad \operatorname*{Max}_{(v,w)\in\mathcal{B}^{\mathcal{L}_K}} B^T v,$$
$$\mathcal{L}_K = \{(v,w) \in \mathbb{R}^m_+ \times \operatorname{int}(\mathbb{R}^k_+) : A^T v + L^T w \in \mathbb{R}^n_+\}$$

where $A \in \mathbb{R}^{m\times n}$, $B \in \mathbb{R}^{m\times k}$ and $L \in \mathbb{R}^{k\times n}$. Although in [118] properly efficient solutions to these vector optimization problems were taken into consideration, by Theorem 5.5.3 we know that it this framework they coincide with the efficient ones of the same problems. As proven in [95], a pair $(\bar{x},\bar{y})$ is efficient to $(PV^{\mathcal{L}_K})$ if and only if there is a $\overline{D} \in \mathbb{R}^{k\times p} \in \mathcal{A}^{\mathcal{L}}_{GKT}$ such that $(\overline{D},\bar{x},\bar{y})$ is efficient to $(PV^{\mathcal{L}_{GKT}})$ and, regarding the vector duals, a pair $(\bar{x},\bar{y})$ is efficient to $(DV^{\mathcal{L}_K})$ if and only if there is a $\overline{V} \in \mathbb{R}^{k\times p} \in \mathcal{B}^{\mathcal{L}}_{GKT}$ such that $(\overline{V},\bar{x},\bar{y})$ is efficient to $(DV^{\mathcal{L}_{GKT}})$.

To overcome the duality gaps signaled in the literature for the problems $(DV^{\mathcal{L}_J})$ and $(DV^{\mathcal{L}_I})$ when $b = 0$ (see also Example 5.5.1) in [84] was recently proposed a new dual problem to $(PV^{\mathcal{L}})$, which, slightly rewritten, is

$$(DV^{\mathcal{L}_H}) \quad \operatorname*{Max}_{U\in\mathcal{B}^{\mathcal{L}_H}} h^{\mathcal{L}_H}(U),$$

where

$$\mathcal{B}^{\mathcal{L}_H} = \left\{U \in \mathbb{R}^{k\times m} : \nexists x \in \mathbb{R}^n_+ \text{ such that } (L - UA)x \leq 0\right\}$$

and

$$h^{\mathcal{L}_H}(U) = Ub + \operatorname{Min}\left((L - UA)(\mathbb{R}^n_+), \mathbb{R}^k_+\right).$$

Originally this dual was introduced in a more general framework, namely by considering an arbitrary nontrivial pointed convex closed cone K in $\mathbb{R}^k$, instead of $\mathbb{R}^k_+$. In the following we put this vector dual problem in relation to $(DV^{\mathcal{L}})$, which is the vector dual introduced and investigated in section 5.2 applied to $(PV^{\mathcal{L}})$.

Proposition 5.5.6. *It holds* $h^{\mathcal{L}_H}(\mathcal{B}^{\mathcal{L}_H}) \subseteq h^{\mathcal{L}}(\mathcal{B}^{\mathcal{L}})$.

Proof. Let $d \in h^{\mathcal{L}_H}(\mathcal{B}^{\mathcal{L}_H})$. Then there are some $U \in \mathcal{B}^{\mathcal{L}_H}$ and $\bar{x} \in \mathbb{R}^n_+$ such that $d = Ub + (L - UA)\bar{x}$ and $(L - UA)\bar{x} \in \operatorname{Min}\left((L - UA)(\mathbb{R}^n_+), \mathbb{R}^k_+\right)$. Denote $t := (t_1,\ldots,t_k)^T = (L - UA)\bar{x}$. Due to the minimality of t in $(L - UA)(\mathbb{R}^n_+)$, there is no $x \in \mathbb{R}^n_+$ such that $(L - UA)x \leq t$. Thus, for $i = 1,\ldots,k$, the system of inequalities

$$\begin{cases} x \geqq 0, \\ (L - UA)x - t \leqq 0, \\ (L - UA)_i x - t_i < 0, \end{cases}$$

has no solution. Similarly to the proof of Proposition 5.5.4 it follows

$$\inf_{\substack{-x \leqq 0, \\ (L-UA)x-t \leqq 0}} \{(L-UA)_i x - t_i\} \geq 0 \ \forall i = 1, \ldots, k,$$

and because of the second inequality constraint all these infima turn out to be equal to 0. Due to Theorem 3.2.14 (see also Remark 3.2.6), there are some $\nu^i \in \mathbb{R}^n_+$ and $\mu^i \in \mathbb{R}^k_+$ such that for $i = 1, \ldots, k$, one has

$$0 = \inf_{\substack{-x \leqq 0, \\ (L-UA)x-t \leqq 0}} \{(L-UA)_i x - t_i\} =$$

$$\inf_{x \in \mathbb{R}^n} \{x^T ((L-UA)_i - \nu^i + (\mu^i)^T (L-UA)) - t_i - (\mu^i)^T t\}.$$

Consequently, $(L-UA)_i - \nu^i + (\mu^i)^T (L-UA) = 0$ and $t_i + (\mu^i)^T t = 0$ for $i = 1, \ldots, k$. Taking $\lambda = \sum_{i=1}^k \mu^i + e$, we obtain directly that $\sum_{i=1}^k \lambda_i t_i = 0$ and $(L-UA)^T \lambda = \sum_{i=1}^k \nu^i \geqq 0$, which means that $(\lambda, U, t) \in \mathcal{B}^{\mathcal{L}}$. As, moreover, $d = Ub + (L-UA)\bar{x} = Ub + t$, we get $h^{\mathcal{L}_H}(\mathcal{B}^{\mathcal{L}_H}) \subseteq h^{\mathcal{L}}(\mathcal{B}^{\mathcal{L}})$. □

As the following example shows, the inclusion opposite to the one proven above does not always hold.

Example 5.5.1. (cf. [40], see also [84]) Let $L = (1, -1)^T$, $n = 1$, $k = 2$, $A = 0$ and $b = 0$. The classical linear vector optimization primal problem is now

$$(PV^{\mathcal{L}}) \quad \operatorname*{Min}_{x \in \mathbb{R}_+} Lx.$$

It is not difficult to note that $(DV^{\mathcal{L}_H})$ actually coincides with $(PV^{\mathcal{L}})$, therefore $h^{\mathcal{L}_H}(\mathcal{B}^{\mathcal{L}_H}) = \{(x, -x) : x \in \mathbb{R}_+\}$. On the other hand, $(DV^{\mathcal{L}})$ turns into

$$(DV^{\mathcal{L}}) \quad \operatorname*{Max}_{\substack{\lambda_1 \geq \lambda_2 > 0, \\ v \in \mathbb{R}}} \begin{pmatrix} -\frac{\lambda_2}{\lambda_1} \\ 1 \end{pmatrix} v.$$

It is clear that $h^{\mathcal{L}}(\mathcal{B}^{\mathcal{L}})$ is larger than $h^{\mathcal{L}_H}(\mathcal{B}^{\mathcal{L}_H})$, as for instance $(-1/2, 1)^T \in h^{\mathcal{L}}(\mathcal{B}^{\mathcal{L}}) \backslash h^{\mathcal{L}_H}(\mathcal{B}^{\mathcal{L}_H})$.

Proposition 5.5.7. *One has* $\operatorname{Max}(h^{\mathcal{L}}(\mathcal{B}^{\mathcal{L}}), \mathbb{R}^k_+) \subseteq \operatorname{Max}(h^{\mathcal{L}_H}(\mathcal{B}^{\mathcal{L}_H}), \mathbb{R}^k_+)$.

Proof. First we show that $\operatorname{Max}(h^{\mathcal{L}}(\mathcal{B}^{\mathcal{L}}), \mathbb{R}^k_+)$ is a subset of $h^{\mathcal{L}_H}(\mathcal{B}^{\mathcal{L}_H})$. Let d be a maximal element of $h^{\mathcal{L}}(\mathcal{B}^{\mathcal{L}})$, i.e. $(DV^{\mathcal{L}})$ has an efficient solution $(\bar{\lambda}, \overline{U}, \bar{v}) \in \mathcal{B}^{\mathcal{L}}$ and $d = \overline{U}b + \bar{v}$. Assume that there is some $x \in \mathbb{R}^n_+$ such that $(L - \overline{U}A)x \leq \bar{v}$. Taking into consideration the definition of $\mathcal{B}^{\mathcal{L}}$, it follows

$$0 = \bar{\lambda}^T \bar{v} > \bar{\lambda}^T (L - \overline{U}A)x \geq 0,$$

which cannot happen. Thus there is no $x \in \mathbb{R}^n_+$ fulfilling $(L - \overline{U}A)x \leq \bar{v}$.

Suppose that there is no $x \in \mathbb{R}^n_+$ fulfilling $(L - \overline{U}A)x \leqq \bar{v}$, too. Then by Gale's theorem of the alternative (see [127, page 35]) it follows that the system

$$\begin{cases} \lambda \geq 0, \\ \lambda^T \bar{v} < 0, \\ (L - \overline{U}A)^T \lambda \geq 0, \end{cases}$$

has a solution $\lambda = (\lambda_1, \ldots, \lambda_k)^T \in \mathbb{R}^k_+$. Denote $\tilde{\lambda} = (1/2)(\bar{\lambda} + \lambda)$. Clearly $\tilde{\lambda} \in \operatorname{int}(\mathbb{R}^k_+)$ and it fulfills also $\sum_{i=1}^k \tilde{\lambda}_i \bar{v}_i < 0$ and $(L - \overline{U}A)^T \tilde{\lambda} \geqq 0$. Therefore there is some $\tilde{v} \geq \bar{v}$ for which $\sum_{i=1}^k \tilde{\lambda}_i \tilde{v}_i = 0$. Consequently, $\overline{U}b + \tilde{v} \geq \overline{U}b + \bar{v} = d$. Noting that $(\tilde{\lambda}, \overline{U}, \tilde{v}) \in \mathcal{B}^{\mathcal{L}}$, we reached a contradiction to the maximality of d in $h^{\mathcal{L}}(\mathcal{B}^{\mathcal{L}})$. Therefore, the existence of some $\bar{x} \in \mathbb{R}^n_+$ for which $(L - \overline{U}A)\bar{x} \leqq \bar{v}$ is granted, and by what we have proven in the beginning of the proof it follows $\bar{v} = (L - \overline{U}A)\bar{x} \in \operatorname{Min}\left((L - UA)(\mathbb{R}^n_+), \mathbb{R}^k_+\right)$

As $(L - \overline{U}A)\bar{x} = \bar{v}$, it follows that $\overline{U} \in \mathcal{B}^{\mathcal{L}_H}$. Indeed, assuming the contrary, there would exist $x \in \mathbb{R}^n_+$ with $(L - \overline{U}A)x \leq 0$ and consequently $(L - \overline{U}A)(\bar{x} + x) \leq \bar{v}$. As we have seen in the first part of the proof, this is impossible. Consequently, $\overline{U}$ is feasible to $(DV^{\mathcal{L}_H})$ and $\overline{U}b + (L - \overline{U}A)\bar{x} = \overline{U}b + \bar{v} = h^{\mathcal{L}}(\bar{\lambda}, \overline{U}, \bar{v}) \in h^{\mathcal{L}_H}(\overline{U})$. This means that

$$\operatorname{Max}\left(h^{\mathcal{L}}(\mathcal{B}^{\mathcal{L}}), \mathbb{R}^k_+\right) \subseteq h^{\mathcal{L}_H}(\mathcal{B}^{\mathcal{L}_H}). \tag{5.6}$$

Assuming that there is some $\bar{d} \in \operatorname{Max}\left(h^{\mathcal{L}}(\mathcal{B}^{\mathcal{L}}), \mathbb{R}^k_+\right) \setminus \operatorname{Max}\left(h^{\mathcal{L}_H}(\mathcal{B}^{\mathcal{L}_H}), \mathbb{R}^k_+\right)$, we get that there is some $\tilde{d} \in h^{\mathcal{L}_H}(\mathcal{B}^{\mathcal{L}_H})$, fulfilling $\bar{d} \leq \tilde{d}$. By (5.6) and Proposition 5.5.6, it follows $\tilde{d} \in h^{\mathcal{L}}(\mathcal{B}^{\mathcal{L}})$. As $\bar{d} \leq \tilde{d}$, we obtained a contradiction to the maximality of $\bar{d}$ in this set. Therefore $\operatorname{Max}\left(h^{\mathcal{L}}(\mathcal{B}^{\mathcal{L}}), \mathbb{R}^k_+\right) \subseteq \operatorname{Max}\left(h^{\mathcal{L}_H}(\mathcal{B}^{\mathcal{L}_H}), \mathbb{R}^k_+\right)$. □

Remark 5.5.3. The reverse inclusion to the one given in Proposition 5.5.7 follows via [84, Theorem 3.14] and Theorem 5.5.5. Consequently,

$$\operatorname{Max}\left(h^{\mathcal{L}}(\mathcal{B}^{\mathcal{L}}), \mathbb{R}^k_+\right) = \operatorname{Max}\left(h^{\mathcal{L}_H}(\mathcal{B}^{\mathcal{L}_H}), \mathbb{R}^k_+\right).$$

Therefore, the schemes employing the vector duals to $(PV^{\mathcal{L}})$ considered earlier in this section can be completed, using also Proposition 5.5.4 and Theorem 5.5.5, as follows

$$\begin{aligned} h^{\mathcal{L}_I}(\mathcal{B}^{\mathcal{L}_I}) &= h^{\mathcal{L}_J}(\mathcal{B}^{\mathcal{L}_J}) \subseteq h^{\mathcal{L}_H}(\mathcal{B}^{\mathcal{L}_H}) \subseteq h^{\mathcal{L}_M}(\mathcal{B}^{\mathcal{L}_M}) \\ &= h^{\mathcal{L}}(\mathcal{B}^{\mathcal{L}}) \subseteq h^{\mathcal{L}_L}(\mathcal{B}^{\mathcal{L}_L}) = h^{\mathcal{L}_P}(\mathcal{B}^{\mathcal{L}_P}) \subseteq h^{\mathcal{L}_N}(\mathcal{B}^{\mathcal{L}_N}), \end{aligned}$$

and, respectively, for the sets of maximal elements,

$$\operatorname{Max}\left(h^{\mathcal{L}_I}(\mathcal{B}^{\mathcal{L}_I}), \mathbb{R}^k_+\right) = \operatorname{Max}\left(h^{\mathcal{L}_J}(\mathcal{B}^{\mathcal{L}_J}), \mathbb{R}^k_+\right) \subseteq \operatorname{Min}\left(L(\mathcal{A}^{\mathcal{L}}), \mathbb{R}^k_+\right) =$$

$$\operatorname{Max}\left(h^{\mathcal{L}_H}(\mathcal{B}^{\mathcal{L}_H}), \mathbb{R}^k_+\right) = \operatorname{Max}\left(h^{\mathcal{L}_M}(\mathcal{B}^{\mathcal{L}_M}), \mathbb{R}^k_+\right) = \operatorname{Max}\left(h^{\mathcal{L}}(\mathcal{B}^{\mathcal{L}}), \mathbb{R}^k_+\right) =$$

$$\operatorname{Max}\left(h^{\mathcal{L}_L}(\mathcal{B}^{\mathcal{L}_L}), \mathbb{R}^k_+\right) = \operatorname{Max}\left(h^{\mathcal{L}_P}(\mathcal{B}^{\mathcal{L}_P}), \mathbb{R}^k_+\right) = \operatorname{Max}\left(h^{\mathcal{L}_N}(\mathcal{B}^{\mathcal{L}_N}), \mathbb{R}^k_+\right).$$

When $b \neq 0$ the inclusion in the latter relation becomes an equality, i.e. all the sets in the scheme above coincide.

5.5.2 Duality with respect to weakly efficient solutions

Consider now $(PVF_w^{\mathcal{L}})$ in the particular framework when $X = \mathbb{R}^n$, $C = \{0\}$ and $S = \mathbb{R}^n_+$, namely

$$(PV_w^{\mathcal{L}}) \quad \underset{x \in \mathcal{A}^{\mathcal{L}}}{\operatorname{WMin}} \, Lx.$$
$$\mathcal{A}^{\mathcal{L}} = \{x \in \mathbb{R}^n_+ : Ax = b\}$$

In this situation the vector dual problems considered to it in chapter 4 and subsection 5.2.2 can be written as follows. We begin with the one from [101, 104]

$$(DV_w^{\mathcal{L}_J}) \quad \underset{(\lambda,U) \in \mathcal{B}_w^{\mathcal{L}_J}}{\operatorname{WMax}} \, h_w^{\mathcal{L}_J}(\lambda, U),$$

where

$$\mathcal{B}_w^{\mathcal{L}_J} = \left\{(\lambda, U) \in (\mathbb{R}^k_+ \backslash \{0\}) \times \mathbb{R}^{k \times m} : (L - UA)^T \lambda \in \mathbb{R}^n_+\right\}$$

and

$$h_w^{\mathcal{L}_J}(\lambda, U) = Ub,$$

followed by the duals introduced in this chapter

$$(DV_w^{\mathcal{L}_M}) \quad \underset{(\lambda,z,v) \in \mathcal{B}_w^{\mathcal{L}_M}}{\operatorname{WMax}} \, h_w^{\mathcal{L}_M}(\lambda, z, v),$$

where

$$\mathcal{B}_w^{\mathcal{L}_M} = \left\{(\lambda, z, v) \in (\mathbb{R}^k_+ \backslash \{0\}) \times \mathbb{R}^m \times \mathbb{R}^k : \lambda^T v = 0 \text{ and } L^T \lambda - A^T z \in \mathbb{R}^n_+\right\}$$

and

$$h_w^{\mathcal{L}_M}(\lambda, z, v) = \frac{1}{\sum\limits_{i=1}^{k} \lambda_i} z^T b + v$$

and, respectively,

$$(DV_w^{\mathcal{L}}) \quad \underset{(\lambda,U,v) \in \mathcal{B}_w^{\mathcal{L}}}{\operatorname{WMax}} \, h_w^{\mathcal{L}}(\lambda, U, v),$$

where

$$\mathcal{B}_w^{\mathcal{L}} = \left\{(\lambda, U, v) \in (\mathbb{R}^k_+ \backslash \{0\}) \times \mathbb{R}^{k \times m} \times \mathbb{R}^k : \lambda^T v = 0 \text{ and } (L - UA)^T \lambda \in \mathbb{R}^n_+\right\}$$

and

$$h_w^{\mathcal{L}}(\lambda, U, v) = Ub + v.$$

We also have the Lagrange type one, which as observed before is equivalent to the Fenchel-Lagrange type ones,

$$(DV_w^{\mathcal{L}_L}) \quad \underset{(\lambda,z,v) \in \mathcal{B}_w^{\mathcal{L}_L}}{\operatorname{WMax}} \, h_w^{\mathcal{L}_L}(\lambda, z, v),$$

where

$$\mathcal{B}_w^{\mathcal{L}_L} = \big\{(\lambda, z, v) \in (\mathbb{R}^k_+ \setminus \{0\}) \times \mathbb{R}^m \times \mathbb{R}^k : \lambda^T v - z^T b \leq 0 \text{ and } L^T\lambda - A^T z \in \mathbb{R}^n_+\big\}$$

and

$$h_w^{\mathcal{L}_L}(\lambda, z, v) = v,$$

then

$$(DV_w^{\mathcal{L}_P}) \quad \underset{(\lambda,v)\in\mathcal{B}_w^{\mathcal{L}_P}}{\mathrm{WMax}} \; h_w^{\mathcal{L}_P}(\lambda, v),$$

where

$$\mathcal{B}_w^{\mathcal{L}_P} = \big\{(\lambda, v) \in (\mathbb{R}^k_+ \setminus \{0\}) \times \mathbb{R}^k : \lambda^T v \leq \inf_{x\in\mathcal{A}^{\mathcal{L}}} \lambda^T Lx\big\}$$

and

$$h_w^{\mathcal{L}_P}(\lambda, v) = v,$$

which is equivalent to the Fenchel type dual, while in section 4.2 we considered Nakayama's vector dual

$$(DV_w^{\mathcal{L}_N}) \quad \underset{(U,v)\in\mathcal{B}_w^{\mathcal{L}_N}}{\mathrm{WMax}} \; h_w^{\mathcal{L}_N}(U, v),$$

where

$$\mathcal{B}_w^{\mathcal{L}_N} = \big\{(U, v) \in \mathbb{R}^{k\times m}_+ \times \mathbb{R}^k : \nexists x \in \mathbb{R}^n_+ \text{ such that } v - Ub > (L - UA)x\big\}$$

and

$$h_w^{\mathcal{L}_N}(U, v) = v.$$

Remark 5.5.4. The scheme of inclusions of the images of the feasible sets through their objective functions of the vector duals with respect to weakly efficient solutions which follows from Remark 5.4.4 in the particular setting we work now in is

$$h_w^{\mathcal{L}_J}(\mathcal{B}_w^{\mathcal{L}_J}) \subseteq h_w^{\mathcal{L}_M}(\mathcal{B}_w^{\mathcal{L}_M}) = h_w^{\mathcal{L}}(\mathcal{B}_w^{\mathcal{L}}) \subseteq h_w^{\mathcal{L}_L}(\mathcal{B}_w^{\mathcal{L}_L}) = h_w^{\mathcal{L}_P}(\mathcal{B}_w^{\mathcal{L}_P}) = h_w^{\mathcal{L}_N}(\mathcal{B}_w^{\mathcal{L}_N}).$$

For the weakly maximal elements of these sets we have then

$$\mathrm{WMax}\big(h_w^{\mathcal{L}_J}(\mathcal{B}_w^{\mathcal{L}_J}), \mathbb{R}^k_+\big) \subseteq \mathrm{WMax}\big(h_w^{\mathcal{L}_M}(\mathcal{B}_w^{\mathcal{L}_M}), \mathbb{R}^k_+\big) = \mathrm{WMax}\big(h_w^{\mathcal{L}}(\mathcal{B}_w^{\mathcal{L}}), \mathbb{R}^k_+\big) =$$

$$\mathrm{WMax}\big(h_w^{\mathcal{L}_L}(\mathcal{B}_w^{\mathcal{L}_L}), \mathbb{R}^k_+\big) = \mathrm{WMax}\big(h_w^{\mathcal{L}_P}(\mathcal{B}_w^{\mathcal{L}_P}), \mathbb{R}^k_+\big) = \mathrm{WMax}\big(h_w^{\mathcal{L}_N}(\mathcal{B}_w^{\mathcal{L}_N}), \mathbb{R}^k_+\big).$$

When $b \neq 0$ the first inclusion turns into equality in each of these schemes, while when $b = 0$, either $\mathrm{WMax}\big(h^{\mathcal{L}_J}(\mathcal{B}^{\mathcal{L}_J}), \mathbb{R}^k_+\big) = \{0\}$ or this set is empty.

For $(PV_w^{\mathcal{L}})$ and the vector duals mentioned above the weak, strong and converse duality statements can be derived directly from the ones given before in more general frameworks, but, using the advantages of the linear duality stressed in the comments before Theorem 5.5.5, it is possible to prove them under weaker hypotheses as follows. Using Remark 5.5.4, it is clear that it is enough to give only the following duality statement, which holds for all the vector duals with respect to weakly efficient solutions we considered here, with the notable exception of $(DV_w^{\mathcal{L}_J})$ in case $b=0$.

Theorem 5.5.8. *(a) There is no $x \in \mathcal{A}^{\mathcal{L}}$ and no $(\lambda, U, v) \in \mathcal{B}_w^{\mathcal{L}}$ such that $Lx < Ub + v$.*
(b) If $\bar{x} \in \mathcal{A}^{\mathcal{L}}$ is a weakly efficient solution to $(PV_w^{\mathcal{L}})$, then there exists $(\bar{\lambda}, \overline{U}, \bar{v}) \in \mathcal{B}_w^{\mathcal{L}}$, a weakly efficient solution to $(DV_w^{\mathcal{L}})$, fulfilling $L\bar{x} = \overline{U}b + \bar{v}$.
(c) If $(\bar{\lambda}, \overline{U}, \bar{v}) \in \mathcal{B}_w^{\mathcal{L}}$ is a weakly efficient solution to $(DV_w^{\mathcal{L}})$, then $\overline{U}b + \bar{v} \in \operatorname{WMin}\big(L(\mathcal{A}^{\mathcal{L}}) + \mathbb{R}^k_+, \mathbb{R}^k_+\big)$.

Remark 5.5.5. Combining Remark 5.5.4 and Theorem 5.5.8, one has when $b \neq 0$

$$\operatorname{WMin}\big(L(\mathcal{A}^{\mathcal{L}}), \mathbb{R}^k_+\big) \subseteq \operatorname{WMax}\big(h_w^{\mathcal{L}_J}(\mathcal{B}_w^{\mathcal{L}_J}), \mathbb{R}^k_+\big) = \operatorname{WMax}\big(h_w^{\mathcal{L}_M}(\mathcal{B}_w^{\mathcal{L}_M}), \mathbb{R}^k_+\big) =$$

$$\operatorname{WMax}\big(h_w^{\mathcal{L}}(\mathcal{B}_w^{\mathcal{L}}), \mathbb{R}^k_+\big) = \operatorname{WMax}\big(h_w^{\mathcal{L}_L}(\mathcal{B}_w^{\mathcal{L}_L}), \mathbb{R}^k_+\big) = \operatorname{WMax}\big(h_w^{\mathcal{L}_P}(\mathcal{B}_w^{\mathcal{L}_P}), \mathbb{R}^k_+\big)$$
$$= \operatorname{WMax}\big(h_w^{\mathcal{L}_N}(\mathcal{B}_w^{\mathcal{L}_N}), \mathbb{R}^k_+\big) \subseteq \operatorname{WMin}\big(L(\mathcal{A}^{\mathcal{L}}) + \mathbb{R}^k_+, \mathbb{R}^k_+\big),$$

while when $b=0$ it holds

$$\operatorname{WMax}\big(h_w^{\mathcal{L}_J}(\mathcal{B}_w^{\mathcal{L}_J}), \mathbb{R}^k_+\big) \subseteq \operatorname{WMin}\big(L(\mathcal{A}^{\mathcal{L}}), \mathbb{R}^k_+\big) \subseteq \operatorname{WMax}\big(h_w^{\mathcal{L}_M}(\mathcal{B}_w^{\mathcal{L}_M}), \mathbb{R}^k_+\big) =$$

$$\operatorname{WMax}\big(h_w^{\mathcal{L}}(\mathcal{B}_w^{\mathcal{L}}), \mathbb{R}^k_+\big) = \operatorname{WMax}\big(h_w^{\mathcal{L}_L}(\mathcal{B}_w^{\mathcal{L}_L}), \mathbb{R}^k_+\big) = \operatorname{WMax}\big(h_w^{\mathcal{L}_P}(\mathcal{B}_w^{\mathcal{L}_P}), \mathbb{R}^k_+\big)$$
$$= \operatorname{WMax}\big(h_w^{\mathcal{L}_N}(\mathcal{B}_w^{\mathcal{L}_N}), \mathbb{R}^k_+\big) \subseteq \operatorname{WMin}\big(L(\mathcal{A}^{\mathcal{L}}) + \mathbb{R}^k_+, \mathbb{R}^k_+\big).$$

Bibliographical notes

As mentioned in the bibliographical notes of chapter 4, the works where vector duality based on the classical scalar Fenchel duality is considered are quite seldom. The paper where the approach used by us in section 5.1 was introduced is [28], where all the spaces involved were taken finite dimensional.

The scalar Fenchel-Lagrange duality, introduced in [186], was quickly applied to vector duality, too, in the sense that it was considered for the scalarized problems attached to vector optimization problems and the vector duals attached to these contained the scalar duals not only in the feasible sets like in the cases attached in chapter 4, but also in the objective functions. This was first done by Boţ and Wanka for linearly constrained vector optimization problems in [182,183], and soon after for cone constrained ones in [184]. These investigations were continued in [24,36,37,185], where a whole class of Fenchel-Lagrange type vector dual problems was assigned to a cone constrained vector optimization problem and comparisons involving the image sets through the objective functions of these duals and of some other vector duals known in the literature and the corresponding maximal sets were performed.

Among the first papers on linear vector duality in finite dimensional spaces belong [70] of Gale, Kuhn and Tucker and Kornbluth's [118]. Then Isermann has done intensive research on linear vector duality in finite dimensional spaces, introducing in [97] a new vector dual to the primal linear vector optimization problem and comparing his result to the previously mentioned ones. He also proved in [98] that proper efficiency and efficiency coincide for primal linear vector optimization problems in finite dimensional spaces. Later, Jahn has shown in [101] that the vector dual he proposed for linear vector optimization problems turns out to be equivalent to Isermann's one, noting for the latter that it has as major drawback the fact that when $b = 0$ duality could not be established. This issue was claimed to be solved in [84], where a new vector dual for the classical linear vector optimization problem was introduced by considering as objective function a set-valued mapping. However, the Fenchel-Lagrange type vector duality introduced earlier in [24, 36] works fine in the linear case when $b = 0$, too.

6

Wolfe and Mond-Weir duality concepts

In this chapter we present scalar and vector duality based on the classical *Wolfe* and *Mond-Weir duality concepts.* As the field is very vast, especially because of different generalizations of the notion of convexity for the functions employed, we limited our exposition to a reasonable framework, large enough to present the most relevant facts in the area. We shall work in parallel with the two mentioned duality concepts. Note that the properly efficient solutions that appear in this chapter are considered in the sense of linear scalarization (see Definition 2.4.12), unless otherwise specified.

6.1 Classical scalar Wolfe and Mond-Weir duality

For the beginnings of Wolfe duality the reader is referred to [202], while the first paper on Mond-Weir duality is considered to be [138]. In both of them the functions involved were taken differentiable, but afterwards both these duality concepts were extended to nondifferentiable functions by making use of convexity. We begin our presentation with the convex case, treating after that the situation when the functions involved are assumed moreover differentiable.

6.1.1 Scalar Wolfe and Mond-Weir duality: nondifferentiable case

Like in section 3.1.3, let X and Z be Hausdorff locally convex spaces, the latter partially ordered by the convex cone $C \subseteq Z$, and consider the nonempty set $S \subseteq X$ and the proper functions $f : X \to \overline{\mathbb{R}}$ and $g : X \to \overline{Z}$, fulfilling $\operatorname{dom} f \cap S \cap g^{-1}(-C) \neq \emptyset$. The primal problem we treat further is

$$(P^C) \quad \inf_{\substack{x \in \mathcal{A} \\ \mathcal{A} = \{x \in S : g(x) \in -C\}}} f(x).$$

To it we attach the *Wolfe dual* problem

R.I. Boţ et al., *Duality in Vector Optimization*, Vector Optimization,
DOI 10.1007/978-3-642-02886-1_6,

$$(D_W^C) \qquad \sup_{\substack{u\in S, z^*\in C^*,\\ 0\in\partial f(u)+\partial(z^*g)(u)+N(S,u)}} \{f(u)+\langle z^*, g(u)\rangle\}$$

and the *Mond-Weir dual* problem

$$(D_{MW}^C) \qquad \sup_{\substack{u\in S, z^*\in C^*, \langle z^*, g(u)\rangle\geq 0,\\ 0\in\partial f(u)+\partial(z^*g)(u)+N(S,u)}} f(u).$$

Note that the feasible set of (D_{MW}^C) is included in the one of (D_W^C). The weak and strong duality statements follow.

Theorem 6.1.1. *One has* $v(D_{MW}^C) \leq v(D_W^C) \leq v(P^C)$.

Proof. We distinguish two cases. If the feasible set of (D_W^C) is empty, then so is the one of (D_{MW}^C), in which case the optimal objective values of these problems are both equal to $-\infty$, which is clearly less than or equal to $v(P^C)$.

Otherwise, let $u \in S$ and $z^* \in C^*$, fulfilling $0 \in \partial f(u) + \partial(z^*g)(u) + N(S,u)$. If $\langle z^*, g(u)\rangle \geq 0$ then (u, z^*) is feasible to (D_{MW}^C) and $f(u) \leq f(u) + \langle z^*, g(u)\rangle$. Taking now in both sides of this inequality the suprema regarding all pairs (u, z^*) feasible to (D_{MW}^C) we obtain in the left-hand side $v(D_{MW}^C)$, while in the right-hand side there is the supremum of the objective function of (D_W^C) concerning only some of the feasible solutions to this problem. Consequently, $v(D_{MW}^C) \leq v(D_W^C)$.

Since $0 \in \partial f(u) + \partial(z^*g)(u) + N(S,u)$, by (2.8) follows $0 \in \partial(f + (z^*g) + \delta_S)(u)$, i.e. for all $x \in S$ one has $f(x) + \langle z^*, g(x)\rangle \geq f(u) + \langle z^*, g(u)\rangle$. Taking in the left-hand side of this inequality the infimum regarding all $x \in S$ for which $g(x) \in -C$, we obtain there a value less than $v(P^C)$. Considering then in the right-hand side of the same inequality the supremum regarding all pairs (u, z^*) feasible to (D_W^C), it follows $v(D_W^C) \leq v(P^C)$. □

Theorem 6.1.2. *Assume that $S \subseteq X$ is a convex set, $f : X \to \overline{\mathbb{R}}$ a proper and convex function and $g : X \to \overline{Z}$ a proper and C-convex function such that* $\operatorname{dom} f \cap S \cap g^{-1}(-C) \neq \emptyset$. *If one of the regularity conditions $(RC_i^{C_L})$, $i \in \{1,2,3,4\}$ is fulfilled, (P^C) has an optimal solution $\bar{x} \in \mathcal{A}$ and one of the following additional conditions*

(i) f *and* g *are continuous at a point in* $\operatorname{dom} f \cap \operatorname{dom} g \cap S$;
(ii) $\operatorname{dom} f \cap \operatorname{dom} g \cap \operatorname{int}(S) \neq \emptyset$ *and* f *or* g *is continuous at a point in* $\operatorname{dom} f \cap \operatorname{dom} g$;
(iii) X *is a Fréchet space,* f *is lower semicontinuous,* g *is star* C*-lower semicontinuous,* S *is closed and* $0 \in \operatorname{sqri}(\operatorname{dom} f \times \operatorname{dom} g \times S - \Delta_{X^3})$;
(iv) $\dim(\operatorname{lin}(\operatorname{dom} f \times \operatorname{dom} g \times S - \Delta_{X^3})) < +\infty$ *and* $0 \in \operatorname{ri}(\operatorname{dom} f \times \operatorname{dom} g \times S - \Delta_{X^3})$;

is fulfilled, then $v(P^C) = v(D_W^C) = v(D_{MW}^C)$ *and there is some* $\bar{z}^* \in C^*$ *with* $(\bar{z}^*g)(\bar{x}) = 0$ *such that* $(\bar{x}, \bar{z}^*)$ *is an optimal solution to each of the duals.*

Proof. By Theorem 3.3.16 the hypotheses ensure the existence of some $\bar{z}^* \in C^*$ for which $f(\bar{x}) = \inf_{x \in S}\{f(x) + (\bar{z}^* g)(x)\}$ and $(\bar{z}^* g)(\bar{x}) = 0$. These yield $0 \in \partial(f + (\bar{z}^* g) + \delta_S)(\bar{x})$, which, when any of the additional conditions $(i)-(iv)$ holds, means actually, via Theorem 3.5.8, $0 \in \partial f(\bar{x}) + \partial(\bar{z}^* g)(\bar{x}) + N(S, \bar{x})$. Therefore $(\bar{x}, \bar{z}^*)$ is feasible to both (D_W^C) and (D_{MW}^C). Note also that $f(\bar{x}) = f(\bar{x}) + (\bar{z}^* g)(\bar{x})$, which implies, via Theorem 6.1.1, $v(P^C) = v(D_W^C) = v(D_{MW}^C)$ and that both duals in discussion have $(\bar{x}, \bar{z}^*)$ among their optimal solutions. □

Remark 6.1.1. If $X = \mathbb{R}^n$, $Z = \mathbb{R}^m$, $C = \mathbb{R}_+^m$, $f : \mathbb{R}^n \to \mathbb{R}$ and $g = (g_1, \ldots, g_m)^T : \mathbb{R}^n \to \mathbb{R}^m$, then (D_W^C) turns out to be the Wolfe dual problem mentioned in [109] (see also [99, 121]). In this case there is no need to assume in Theorem 6.1.2 the fulfillment of the additional conditions $(i) - (iv)$.

Remark 6.1.2. In the framework of Remark 6.1.1 assuming moreover that $S = \mathbb{R}^n$ the dual (D_W^C) turns out to be the classical *nondifferentiable Wolfe dual problem* introduced by Schechter in [166]

$$(D_W^C) \qquad \sup_{\substack{u \in \mathbb{R}^n, z^* = (z_1^*, \ldots, z_m^*)^T \in \mathbb{R}_+^m, \\ 0 \in \partial f(u) + \sum_{j=1}^m z_j^* \partial g_j(u)}} \left\{ f(u) + z^{*T} g(u) \right\},$$

while (D_{MW}^C) becomes the *nondifferentiable Mond-Weir dual problem*

$$(D_{MW}^C) \qquad \sup_{\substack{u \in \mathbb{R}^n, z^* = (z_1^*, \ldots, z_m^*)^T \in \mathbb{R}_+^m, \ z^{*T} g(u) \geq 0, \\ 0 \in \partial f(u) + \sum_{j=1}^m z_j^* \partial g_j(u)}} f(u).$$

6.1.2 Scalar Wolfe and Mond-Weir duality: differentiable case

Very important for both Wolfe and Mond-Weir duality concepts is the differentiable case, i.e. the situation when the sets involved are taken open and the functions differentiable.

Like in the previous subsection let X be a Hausdorff locally convex space and take $Z = \mathbb{R}^m$ partially ordered by the nonnegative orthant $\mathbb{R}_+^m$. Consider the nonempty open set $S \subseteq X$ and the real-valued functions $f : S \to \mathbb{R}$ and $g_j : S \to \mathbb{R}$, $j = 1, \ldots, m$, all Gâteaux differentiable on S and fulfilling $S \cap g^{-1}(-\mathbb{R}_+^m) \neq \emptyset$, where $g = (g_1, \ldots, g_m)^T : S \to \mathbb{R}^m$. Notice that one can consider also here the framework of subsection 6.1.1 by working with the extensions to the whole space of the functions involved, $\tilde{f} : X \to \overline{\mathbb{R}}$, $\tilde{f} = f + \delta_S$ and $\tilde{g} : X \to \overline{\mathbb{R}^m}$, $\tilde{g} = g + \delta_S^{\mathbb{R}^m}$. To the primal problem

$$(P^C) \quad \inf_{x \in \mathcal{A}} f(x)$$
$$\mathcal{A} = \{x \in S : g(x) \in -\mathbb{R}_+^m\}$$

we attach in this section the *differentiable Wolfe dual* problem

$$(DD_W^C) \qquad \sup_{\substack{u\in S, z^*=(z_1^*,\dots,z_m^*)^T\in\mathbb{R}_+^m,\\ \nabla f(u)+\nabla(z^{*T}g)(u)=0}} \left\{f(u)+z^{*T}g(u)\right\}$$

and the *differentiable Mond-Weir dual* problem

$$(DD_{MW}^C) \qquad \sup_{\substack{u\in S, z^*=(z_1^*,\dots,z_m^*)^T\in\mathbb{R}_+^m,\\ \nabla f(u)+\nabla(z^{*T}g)(u)=0, z^{*T}g(u)\geq 0}} f(u).$$

When the functions involved are taken Fréchet differentiable, the formulation of these duals is formally the same. Note that the the feasible set of (DD_{MW}^C) is included in the one of (DD_W^C), which has as consequence that $v(DD_{MW}^C) \leq v(DD_W^C)$. The weak and strong duality statements for the case when convexity is additionally assumed follow.

Theorem 6.1.3. *Assume that S is moreover convex and the functions f and g_j, $j=1,\dots,m$, are convex on S. Then it holds $v(DD_{MW}^C) \leq v(DD_W^C) \leq v(P^C)$.*

Proof. For $z^*=(z_1^*,\dots,z_m^*)^T \in \mathbb{R}_+^m$ and $u\in S$ there is, via Theorem 3.5.8, $\partial(z^*g)(u)=\sum_{j=1}^m \partial(z_j^*g_j)(u)=\sum_{j=1}^m z_j^*\partial g_j(u)$. By Proposition 2.3.19$(a)$ it follows that for all $u\in S$ one has $\partial f(u)=\{\nabla f(u)\}$ and $\partial g_j(u)=\{\nabla g_j(u)\}$, $j=1,\dots,m$. Moreover, $N(S,u)=\{0\}$, and we get that (DD_{MW}^C) and (DD_W^C) are actually (D_{MW}^C) and (DD_W^C), respectively. The conclusion follows by Theorem 6.1.1. □

Theorem 6.1.4. *Assume that S is moreover convex, the functions f and g_j, $j=1,\dots,m$, are convex on S, and that $0\in \operatorname{ri}(g(S)+\mathbb{R}_+^k)$. If (P^C) has an optimal solution $\bar{x}$, then $v(P^C)=v(DD_W^C)=v(DD_{MW}^C)$ and there is some $\bar{z}^*\in\mathbb{R}_+^m$ such that $(\bar{x},\bar{z}^*)$ is an optimal solution to each of the duals.*

Proof. As noted in the proof of Theorem 6.1.3, the differentiable duals to (P^C) introduced in this subsection coincide under the present hypotheses with the dual problems introduced in subsection 6.1.1, namely (D_W^C) and (D_{MW}^C), respectively. The conclusion follows directly from Theorem 6.1.2, as condition (i) from there is fulfilled. □

Remark 6.1.3. The regularity condition considered in Theorem 6.1.4 is actually $(RC_3^{C_L})$, introduced in subsection 3.2.3, written in the particular setting considered here. Note that the conclusion of the theorem remains valid if one replaces the regularity condition by any of $(RC_i^{C_L})$, $i\in\{1,2,4\}$, as all these conditions imply $(RC_3^{C_L})$ in this context. Note also that if $X=\mathbb{R}^n$ as a regularity condition in Theorem 6.1.4 one can use also $(\widetilde{RC}^{C_L})$ from Remark 3.2.6.

Remark 6.1.4. As it will be noted in the next subsection, the convexity assumptions from the hypotheses of Theorem 6.1.3 and Theorem 6.1.4 can be relaxed to some generalized convexity ones. Though, as the next example from [127] shows, such hypotheses cannot be completely removed without losing the duality for (P^C) and (DD^C_W) (or (DD^C_{MW})). Other examples sustaining this are available in [138].

Example 6.1.1. Take $X = \mathbb{R}$, $S = \mathbb{R}$ and $f, g : \mathbb{R} \to \mathbb{R}$ defined by $f(x) = -e^{-x^2}$ and $g(x) = 1 - x$. Note that f is not convex and that $v(P^C) = -1/e$. Then the optimal objective value of the differentiable Wolfe dual is

$$v(DD^C_W) = \sup_{\substack{u\in\mathbb{R}, z^*\in\mathbb{R}_+,\\ 2ue^{-u^2}=z^*}} \{-e^{-u^2} - uz^* + z^*\} = \sup_{u\in\mathbb{R}_+} \{e^{-u^2}(-1+2u-2u^2)\} = 0,$$

and this supremum is nowhere attained. Therefore, not even the weak duality for this pair of primal-dual optimization problems is guaranteed.

Remark 6.1.5. If $X = \mathbb{R}^n$ the dual (DD^C_W) turns out to be the Wolfe dual problem mentioned in [126, 127]. When moreover $S = \mathbb{R}^n_+$, (DD^C_{MW}) and (DD^C_W) are actually the dual problems treated in [138, section 3]. When $X = S = \mathbb{R}^n$, (DD^C_W) becomes the *classical (differentiable) Wolfe dual* problem introduced in [202] and mentioned also in [11], while (DD^C_{MW}) is the *classical (differentiable) Mond-Weir dual* problem introduced in [138].

Besides the ones already presented here, there were proposed in the literature other dual problems which are based on the Wolfe and Mond-Weir duality concepts, respectively. Take $X = \mathbb{R}^n$. In [126] the following Wolfe type dual problem to (P^C)

$$(DD^C_{\widetilde{W}}) \quad \sup_{\substack{u\in S, z^*=(z_1^*,\ldots,z_m^*)^T\in\mathbb{R}^m_+,\\ \nabla f(u)+\nabla\left(z^{*T}g\right)(u)\in -N(\mathcal{A},u)}} \left\{f(u) + z^{*T}g(u)\right\}$$

was treated, while in [138] were considered some different dual problems to (P^C) for the case $S = \mathbb{R}^n$. These can be formulated actually for an arbitrary nonempty open set $S \subseteq \mathbb{R}^n$, as follows. A Mond-Weir type dual is

$$(DD^C_{\widetilde{MW}}) \quad \sup_{\substack{u\in S, z^*=(z_1^*,\ldots,z_m^*)^T\in\mathbb{R}^m_+,\\ z_j^* g_j(u)\geq 0, j=1,\ldots,m,\\ \nabla f(u)+\nabla\left(z^{*T}g\right)(u)=0}} f(u),$$

while by considering the disjoint sets $J_l \subseteq \{1, \ldots, m\}$, $l = 0, \ldots, s$, such that $\cup_{l=0}^s J_l = \{1, \ldots, m\}$, the following dual problem to (P^C) was introduced,

$$(DD^C_{W-MW}) \quad \sup_{\substack{u\in S, z^*=(z_1^*,\ldots,z_m^*)^T\in\mathbb{R}^m_+,\\ \sum\limits_{j\in J_l} z_j^* g_j(u)\geq 0,\ l=1,\ldots,s,\\ \nabla f(u)+\nabla\left(z^{*T}g\right)(u)=0}} \left\{f(u) + \sum_{j\in J_0} z_j^* g_j(u)\right\}.$$

The latter dual is constructed as a "combination" of the Wolfe and Mond-Weir dual problems, which can be rediscovered as special instances of it. Taking $J_0 = \{1,\dots,m\}$, (DD^C_{W-MW}) turns into (DD^C_W), while if $J_0 = \emptyset$ and $s = 1$ it becomes (DD^C_{MW}). Also $(DD^C_{\widetilde{MW}})$ can be obtained as a particular case of (DD^C_{W-MW}) by taking $J_0 = \emptyset$, $s = m$ and each J_l, $l \in \{1,\dots,m\}$ to be a singleton. Besides these, other similar Wolfe or Mond-Weir type duals were proposed in literature to (P^C), see for instance [137], while in papers like [126, 135, 138] Wolfe or Mond-Weir type duals were considered to the problem obtained by attaching to (P^C) affine equality constraints. Since the so-obtained primal is actually a special case of (P^C) and most of the duals to it can be obtained as particular instances of (DD^C_{W-MW}) we will not mention them here.

Regarding the optimal objective values of the duals introduced above, taking into account only the way they are defined one can prove the following inequalities.

Proposition 6.1.5. *It holds*

$$v(DD^C_{\widetilde{MW}}) \leq \begin{matrix} v(DD^C_{W-MW}) \\ v(DD^C_{MW}) \end{matrix} \leq v(DD^C_W) \leq v(DD^C_{\widetilde{W}}).$$

Remark 6.1.6. For all these duals weak duality statements are not always valid in the most general case, being usually proven under some generalized convexity assumptions, as it done within the next subsection.

6.1.3 Scalar Wolfe and Mond-Weir duality under generalized convexity hypotheses

We work further in the framework of the second part of subsection 6.1.2, namely when $X = \mathbb{R}^n$, $S \subseteq \mathbb{R}^n$ is a nonempty open set, $Z = \mathbb{R}^m$ is partially ordered by the nonnegative orthant $\mathbb{R}^m_+$ and the real-valued functions $f : S \to \mathbb{R}$ and $g_j : S \to \mathbb{R}$, $j = 1,\dots,m$, are Fréchet differentiable on S and fulfill $S \cap g^{-1}(-\mathbb{R}^m_+) \neq \emptyset$, where $g = (g_1,\dots,g_m)^T$. As there is no possibility of confusion, we denote the Fréchet differential of f at x by $\nabla f(x)$, too. In order to prove duality statements for the dual optimization problems attached in subsection 6.1.2 to

$$(P^C) \quad \inf_{x\in\mathcal{A}} f(x), \qquad \mathcal{A} = \{x \in S : g(x) \leqq 0\}$$

we need to introduce some generalizations of convexity for the functions involved. We refer to [127] for the proofs of the characterizations and the existing relations between these notions.

For a nonempty convex set $U \subseteq X$ and $f : U \to \mathbb{R}$ we say that f is *quasiconvex on* U if for all $x, y \in U$ such that $f(x) < f(y)$ and all $\lambda \in (0,1)$

there is $f(\lambda x + (1-\lambda)y) < f(y)$. When $f : \mathbb{R}^n \to \mathbb{R}$ is quasiconvex on $\mathbb{R}^n$ we call it *quasiconvex*.

If $U \subseteq X$ is, additionally, open and $f : U \to \mathbb{R}$ is Fréchet differentiable, then f is quasiconvex on U if and only if for all $x, y \in U$ such that $f(y) \geq f(x)$ there is $\nabla f(y)^T(y-x) \geq 0$.

Given a nonempty open set $U \subseteq \mathbb{R}^n$, a Fréchet differentiable function $f : U \to \mathbb{R}$ is said to be *pseudoconvex on* U if for all $x, y \in U$ such that $\nabla f(x)^T(y-x) \geq 0$ one has $f(y) \geq f(x)$. When $f : \mathbb{R}^n \to \mathbb{R}$ is pseudoconvex on $\mathbb{R}^n$ we call it *pseudoconvex*. If $-f$ is pseudoconvex we call the function f *pseudoconcave*.

If $U \subseteq \mathbb{R}^n$ is nonempty, convex and open and the Fréchet differentiable function $f : U \to \mathbb{R}$ is pseudoconvex on U, then f is quasiconvex on U, too.

If $U \subseteq \mathbb{R}^n$ is an open set and a function $\eta : U \times U \to \mathbb{R}^n$ is given, a function $f : U \to \mathbb{R}$, Fréchet differentiable on U, is called *invex with respect to* η *on* U if $f(x) - f(u) \geq \nabla f(u)^T \eta(x,u)$ for all $x, u \in U$. When $f : \mathbb{R}^n \to \mathbb{R}$ is invex with respect to η on $\mathbb{R}^n$ we call it *invex with respect to* η.

Remark 6.1.7. Let be the convex set $U \subseteq \mathbb{R}^n$ and $f : U \to \mathbb{R}$ a given function. If f is convex on U, then f is quasiconvex on U, too. If U is an open set and f is convex and Fréchet differentiable on U, then it is pseudoconvex on U and also invex with respect to $\eta : U \times U \to \mathbb{R}$, $\eta(x,u) = x - u$. In general, the reverse implications fail to hold.

Now we give duality statements involving the duals mentioned in subsection 6.1.2, where the generalizations of convexity introduced above play important roles. As underlined by Example 6.1.1, in the differentiable case weak duality requires different hypotheses for the functions involved, unlike what happens when not necessarily differentiable functions are taken into consideration and the gradients are replaced by convex subgradients. Note that in many papers on Wolfe duality and Mond-Weir duality, respectively, the importance of the regularity conditions is minimal, in some of them they being not even stated or named. When convexity is not assumed, the classical regularity condition due to Kuhn and Tucker is largely used in the differentiable case. To state it consider $\bar{x} \in \mathcal{A}$ and denote the set of *active indices of* g *at* $\bar{x}$ by $I(\bar{x}) = \{j \in \{1,\ldots,m\} : g_j(\bar{x}) = 0\}$. The *Kuhn-Tucker regularity condition at* $\bar{x} \in \mathcal{A}$ is

$(RC_{KT}^C)(\bar{x})$	for all $d \in \mathbb{R}^n$ such that $\nabla g_i(\bar{x})^T d \leq 0$ for all $i \in I(\bar{x})$, there exists $\varphi : [0,1] \to \mathbb{R}^n$ differentiable at 0 such that $\varphi(0) = \bar{x}$, $\varphi([0,1]) \subseteq \mathcal{A}$ and $\nabla\varphi(0) = td$ for some $t > 0$.

As seen in the following statement (cf. [127, Theorem 7.3.7]), this regularity condition (and others that can be found in [127, Chapter 7] and in Remark 6.1.8) is sufficient for providing the Karush-Kuhn-Tucker optimality conditions for (P^C).

Lemma 6.1.6. *Assume that $\bar{x} \in \mathcal{A}$ is an optimal solution to (P^C) and that $(RC_{KT}^C)(\bar{x})$ is fulfilled. Then there exists $\bar{z}^* \in \mathbb{R}_+^m$ such that $\nabla f(\bar{x}) + \nabla\big(\bar{z}^{*T} g\big)(\bar{x}) = 0$ and $\bar{z}^{*T} g(\bar{x}) = 0$.*

Remark 6.1.8. Let us recall other regularity conditions used in the literature when dealing with duality for problems involving differentiable functions. For a given $\bar{x} \in \mathcal{A}$ we have the *Abadie constraint qualification*

$$(ACQ)(\bar{x}) \mid T(\mathcal{A}, \bar{x}) = \big\{d \in \mathbb{R}^n : \nabla g_i(\bar{x})^T d \leq 0, i \in I(\bar{x})\big\},$$

the *Mangasarian-Fromovitz constraint qualification*

$$(MFCQ)(\bar{x}) \mid \exists d \in \mathbb{R}^n : \nabla g_i(\bar{x})^T d < 0 \ \forall i \in I(\bar{x}),$$

and the *Linear Independence constraint qualification*

$$(LICQ)(\bar{x}) \mid \nabla g_i(\bar{x}),\ i \in I(\bar{x}), \text{ are linear independent.}$$

If f and g_j, $j = 1, \ldots, m$, are Fréchet continuously differentiable, one has $(LICQ)(\bar{x}) \Rightarrow (MFCQ)(\bar{x}) \Rightarrow (ACQ)(\bar{x})$ and all of them guarantee, if $\bar{x}$ is a local minimum to (P^C), the existence of a $\bar{z}^* \in \mathbb{R}_+^m$ such that $\nabla f(\bar{x}) + \nabla\big(\bar{z}^{*T} g\big)(\bar{x}) = 0$ and $\bar{z}^{*T} g(\bar{x}) = 0$. Though, we work further with $(RC_{KT}^C)(\bar{x})$ since the hypotheses it requires to work, i.e. f and g_j, $j = 1, \ldots, m$, are Fréchet differentiable on the open set S, are minimal.

In the light of Lemma 6.1.6, one can conclude that in order to give a complete duality scheme in the setting considered in this section one has only to ensure the existence of weak duality, since strong duality follows under $(RC_{KT}^C)(\bar{x})$ automatically. Next we prove the existence of weak duality and also strong duality for problems where the convexity is replaced by pseudoconvexity and quasiconvexity.

Theorem 6.1.7. *Assume that for each (u, z^*) feasible to the dual $\big(DD_{\widetilde{W}}^C\big)$ the function $f + z^{*T} g$ is pseudoconvex on S. Then $v(P^C) \geq v\big(DD_{\widetilde{W}}^C\big)$.*

Proof. Let $x \in \mathcal{A}$ and (u, z^*) be feasible to $\big(DD_{\widetilde{W}}^C\big)$. Then one has $\big(\nabla(f + z^{*T} g)(u)\big)^T (x - u) \geq 0$. Using the pseudoconvexity of $f + z^{*T} g$ on S we get $f(x) + z^{*T} g(x) \geq f(u) + z^{*T} g(u)$, which yields, taking into account that $z^* \in \mathbb{R}_+^m$ and $g(x) \in -\mathbb{R}_+^m$, $f(x) \geq f(u) + z^{*T} g(u)$. As the feasible points were chosen arbitrarily, there is weak duality for the problems in discussion. □

Theorem 6.1.8. *Assume that for each (u, z^*) feasible to the dual (DD_W^C) the function $f + z^{*T} g$ is pseudoconvex on S. Then $v(P^C) \geq v(DD_W^C)$.*

Proof. Let $x \in \mathcal{A}$ and (u, z^*) be feasible to (DD_W^C). Then $\big(\nabla(f + z^{*T} g)(u)\big)^T (x - u) = 0$ and the pseudoconvexity of $f + z^{*T} g$ on S yields $f(x) + z^{*T} g(x) \geq f(u) + z^{*T} g(u)$, which implies, taking into account that $z^* \in \mathbb{R}_+^m$ and $g(x) \in -\mathbb{R}_+^m$, $f(x) \geq f(u) + z^{*T} g(u)$. As the feasible points were chosen arbitrarily, there is weak duality for the problems in discussion. □

Analogously one can prove also the following weak duality statements.

Theorem 6.1.9. *(a) If for each (u, z^*) feasible to the dual (DD^C_{MW}) the function $f + z^{*T}g$ is pseudoconvex on S, then $v(P^C) \geq v(DD^C_{MW})$.*
(b) If for each (u, z^*) feasible to the dual $\big(DD^C_{\widetilde{MW}}\big)$ the function $f + z^{*T}g$ is pseudoconvex on S, then $v(P^C) \geq v\big(DD^C_{\widetilde{MW}}\big)$.
(c) If for each (u, z^*) feasible to the dual (DD^C_{W-MW}) the function $f + z^{*T}g$ is pseudoconvex on S, then $v(P^C) \geq v(DD^C_{W-MW})$.

Now we give the corresponding strong duality statements.

Theorem 6.1.10. *Assume that for each (u, z^*) feasible to the dual (DD^C_W) the function $f + z^{*T}g$ is pseudoconvex on S. If $\bar{x} \in \mathcal{A}$ is an optimal solution to (P^C) and the regularity condition $(RC^C_{KT})(\bar{x})$ is fulfilled, then $v(P^C) = v(DD^C_W)$ and there is a $\bar{z}^* \in \mathbb{R}^m_+$ such that $(\bar{x}, \bar{z}^*)$ is an optimal solution to the dual.*

Proof. The fulfillment of $(RC^C_{KT})(\bar{x})$ guarantees, via Lemma 6.1.6, that there is some $\bar{z}^* \in \mathbb{R}^m_+$ such that $\nabla f(\bar{x}) + \nabla\big(\bar{z}^{*T}g\big)(\bar{x}) = 0$ and $\bar{z}^{*T}g(\bar{x}) = 0$. Consequently, $(\bar{x}, \bar{z}^*)$ is feasible to (DD^C_W) and $f(\bar{x}) = f(\bar{x}) + \bar{z}^{*T}g(\bar{x})$. Using Theorem 6.1.8 we obtain $v(P^C) = v(DD^C_W)$ and that $(\bar{x}, \bar{z}^*)$ is an optimal solution to the dual. □

Analogously one can prove the following strong duality statements.

Theorem 6.1.11. *Assume that for each (u, z^*) feasible to the dual (DD^C_{MW}) the function $f + z^{*T}g$ is pseudoconvex on S. If $\bar{x} \in \mathcal{A}$ is an optimal solution to (P^C) and the regularity condition $(RC^C_{KT})(\bar{x})$ is fulfilled, then $v(P^C) = v(DD^C_{MW})$ and there is a $\bar{z}^* \in \mathbb{R}^m_+$ such that $(\bar{x}, \bar{z}^*)$ is an optimal solution to the dual.*

Theorem 6.1.12. *Assume that $\bar{x} \in \mathcal{A}$ is an optimal solution to (P^C) and the regularity condition $(RC^C_{KT})(\bar{x})$ is fulfilled.*

(a) If for each (u, z^*) feasible to the dual $\big(DD^C_{\widetilde{W}}\big)$ the function $f + z^{*T}g$ is pseudoconvex on S, then $v(P^C) = v\big(DD^C_{\widetilde{W}}\big)$ and there is a $\bar{z}^* \in \mathbb{R}^m_+$ such that $(\bar{x}, \bar{z}^*)$ is an optimal solution to the dual.
(b) If for each (u, z^*) feasible to the dual $\big(DD^C_{\widetilde{MW}}\big)$ the function $f + z^{*T}g$ is pseudoconvex on S, then $v(P^C) = v\big(DD^C_{\widetilde{MW}}\big)$ and there is a $\bar{z}^* \in \mathbb{R}^m_+$ such that $(\bar{x}, \bar{z}^*)$ is an optimal solution to the dual.
(c) If for each (u, z^*) feasible to the dual (DD^C_{W-MW}) the function $f+z^{*T}g$ is pseudoconvex on S, then $v(P^C) = v(DD^C_{W-MW})$ and there is a $\bar{z}^* \in \mathbb{R}^m_+$ such that $(\bar{x}, \bar{z}^*)$ is an optimal solution to the dual.

Other duality statements involving pseudoconvexity and quasiconvexity, valid only for the Mond-Weir type duals, follow.

Theorem 6.1.13. *Assume that the set S is moreover convex.*

(a) *If f is pseudoconvex on S and $z^{*T}g$ is quasiconvex on S whenever (u, z^*) is feasible to (DD^C_{MW}), then $v(P^C) \geq v(DD^C_{MW})$.*
(b) *If f is pseudoconvex on S and g_j is quasiconvex on S for $j = 1, \ldots, m$, then $v(P^C) \geq v\big(DD^C_{\widetilde{MW}}\big)$.*
(c) *If $f + \sum_{l \in J_0} z_l^* g_l(u)$ is pseudoconvex on S and $\sum_{j \in J_l} z_j^* g_j$ is quasiconvex on S for $l = 1, \ldots, s$, whenever (u, z^*) is feasible to (DD^C_{W-MW}), then $v(P^C) \geq v(DD^C_{W-MW})$.*

Proof. We prove only (a), the other weak duality statements following analogously. Let $x \in \mathcal{A}$ and a pair (u, z^*) feasible to (DD^C_{MW}). Then $z^{*T}(g(u) - g(x)) \geq 0$ and by the quasiconvexity on S of $z^{*T}g$ one gets $\nabla(z^{*T}g)(u)^T(u - x) \geq 0$. This is nothing but $\nabla f(u)^T(x - u) \geq 0$, which, because f is pseudoconvex on S, yields $f(x) \geq f(u)$. As the mentioned feasible points were arbitrarily chosen, by taking in the left-hand side of the last inequality the infimum after $x \in \mathcal{A}$ and in the right-hand side the supremum regarding all the pairs (u, z^*) feasible to (DD^C_{MW}), one gets $v(P^C) \geq v(DD^C_{MW})$. □

Remark 6.1.9. The hypotheses used in the latter theorem do not always ensure also the weak duality for (P^C) and (DD^C_W) or $\big(DD^C_{\widetilde{W}}\big)$. See Example 6.1.1 where the objective function is pseudoconvex and the constraint function is quasiconvex or [138] for other counter-examples.

Theorem 6.1.14. *Assume that the set S is moreover convex, $\bar{x} \in \mathcal{A}$ is an optimal solution to (P^C) and the regularity condition $(RC^C_{KT})(\bar{x})$ is fulfilled.*

(a) *If f is pseudoconvex on S and $z^{*T}g$ is quasiconvex on S whenever (u, z^*) is feasible to (DD^C_{MW}), then $v(P^C) = v(DD^C_{MW})$ and there is some $\bar{z}^* \in \mathbb{R}^m_+$ such that $(\bar{x}, \bar{z}^*)$ is an optimal solution to the dual.*
(b) *f is pseudoconvex on S and g_j is quasiconvex on S for $j = 1, \ldots, m$, then $v(P^C) = v\big(DD^C_{\widetilde{MW}}\big)$ and there is some $\bar{z}^* \in \mathbb{R}^m_+$ such that $(\bar{x}, \bar{z}^*)$ is an optimal solution to the dual.*
(c) *If $f + \sum_{l \in J_0} z_l^* g_l(u)$ is pseudoconvex on S and $\sum_{j \in J_l} z_j^* g_j$ is quasiconvex on S for $l = 1, \ldots, s$, whenever (u, z^*) is feasible to (DD^C_{W-MW}), then $v(P^C) = v(DD^C_{W-MW})$ and there is some $\bar{z}^* \in \mathbb{R}^m_+$ such that $(\bar{x}, \bar{z}^*)$ is an optimal solution to the dual.*

Proof. We prove only (a), the other strong duality statements following analogously. The fulfillment of $(RC^C_{KT})(\bar{x})$ guarantees via Lemma 6.1.6 that there is some $\bar{z}^* \in \mathbb{R}^m_+$ such that $\nabla f(\bar{x}) + \nabla\big(\bar{z}^{*T}g\big)(\bar{x}) = 0$ and $\bar{z}^{*T}g(\bar{x}) = 0$. Consequently, $(\bar{x}, \bar{z}^*)$ is feasible to (DD^C_{MW}) and $f(\bar{x})$ is a value taken by both objective functions of the primal and dual, in the corresponding feasible set, respectively. Due to Theorem 6.1.13, we obtain $v(P^C) = v(DD^C_{MW})$ and consequently $(\bar{x}, \bar{z}^*)$ is an optimal solution to the dual. □

Remark 6.1.10. The hypotheses used in the latter theorem do not always guarantee also the strong duality for (P^C) and (DD_W^C) or $(DD_{\widetilde{W}}^C)$. The regularity condition ensures the existence of a common value taken by the objective functions of the primal and dual but, as pointed out in Remark 6.1.9, the optimal objective value of the dual can surpass the one of the primal.

Another class of generalizations of convexity employed in Wolfe and Mond-Weir duality is invexity. The classical invexity was extended in many ways during the recent years, but the lack of examples justifying these notions and the complexity of the way they are introduced made us to remain in a reasonable framework by not considering them here.

Theorem 6.1.15. *Assume that the functions f and g_j, $j = 1, \ldots, m$, are invex with respect to the given function $\eta : S \times S \to \mathbb{R}^n$ on S. Then $v(P^C) \geq v(DD_W^C)$.*

Proof. Let be $x \in \mathcal{A}$ and (u, z^*) feasible to (DD_W^C). Then one has $\nabla f(u) = -\nabla(z^{*T}g)(u)$ and $z^{*T}g(x) \leq 0$. Due to the invexity of f on S there is $f(x) - (f(u) + z^{*T}g(u)) \geq \nabla f(u)^T \eta(x,u) - z^{*T}g(u) = -\nabla(z^{*T}g)(u)^T\eta(x,u) - z^{*T}g(u)$. Using now the invexity of g_j on S, $j = 1, \ldots, m$, we obtain $-\nabla(z^{*T}g)(u)^T\eta(x,u) - z^{*T}g(u) \geq -z^{*T}g(x) \geq 0$. Consequently, $f(x) - (f(u) + z^{*T}g(u)) \geq 0$. Since the feasible points were chosen arbitrarily, there is weak duality for the problems in discussion. □

Analogously one can prove the following statement.

Theorem 6.1.16. *Assume that the functions f and g_j, $j = 1, \ldots, m$, are invex with respect to the function $\eta : S \times S \to \mathbb{R}^n$, $\eta(x,u) = x - u$, on S. Then $v(P^C) \geq (DD_{\widetilde{W}}^C)$.*

Via Proposition 6.1.5, Theorem 6.1.15 yields the following consequences.

Theorem 6.1.17. *Assume that the functions f and g_j, $j = 1, \ldots, m$, are invex with respect to the given function $\eta : S \times S \to \mathbb{R}^n$ on S. Then $v(P^C) \geq v(DD_{MW}^C)$, $v(P^C) \geq (DD_{\widetilde{MW}}^C)$ and $v(P^C) \geq (DD_{W-MW}^C)$.*

To obtain strong duality we employ again the regularity condition due to Kuhn and Tucker, alongside the invexity hypotheses used for weak duality.

Theorem 6.1.18. *Assume that the functions f and g_j, $j = 1, \ldots, m$, are invex with respect to the given function $\eta : S \times S \to \mathbb{R}^n$ on S. If $\bar{x} \in \mathcal{A}$ is an optimal solution to (P^C) and the regularity condition $(RC_{KT}^C)(\bar{x})$ is fulfilled, then $v(P^C) = v(DD_W^C)$ and there is a $\bar{z}^* \in \mathbb{R}_+^m$ such that $(\bar{x}, \bar{z}^*)$ is an optimal solution to the dual.*

Proof. The fulfillment of $(RC_{KT}^C)(\bar{x})$ guarantees via Lemma 6.1.6 that there is some $\bar{z}^* \in \mathbb{R}_+^m$ such that $\nabla f(\bar{x}) + \nabla(\bar{z}^{*T}g)(\bar{x}) = 0$ and $\bar{z}^{*T}g(\bar{x}) = 0$. Consequently, $(\bar{x}, \bar{z}^*)$ is feasible to (DD_W^C) and $f(\bar{x})$ is a value taken by both objective functions, of the primal and dual, in the corresponding feasible set, respectively. Due to Theorem 6.1.15, we obtain $v(P^C) = v(DD_W^C)$ and that $(\bar{x}, \bar{z}^*)$ is an optimal solution to the dual. □

Theorem 6.1.19. *Assume that the functions f and g_j, $j = 1, \ldots, m$, are invex with respect to the function $\eta : S \times S \to \mathbb{R}^n$, $\eta(x, u) = x - u$, on S. If $\bar{x} \in \mathcal{A}$ is an optimal solution to (P^C) and the regularity condition $(RC^C_{KT})(\bar{x})$ is fulfilled, then $v(P^C) = v\big(DD^C_{\widetilde{W}}\big)$ and there is a $\bar{z}^* \in \mathbb{R}^m_+$ such that $(\bar{x}, \bar{z}^*)$ is an optimal solution to the dual.*

Theorem 6.1.20. *Assume that the functions f and g_j, $j = 1, \ldots, m$, are invex with respect to the given function $\eta : S \times S \to \mathbb{R}^n$ on S. If $\bar{x} \in \mathcal{A}$ is an optimal solution to (P^C) and the regularity condition $(RC^C_{KT})(\bar{x})$ is fulfilled, then $v(P^C) = v(DD^C_{MW}) = v\big(DD^C_{\widetilde{MW}}\big) = v(DD^C_{W-MW})$ and there is $\bar{z}^* \in \mathbb{R}^m_+$ such that $(\bar{x}, \bar{z}^*)$ is an optimal solution to each of the duals.*

Finally, we give without proofs a pair of weak and strong duality statements stated in [135] for Wolfe duality where neither convexity nor one of its generalizations are assumed for any of the sets and functions involved, being replaced by a condition imposed on the objective function of the dual.

Theorem 6.1.21. *If for each (u, z^*) feasible to (DD^C_W) there is $f(u) + z^{*T} g(u) = \inf_{x \in S} \{f(x) + z^{*T} g(x)\}$, then $v(P^C) \geq v(DD^C_W)$.*

Theorem 6.1.22. *If $\bar{x} \in \mathcal{A}$ is an optimal solution to (P^C), the regularity condition $(RC^C_{KT})(\bar{x})$ is fulfilled and for each (u, z^*) feasible to (DD^C_W) there is $f(u) + z^{*T} g(u) = \inf_{x \in S} \{f(x) + z^{*T} g(x)\}$, then $v(P^C) = v(DD^C_W)$ and there is a $\bar{z}^* \in \mathbb{R}^m_+$ such that $(\bar{x}, \bar{z}^*)$ is an optimal solution to the dual.*

Remark 6.1.11. Some of the weak and strong duality statements given in this subsection were collected from [11, 85, 126, 138]. Note also that in [128] equivalent characterizations of weak and strong duality for (P^C) and (DD^C_W) via different types of generalized invexities were given.

Remark 6.1.12. Assuming the functions f and g_j, $j = 1, \ldots, m$, Fréchet continuously differentiable on S, the strong duality statements given within this subsection remain valid when replacing the regularity condition $(RC^C_{KT})(\bar{x})$ with any of the regularity conditions considered in Remark 6.1.8.

6.2 Classical vector Wolfe and Mond-Weir duality

The Wolfe and Mond-Weir duality concepts were employed in vector optimization, too. Like in the previous section we begin with the nondifferentiable case, turning then our attention to the situation when the functions involved are assumed to be differentiable. Duality theorems are given under both convexity and generalized convexity assumptions. Note that the proper efficiency is considered in the sense of linear scalarization.

6.2.1 Vector Wolfe and Mond-Weir duality: nondifferentiable case

Let X and Z be Hausdorff locally convex spaces, the latter partially ordered by the convex cone $C \subseteq Z$, and consider the nonempty convex set $S \subseteq X$, the proper and convex functions $f_i : X \to \overline{\mathbb{R}}$, $i = 1, \ldots, k$, and the proper and C-convex function $g : X \to \overline{Z}$, fulfilling $\cap_{i=1}^k \operatorname{dom} f_i \cap S \cap g^{-1}(-C) \neq \emptyset$. Consider the vector function

$$f : X \to \overline{\mathbb{R}^k}, f(x) = \begin{cases} (f_1(x), \ldots, f_k(x))^T, & \text{if } x \in \bigcap\limits_{i=1}^k \operatorname{dom} f_i, \\ +\infty_{\mathbb{R}^k_+}, & \text{otherwise.} \end{cases}$$

Due to the hypotheses on the functions f_i, $i = 1, \ldots, k$, f is proper and $\mathbb{R}^k_+$-convex. Let be the primal vector optimization problem with geometric and cone constraints

$$(PV^C) \quad \operatorname*{Min}_{x \in \mathcal{A}} f(x).$$
$$\mathcal{A} = \{x \in S : g(x) \in -C\}$$

According to Definition 2.5.4 we say that an element $\bar{x} \in \mathcal{A}$ is properly efficient solution to (PV^C) in the sense of linear scalarization if $\bar{x} \in \cap_{i=1}^k \operatorname{dom} f_i$ and $f(\bar{x}) \in \operatorname{PMin}_{LSc}(f(\cap_{i=1}^k \operatorname{dom} f_i \cap \mathcal{A}), \mathbb{R}^k_+)$. To (PV^C) we attach the *Wolfe vector dual problem* with respect to properly efficient solutions

$$(DV^C_W) \quad \operatorname*{Max}_{(\lambda, u, z^*) \in \mathcal{B}^C_W} h^C_W(\lambda, u, z^*),$$

where

$$\mathcal{B}^C_W = \Bigg\{(\lambda, u, z^*) \in \operatorname{int}(\mathbb{R}^k_+) \times S \times C^* : \lambda = (\lambda_1, \ldots, \lambda_k)^T,$$
$$\sum_{i=1}^k \lambda_i = 1, 0 \in \partial\Bigg(\sum_{i=1}^k \lambda_i f_i\Bigg)(u) + \partial(z^* g)(u) + N(S, u)\Bigg\}$$

and

$$h^C_W(\lambda, u, z^*) = \begin{pmatrix} f_1(u) + \langle z^*, g(u)\rangle \\ \vdots \\ f_k(u) + \langle z^*, g(u)\rangle \end{pmatrix}$$

and the *Mond-Weir vector dual problem* with respect to properly efficient solutions

$$(DV^C_{MW}) \quad \operatorname*{Max}_{(\lambda, u, z^*) \in \mathcal{B}^C_{MW}} h^C_{MW}(\lambda, u, z^*),$$

where

$$\mathcal{B}^C_{MW} = \Bigg\{(\lambda, u, z^*) \in \operatorname{int}(\mathbb{R}^k_+) \times S \times C^* : \lambda = (\lambda_1, \ldots, \lambda_k)^T, \sum_{i=1}^k \lambda_i = 1,$$
$$(z^* g)(u) \geq 0, 0 \in \partial\Bigg(\sum_{i=1}^k \lambda_i f_i\Bigg)(u) + \partial(z^* g)(u) + N(S, u)\Bigg\}$$

and

$$h^C_{MW}(\lambda, u, z^*) = f(u).$$

Note that $\mathcal{B}^C_{MW} \subseteq \mathcal{B}^C_W$ and both $h^C_{MW}(\mathcal{B}^C_{MW})$ and $h^C_W(\mathcal{B}^C_W)$ are subsets of $\mathbb{R}^k$. Weak and strong duality statements follow.

Theorem 6.2.1. *There is no $x \in \mathcal{A}$ and no $(\lambda, u, z^*) \in \mathcal{B}^C_W$ such that $f_i(x) \leq h^C_{Wi}(\lambda, u, z^*)$ for all $i = 1, \dots, k$, and $f_j(x) < h^C_{Wj}(\lambda, u, z^*)$ for at least one $j \in \{1, \dots, k\}$.*

Proof. Assume that there are some $x \in \mathcal{A}$ and $(\lambda, u, z^*) \in \mathcal{B}^C_W$ such that $f_i(x) \leq h^C_{Wi}(\lambda, u, z^*)$ for all $i = 1, \dots, k$, and $f_j(x) < h^C_{Wj}(\lambda, u, z^*)$ for at least one $j \in \{1, \dots, k\}$. From here we obtain immediately $\sum_{i=1}^k \lambda_i f_i(x) < \sum_{i=1}^k \lambda_i h^C_{Wi}(\lambda, u, z^*)$. On the other hand, from the way the feasible set of the dual is defined one gets, via (2.8), $\sum_{i=1}^k \lambda_i(f_i(x) + (z^*g)(x) + \delta_S(x)) - \sum_{i=1}^k \lambda_i(f_i(u) + (z^*g)(u) + \delta_S(u)) \geq 0$. Taking into consideration that $x, u \in S$ and that $(z^*g)(x) \leq 0$, we obtain $\sum_{i=1}^k \lambda_i f_i(x) \geq \sum_{i=1}^k \lambda_i f_i(u) + (z^*g)(u) = \sum_{i=1}^k \lambda_i\, h^C_{Wi}(\lambda, u, z^*)$, which contradicts the inequality obtained in the beginning of the proof. Thus the supposition we made is false and there is weak duality for the problems in discussion. □

Note that the convexity assumptions on S, f_i, $i = 1, \dots, k$, and g are not necessary for proving the weak duality statement.

Concerning the solutions concepts we use here, recall that $(\lambda, u, z^*) \in \mathcal{B}^C_W$ is efficient to (DV^C_W) if $h^C_W(\lambda, u, z^*) \in \operatorname{Min}(h^C_W(\mathcal{B}^C_W), \mathbb{R}^k_+)$ and $(\lambda, u, z^*) \in \mathcal{B}^C_{MW}$ is efficient to (DV^C_{MW}) if $h^C_{MW}(\lambda, u, z^*) \in \operatorname{Min}(h^C_{MW}(\mathcal{B}^C_{MW}), \mathbb{R}^k_+)$, respectively.

For strong duality we use the following regularity condition (see also section 4.2)

$$(RCV^{C_L}) \;\Big|\; \exists x' \in \bigcap_{i=1}^k \operatorname{dom} f_i \cap S \text{ such that } g(x') \in -\operatorname{int}(C).$$

Theorem 6.2.2. *Assume that the regularity condition (RCV^{C_L}) is fulfilled. If $\bar{x}$ is a properly efficient solution to (PV^C) and one of the following additional conditions*

(i) f_i, $i = 1, \dots, k$, and g are continuous at a point in $\cap_{i=1}^k \operatorname{dom} f_i \cap \operatorname{dom} g \cap S$;
(ii) $\cap_{i=1}^k \operatorname{dom} f_i \cap \operatorname{dom} g \cap \operatorname{int}(S) \neq \emptyset$ and (f_i is continuous at a point in $\cap_{i=1}^k \operatorname{dom} f_i \cap \operatorname{dom} g$ for $i = 1, \dots, k$, or g is continuous at a point in $\cap_{i=1}^k \operatorname{dom} f_i \cap \operatorname{dom} g$);

is fulfilled, then there exists $(\bar{\lambda}, \bar{z}^*) \in \operatorname{int}(\mathbb{R}^k_+) \times C^*$ such that $(\bar{\lambda}, \bar{x}, \bar{z}^*)$ is an efficient solution to (DV^C_W) and $f_i(\bar{x}) = h^C_{Wi}(\bar{\lambda}, \bar{x}, \bar{z}^*)$ for $i = 1, \dots, k$.

Proof. Since $\bar{x}$ is a properly efficient solution to (PV^C), there is a $\tilde{\lambda} = (\tilde{\lambda}_1, \dots, \tilde{\lambda}_k)^T \in \operatorname{int}(\mathbb{R}^k_+)$ such that $\bar{x}$ is an optimal solution to the scalar optimization problem

$$\inf_{x \in \mathcal{A}} \sum_{i=1}^{k} \tilde{\lambda}_i f_i(x).$$

Denoting $\bar{\lambda} := (1/(\sum_{i=1}^k \tilde{\lambda}_i))\tilde{\lambda}$, it is obvious that $\bar{\lambda} = (\bar{\lambda}_1, \ldots, \bar{\lambda}_k)^T \in \operatorname{int}(\mathbb{R}^k_+)$, $\sum_{i=1}^k \bar{\lambda}_i = 1$ and $\bar{x}$ is an optimal solution to the scalar optimization problem

$$\inf_{x \in \mathcal{A}} \sum_{i=1}^{k} \bar{\lambda}_i f_i(x).$$

Now Theorem 3.3.16 ensures the existence of some $\bar{z}^* \in C^*$ for which $\sum_{i=1}^k \bar{\lambda}_i f_i\ (\bar{x}) = \inf_{x \in S}\{\sum_{i=1}^k \bar{\lambda}_i f_i(x) + (\bar{z}^* g)(x)\}$ and $(\bar{z}^* g)(\bar{x}) = 0$. These yield $0 \in \partial(\sum_{i=1}^k \bar{\lambda}_i f_i + (\bar{z}^* g) + \delta_S)(\bar{x})$, which, when one of the additional conditions $(i) - (ii)$ is fulfilled, means actually $0 \in \partial(\sum_{i=1}^k \bar{\lambda}_i f_i)(\bar{x}) + \partial(\bar{z}^* g)(\bar{x}) + N(S, \bar{x})$. Therefore $(\bar{\lambda}, \bar{x}, \bar{z}^*)$ is feasible to (DV_W^C). Note also that $f_i(\bar{x}) = f_i(\bar{x}) + (\bar{z}^* g)(\bar{x})$ for $i = 1, \ldots, k$. The efficiency of $(\bar{\lambda}, \bar{x}, \bar{z}^*)$ to (DV_W^C) follows by Theorem 6.2.1. □

Analogously one can prove similar duality statements for the Mond-Weir vector dual to (PV^C).

Theorem 6.2.3. *There is no $x \in \mathcal{A}$ and no $(\lambda, u, z^*) \in \mathcal{B}_{MW}^C$ such that $f_i(x) \leq h_{MWi}^C(\lambda, u, z^*)$ for $i = 1, \ldots, k$, and $f_j(x) < h_{MWj}^C(\lambda, u, z^*)$ for at least one $j \in \{1, \ldots, k\}$.*

Theorem 6.2.4. *Assume that the regularity condition (RCV^{C_L}) is fulfilled. If $\bar{x}$ is a properly efficient solution to (PV^C) and one of the additional conditions $(i) - (ii)$ from Theorem 6.2.2 is satisfied, then there exists $(\bar{\lambda}, \bar{z}^*) \in \operatorname{int}(\mathbb{R}^k_+) \times C^*$ such that $(\bar{\lambda}, \bar{x}, \bar{z}^*)$ is an efficient solution to (DV_{MW}^C) and $f_i(\bar{x}) = h_{MWi}^C(\bar{\lambda}, \bar{x}, \bar{z}^*)$ for $i = 1, \ldots, k$.*

Remark 6.2.1. (a) The conclusions of Theorem 6.2.2 and Theorem 6.2.4, respectively, remain valid if one replaces the regularity condition (RCV^{C_L}) by any condition that ensures the stability of the scalar optimization problem

$$\inf_{x \in \mathcal{A}} \sum_{i=1}^{k} \bar{\lambda}_i f_i(x)$$

with respect to its Lagrange dual. For instance (see also Remark 4.3.5), in case X and Z are Fréchet spaces, S is closed, f_i is lower semicontinuous, $i = 1, ..., k$, and g is C-epi closed, one can use the condition $0 \in \operatorname{sqri}(g(\cap_{i=1}^k \operatorname{dom} f_i \cap S \cap \operatorname{dom} g) + C)$, while if $\operatorname{lin}(g(\cap_{i=1}^k \operatorname{dom} f_i \cap S \cap \operatorname{dom} g) + C)$ is a finite dimensional linear subspace one can assume instead that $0 \in \operatorname{ri}(g(\cap_{i=1}^k \operatorname{dom} f_i \cap S \cap \operatorname{dom} g) + C)$.

(b) Instead of the additional conditions $(i) - (ii)$ one can consider in the last strong duality theorems, for instance, conditions similar to $(iii) - (iv)$ in Theorem 6.1.2, like (iii') X is a Fréchet space, S is closed, f_i is lower semicontinuous, $i = 1, ..., k$, g is star C-lower semicontinuous and $0 \in \operatorname{sqri}(\cap_{i=1}^k \operatorname{dom} f_i \times \operatorname{dom} g \times S - \Delta_{X^3})$;

(iv') $\dim(\operatorname{lin}(\cap_{i=1}^k \operatorname{dom} f_i \times \operatorname{dom} g \times S - \Delta_{X^3})) < +\infty$ and $0 \in \operatorname{ri}(\cap_{i=1}^k \operatorname{dom} f_i \times \operatorname{dom} g \times S - \Delta_{X^3})$.

Remark 6.2.2. When $k = 1$ the duals and the duality statements from the this subsection collapse into the corresponding ones from the scalar case.

6.2.2 Vector Wolfe and Mond-Weir duality: differentiable case

Next we assume that the functions involved in the formulation of the primal vector optimization problem (PV^C) are differentiable. Let X be a Hausdorff locally convex space, take $Z = \mathbb{R}^m$, partially ordered by the nonnegative orthant $\mathbb{R}^m_+$ and consider the nonempty open and not necessarily convex set $S \subseteq X$, while the functions $f_i : S \to \mathbb{R}$, $i = 1, \dots, k$, and $g_j : S \to \mathbb{R}$, $j = 1, \dots, m$, are assumed to be Gâteaux differentiable on S, but not necessarily convex on S. Denote $f = (f_1, \dots, f_k)^T$ and $g = (g_1, \dots, g_m)^T$ and assume that they fulfill $S \cap g^{-1}(-\mathbb{R}^m_+) \neq \emptyset$. To the primal vector optimization problem (PV^C)

$$(PV^C) \quad \operatorname*{Min}_{x \in \mathcal{A}} f(x)$$
$$\mathcal{A} = \{x \in S : g(x) \leqq 0\}$$

we attach the *differentiable Wolfe vector dual* problem with respect to properly efficient solutions

$$(DDV_W^C) \quad \operatorname*{Max}_{(\lambda,u,z^*) \in \mathcal{B}_W^{DC}} h_W^{DC}(\lambda, u, z^*),$$

where

$$\mathcal{B}_W^{DC} = \Bigg\{(\lambda, u, z^*) \in \operatorname{int}(\mathbb{R}^k_+) \times S \times \mathbb{R}^m_+ : \lambda = (\lambda_1, \dots, \lambda_k)^T,$$
$$z^* = (z_1^*, \dots, z_m^*)^T, \sum_{i=1}^k \lambda_i = 1,$$
$$\nabla\Big(\sum_{i=1}^k \lambda_i f_i\Big)(u) + \nabla\Big(\sum_{j=1}^m z_j^* g_j\Big)(u) = 0\Bigg\}$$

and

$$h_W^{DC}(\lambda, y^*, u) = \begin{pmatrix} f_1(u) + \sum\limits_{j=1}^m z_j^* g_j(u) \\ \vdots \\ f_k(u) + \sum\limits_{j=1}^m z_j^* g_j(u) \end{pmatrix},$$

and the *differentiable Mond-Weir vector dual* problem with respect to properly efficient solutions

$$(DDV^C_{MW}) \qquad \underset{(\lambda,u,z^*)\in \mathcal{B}^{DC}_{MW}}{\text{Max}} h^{DC}_{MW}(\lambda, u, z^*),$$

where

$$\mathcal{B}^{DC}_{MW} = \Bigg\{ (\lambda, u, z^*) \in \text{int}(\mathbb{R}^k_+) \times S \times \mathbb{R}^m_+ : \lambda = (\lambda_1, \dots, \lambda_k)^T, \sum_{i=1}^k \lambda_i = 1,$$
$$z^* = (z^*_1, \dots, z^*_m)^T, \sum_{j=1}^m z^*_j g_j(u) \geq 0,$$
$$\nabla\Big(\sum_{i=1}^k \lambda_i f_i\Big)(u) + \nabla\Big(\sum_{j=1}^m z^*_j g_j\Big)(u) = 0 \Bigg\}$$

and

$$h^{DC}_{MW}(\lambda, u, z^*) = f(u).$$

When the functions involved are taken Fréchet differentiable, the formulation of these duals is formally the same. Note that $\mathcal{B}^{DC}_{MW} \subseteq \mathcal{B}^{DC}_W$ and $h^{DC}_{MW}(\mathcal{B}^{DC}_{MW}), h^{DC}_W(\mathcal{B}^{DC}_W) \subseteq \mathbb{R}^k$. The weak and strong duality statements follow, first for the convex case, then for the situation when the functions involved have only some generalized convexity properties. The proofs in the convex case follow from the ones given in subsection 6.2.1.

Theorem 6.2.5. *Assume that the set S is moreover convex and the functions $f_i : S \to \mathbb{R}$, $i = 1, \dots, k$, and $g_j : S \to \mathbb{R}$, $j = 1, \dots, m$, are convex on S. Then there is no $x \in \mathcal{A}$ and no $(\lambda, u, z^*) \in \mathcal{B}^{DC}_W$ such that $f_i(x) \leq h^{DC}_{Wi}(\lambda, u, z^*)$ for $i = 1, \dots, k$, and $f_j(x) < h^{DC}_{Wj}(\lambda, u, z^*)$ for at least one $j \in \{1, \dots, k\}$.*

Theorem 6.2.6. *Assume that the set S is moreover convex, the functions $f_i : S \to \mathbb{R}$, $i = 1, \dots, k$, and $g_j : S \to \mathbb{R}$, $j = 1, \dots, m$, are convex on S and the regularity condition (RCV^{C_L}) is fulfilled. If $\bar{x}$ is a properly efficient solution to (PV^C), then there exists $(\bar{\lambda}, \bar{z}^*) \in \text{int}(\mathbb{R}^k_+) \times \mathbb{R}^m_+$ such that $(\bar{\lambda}, \bar{x}, \bar{z}^*)$ is an efficient solution to (DDV^C_W) and $f_i(\bar{x}) = h^{DC}_{Wi}(\bar{\lambda}, \bar{x}, \bar{z}^*)$ for $i = 1, \dots, k$.*

Theorem 6.2.7. *Assume that the set S is moreover convex and the functions $f_i : S \to \mathbb{R}$, $i = 1, \dots, k$, and $g_j : S \to \mathbb{R}$, $j = 1, \dots, m$, are convex on S. Then there is no $x \in \mathcal{A}$ and no $(\lambda, u, z^*) \in \mathcal{B}^{DC}_{MW}$ such that $f_i(x) \leq h^{DC}_{MWi}(\lambda, u, z^*)$ for $i = 1, \dots, k$, and $f_j(x) < h^{DC}_{MWj}(\lambda, u, z^*)$ for at least one $j \in \{1, \dots, k\}$.*

Theorem 6.2.8. *Assume that the set S is moreover convex, the functions $f_i : S \to \mathbb{R}$, $i = 1, \dots, k$, and $g_j : S \to \mathbb{R}$, $j = 1, \dots, m$, are convex on S and the regularity condition (RCV^{C_L}) is fulfilled. If $\bar{x}$ is a properly efficient solution to (PV^C), then there exists $(\bar{\lambda}, \bar{z}^*) \in \text{int}(\mathbb{R}^k_+) \times \mathbb{R}^m_+$ such that $(\bar{\lambda}, \bar{x}, \bar{z}^*)$ is an efficient solution to (DDV^C_{MW}) and $f_i(\bar{x}) = h^{DC}_{MWi}(\bar{\lambda}, \bar{x}, \bar{z}^*)$ for $i = 1, \dots, k$.*

Remark 6.2.3. Note that in both Theorem 6.2.6 and Theorem 6.2.8 the regularity condition can be replaced according to Remark 6.2.1(a).

Now let us focus on vector duality for the pairs of vector problems we considered so far when generalized convexity properties are considered for the functions in discussion. For this, let $X = \mathbb{R}^n$, $Z = \mathbb{R}^m$ and $C = \mathbb{R}^m_+$ and take the functions $f_i : S \to \mathbb{R}$, $i = 1, \dots, k$, and $g_j : S \to \mathbb{R}$, $j = 1, \dots, m$, be Fréchet differentiable on S. First, only pseudoconvexity is used.

Theorem 6.2.9. *Assume that S is moreover convex and for each $(\lambda, u, z^*) \in \mathcal{B}_W^{DC}$ the function $f_i + z^{*T} g$ is pseudoconvex on S, for $i = 1, \dots, k$. Then there is no $x \in \mathcal{A}$ and no $(\lambda, u, z^*) \in \mathcal{B}_W^{DC}$ such that $f_i(x) \leq h_{Wi}^{DC}(\lambda, u, z^*)$ for $i = 1, \dots, k$, and $f_j(x) < h_{Wj}^{DC}(\lambda, u, z^*)$ for at least one $j \in \{1, \dots, k\}$.*

Proof. Assume the contrary, i.e. that there are some $x \in \mathcal{A}$ and $(\lambda, u, z^*) \in \mathcal{B}_W^{DC}$ with $f_i(x) \leq h_{Wi}^{DC}(\lambda, u, z^*)$ for $i = 1, \dots, k$, and $f_j(x) < h_{Wj}^{DC}(\lambda, u, z^*)$ for at least one $j \in \{1, \dots, k\}$. The functions $f_i + z^{*T} g$, $i = 1, \dots, k$, being pseudoconvex on S, thus also quasiconvex on S, we get $\nabla(f_i + z^{*T} g)(u)^T (x - u) \leq 0$, $i = 1, \dots, k$. On the other hand, the pseudoconvexity on S of $f_j + z^{*T} g$ yields, since $f_j(x) < h_{Wj}^{DC}(\lambda, u, z^*)$, $\nabla(f_j + z^{*T} g)(u)^T (x - u) < 0$. Consequently, $\nabla(\sum_{i=1}^k \lambda_i f_i + z^{*T} g)(u)^T (x - u) < 0$, which contradicts one of the constraints of the vector dual problem. Consequently, there is weak duality for the problems in discussion. □

Theorem 6.2.10. *Assume that the set S is moreover convex and for each $(\lambda, u, z^*) \in \mathcal{B}_W^{DC}$ the function $f_i + z^{*T} g$ is pseudoconvex on S for $i = 1, \dots, k$. If $\bar{x}$ is a properly efficient solution to (PV^C) and the regularity condition $(RC_{KT}^C)(\bar{x})$ is fulfilled, then there exists $(\bar{\lambda}, \bar{z}^*) \in \operatorname{int}(\mathbb{R}^k_+) \times \mathbb{R}^m_+$ such that $(\bar{\lambda}, \bar{x}, \bar{z}^*)$ is an efficient solution to (DDV_W^C) and $f_i(\bar{x}) = h_{Wi}^{DC}(\bar{\lambda}, \bar{x}, \bar{z}^*)$ for $i = 1, \dots, k$.*

Proof. The satisfaction of the regularity condition ensures the existence of some $\bar{\lambda} \in \operatorname{int}(\mathbb{R}^k_+)$ and $\bar{z}^* \in \mathbb{R}^m_+$ such that $\sum_{i=1}^k \bar{\lambda}_i = 1$, $\nabla(\sum_{i=1}^k \bar{\lambda}_i f_i)(\bar{x}) + \nabla(\bar{z}^{*T} g)(\bar{x}) = 0$ and $\bar{z}^{*T} g(\bar{x}) = 0$. Thus $(\bar{\lambda}, \bar{x}, \bar{z}^*) \in \mathcal{B}_W^{DC}$. The efficiency of this element to the vector dual follows via Theorem 6.2.9. □

Similar duality statements for the differentiable Mond-Weir vector dual can be proven analogously.

Theorem 6.2.11. *Assume that the set S is moreover convex and for each $(\lambda, u, z^*) \in \mathcal{B}_{MW}^{DC}$ the function $f_i + z^{*T} g$ is pseudoconvex on S for $i = 1, \dots, k$. Then there is no $x \in \mathcal{A}$ and no $(\lambda, u, z^*) \in \mathcal{B}_{MW}^{DC}$ such that $f_i(x) \leq h_{MWi}^{DC}(\lambda, u, z^*)$ for $i = 1, \dots, k$, and $f_j(x) < h_{MWj}^{DC}(\lambda, u, z^*)$ for at least one $j \in \{1, \dots, k\}$.*

Theorem 6.2.12. *Assume that the set S is moreover convex and for each $(\lambda, u, z^*) \in \mathcal{B}_{MW}^{DC}$ the function $f_i + z^{*T} g$ is pseudoconvex on S for $i = 1, \dots, k$. If $\bar{x}$ is a properly efficient solution to (PV^C) and the regularity condition $(RC_{KT}^C)(\bar{x})$ is fulfilled, then there exists $(\bar{\lambda}, \bar{z}^*) \in \operatorname{int}(\mathbb{R}^k_+) \times \mathbb{R}^m_+$ such that $(\bar{\lambda}, \bar{x}, \bar{z}^*)$ is an efficient solution to (DDV_{MW}^C) and $f_i(\bar{x}) = h_{MWi}^{DC}(\bar{\lambda}, \bar{x}, \bar{z}^*)$ for $i = 1, \dots, k$.*

Duality statements for the differentiable Mond-Weir vector dual can be obtained under weaker hypotheses by employing also quasiconvexity.

Theorem 6.2.13. *Assume that the set S is moreover convex, f_i is pseudoconvex on S, $i = 1, \ldots, k$, and for each $(\lambda, u, z^*) \in \mathcal{B}_{MW}^{DC}$ the function $z^{*T} g$ is quasiconvex on S. Then there is no $x \in \mathcal{A}$ and no $(\lambda, u, z^*) \in \mathcal{B}_{MW}^{DC}$ such that $f_i(x) \leq h_{MWi}^{DC}(\lambda, u, z^*)$ for $i = 1, \ldots, k$, and $f_j(x) < h_{MWj}^{DC}(\lambda, u, z^*)$ for at least one $j \in \{1, \ldots, k\}$.*

Proof. Assume the contrary, i.e. that there are some $x \in \mathcal{A}$ and $(\lambda, u, z^*) \in \mathcal{B}_{MW}^{DC}$ such that $f_i(x) \leq h_{MWi}^{DC}(\lambda, u, z^*)$ for $i = 1, \ldots, k$, and $f_j(x) < h_{MWj}^{DC}(\lambda, u, z^*)$ for at least one $j \in \{1, \ldots, k\}$. The functions f_i, $i = 1, \ldots, k$, being pseudoconvex on S, thus also quasiconvex on S, we get $\nabla f_i(u)^T (x - u) \leq 0$, $i = 1, \ldots, k$. Moreover, the pseudoconvexity on S of f_j implies $\nabla f_j(u)^T (x - u) < 0$ and consequently $\nabla(\sum_{i=1}^k \lambda_i f_i)(u)^T (x - u) < 0$, that yields, using the construction of $\mathcal{B}_{MW}^{DC}$, $\nabla\big(z^{*T} g\big)(u)^T (x - u) > 0$. On the other hand, the way the feasible sets of the pair of problems in discussion are built yields $z^{*T} g(x) \leq z^{*T} g(u)$. Employing here the quasiconvexity on S of $z^{*T} g$ it follows $\nabla\big(z^{*T} g\big)(u)^T (x - u) \leq 0$, which contradicts a previously obtained inequality. Consequently, there is weak duality for the problems in discussion. □

The following strong duality statement can be proven similarly to the corresponding assertion involving Wolfe vector duality, via Theorem 6.2.13.

Theorem 6.2.14. *Assume that the set S is moreover convex, f_i is pseudoconvex on S, $i = 1, \ldots, k$, and for each $(\lambda, u, z^*) \in \mathcal{B}_{MW}^{DC}$ the function $z^{*T} g$ is quasiconvex on S. If $\bar{x}$ is a properly efficient solution to (PV^C) and the regularity condition $(RC_{KT}^C)(\bar{x})$ is fulfilled, then there exists $(\bar{\lambda}, \bar{z}^*) \in \operatorname{int}(\mathbb{R}_+^k) \times \mathbb{R}_+^m$ such that $(\bar{\lambda}, \bar{x}, \bar{z}^*)$ is an efficient solution to (DDV_{MW}^C) and $f_i(\bar{x}) = h_{MWi}^{DC}(\bar{\lambda}, \bar{x}, \bar{z}^*)$ for $i = 1, \ldots, k$.*

Remark 6.2.4. In Theorem 6.2.13 and Theorem 6.2.14 the hypothesis of pseudoconvexity on S for f_i, $i = 1, \ldots, k$, can be replaced by assuming that $\sum_{i=1}^k \lambda_i f_i$ is pseudoconvex on S whenever $(\lambda, u, z^*) \in \mathcal{B}_{MW}^{DC}$.

Vector duality statements of Wolfe type and Mond-Weir type when the functions involved are invex with respect to the same function on S are considered, too. The skipped proofs are analogous to the ones just presented in this subsection.

Theorem 6.2.15. *Assume that the functions f_i, $i = 1, \ldots, k$, and g_j, $j = 1, \ldots, m$, are invex with respect to the same $\eta : S \times S \to \mathbb{R}^n$ on S. Then there is no $x \in \mathcal{A}$ and no $(\lambda, u, z^*) \in \mathcal{B}_W^{DC}$ such that $f_i(x) \leq h_{Wi}^{DC}(\lambda, u, z^*)$ for $i = 1, \ldots, k$, and $f_j(x) < h_{Wj}^{DC}(\lambda, u, z^*)$ for at least one $j \in \{1, \ldots, k\}$.*

Proof. Assume the contrary, i.e. that there are some $x \in \mathcal{A}$ and $(\lambda, u, z^*) \in \mathcal{B}_W^{DC}$ with $f_i(x) \leq h_{Wi}^{DC}(\lambda, u, z^*)$ for $i = 1, \dots, k$, and $f_j(x) < h_{Wj}^{DC}(\lambda, u, z^*)$ for at least one $j \in \{1, \dots, k\}$. Then $\sum_{i=1}^k \lambda_i(f_i(x) - (f_i(u) + z^{*T} g(u))) < 0$. On the other hand, by using the invexity hypotheses, we get $\sum_{i=1}^k \lambda_i(f_i(x) - (f_i(u) + z^{*T} g(u))) \geq \eta(x,u)^T(\nabla(\sum_{i=1}^k \lambda_i f_i)(u) + \nabla(z^{*T} g)(u)) - z^{*T} g(x) = -z^{*T} g(x) \geq 0$, which contradicts the first obtained inequality. Consequently, there is weak duality for the problems in discussion. □

Theorem 6.2.16. *Assume that the functions f_i, $i = 1, \dots, k$, and g_j, $j = 1, \dots, m$, are invex with respect to the same $\eta : S \times S \to \mathbb{R}^n$ on S. If $\bar{x}$ is a properly efficient solution to (PV^C) and the regularity condition $(RC_{KT}^C)(\bar{x})$ is fulfilled, then there exists $(\bar{\lambda}, \bar{z}^*) \in \operatorname{int}(\mathbb{R}_+^k) \times \mathbb{R}_+^m$ such that $(\bar{\lambda}, \bar{x}, \bar{z}^*)$ is an efficient solution to (DDV_W^C) and $f_i(\bar{x}) = h_{Wi}^{DC}(\bar{\lambda}, \bar{x}, \bar{z}^*)$ for $i = 1, \dots, k$.*

Analogously, one can prove the similar statements for the Mond-Weir vector dual.

Theorem 6.2.17. *Assume that the functions f_i, $i = 1, \dots, k$, and g_j, $j = 1, \dots, m$, are invex with respect to the same $\eta : S \times S \to \mathbb{R}^n$ on S. Then there is no $x \in \mathcal{A}$ and no $(\lambda, u, z^*) \in \mathcal{B}_{MW}^{DC}$ such that $f_i(x) \leq h_{MWi}^{DC}(\lambda, u, z^*)$ for $i = 1, \dots, k$, and $f_j(x) < h_{MWj}^{DC}(\lambda, u, z^*)$ for at least one $j \in \{1, \dots, k\}$.*

Theorem 6.2.18. *Assume that the functions f_i, $i = 1, \dots, k$, and g_j, $j = 1, \dots, m$, are invex with respect to the same $\eta : S \times S \to \mathbb{R}^n$ on S. If $\bar{x}$ is a properly efficient solution to (PV^C) and the regularity condition $(RC_{KT}^C)(\bar{x})$ is fulfilled, then there exists $(\bar{\lambda}, \bar{z}^*) \in \operatorname{int}(\mathbb{R}_+^k) \times \mathbb{R}_+^m$ such that $(\bar{\lambda}, \bar{x}, \bar{z}^*)$ is an efficient solution to (DDV_{MW}^C) and $f_i(\bar{x}) = h_{MWi}^{DC}(\bar{\lambda}, \bar{x}, \bar{z}^*)$ for $i = 1, \dots, k$.*

Remark 6.2.5. The statements on Wolfe and Mond-Weir vector duality from above remain valid when the invexity hypotheses are replaced by asking $\sum_{i=1}^k \lambda_i f_i$ and $z^{*T} g$ to be invex with respect to the same η on S for all (λ, u, z^*) feasible to the corresponding vector dual.

Remark 6.2.6. In papers like [62, 191, 192] it is claimed to be proven strong duality statements for (PV^C) and its Wolfe and Mond-Weir vector dual problems where properly efficient solutions in the sense of Geoffrion are obtained for the vector duals under hypotheses similar to the ones considered in this section. We doubt that the proofs of those results are correct.

Remark 6.2.7. In papers like [61, 99, 195] there are given strong duality statements for (PV^C) and its Wolfe and Mond-Weir vector dual problems where efficient solutions are considered for the primal and efficient solutions are obtained for the vector duals. These assertions are based on the fact that $\bar{x} \in \mathcal{A}$ is efficient to (PV^C) if and only if it is an optimal solution to each of the scalar optimization problems

$$\inf_{\substack{x\in\mathcal{A},\\ f_j(x)\leq f_j(\bar{x}),\\ j\in\{1,\ldots,k\}\setminus\{i\}}} f_i(x)$$

where $i = 1,\ldots,k$. In this case the regularity conditions that ensure strong vector duality are too demanding, namely asking the stability of each of these scalar optimization problems. This is why we omit these investigations from our presentation.

Remark 6.2.8. In [6] a Mond-Weir vector dual problem to (PV^C) is proposed, where the constraint $\sum_{j=1}^m z_j^* g_j(u) \geq 0$ is replaced by $z_j^* g_j(u) \geq 0$ for $j = 1,\ldots,m$. Weak and strong duality are proven under invexity hypotheses.

Remark 6.2.9. Assuming the functions f_i, $i = 1,\ldots,k$, and g_j, $j = 1,\ldots,m$, to be Fréchet continuously differentiable on S, the strong duality statements given within this subsection remain valid when replacing the regularity condition $(RC_{KT}^C)(\bar{x})$ with any of the regularity conditions considered in Remark 6.1.8.

Remark 6.2.10. When $k = 1$ the duals and the duality statements from the this subsection collapse into the corresponding ones from the scalar case.

6.2.3 Vector Wolfe and Mond-Weir duality with respect to weakly efficient solutions

Let X and Z be Hausdorff locally convex spaces, the latter partially ordered by the convex cone $C \subseteq Z$, and consider the nonempty convex set $S \subseteq X$, the proper and convex functions $f_i : X \to \overline{\mathbb{R}}$, $i = 1,\ldots,k$, and the proper and C-convex function $g : X \to \overline{Z}$ fulfilling $\cap_{i=1}^k \operatorname{dom} f_i \cap S \cap g^{-1}(-C) \neq \emptyset$. Further, consider the vector function

$$f : X \to \overline{\mathbb{R}^k}, f(x) = \begin{cases} (f_1(x),\ldots,f_k(x))^T, & \text{if } x \in \bigcap\limits_{i=1}^k \operatorname{dom} f_i, \\ +\infty_{\mathbb{R}^k_+}, & \text{otherwise,} \end{cases}$$

which is proper and $\mathbb{R}^k_+$-convex. Let be the primal vector optimization problem with geometric and cone constraints

$$(PV_w^C) \quad \underset{x\in\mathcal{A}}{\operatorname{WMin}} f(x).$$
$$\mathcal{A} = \{x \in S : g(x) \in -C\}$$

According to Definition 2.5.1, an element $\bar{x} \in \mathcal{A}$ is said to be a weakly efficient solution to (PV_w^C) if $\bar{x} \in \cap_{i=1}^k \operatorname{dom} f_i$ and $f(\bar{x}) \in \operatorname{WMin}(f(\cap_{i=1}^k \operatorname{dom} f_i \cap \mathcal{A}), \mathbb{R}^k_+)$. To (PV_w^C) we attach the *Wolfe vector dual* problem with respect to weakly efficient solutions

$$(DV_{Ww}^C) \quad \underset{(\lambda,u,z^*)\in\mathcal{B}_{Ww}^C}{\operatorname{Max}} h_{Ww}^C(\lambda,u,z^*),$$

where

$$\mathcal{B}^C_{Ww} = \left\{(\lambda, u, z^*) \in (\mathbb{R}^k_+\backslash\{0\}) \times S \times C^* : \lambda = (\lambda_1, \ldots, \lambda_k)^T, \sum_{i=1}^k \lambda_i = 1, \right.$$
$$\left. 0 \in \partial\left(\sum_{i=1}^k \lambda_i f_i\right)(u) + \partial(z^* g)(u) + N(S, u)\right\}$$

and

$$h^C_{Ww}(\lambda, u, z^*) = \begin{pmatrix} f_1(u) + \langle z^*, g(u)\rangle \\ \vdots \\ f_k(u) + \langle z^*, g(u)\rangle \end{pmatrix},$$

and the *Mond-Weir vector dual* problem with respect to weakly efficient solutions

$$(DV^C_{MWw}) \quad \underset{(\lambda,u,z^*)\in\mathcal{B}^C_{MWw}}{\operatorname{Max}} \quad h^C_{MWw}(\lambda, u, z^*),$$

where

$$\mathcal{B}^C_{MWw} = \left\{(\lambda, u, z^*) \in (\mathbb{R}^k_+\backslash\{0\}) \times S \times C^* : \lambda = (\lambda_1, \ldots, \lambda_k)^T, \sum_{i=1}^k \lambda_i = 1, \right.$$
$$\left. (z^*g)(u) \geq 0, 0 \in \partial\left(\sum_{i=1}^k \lambda_i f_i\right)(u) + \partial(z^* g)(u) + N(S, u)\right\}$$

and

$$h^C_{MWw}(\lambda, u, z^*) = f(u).$$

Note that $\mathcal{B}^C_{MWw} \subseteq \mathcal{B}^C_{Ww}$, while $h^C_{Ww}(\mathcal{B}^C_{Ww})$ and $h^C_{MWw}(\mathcal{B}^C_{MWw})$ are subsets of $\mathbb{R}^k$. The weak and strong duality statements follow. Some proofs are skipped, being similar to the ones in subsection 6.2.1 and subsection 6.2.2

Theorem 6.2.19. *There is no $x \in \mathcal{A}$ and no $(\lambda, u, z^*) \in \mathcal{B}^C_{Ww}$ such that $f_i(x) < h^C_{Wwi}(\lambda, u, z^*)$ for all $i = 1, \ldots, k$.*

Theorem 6.2.20. *If $\bar{x}$ is a weakly efficient solution to (PV^C_w), the regularity condition (RCV^{C_L}) is fulfilled and one of the following additional conditions $(i) - (ii)$ from Theorem 6.2.2 is fulfilled, then there exists $(\bar{\lambda}, \bar{z}^*) \in (\mathbb{R}^k_+ \setminus \{0\}) \times C^*$ such that $(\bar{\lambda}, \bar{x}, \bar{z}^*)$ is a weakly efficient solution to (DV^C_{Ww}) and $f_i(\bar{x}) = h^C_{Wwi}(\bar{\lambda}, \bar{x}, \bar{z}^*)$ for $i = 1, \ldots, k$.*

Proof. Since $f(\bar{x}) \in \operatorname{WMin}(f(\cap_{i=1}^k \operatorname{dom} f_i \cap \mathcal{A}), \mathbb{R}^k_+)$, there is a $\bar{\lambda} \in \mathbb{R}^k_+\backslash\{0\}$ such that $\bar{x}$ is an optimal solution to the scalar optimization problem

$$\inf_{x\in\mathcal{A}} \sum_{i=1}^k \bar{\lambda}_i f_i(x).$$

Like in the proof of Theorem 6.2.2, one can take without loss of generality $\sum_{i=1}^k \bar{\lambda}_i = 1$. Denoting $I(\bar{\lambda}) = \{i \in \{1,\ldots,k\} : \bar{\lambda}_i > 0\} \neq \emptyset$, we obtain that $\bar{x}$ is an optimal solution to the scalar optimization problem

$$\inf_{x \in \mathcal{A}} \Big\{ \sum_{i \in I(\bar{\lambda})} \bar{\lambda}_i f_i(x) + \delta_{\bigcap\limits_{i \notin I(\bar{\lambda})} \operatorname{dom} f_i}(x) \Big\}.$$

Analogously to the proof of Theorem 6.2.2, by Theorem 3.3.16 there is $\bar{z}^* \in C^*$ for which $(\bar{z}^* g)(\bar{x}) = 0$ and $0 \in \partial\big(\sum_{i \in I(\bar{\lambda})} \bar{\lambda}_i f_i + \delta_{\cap_{i \notin I(\bar{\lambda})} \operatorname{dom} f_i} + (\bar{z}^* g) + \delta_S\big)(\bar{x})$. By one of the additional conditions (i) – (ii), one gets further $0 \in \partial\big(\sum_{i \in I(\bar{\lambda})} \bar{\lambda}_i f_i + \delta_{\cap_{i \notin I(\bar{\lambda})} \operatorname{dom} f_i}\big)(\bar{x}) + \partial(\bar{z}^* g)(\bar{x}) + N(S, \bar{x}) = \partial\big(\sum_{i=1}^k \bar{\lambda}_i f_i\big)(\bar{x}) + \partial(\bar{z}^* g)(\bar{x}) + N(S, \bar{x})$. Therefore $(\bar{\lambda}, \bar{x}, \bar{z}^*)$ is feasible to (DV_W^C). Note also that $f_i(\bar{x}) = f_i(\bar{x}) + (\bar{z}^* g)(\bar{x})$ for $i = 1,\ldots,k$. The weak efficiency of $(\bar{\lambda}, \bar{x}, \bar{z}^*)$ to (DV_W^C) follows by Theorem 6.2.19. □

Theorem 6.2.21. *There is no $x \in \mathcal{A}$ and no $(\lambda, u, z^*) \in \mathcal{B}_{MWw}^C$ such that $f_i(x) < h_{MWwi}^C(\lambda, u, z^*)$ for all $i = 1,\ldots,k$.*

Theorem 6.2.22. *If $\bar{x}$ is a weakly efficient solution to (PV_w^C), the regularity condition (RCV^{C_L}) is fulfilled and one of the following additional conditions (i) – (ii) from Theorem 6.2.2 is fulfilled, then there exists $(\bar{\lambda}, \bar{z}^*) \in (\mathbb{R}_+^k \setminus \{0\}) \times C^*$ such that $(\bar{\lambda}, \bar{x}, \bar{z}^*)$ is a weakly efficient solution to (DV_{MWw}^C) and $f_i(\bar{x}) = h_{MWwi}^C(\bar{\lambda}, \bar{x}, \bar{z}^*)$ for $i = 1,\ldots,k$.*

Remark 6.2.11. (a) Note that in both Theorem 6.2.20 and Theorem 6.2.22 the regularity condition (RCV^{C_L}) can be replaced according to Remark 6.2.1(a), while as additional condition one can assume (iv').

(b) It is worth noticing that in the strong duality theorems mentioned above one cannot use directly the eventual fulfillment of the additional condition (iii') from Remark 6.2.1, since $\delta_{\cap_{i \notin I(\bar{\lambda})} \operatorname{dom} f_i}$ is only proper and convex, but not necessarily lower semicontinuous, when f_i, $i = 1,\ldots,k$, are proper, convex and lower semicontinuous. When X and Z are Fréchet spaces, to consider additional conditions similar to $(iii)'$ from Remark 6.2.1 the reader is referred to [207].

Remark 6.2.12. In [200], where all the spaces involved are taken normed and the functions f and g are cone-convex, vector duality of Wolfe type is considered with respect to weakly efficient solutions.

For the remainder of this subsection we take $Z = \mathbb{R}^m$, $C = \mathbb{R}_+^m$ and $S \subseteq X$ an open set, not necessarily convex. First let the functions $f_i : S \to \mathbb{R}$, $i = 1,\ldots,k$, and $g_j : S \to \mathbb{R}$, $j = 1,\ldots,m$, be Gâteaux differentiable on S and not necessarily convex on S and denote $g = (g_1,\ldots,g_m)^T$.

The *differentiable Wolfe vector dual* problem with respect to weakly efficient solutions is

$$(DDV_W^C) \quad \operatorname*{Max}_{(\lambda,u,z^*)\in\mathcal{B}_{Ww}^{DC}} h_{Ww}^{DC}(\lambda,u,z^*),$$

where

$$\mathcal{B}_{Ww}^{DC} = \Bigg\{(\lambda,u,z^*) \in (\mathbb{R}_+^k\backslash\{0\}) \times S \times \mathbb{R}_+^m : \lambda = (\lambda_1,\ldots,\lambda_k)^T,$$
$$z^* = (z_1^*,\ldots,z_m^*)^T, \sum_{i=1}^k \lambda_i = 1,$$
$$\nabla\Big(\sum_{i=1}^k \lambda_i f_i\Big)(u) + \nabla\Big(\sum_{j=1}^m z_j^* g_j\Big)(u) = 0\Bigg\}$$

and

$$h_{Ww}^{DC}(\lambda,y^*,u) = \begin{pmatrix} f_1(u) + \sum\limits_{j=1}^m z_j^* g_j(u) \\ \vdots \\ f_k(u) + \sum\limits_{j=1}^m z_j^* g_j(u) \end{pmatrix},$$

and the *differentiable Mond-Weir vector dual* problem with respect to weakly efficient solutions

$$(DDV_{MW}^C) \quad \operatorname*{Max}_{(\lambda,u,z^*)\in\mathcal{B}_{MWw}^{DC}} h_{MWw}^{DC}(\lambda,u,z^*),$$

where

$$\mathcal{B}_{MWw}^{DC} = \Bigg\{(\lambda,u,z^*) \in (\mathbb{R}_+^k\backslash\{0\}) \times S \times \mathbb{R}_+^m : \lambda = (\lambda_1,\ldots,\lambda_k)^T, \sum_{i=1}^k \lambda_i = 1,$$
$$z^* = (z_1^*,\ldots,z_m^*)^T, \sum_{j=1}^m z_j^* g_j(u) \geq 0,$$
$$\nabla\Big(\sum_{i=1}^k \lambda_i f_i\Big)(u) + \nabla\Big(\sum_{j=1}^m z_j^* g_j\Big)(u) = 0\Bigg\}$$

and

$$h_{MWw}^{DC}(\lambda,u,z^*) = f(u).$$

When the functions involved are taken Fréchet differentiable the formulation of these duals is formally the same. Note that $\mathcal{B}_{MWw}^{DC} \subseteq \mathcal{B}_{Ww}^{DC}$ and $h_{MWw}^{DC}(\mathcal{B}_{MWw}^{DC}), h_{Ww}^{DC}(\mathcal{B}_{Ww}^{DC}) \subseteq \mathbb{R}^k$. Weak and strong duality statements for these vector duals follow.

Theorem 6.2.23. *Assume that the set S is moreover convex and the functions f_i, $i = 1,\ldots,k$, and g_j, $j = 1,\ldots,m$, are convex on S. Then there is no $x \in \mathcal{A}$ and no $(\lambda,u,z^*) \in \mathcal{B}_{Ww}^{DC}$ such that $f_i(x) < h_{Wwi}^{DC}(\lambda,u,z^*)$ for $i = 1,\ldots,k$.*

Theorem 6.2.24. *Assume that the set S is moreover convex and the functions f_i, $i = 1, \dots, k$, and g_j, $j = 1, \dots, m$, are convex on S. If $\bar{x}$ is a weakly efficient solution to (PV_w^C) and the regularity condition (RCV^{C_L}) is fulfilled, then there exists $(\bar{\lambda}, \bar{z}^*) \in (\mathbb{R}_+^k \setminus \{0\}) \times \mathbb{R}_+^m$ such that $(\bar{\lambda}, \bar{x}, \bar{z}^*)$ is a weakly efficient solution to (DDV_{Ww}^C) and $f_i(\bar{x}) = h_{Wwi}^{DC}(\bar{\lambda}, \bar{x}, \bar{z}^*)$ for $i = 1, \dots, k$.*

Theorem 6.2.25. *Assume that the set S is moreover convex and the functions f_i, $i = 1, \dots, k$, and g_j, $j = 1, \dots, m$, are convex on S. Then there is no $x \in \mathcal{A}$ and no $(\lambda, u, z^*) \in \mathcal{B}_{MWw}^{DC}$ such that $f_i(x) < h_{MWwi}^{DC}(\lambda, u, z^*)$ for $i = 1, \dots, k$.*

Theorem 6.2.26. *Assume that the set S is moreover convex and the functions f_i, $i = 1, \dots, k$, and g_j, $j = 1, \dots, m$, are convex on S. If $\bar{x}$ is a weakly efficient solution to (PV_w^C) and the regularity condition (RCV^{C_L}) is fulfilled, then there exists $(\bar{\lambda}, \bar{z}^*) \in (\mathbb{R}_+^k \setminus \{0\}) \times \mathbb{R}_+^m$ such that $(\bar{\lambda}, \bar{x}, \bar{z}^*)$ is a weakly efficient solution to (DV_{MWw}^{DC}) and $f_i(\bar{x}) = h_{MWwi}^{DC}(\bar{\lambda}, \bar{x}, \bar{z}^*)$ for $i = 1, \dots, k$.*

Remark 6.2.13. Note that in both Theorem 6.2.24 and Theorem 6.2.26 the regularity conditions can be replaced according to Remark 6.2.1(a).

Generalized convexity properties can be considered for the functions involved without losing the vector duality statements. Further let be $X = \mathbb{R}^n$ and the functions $f_i : S \to \mathbb{R}$, $i = 1, \dots, k$, and $g_j : S \to \mathbb{R}$, $j = 1, \dots, m$, Fréchet differentiable on S and not necessarily convex on S. We begin with the duality statements for the differentiable Wolfe vector dual.

Theorem 6.2.27. *Assume that for each $(\lambda, u, z^*) \in \mathcal{B}_{Ww}^{DC}$ the function $f_i + z^{*T} g$ is pseudoconvex on S for $i = 1, \dots, k$. Then there is no $x \in \mathcal{A}$ and no $(\lambda, u, z^*) \in \mathcal{B}_{Ww}^{DC}$ such that $f_i(x) < h_{Wwi}^{DC}(\lambda, u, z^*)$ for $i = 1, \dots, k$.*

Theorem 6.2.28. *Assume that for each $(\lambda, u, z^*) \in \mathcal{B}_{Ww}^{DC}$ the function $f_i + z^{*T} g$ is pseudoconvex on S for $i = 1, \dots, k$. If $\bar{x}$ is a weakly efficient solution to (PV_w^C) and the regularity condition $(RC_{KT}^C)(\bar{x})$ is satisfied, then there exists $(\bar{\lambda}, \bar{z}^*) \in (\mathbb{R}_+^k \setminus \{0\}) \times \mathbb{R}_+^m$ such that $(\bar{\lambda}, \bar{x}, \bar{z}^*)$ is a weakly efficient solution to (DDV_{Ww}^C) and $f_i(\bar{x}) = h_{Wwi}^{DC}(\bar{\lambda}, \bar{x}, \bar{z}^*)$ for $i = 1, \dots, k$.*

The duality statements for the differentiable Mond-Weir vector dual follow.

Theorem 6.2.29. *Assume that for each $(\lambda, u, z^*) \in \mathcal{B}_{MWw}^{DC}$ the function $f_i + z^{*T} g$ is pseudoconvex on S for $i = 1, \dots, k$. Then there is no $x \in \mathcal{A}$ and no $(\lambda, u, z^*) \in \mathcal{B}_{MWw}^{DC}$ such that $f_i(x) < h_{MWwi}^{DC}(\lambda, u, z^*)$ for $i = 1, \dots, k$.*

Theorem 6.2.30. *Assume that for each $(\lambda, u, z^*) \in \mathcal{B}_{MWw}^{DC}$ the function $f_i + z^{*T} g$ is pseudoconvex on S for $i = 1, \dots, k$. If $\bar{x}$ is a weakly efficient solution to (PV_w^C) and the regularity condition $(RC_{KT}^C)(\bar{x})$ is satisfied, then there exists $(\bar{\lambda}, \bar{z}^*) \in (\mathbb{R}_+^k \setminus \{0\}) \times \mathbb{R}_+^m$ such that $(\bar{\lambda}, \bar{x}, \bar{z}^*)$ is a weakly efficient solution to (DDV_{MWw}^C) and $f_i(\bar{x}) = h_{MWwi}^{DC}(\bar{\lambda}, \bar{x}, \bar{z}^*)$ for $i = 1, \dots, k$.*

Note that, unlike the corresponding statements from the previous subsection, in the last four assertions it was not necessary to impose the convexity of the set S. Duality statements involving the differentiable Mond-Weir vector dual can be obtained under weaker hypotheses by employing also quasiconvexity.

Theorem 6.2.31. *Assume that that the set S is moreover convex, f_i is pseudoconvex on S, $i = 1, \dots, k$, and $z^{*T} g$ is quasiconvex on S for each $(\lambda, u, z^*) \in \mathcal{B}^{DC}_{MWw}$. Then there is no $x \in \mathcal{A}$ and no $(\lambda, u, z^*) \in \mathcal{B}^{DC}_{MWw}$ such that $f_i(x) < h^{DC}_{MWwi}(\lambda, u, z^*)$ for $i = 1, \dots, k$.*

Theorem 6.2.32. *Assume that that the set S is moreover convex, f_i is pseudoconvex on S, $i = 1, \dots, k$, and $z^{*T} g$ is quasiconvex on S for each $(\lambda, u, z^*) \in \mathcal{B}^{DC}_{MWw}$. If $\bar{x}$ is a weakly efficient solution to (PV^C_w) and the regularity condition $(RC^C_{KT})(\bar{x})$ is satisfied, then there exists $(\bar{\lambda}, \bar{z}^*) \in (\mathbb{R}^k_+ \setminus \{0\}) \times \mathbb{R}^m_+$ such that $(\bar{\lambda}, \bar{x}, \bar{z}^*)$ is a weakly efficient solution to (DDV^C_{MWw}) and $f_i(\bar{x}) = h^{DC}_{MWwi}(\bar{\lambda}, \bar{x}, \bar{z}^*)$ for $i = 1, \dots, k$.*

Remark 6.2.14. In Theorem 6.2.31 and Theorem 6.2.32 the hypothesis of pseudoconvexity on S for f_i, $i = 1, \dots, k$, can be replaced by assuming that $\sum_{i=1}^k \lambda_i f_i$ is pseudoconvex on S for each $(\lambda, u, z^*) \in \mathcal{B}^{DC}_{MWw}$.

Remark 6.2.15. In [197], in the case $S = \mathbb{R}^n$, Wolfe vector duality was proven under the generalized convexity hypotheses of Theorem 6.2.27 and Mond-Weir vector duality under the ones of Theorem 6.2.31, both with respect to weakly efficient solutions.

Invexity can be employed to deliver vector duality statements in this subsection, too, and here it is not necessary to have the set S moreover convex.

Theorem 6.2.33. *Assume that the functions f_i, $i = 1, \dots, k$, and g_j, $j = 1, \dots, m$, are invex with respect to the same $\eta : S \times S \to \mathbb{R}^n$ on S. Then there is no $x \in \mathcal{A}$ and no $(\lambda, u, z^*) \in \mathcal{B}^{DC}_{Ww}$ such that $f_i(x) < h^{DC}_{Wwi}(\lambda, u, z^*)$ for $i = 1, \dots, k$.*

Theorem 6.2.34. *Assume that the functions f_i, $i = 1, \dots, k$, and g_j, $j = 1, \dots, m$, are invex with respect to the same $\eta : S \times S \to \mathbb{R}^n$ on S. If $\bar{x}$ is a weakly efficient solution to (PV^C_w) and the regularity condition $(RC^C_{KT})(\bar{x})$ is fulfilled, then there exists $(\bar{\lambda}, \bar{z}^*) \in (\mathbb{R}^k_+ \setminus \{0\}) \times \mathbb{R}^m_+$ such that $(\bar{\lambda}, \bar{x}, \bar{z}^*)$ is a weakly efficient solution to (DDV^C_{Ww}) and $f_i(\bar{x}) = h^{DC}_{Wwi}(\bar{\lambda}, \bar{x}, \bar{z}^*)$ for $i = 1, \dots, k$.*

Theorem 6.2.35. *Assume that the functions f_i, $i = 1, \dots, k$, and g_j, $j = 1, \dots, m$, are invex with respect to the same $\eta : S \times S \to \mathbb{R}^n$ on S. Then there is no $x \in \mathcal{A}$ and no $(\lambda, u, z^*) \in \mathcal{B}^{DC}_{MWw}$ such that $f_i(x) < h^{DC}_{MWwi}(\lambda, u, z^*)$ for $i = 1, \dots, k$.*

Theorem 6.2.36. *Assume that the functions f_i, $i = 1, \dots, k$, and g_j, $j = 1, \dots, m$, are invex with respect to the same $\eta : S \times S \to \mathbb{R}^n$ on S. If $\bar{x}$ is a weakly efficient solution to (PV_w^C) and the regularity condition $(RC_{KT}^C)(\bar{x})$ is fulfilled, then there exists $(\bar{\lambda}, \bar{z}^*) \in (\mathbb{R}_+^k \setminus \{0\}) \times \mathbb{R}_+^m$ such that $(\bar{\lambda}, \bar{x}, \bar{z}^*)$ is a weakly efficient solution to $(DDV_{MW_w}^C)$ and $f_i(\bar{x}) = h_{MWwi}^{DC}(\bar{\lambda}, \bar{x}, \bar{z}^*)$ for $i = 1, \dots, k$.*

Remark 6.2.16. The statements on Wolfe and Mond-Weir vector duality from above remain valid when the invexity hypotheses are replaced by asking $\sum_{i=1}^k \lambda_i f_i$ and $\sum_{j=1}^m z_j^* g_j$ to be invex with respect to the same $\eta : S \times S \to \mathbb{R}$ on S for all (λ, u, z^*) feasible to the corresponding dual.

Remark 6.2.17. In [6] another Mond-Weir vector dual problem to (PV_w^C) with respect to weakly efficient solutions is proposed, where the constraint $\sum_{j=1}^m z_j^* g_j \geq 0$ is replaced by $z_j^* g_j(u) \geq 0$ for $j = 1, \dots, m$. Weak and strong duality are proven under invexity hypotheses.

Remark 6.2.18. Assuming the functions f_i, $i = 1, \dots, k$, and g_j, $j = 1, \dots, m$, Fréchet continuously differentiable on S, the strong duality statements for the differentiable duals given within this subsection remain valid when replacing the regularity condition $(RC_{KT}^C)(\bar{x})$ with any of the regularity conditions considered in Remark 6.1.8.

Remark 6.2.19. When $k = 1$ the duals and the duality statements from the this subsection collapse into the corresponding ones from the scalar case.

Remark 6.2.20. Other vector duals of Wolfe and Mond-Weir types or closely related to them were considered in the literature, too. In [134] there is mentioned a vector dual problem to (PV^C) constructed in a similar way to the scalar dual problem $(DD_{\widetilde{MW}}^C)$. In [197] a so-called *Wolfe-Mond-Weir type vector dual* is proposed, while in [203] a similar one is proposed to a special case of (PV^C). On the other hand, in [53], working in normed spaces, a Wolfe type vector dual is considered, where the functionals that bring the constraint vector function of the primal vector minimization problem in the objective function of the dual, in our case z^*, are only linear, not also continuous. Something similar can be found for $V = \mathbb{R}^k$ and $Z = \mathbb{R}^m$ in [55], where the mentioned functionals are actually linear maps mapping C into a cone that contains K, and in [145]. Vector dual problems to (PV^C) constructed in a similar way to the scalar dual problem (DD_{W-MW}^C) can be found also in [44, 134, 191, 197].

6.3 Other Wolfe and Mond-Weir type duals and special cases

In this section we present some applications of the Wolfe and Mond-Weir duality concepts, first for constructing new dual problems for which the strong duality occurs *without the fulfillment of any regularity condition*, then for introducing dual problems whose duals are actually their corresponding primal problems, obtaining the so-called *symmetric duality*.

6.3.1 Scalar Wolfe and Mond-Weir duality without regularity conditions

Sometimes the validity of regularity conditions is not so easy to verify and different methods were proposed in order to avoid this situation without losing the strong duality for the problem in discussion. A way to overcome this difficulty is to consider stronger regularity conditions that are easier verifiable, but this method has as drawback the fact that there are situations when these stronger regularity condition are not fulfilled. Another possibility is the one presented in the following, namely to assign to the primal problem a dual for which the strong duality is automatically valid, without any additional assumption.

The scalar primal problem we investigate here is a particular case of (P^C), obtained for $X = \mathbb{R}^n$, $Z = \mathbb{R}^m$, $C = \mathbb{R}^m_+$, $S = \mathbb{R}^n$ and the convex functions $f : \mathbb{R}^n \to \mathbb{R}$ and $g_j : \mathbb{R}^n \to \mathbb{R}$, $j = 1, \ldots, m$, with $g^{-1}(\mathbb{R}^m_+) \neq \emptyset$. Denote $g = (g_1, \ldots, g_m)^T$. Then the primal problem becomes

$$(P^C) \quad \inf_{x \in \mathcal{A}} f(x).$$
$$\mathcal{A} = \{x \in \mathbb{R}^n : g(x) \leqq 0\}$$

We give first some Wolfe and Mond-Weir type duals for it for which there is strong duality without the fulfillment of a regularity condition. To do this, let be the *set of binding constraints* $Z(g) := \cap_{x \in \mathcal{A}} I(x) = \{j \in \{1, \ldots, m\} : g_j(x) = 0 \ \forall x \in \mathcal{A}\}$ and the following so-called *set of constant directions of* $Z(g)$ *at* $x \in \mathcal{A}$

$$D^{=}_{Z(g)}(x) := \{d \in \mathbb{R}^n | \exists t > 0 : g_j(x + sd) = g_j(x) \ \forall s \in [0, t) \ \forall j \in Z(g)\}.$$

For any $x \in \mathcal{A}$ the set $D^{=}_{Z(g)}(x)$ is a convex cone.

To (P^C) we attach a *Wolfe type dual* problem (cf. [140])

$$(D^{WC}_W) \quad \sup_{\substack{u \in \mathbb{R}^n, z^* = (z_1^*, \ldots, z_m^*)^T \in \mathbb{R}^m_+, \\ g_j(u) = 0 \ \forall j \in Z(g), \\ 0 \in \partial f(u) + \sum_{j=1}^m z_j^* \partial g_j(u) - (D^{=}_{Z(g)}(u))^*}} \left\{ f(u) + \sum_{j=1}^m z_j^* g_j(u) \right\},$$

and a *Mond-Weir type dual*

$$(D^{WC}_{MW}) \quad \sup_{\substack{u \in \mathbb{R}^n, z^* = (z_1^*, \ldots, z_m^*)^T \in \mathbb{R}^m_+, \\ g_j(u) = 0 \ \forall j \in Z(g), \ \sum_{j=1}^m z_j^* g_j(u) \geq 0 \\ 0 \in \partial f(u) + \sum_{j=1}^m z_j^* \partial g_j(u) - (D^{=}_{Z(g)}(u))^*}} f(u).$$

Note that the feasible set of (D_{MW}^{WC}) is included in the one of (D_W^{WC}). To give duality statements for (P^C) and these duals the following preliminary results are needed (cf. [140]).

Lemma 6.3.1. *The set* $\{x \in \mathbb{R}^n : g_j(x) = 0 \ \forall j \in Z(g)\}$ *is convex.*

Proof. Take $x, y \in \mathbb{R}^n$ such that $g_j(x) = g_j(y) = 0$ for all $j \in Z(g)$ and an arbitrary $t \in (0,1)$. Denote $z := tx + (1-t)y$. Showing that $g_j(z) = 0$ for all $j \in Z(g)$ would yield the desired convexity.

Assume to the contrary that $g_j(z) \neq 0$ for some $j \in Z(g)$. Due to the convexity of g_j we must have $g_j(z) < 0$. Thus $z \notin \mathcal{A}$. Take $w \in \operatorname{ri}(\mathcal{A})$ and construct the half-line $W := \{z + t(w-z) : t \geq 0\}$. Surely w belongs to the set $\mathcal{A} \cap W$.

Were there other points in this set, one would have $g_i(s) = 0$ for all $i \in Z(g)$ and all $s \in \mathcal{A} \cap W$. Thus g_j would take the value 0 along a segment of a line which contains z and there is also $g_j(z) < 0$. But this cannot happen for the convex function g_j, consequently $\mathcal{A} \cap W = \{w\}$. As $z \notin \mathcal{A}$, there is some $l \in \{1, \ldots, m\}$ such that $g_l(w) \leq 0$ and $g_l(z) > 0$. We cannot have $l \in Z(g)$, since this would yield $g_l(x) = g_l(y) = 0$, which together with $g_l(z) > 0$ contradicts the convexity of g_l. Then $l \in \{1, \ldots, m\} \backslash Z(g)$.

Assume now that $g_l(w) < 0$. As g_l is convex, there is some $t_l \in (0,1)$ such that $g_l(t_l w + (1-t_l)z) = 0$. From all the l's obtained as above, choose the one which delivers the largest t_l and denote it by $\bar{l}$. Then $g_l(t_{\bar{l}} w + (1-t_{\bar{l}})z) \leq 0$ for all these l's. As for any other $i \in \{1, \ldots, m\}$ we have $g_i(z) \leq 0$ and $g_i(w) \leq 0$, it follows $g_i(t_{\bar{l}} w + (1-t_{\bar{l}})z) \leq 0$, therefore $t_{\bar{l}} w + (1-t_{\bar{l}})z \in \mathcal{A}$ and, since $t_{\bar{l}} w + (1-t_{\bar{l}})z \in W$, there is another point in $\mathcal{A} \cap W$ besides w. As this cannot happen, it follows $g_l(w) = 0$. As $l \in \{1, \ldots, m\} \backslash Z(g)$, there is some point $\bar{w} \in \mathcal{A}$ such that $g_l(\bar{w}) < 0$ and, since $w \in \operatorname{ri}(\mathcal{A})$, there is a nontrivial segment on the line containing both w and $\bar{w}$ completely contained in $\operatorname{ri}(\mathcal{A})$, on which g_l takes everywhere the value 0. But g_l is convex, thus it cannot happen to take the value 0 on a segment of a line and a negative value at another point of the line. Consequently our initial assumptions is false, thus $g_i(z) = 0$ for $i \in Z(g)$. □

Lemma 6.3.2. *If* $x \in \mathcal{A}$ *and* $u \in \mathbb{R}^n$ *are such that* $g_j(u) = 0$ *for* $j \in Z(g)$, *then* $d^{*T}(x-u) \geq 0$ *whenever* $d^* \in (D^{=}_{Z(g)}(u))^*$.

Proof. It is enough to show that $g_i(u + t(x-u)) = g_i(tx + (1-t)u) = 0$ for $i \in Z(g)$ and all $t \in (0,1)$ as this yields $(x-u) \in D^{=}_{Z(g)}(u)$, fact that leads to the desired conclusion. Were it not true, then there would exist $j \in Z(g)$ and $\bar{t} \in (0,1)$ such that $g_j(u + \bar{t}(x-u)) < 0$, which contradicts the convexity of the set $\{x \in \mathbb{R}^n : g_j(x) = 0 \ \forall j \in Z(g)\}$ proven in Lemma 6.3.1. □

The proof of the following statement can be found in [14].

Lemma 6.3.3. *A necessary and sufficient condition for* $\bar{x} \in \mathcal{A}$ *to be an optimal solution to* (P^C) *is the existence of some* $z_j^* \in \mathbb{R}_+$, *where* $j \in J(g, \bar{x}) :=$

$\{i \in \{1,\dots,m\}\backslash Z(g) : g_i(\bar{x}) = 0\}$, *for which* $\partial f(\bar{x}) + \sum_{j\in J(g,\bar{x})} z_j^* \partial g_j(\bar{x}) \subseteq (D^{=}_{Z(g)}(\bar{x}))^*$.

The weak and strong duality statements follow.

Theorem 6.3.4. *One has* $v(D^{WC}_{MW}) \leq v(D^{WC}_{W}) \leq v(P^C)$.

Proof. Let be $u \in \mathbb{R}^n$ such that $g_j(u) = 0$ for $j \in Z(g)$ and $z^* = (z_1^*,\dots,z_m^*)^T \in \mathbb{R}^m_+$, fulfilling $\partial f(u) + \sum_{j=1}^m z_j^* \partial g_j(u) \cap (D^{=}_{Z(g)}(u))^* \neq \emptyset$.

If $\sum_{j=1}^m z_j^* g_j(u) \geq 0$ then (u, z^*) is feasible to (D^{WC}_{MW}) and $f(u) \leq f(u) + \sum_{j=1}^m z_j^* g_j(u)$. Taking now in both sides of this inequality the supremum regarding all pairs (u, z^*) feasible to (D^{WC}_{MW}) we obtain in the left-hand side $v(D^{WC}_{MW})$, while in the right-hand side there is the supremum of the objective function of (D^{WC}_{W}) concerning only some of the feasible solutions to this problem. Consequently, $v(D^{WC}_{MW}) \leq v(D^{WC}_{W})$.

Take now an element (u, z^*) feasible to (D^{WC}_{W}). Then there are some $u^* \in \partial f(u)$, $w^{j*} \in \partial g_j(u)$, $j = 1,\dots,m$, and $d^* \in (D^{=}_{Z(g)}(u))^*$ such that $u^* + \sum_{j=1}^m z_j^* w^{j*} = d^*$. Then for all $x \in \mathcal{A}$ we have

$$f(x) - \left(f(u) + \sum_{j=1}^m z_j^* g_j(u)\right) \geq u^{*T}(x-u) - \sum_{j=1}^m z_j^* g_j(u)$$

$$= -\sum_{j=1}^m z_j^* {w^{j*}}^T (x-u) - \sum_{j=1}^m z_j^* g_j(u) + d^{*T}(x-u)$$

$$\geq \sum_{j=1}^m z_j^*(g_j(u) - g_j(x)) - \sum_{j=1}^m z_j^* g_j(u) = -\sum_{j=1}^m z_j^* g_j(x) \geq 0.$$

As the feasible points were arbitrarily chosen, we get $v(D^{WC}_{W}) \leq v(P^C)$. □

Theorem 6.3.5. *If* $\bar{x} \in \mathcal{A}$ *is an optimal solution to* (P^C), *then* $v(P^C) = v(D^{WC}_{W})$ *and there exists* $\bar{z}^* \in \mathbb{R}^m_+$ *such that* $(\bar{x}, \bar{z}^*)$ *is an optimal solution to* (D^{WC}_{W}).

Proof. As $\bar{x}$ solves (P^C), Lemma 6.3.3 ensures the existence of some $\bar{z}_j^* \in \mathbb{R}_+$, $j \in J(g,\bar{x})$, for which $0 \in \partial f(\bar{x}) + \sum_{j\in J(g,\bar{x})} \bar{z}_j^* \partial g_j(\bar{x}) - (D^{=}_{Z(g)}(\bar{x}))^*$. Take $\bar{z}_j^* = 0$ for $j \in \{1,\dots,m\}\backslash J(g,\bar{x})$. Thus $0 \in \partial f(\bar{x}) + \sum_{j=1}^m \bar{z}_j^* \partial g_j(\bar{x}) - (D^{=}_{Z(g)}(\bar{x}))^*$ and we obtained a $\bar{z}^* \in \mathbb{R}^m_+$ such that $(\bar{x}, \bar{z}^*)$ is feasible to (D^{WC}_{W}) and $\sum_{j=1}^m \bar{z}_j^* g_j(\bar{x}) = 0$. Then $f(\bar{x}) = f(\bar{x}) + \sum_{j=1}^m \bar{z}_j^* g_j(\bar{x})$, i.e. the objective functions of the primal and dual take a common value. Employing Theorem 6.3.4 we obtain the desired conclusion. □

Theorem 6.3.6. *If* $\bar{x} \in \mathcal{A}$ *is an optimal solution to* (P^C), *then* $v(P^C) = v(D^{WC}_{MW})$ *and there exists* $\bar{z}^* \in \mathbb{R}^m_+$ *such that* $(\bar{x}, \bar{z}^*)$ *is an optimal solution to* (D^{WC}_{MW}).

Proof. Analogously to the proof of Theorem 6.3.5 we obtain a $\bar{z}^* \in \mathbb{R}^m_+$ such that $(\bar{x}, \bar{z}^*)$ is feasible to (D^{WC}_{MW}). Then the objective functions of the primal and dual take a common value, $f(\bar{x})$. Employing Theorem 6.3.4 we obtain again the conclusion. □

Remark 6.3.1. According to [140], the constraints $g_j(u) = 0$ for $j \in Z(g)$ can be replaced in the formulations of (D^{WC}_{W}) and (D^{WC}_{MW}), respectively, by $g_j(u) \leq 0$ for $j \in Z(g)$ without affecting the duality statements.

Similar assertions can be made also when the functions $f : \mathbb{R}^n \to \mathbb{R}$ and $g_j : \mathbb{R}^n \to \mathbb{R}$, $j = 1, \ldots, m$, are moreover Gâteaux differentiable. Then the subdifferentials that appear in the formulations of the duals introduced in this subsection turn into gradients and both of these duals can be obtained as special cases of the dual considered to (P^C) in [196], as an analogous to (DD^C_{W-MW}) from subsection 6.1.2. To give it, consider the disjoint sets $J_l \subseteq \{1, \ldots, m\}$, $l = 0, \ldots, s$, such that $\cup_{l=0}^s J_l = \{1, \ldots, m\}$. The dual in discussion, a *"combination" of the Wolfe and Mond-Weir duality concepts*, is

$$(DD^{WC}_{W-MW}) \quad \sup_{\substack{u \in \mathbb{R}^n, z^* = (z_1^*, \ldots, z_m^*)^T \in \mathbb{R}^m_+, \\ g_j(u) = 0 \ \forall j \in Z(g), \ \sum\limits_{j \in J_l} z_j^* g_j(u) \geq 0, \ l = 1, \ldots, s, \\ \nabla f(u) + \nabla \left(\sum\limits_{j=1}^m z_j^* g_j \right)(u) \in (D^{=}_{Z(g)}(u))^*}} \left\{ f(u) + \sum_{j \in J_0} z_j^* g_j(u) \right\}.$$

Taking $J_0 = \{1, \ldots, m\}$, $(DD^C_{W-MW^n})$ turns into a *differentiable Wolfe type dual* to (P^C),

$$(DD^{WC}_{W}) \quad \sup_{\substack{u \in \mathbb{R}^n, z^* = (z_1^*, \ldots, z_m^*)^T \in \mathbb{R}^m_+, \ g_j(u) = 0 \ \forall j \in Z(g), \\ \nabla f(u) + \nabla \left(\sum\limits_{j=1}^m z_j^* g_j \right)(u) \in (D^{=}_{Z(g)}(u))^*}} \left\{ f(u) + \sum_{j=1}^m z_j^* g_j(u) \right\},$$

while if $J_0 = \emptyset$ we obtain *differentiable Mond-Weir type duals* to (P^C), namely if for some $l \in \{1, \ldots, s\}$ there is $J_l = \{1, \ldots, m\}$ we get

$$(DD^{WC}_{MW}) \quad \sup_{\substack{u \in \mathbb{R}^n, z^* = (z_1^*, \ldots, z_m^*)^T \in \mathbb{R}^m_+, \\ g_j(u) = 0 \ \forall j \in Z(g), \ \sum\limits_{j=1}^m z_j^* g_j(u) \geq 0, \\ \nabla f(u) + \nabla \left(\sum\limits_{j=1}^m z_j^* g_j \right)(u) \in (D^{=}_{Z(g)}(u))^*}} f(u),$$

and when $s = k$ and each J_l, $l \in \{1, \ldots, m\}$ is a singleton, (DD^{WC}_{W-MW}) is

$$(DD^{WC}_{\widetilde{MW}}) \quad \sup_{\substack{u \in \mathbb{R}^n, z^* = (z_1^*, \ldots, z_m^*)^T \in \mathbb{R}^m_+, \\ g_j(u) = 0 \ \forall j \in Z(g), \ z_j^* g_j(u) \geq 0, \ j = 1, \ldots, m, \\ \nabla f(u) + \nabla \left(\sum\limits_{j=1}^m z_j^* g_j \right)(u) \in (D^{=}_{Z(g)}(u))^*}} f(u).$$

Analogously to the proofs of Theorem 6.3.4, Theorem 6.3.5 and Theorem 6.3.6 we get the following weak and strong duality statements for (P^C) and (DD^{WC}_{W-MW}), from which one can deduce weak and strong duality assertions for (DD^{WC}_{W}), (DD^{WC}_{MW}) and $(DD^{WC}_{\widetilde{MW}})$, too.

Theorem 6.3.7. *One has* $v(DD^{WC}_{W-MW}) \leq v(P^C)$.

Theorem 6.3.8. *If* $\bar{x} \in \mathcal{A}$ *is an optimal solution to* (P^C), *then* $v(P^C) = v(DD^{WC}_{W-MW})$ *and there exists* $\bar{z}^* \in \mathbb{R}^m_+$ *such that* $(\bar{x}, \bar{z}^*)$ *is an optimal solution to* (DD^{WC}_{W-MW}).

Remark 6.3.2. In [196] it is claimed to have been proven that Theorem 6.3.7 and Theorem 6.3.8 remain valid also when replacing the convexity hypotheses imposed here on f and g_j, $j = 1, \ldots, m$, by weaker assumptions of pseudoconvexity and quasiconvexity and taking these functions Fréchet differentiable. However, the proof of the weak duality statement uses Lemma 6.3.2, which is demonstrated by making use of Lemma 6.3.1, where the convexity of g_j, $j = 1, \ldots, m$, is decisive.

Remark 6.3.3. When there exists $x' \in \mathbb{R}^n$ such that $g(x') \in -\operatorname{int}(\mathbb{R}^m_+)$, in other words the classical Slater constraint qualification is fulfilled, in [14, 140] is stated that $Z(g) = \emptyset$ and, consequently, $D^{=}_{Z(g)}(u) = \mathbb{R}^n$ and $(D^{=}_{Z(g)}(u))^* = \{0\}$. In this situation the results given in this subsection turn out to collapse into the classical ones from section 6.1.

6.3.2 Vector Wolfe and Mond-Weir duality without regularity conditions

There are Wolfe type and Mond-Weir type vector duals for which the strong duality holds without regularity conditions, too. The way they are presented here has its roots in [63, 198]. The primal vector optimization problem considered there is (PV^C), formulated for $X = \mathbb{R}^n$, $Z = \mathbb{R}^m$, $S = \mathbb{R}^n$, $C = \mathbb{R}^m_+$, $V = \mathbb{R}^k$ partially ordered by the corresponding nonnegative orthant, $f = (f_1, \ldots, f_k)^T : \mathbb{R}^n \to \mathbb{R}^k$ and $g = (g_1, \ldots, g_m)^T : \mathbb{R}^n \to \mathbb{R}^m$, with f_i, $i = 1, \ldots, k$, and g_j, $j = 1, \ldots, m$, convex functions.

We give in the following Wolfe and Mond-Weir type vector duals to (PV^C) with respect to properly efficient solutions where strong duality holds without any regularity condition.

The *Wolfe type vector dual* for (PV^C) with respect to properly efficient solutions we consider here is

$$(DV^{WC}_W) \qquad \underset{(\lambda, u, z^*) \in \mathcal{B}^{WC}_W}{\operatorname{Max}} h^{WC}_W(\lambda, u, z^*),$$

where

$$\mathcal{B}_W^{WC} = \Big\{(\lambda, u, z^*) \in \operatorname{int}(\mathbb{R}^m_+) \times \mathbb{R}^n \times \mathbb{R}^m_+ : \lambda = (\lambda_1, \dots, \lambda_k)^T, \sum_{i=1}^k \lambda_i = 1,$$
$$z^* = (z_1^*, \dots, z_m^*)^T, g_j(u) = 0 \ \forall j \in Z(g),$$
$$0 \in \partial\Big(\sum_{i=1}^k \lambda_i f_i\Big)(u) + \partial\Big(\sum_{j=1}^m z_j^* g_j\Big)(u) - (D^{\equiv}_{Z(g)}(u))^*\Big\}$$

and

$$h_W^{WC}(\lambda, u, z^*) = \begin{pmatrix} f_1(u) + \sum\limits_{j=1}^m z_j^* g_j(u) \\ \vdots \\ f_k(u) + \sum\limits_{j=1}^m z_j^* g_j(u) \end{pmatrix},$$

while the corresponding *Mond-Weir type vector dual* is

$$(DV_{MW}^{WC}) \qquad \operatorname*{Max}_{(\lambda,u,z^*)\in\mathcal{B}_{MW}^{WC}} h_{MW}^{WC}(\lambda, u, z^*),$$

where

$$\mathcal{B}_{MW}^{WC} = \Big\{(\lambda, u, z^*) \in \operatorname{int}(\mathbb{R}^m_+) \times \mathbb{R}^n \times \mathbb{R}^m_+ : \lambda = (\lambda_1, \dots, \lambda_k)^T, \sum_{i=1}^k \lambda_i = 1,$$
$$z^* = (z_1^*, \dots, z_m^*)^T, g_j(u) = 0 \forall j \in Z(g), \sum_{j=1}^m z_j^* g_j(u) \geq 0,$$
$$0 \in \partial\Big(\sum_{i=1}^k \lambda_i f_i\Big)(u) + \partial\Big(\sum_{j=1}^m z_j^* g_j\Big)(u) - (D^{\equiv}_{Z(g)}(u))^*\Big\}$$

and

$$h_{MW}^{WC}(\lambda, u, z^*) = f(u).$$

The weak and strong duality statements for (PV^C) and these dual problems follow.

Theorem 6.3.9. *(a) There is no $x \in \mathcal{A}$ and no $(\lambda, u, z^*) \in \mathcal{B}_W^{WC}$ such that $f_i(x) \leq h_{Wi}^{WC}(\lambda, u, z^*)$ for $i = 1, \dots, k$, and $f_j(x) < h_{Wj}^{WC}(\lambda, u, z^*)$ for at least one $j \in \{1, \dots, k\}$.*

(b) If $\bar{x}$ is a properly efficient solution to (PV^C), then there exists $(\bar{\lambda}, \bar{z}^*) \in \operatorname{int}(\mathbb{R}^k_+) \times \mathbb{R}^m_+$ such that $(\bar{\lambda}, \bar{x}, \bar{z}^*)$ is an efficient solution to (DV_W^{WC}) and $f_i(\bar{x}) = h_{Wi}^{WC}(\bar{\lambda}, \bar{x}, \bar{z}^*)$ for $i = 1, \dots, k$.

Proof. (a) Assume that there are $x \in \mathcal{A}$ and $(\lambda, u, z^*) \in \mathcal{B}_W^{WC}$ for which $f_i(x) \leq h_{Wi}^{WC}(\lambda, u, z^*)$ for $i = 1, \dots, k$, and $f_j(x) < h_{Wj}^{WC}(\lambda, u, z^*)$ for at least one $j \in \{1, \dots, k\}$. Then $\sum_{i=1}^k \lambda_i f_i(x) < \sum_{i=1}^k \lambda_i f_i(u) + \sum_{j=1}^m z_j^* g_j(u)$. On the other hand, using the way the feasible set of the dual problem is constructed and Lemma 6.3.2, we get, like in the proof of Theorem 6.3.4,

$$\sum_{i=1}^{k} \lambda_i f_i(x) \geq \sum_{i=1}^{k} \lambda_i f_i(u) + \sum_{j=1}^{m} z_j^* g_j(u),$$

which contradicts the inequality obtained above. Consequently, the initial supposition turns out to be false and there is the desired weak duality.

(*b*) The proper efficiency of $\bar{x}$ to (PV^C) delivers a $\bar{\lambda} = (\bar{\lambda}_1, \ldots, \bar{\lambda}_k)^T \in \operatorname{int}(\mathbb{R}^k_+)$, which can be taken to fulfill $\sum_{i=1}^k \bar{\lambda}_i = 1$, such that $\bar{x}$ is an optimal solution to the problem

$$\inf_{x \in \mathcal{A}} \sum_{i=1}^{k} \bar{\lambda}_i f_i(x),$$

while by using Lemma 6.3.3 we can construct an element $\bar{z}^* = (\bar{z}_1^*, \ldots, \bar{z}_m^*)^T \in \mathbb{R}^m_+$ such that $0 \in \partial(\sum_{i=1}^k \bar{\lambda}_i f_i)(\bar{x}) + \sum_{j \in J(g,\bar{x})} \bar{z}_j^* \partial g_j(\bar{x}) - (D^{=}_{\overline{Z(g)}}(\bar{x}))^*$ and $\bar{z}_j^* = 0$, $j \in \{1, \ldots, m\} \backslash J(g, \bar{x})$. Thus $0 \in \partial(\sum_{i=1}^k \bar{\lambda}_i f_i)(\bar{x}) + \sum_{j=1}^m \bar{z}_j^* \partial g_j(\bar{x}) - (D^{=}_{\overline{Z(g)}}(\bar{x}))^*$ and $(\bar{\lambda}, \bar{x}, \bar{z}^*)$ is feasible to (DV_W^{WC}).

The efficiency of $(\bar{\lambda}, \bar{x}, \bar{z}^*)$ to (DV_W^{WC}) follows via (*a*). □

Analogously one can prove the following statement.

Theorem 6.3.10. *(a) There is no $x \in \mathcal{A}$ and no $(\lambda, u, z^*) \in \mathcal{B}_{MW}^{WC}$ such that $f_i(x) \leq h_{MWi}^{WC}(\lambda, u, z^*)$ for $i = 1, \ldots, k$, and $f_j(x) < h_{MWj}^{WC}(\lambda, u, z^*)$ for at least one $j \in \{1, \ldots, k\}$.*
(b) If $\bar{x}$ is a properly efficient solution to (PV^C), then there exists a $(\bar{\lambda}, \bar{z}^*) \in \operatorname{int}(\mathbb{R}^k_+) \times \mathbb{R}^m_+$ such that $(\bar{\lambda}, \bar{x}, \bar{z}^*)$ is an efficient solution to (DV_{MW}^{WC}) and $f_i(\bar{x}) = h_{MWi}^{WC}(\bar{\lambda}, \bar{x}, \bar{z}^*)$ for $i = 1, \ldots, k$.

Remark 6.3.4. When the functions involved are taken moreover Gâteaux differentiable one can formulate, starting eventually from the scalar differentiable duals (DD_W^{WC}), (DD_{MW}^{WC}) and $(DD_{\widetilde{MW}}^{WC})$, differentiable vector duals to (PV^C) for which strong duality holds without asking the fulfillment of any regularity condition. A Mond-Weir type such vector dual is given in [63].

Remark 6.3.5. In [63] there are introduced Wolfe and Mond-Weir type vector duals to (PV^C) with respect to efficient solutions for which strong duality holds without asking the fulfillment of any regularity condition. However, their formulation is more complicated than (DV_W^{WC}) and (DV_{MW}^{WC}), respectively, thus we do not treat them here.

Remark 6.3.6. In [63] it is claimed to prove weak and strong duality statements for the differentiable version of (DV_{MW}^{WC}) under pseudoconvexity and quasiconvexity assumptions for the functions involved. As stated in Remark 6.3.2, we doubt that these results are valid. Moreover, in [198] it is claimed that in Theorem 6.3.9 properly efficient solutions in the sense of Geoffrion to the vector dual can be obtained. We doubt that the proof given there is correct.

Remark 6.3.7. Statements similar to Remark 6.3.1 and Remark 6.3.3 hold in the vector case, too. Note also that when $k = 1$ the duals and the duality statements from the vector case collapse into the corresponding ones from the scalar case.

6.3.3 Scalar Wolfe and Mond-Weir symmetric duality

Wolfe and Mond Weir duality approaches were incorporated also in the so-called *symmetric duality*, as it is known the situation when the dual of the dual problem is the primal problem itself. Mond-Weir symmetric duality was considered since the inception of the Mond-Weir duality in [138]. Unlike the special case treated in subsection 6.3.1, the primal problem in a symmetric primal-dual pair has a special formulation, being no more the classical constrained optimization problem (P^C) or (PV^C), respectively.

We begin with the scalar case. Let be the twice Fréchet differentiable function $f : \mathbb{R}^n \times \mathbb{R}^q \to \mathbb{R}$ and the pointed convex closed cones with nonempty interiors $C_1 \subseteq \mathbb{R}^n$ and $C_2 \subseteq \mathbb{R}^q$. The gradient of f with respect to its first variable is denoted by $\nabla_x f$, while the one with respect to its second variable $\nabla_y f$. Moreover, the Hessian matrices of f with respect to the first and second variable, respectively, are denoted by $\nabla_{xx} f$ and $\nabla_{yy} f$, respectively. The *Jacobi matrix* of $\nabla_x f$ with respect to the second variable is denoted by $\nabla_{xy} f$, while the *Jacobi matrix* of $\nabla_y f$ with respect to the first variable is denoted by $\nabla_{yx} f$.

The *Wolfe type symmetric duality scheme* we propose generalizes the one treated in [136] and it consists of the primal problem

$$(P_W^S) \qquad \inf_{\substack{x \in C_1, y \in \mathbb{R}^q, \\ \nabla_y f(x,y) \in -C_2^*}} \{f(x,y) - y^T \nabla_y f(x,y)\}$$

and the dual problem

$$(D_W^S) \qquad \sup_{\substack{u \in \mathbb{R}^n, z \in C_2, \\ \nabla_x f(u,z) \in C_1^*}} \{f(u,z) - u^T \nabla_x f(u,z)\}.$$

Mond-Weir type symmetric duality schemes were considered in the literature, too. We present here a scheme of type that generalizes the one introduced in [138], consisting of the primal problem

$$(P_{MW}^S) \qquad \inf_{\substack{x \in C_1, y \in \mathbb{R}^q, \\ \nabla_y f(x,y) \in -C_2^*, \\ y^T \nabla_y f(x,y) \geq 0}} f(x,y)$$

and the dual problem

$$(D_{MW}^S) \quad \sup_{\substack{u\in\mathbb{R}^n, z\in C_2,\\ \nabla_x f(u,z)\in C_1^*,\\ u^T\nabla_x f(u,z)\leq 0}} f(u,z).$$

Remark 6.3.8. The pairs of symmetric dual problems proposed in [136, 138] are special situations of the ones considered above, obtainable when the cones C_1 and C_2 are the corresponding nonnegative orthants.

The weak and strong duality statements for these pair of problems follow. We begin with the Wolfe type one.

Theorem 6.3.11. *Assume that $f(\cdot, y)$ is convex for any fixed $y \in \mathbb{R}^q$ and $f(x,\cdot)$ is concave for any fixed $x \in \mathbb{R}^n$. Then $v(D_W^S) \leq v(P_W^S)$.*

Proof. Let (x, y) be feasible to the primal and (u, z) be feasible to the dual. The convexity hypotheses yield $f(x,z) - f(u,z) \geq (x-u)^T\nabla_x f(u,z)$ and $(z-y)^T\nabla_y f(x,y) \geq f(x,z) - f(x,y)$. Summing these relations up, one gets

$$f(x,y)-y^T\nabla_y f(x,y)-(f(u,z)-u^T\nabla_x f(u,z)) \geq x^T\nabla_x f(u,z)-z^T\nabla_y f(x,y),$$

and the term in the right-hand side is nonnegative because of the way the feasible sets of the two problems are constructed. Since the feasible points were arbitrarily chosen, it follows $v(D_W^S) \leq v(P_W^S)$. □

Theorem 6.3.12. *Assume that $f(\cdot, y)$ is convex for any fixed $y \in \mathbb{R}^q$ and $f(x,\cdot)$ is concave for any fixed $x \in \mathbb{R}^n$. If $(\bar{x},\bar{y})$ is an optimal solution to (P_W^S) and $\nabla_{yy} f(\bar{x},\bar{y})$ is positive or negative definite, then $(\bar{x},\bar{y})$ is an optimal solution to (D_W^S), too, and $v(P_W^S) = v(D_W^S)$.*

Proof. Since $(\bar{x},\bar{y})$ is an optimal solution to (P_W^S), by the Fritz John optimality conditions (see, for instance, [56, 57]) there exists a pair $(t,\tau) \in \mathbb{R}_+ \times C_2$, $(t,\tau) \neq 0$, such that

$$\begin{cases} t\nabla_x f(\bar{x},\bar{y}) + (\tau - t\bar{y})^T\nabla_{yx} f(\bar{x},\bar{y}) = 0,\\ (\tau - t\bar{y})^T\nabla_{yy} f(\bar{x},\bar{y}) = 0,\\ \tau^T\nabla_y f(\bar{x},\bar{y}) = 0,\\ \nabla_y f(\bar{x},\bar{y}) \in -C_2^*, \bar{x} \in C_1^*, \bar{y} \in \mathbb{R}^q. \end{cases}$$

As $\nabla_{yy} f(\bar{x},\bar{y})$ is positive or negative definite, multiplying the second equality from above by $\tau - t\bar{y}$ we should get in the left-hand side a nonzero value, unless if $\tau - t\bar{y} = 0$. Thus $t\bar{y} = \tau$. If $t = 0$ then $\tau = 0$, which cannot happen. Thus $t > 0$ and $\bar{y} = (1/t)\tau \in C_2$. Therefore $\bar{y}^T\nabla_y f(\bar{x},\bar{y}) = 0$. More than this, $\nabla_x f(\bar{x},\bar{y}) = 0$ and so $(\bar{x},\bar{y})$ is feasible to (D_W^S). Consequently,

$$f(\bar{x},\bar{y}) - \bar{y}^T\nabla_y f(\bar{x},\bar{y}) = f(\bar{x},\bar{y}) = f(\bar{x},\bar{y}) - \bar{x}^T\nabla_x f(\bar{x},\bar{y}),$$

i.e. at $(\bar{x},\bar{y})$ the values of the the objective functions of the primal and dual coincide. By Theorem 6.3.11 it follows that $(\bar{x},\bar{y})$ is an optimal solution to the dual and, consequently, there is strong duality. □

For weak and strong duality regarding the Mond-Weir type pair of dual problems one can consider the convexity hypotheses of Theorem 6.3.11, but they are valid under generalized convexity assumptions as shown below.

Theorem 6.3.13. *Assume that $f(\cdot, y)$ is pseudoconvex for any $y \in \mathbb{R}^q$ and $f(x, \cdot)$ is pseudoconcave for any fixed $x \in \mathbb{R}^n$. Then $v(D^S_{MW}) \leq v(P^S_{MW})$.*

Proof. Let (x, y) be feasible to the primal and (u, z) be feasible to the dual. Then $x^T \nabla_x f(u, z) \geq 0$ and $z^T \nabla_y f(x, y) \leq 0$, consequently, $(x - u)^T \nabla_x f(u, z) \geq 0$ and $(z - y)^T \nabla_y (-f(x, y)) \geq 0$. The pseudoconvexity hypotheses yield then $f(x, y) \geq f(u, y)$ and, respectively, $-f(u, z) \geq -f(u, y)$. Summing these two inequalities one gets $f(x, y) \geq f(u, z)$. Since the feasible points were arbitrarily chosen, it follows $v(D^S_W) \leq v(P^S_W)$. □

The proof of the strong duality statement is analogous to the one of Theorem 6.3.12, the only difference consisting in the weak duality statement that yields the conclusion.

Theorem 6.3.14. *Assume that $f(\cdot, y)$ is pseudoconvex for any $y \in \mathbb{R}^q$ and $f(x, \cdot)$ is pseudoconcave for any fixed $x \in \mathbb{R}^n$. If $(\bar{x}, \bar{y})$ is an optimal solution to (P^S_{MW}) and $\nabla_{yy} f(\bar{x}, \bar{y})$ is positive or negative definite, then $(\bar{x}, \bar{y})$ is an optimal solution to (D^S_{MW}), too, and $v(D^S_{MW}) = v(P^S_{MW})$.*

6.3.4 Vector Wolfe and Mond-Weir symmetric duality

Now we treat the vector case. Consider the twice Fréchet differentiable function $f = (f_1, \ldots, f_k)^T : \mathbb{R}^n \times \mathbb{R}^q \to \mathbb{R}^k$ and the pointed convex closed cones with nonempty interiors $C_1 \subseteq \mathbb{R}^n$ and $C_2 \subseteq \mathbb{R}^q$. Assume that $f_i(\cdot, y)$ is convex for any fixed $y \in \mathbb{R}^q$ and $f_i(x, \cdot)$ is concave for any fixed $x \in \mathbb{R}^n$ for $i = 1, \ldots, k$.

The *Wolfe type* pair of symmetric primal-dual vector problems with respect to efficient solutions we work with consists of

$$(PV^S_W) \quad \operatorname*{Min}_{(\lambda, x, y) \in \mathcal{A}^S_W} f^S_W(\lambda, x, y),$$

where

$$\mathcal{A}^S_W = \left\{ (\lambda, x, y) \in \operatorname{int}(\mathbb{R}^k_+) \times C_1 \times \mathbb{R}^q : \lambda = (\lambda_1, \ldots, \lambda_k)^T, \right.$$
$$\left. \sum_{i=1}^k \lambda_i = 1, \nabla_y \left(\sum_{i=1}^k \lambda_i f_i \right)(x, y) \in -C_2^* \right\}$$

and

$$f_W^S(\lambda,x,y) = \begin{pmatrix} f_1(x,y) - y^T\nabla_y\Big(\sum\limits_{i=1}^k \lambda_i f_i\Big)(x,y) \\ \vdots \\ f_k(x,y) - y^T\nabla_y\Big(\sum\limits_{i=1}^k \lambda_i f_i\Big)(x,y) \end{pmatrix}$$

and

$$(DV_W^S) \quad \underset{(\lambda,u,z)\in\mathcal{B}_W^S}{\text{Max}} \; h_W^S(\lambda,u,z),$$

where

$$\mathcal{B}_W^S = \Big\{(\lambda,u,z) \in \operatorname{int}(\mathbb{R}_+^k) \times \mathbb{R}^n \times C_2 : \lambda = (\lambda_1,\ldots,\lambda_k)^T, \sum_{i=1}^k \lambda_i = 1, \nabla_x\Big(\sum_{i=1}^k \lambda_i f_i\Big)(u,z) \in C_1^*\Big\}$$

and

$$h_W^S(\lambda,u,z) = \begin{pmatrix} f_1(u,z) - u^T\nabla_x\Big(\sum\limits_{i=1}^k \lambda_i f_i\Big)(u,z) \\ \vdots \\ f_k(u,z) - u^T\nabla_x\Big(\sum\limits_{i=1}^k \lambda_i f_i\Big)(u,z) \end{pmatrix},$$

while the corresponding *Mond-Weir type* pair contains

$$(PV_{MW}^S) \quad \underset{(\lambda,x,y)\in\mathcal{A}_{MW}^S}{\text{Min}} \; f_{MW}^S(\lambda,x,y),$$

where

$$\mathcal{A}_{MW}^S = \Big\{(\lambda,x,y) \in \operatorname{int}(\mathbb{R}_+^k) \times C_1 \times \mathbb{R}^q : \lambda = (\lambda_1,\ldots,\lambda_k)^T, \sum_{i=1}^k \lambda_i = 1, \nabla_y\Big(\sum_{i=1}^k \lambda_i f_i\Big)(x,y) \in -C_2^*, y^T\nabla_y\Big(\sum_{i=1}^k \lambda_i f_i\Big)(x,y) \geq 0\Big\}$$

and

$$f_{MW}^S(\lambda,x,y) = f(x,y)$$

and

$$(DV_{MW}^S) \quad \underset{(\lambda,u,z)\in\mathcal{B}_{MW}^S}{\text{Max}} \; h_{MW}^S(\lambda,u,z),$$

where

$$\mathcal{B}_{MW}^S = \Big\{(\lambda,u,z) \in \operatorname{int}(\mathbb{R}_+^k) \times \mathbb{R}^n \times C_2 : \lambda = (\lambda_1,\ldots,\lambda_k)^T, \sum_{i=1}^k \lambda_i = 1, \nabla_x\Big(\sum_{i=1}^k \lambda_i f_i\Big)(u,z) \in C_1^*, u^T\nabla_x\Big(\sum_{i=1}^k \lambda_i f_i\Big)(u,z) \leq 0\Big\}$$

and

$$h^S_{MW}(\lambda, u, z) = f(u, z).$$

Remark 6.3.9. Different pairs of symmetric vector problems of both Wolfe and Mond-Weir types were considered in the literature, see for instance [111,115, 116,137,139,171,199], in [116] even a "combination" of Wolfe type and Mond-Weir type pairs of symmetric vector dual problems is proposed. Some of these problems are considered with respect to properly efficient solutions, some with respect to efficient solutions and there are pairs of problems considered with respect to weakly efficient solutions, too. We choose to consider only pairs of problems with respect to efficient solutions because in all these papers the main tool for showing the coincidence of the objective values of the primal and dual at some point is vaguely mentioned, making us having doubts in the correctness of some of those results.

Remark 6.3.10. In some of the papers dealing with symmetric Wolfe or Mond-Weir type vector duality the variable λ is fixed before, being not considered as variable in any of the primal and dual symmetric vector optimization problems. However, when it comes to strong duality the proofs are not very accurate.

In the following we state weak and strong duality type statements for the primal-dual pair of Wolfe type symmetric vector problems follow. Due to the special formulation of the symmetric vector optimization problems we deal with, these assertions are not weak and strong vector duality statements as understood anywhere else in this book.

Theorem 6.3.15. *Let be $\lambda \in \text{int}(\mathbb{R}^k_+)$ with $\sum_{i=1}^k \lambda_i = 1$, such that $(\lambda, x, y) \in \mathcal{A}^S_W$ and $(\lambda, u, z) \in \mathcal{B}^S_W$. Then it is not possible to have $f^S_{Wi}(\lambda, x, y) \leq h^S_{Wi}(\lambda, u, z)$ for $i = 1, \dots, k$ and $f^S_{Wj}(\lambda, x, y) < h^S_{Wj}(\lambda, u, z)$ for at least one $j \in \{1, \dots, k\}$.*

Proof. Assume that it is possible to choose $(\lambda, x, y) \in \mathcal{A}^S_W$ and $(\lambda, u, z) \in \mathcal{B}^S_W$ such that $f^S_{Wi}(\lambda, x, y) \leq h^S_{Wi}(\lambda, u, z)$ for $i = 1, \dots, k$ and $f^S_{Wj}(\lambda, x, y) \leq h^S_{Wj}(\lambda, u, z)$ for at least one $j \in \{1, \dots, k\}$. Then

$$\sum_{i=1}^k \lambda_i f_i(x, y) - y^T \nabla_y \Big(\sum_{i=1}^k \lambda_i f_i \Big)(x, y) - \sum_{i=1}^k \lambda_i f_i(u, z) + u^T \nabla_x \Big(\sum_{i=1}^k \lambda_i f_i \Big)(u, z) < 0. \tag{6.1}$$

On the other hand, the convexity hypotheses yield

$$\sum_{i=1}^k \lambda_i f_i(x, z) - \sum_{i=1}^k \lambda_i f_i(u, z) \geq (x - u)^T \nabla_x \left(\sum_{i=1}^k \lambda_i f_i \right)(u, z)$$

and, respectively,

$$(z-y)^T \nabla_y \left(\sum_{i=1}^k \lambda_i f_i \right)(x,y) \geq \sum_{i=1}^k \lambda_i f_i(x,z) - \sum_{i=1}^k \lambda_i f_i(x,y).$$

Summing these relations up, one gets

$$\sum_{i=1}^k \lambda_i f_i(x,y) - y^T \nabla_y \left(\sum_{i=1}^k \lambda_i f_i \right)(x,y) -$$

$$\left(\sum_{i=1}^k \lambda_i f_i(u,z) - u^T \nabla_x \left(\sum_{i=1}^k \lambda_i f_i(u,z) \right) \right) \geq$$

$$x^T \nabla_x \left(\sum_{i=1}^k \lambda_i f_i \right)(u,z) - z^T \nabla_y \left(\sum_{i=1}^k \lambda_i f_i \right)(x,y) \geq 0,$$

where the last inequality follows because of the way the feasible sets of the two problems are constructed. This contradicts (6.1), thus the desired conclusion follows. □

The following result from [171] delivers necessary *Fritz-John optimality conditions* for a vector optimization problem, being useful for the strong duality type statement following it.

Lemma 6.3.16. *Let $S \subseteq \mathbb{R}^n$ be a nonempty convex set, $C \subseteq \mathbb{R}^m$ a convex closed cone with nonempty interior and $f = (f_1, \ldots, f_k)^T : \mathbb{R}^n \to \mathbb{R}^k$ and $g = (g_1, \ldots, g_m)^T : \mathbb{R}^n \to \mathbb{R}^m$ vector-valued functions such that f_i, $i = 1, \ldots, k$, and g_j, $j = 1, \ldots, m$, are Fréchet differentiable. If $\bar{x} \in \mathcal{A}$ is a weakly efficient solution to*

$$(PV_w^C) \quad \underset{x \in \mathcal{A}}{\mathrm{WMin}}\, f(x),$$
$$\mathcal{A} = \{x \in S : g(x) \in -C\}$$

then there exist $\alpha = (\alpha_1, \ldots, \alpha_k)^T \in \mathbb{R}^k_+$ and $\beta = (\beta_1, \ldots, \beta_m)^T \in C^*$ with $(\alpha, \beta) \neq (0,0)$ fulfilling the following optimality conditions

(i) $\nabla\big(\sum_{i=1}^k \alpha_i f_i\big)(\bar{x})^T(x - \bar{x}) + \nabla\big(\sum_{j=1}^m \beta_j g_j\big)(\bar{x})^T(x - \bar{x}) \geq 0$ *for all $x \in S$;*
(ii) $\sum_{j=1}^m \beta_j g_j(\bar{x}) = 0$.

Now we give a strong duality type statement for (PV_W^S) and (DV_W^S).

Theorem 6.3.17. *Let $(\bar{\lambda}, \bar{x}, \bar{y})$ be an efficient solution to (PV_W^S). If the matrix $\nabla_{yy}(\sum_{i=1}^k \bar{\lambda}_i f_i)(\bar{x}, \bar{y})$ is positive or negative definite and the vectors $\{\nabla_y f_i(\bar{x}, \bar{y}) : i = 1, \ldots, k\}$ are linearly independent, then $(\bar{\lambda}, \bar{x}, \bar{y}) \in \mathcal{B}_W^S$ and $f_{Wi}^S(\bar{\lambda}, \bar{x}, \bar{y}) = h_{Wi}^S(\bar{\lambda}, \bar{x}, \bar{y})$ for $i = 1, \ldots, k$.*

Proof. Since $(\bar{\lambda}, \bar{x}, \bar{y})$ is an efficient solution to (PV_W^S), it is also a weakly efficient solution to (PV_W^S). The assumptions made on the cone C_2 ensure that $\text{int}(C_2^*) \neq \emptyset$. We can apply Lemma 6.3.16 and in this way we obtain a pair $(\alpha, \beta) \in \mathbb{R}^k_+ \times C_2 \setminus \{(0,0)\}$ such that

$$\left(\beta - (\alpha^T e)\,\bar{y}\right)^T \nabla_y f(\bar{x},\bar{y})^T(\lambda - \bar{\lambda}) \geq 0 \ \forall \lambda \in \text{int}(\mathbb{R}^k_+) \text{ with } \sum_{i=1}^k \lambda_i = 1, \quad (6.2)$$

$$\left(\alpha^T \nabla_x f(\bar{x},\bar{y})^T + (\beta - (\alpha^T e)\bar{y})^T \nabla_{yx}\Big(\sum_{i=1}^k \bar{\lambda}_i f_i\Big)(\bar{x},\bar{y})^T\right)(x - \bar{x}) \geq 0 \ \forall x \in C_1, \quad (6.3)$$

$$(\alpha - (\alpha^T e)\bar{\lambda})^T \nabla_y f(\bar{x},\bar{y}) + (\beta - (\alpha^T e)\bar{y})^T \nabla_{yy}\Big(\sum_{i=1}^k \bar{\lambda}_i f_i\Big)(\bar{x},\bar{y}) = 0 \quad (6.4)$$

and

$$\beta^T \nabla_y \Big(\sum_{i=1}^k \bar{\lambda}_i f_i\Big)(\bar{x},\bar{y}) = 0. \quad (6.5)$$

Assume that $\alpha = 0$. Then from (6.4) follows that $\beta^T \nabla_{yy}\big(\sum_{i=1}^k \bar{\lambda}_i f_i\big)(\bar{x},\bar{y}) = 0$ and, consequently, $\beta^T \nabla_{yy}\big(\sum_{i=1}^k \bar{\lambda}_i f_i\big)(\bar{x},\bar{y})\beta = 0$. Taking into consideration the assumption made on $\nabla_{yy}(\sum_{i=1}^k \bar{\lambda}_i f_i)(\bar{x},\bar{y})$, it follows that $\beta = 0$ and this leads to a contradiction. Consequently, $\alpha \neq 0$.

One can easily see that from (6.2) we obtain that there exists $\gamma \in \mathbb{R}$ such that $\nabla_y f(\bar{x},\bar{y})\big(\beta - (\alpha^T e)\bar{y}\big) = \gamma e$. Thus multiplying (6.4) from the right with $(\beta - (\alpha^T e)\bar{y})$ it follows that

$$(\beta - (\alpha^T e)\bar{y})^T \nabla_{yy}\Big(\sum_{i=1}^k \bar{\lambda}_i f_i\Big)(\bar{x},\bar{y})(\beta - (\alpha^T e)\bar{y}) = 0,$$

which, because of the positive or negative definiteness of $\nabla_{yy}(\sum_{i=1}^k \bar{\lambda}_i f_i)(\bar{x},\bar{y})$, yields $\beta - (\alpha^T e)\bar{y} = 0$. This guarantees that $\bar{y} \in C_2$ and, via (6.5),

$$\bar{y}^T \nabla_y \Big(\sum_{i=1}^k \bar{\lambda}_i f_i\Big)(\bar{x},\bar{y}) = 0.$$

Further, (6.4) turns out to be $(\alpha - (\alpha^T e)\bar{\lambda})^T \nabla_y f(\bar{x},\bar{y}) = 0$. As the vectors $\{\nabla_y f_i(\bar{x},\bar{y}) : i = 1,\ldots,k\}$ are linearly independent, we must have $\alpha - (\alpha^T e)\bar{\lambda} = 0$. Writing (6.3) after getting these relations, it becomes

$$\nabla_x \Big(\sum_{i=1}^k \bar{\lambda}_i f_i\Big)(\bar{x},\bar{y})^T (x - \bar{x}) \geq 0 \ \forall x \in C_1$$

and from here one has

$$x^T \nabla_x \Big(\sum_{i=1}^k \bar{\lambda}_i f_i\Big)(\bar{x},\bar{y}) \geq 0 \ \forall x \in C_1.$$

Consequently,

$$\bar{x}^T \nabla_x \left(\sum_{i=1}^{k} \bar{\lambda}_i f_i \right)(\bar{x}, \bar{y}) = 0.$$

Since C_1 is a convex closed cone, it yields $\nabla_x(\sum_{i=1}^k \bar{\lambda}_i f_i)(\bar{x},\bar{y}) \in C_1^*$ and so $(\bar{\lambda},\bar{x},\bar{y}) \in \mathcal{B}_W^S$. More than that, as one can easily verify, $f_W^S(\bar{\lambda},\bar{x},\bar{y}) = h_W^S(\bar{\lambda},\bar{x},\bar{y})$. □

For the pair of Mond-Weir type problems the weak and strong duality type statements can be proven analogously.

Theorem 6.3.18. *Let be $\lambda \in \operatorname{int}(\mathbb{R}_+^k)$ with $\sum_{i=1}^k \lambda_i = 1$, such that $(\lambda,x,y) \in \mathcal{A}_{MW}^S$ and $(\lambda,u,z) \in \mathcal{B}_{MW}^S$. Then it is not possible to have $f_{MWi}^S(\lambda,x,y) \leq h_{MWi}^S(\lambda,u,z)$ for $i=1,\dots,k$ and $f_{MWj}^S(\lambda,x,y) < h_{MWj}^S(\lambda,u,z)$ for at least one $j \in \{1,\dots,k\}$.*

Theorem 6.3.19. *Let $(\bar{\lambda},\bar{x},\bar{y})$ be an efficient solution to (PV_{MW}^S). If the matrix $\nabla_{yy}(\sum_{i=1}^k \bar{\lambda}_i f_i)(\bar{x},\bar{y})$ is positive or negative definite and the vectors $\{\nabla_y f_i(\bar{x},\bar{y}) : i=1,\dots,k\}$ are linearly independent, then $(\bar{\lambda},\bar{x},\bar{y}) \in \mathcal{B}_{MW}^S$ and $f_{MWi}^S(\bar{\lambda},\bar{x},\bar{y}) = h_{MWi}^S(\bar{\lambda},\bar{x},\bar{y})$ for $i=1,\dots,k$.*

Remark 6.3.11. A special case of symmetric duality is the so-called *self-duality*, namely the situation when the dual can be rewritten in a form which coincides with the one of the primal problem. This can be obtained for instance when the objective functions of the primal symmetric duals are skew symmetric. We refer to [115,138,171] for more on Wolfe type and Mond-Weir type self-duality.

Remark 6.3.12. Using some remarks from [139,199], one can notice that carefully choosing the function f, the primal-dual pairs considered within this subsection become special cases of the primal-dual pairs of vector problems from subsection 6.2.3.

6.4 Wolfe and Mond-Weir fractional duality

6.4.1 Wolfe and Mond-Weir duality in scalar fractional programming

Wolfe and Mond-Weir duality concepts were used in fractional programming, too, even though not directly due to the special way such problems look like. There are different ways to attach a dual to the primal scalar fractional programming problem

$$(PQ^C) \quad \inf_{x \in \mathcal{A}^Q} \frac{f(x)}{h(x)},$$
$$\mathcal{A}^Q = \{x \in S : g(x) \in -\mathbb{R}_+^m\}$$

where $S \subseteq \mathbb{R}^n$ is a nonempty convex set, $f : \mathbb{R}^n \to \mathbb{R}$ and $g_j : \mathbb{R}^n \to \mathbb{R}$, $j = 1, \ldots, m$, are convex functions and $h : \mathbb{R}^n \to \mathbb{R}$ is a concave function such that $S \cap g^{-1}(-C) \neq \emptyset$, where $g = (g_1, \ldots, g_m)^T$. Assume that for all $x \in S$ it holds $f(x) \geq 0$ and $h(x) > 0$. In general (PQ^C) is not a convex problem. The two main approaches used in the literature to assign dual problems to it are (cf. [13,194]) the one due to Jagannathan and Schaible, which can be applied also for nondifferentiable functions and, respectively, the one of Bector, where differentiability for the functions involved is essential.

Jagannathan (cf. [100]) and Schaible (cf. [164]) have considered a dual based on *Dinkelbach's classical approach* (cf. [59]) to fractional programming problems, namely

$$(DQ^C_{W-JS}) \qquad \sup_{\substack{u \in S, z^* \in \mathbb{R}^m_+, z^* = (z_1^*, \ldots, z_m^*)^T, t \in \mathbb{R}_+, \\ f(u) - th(u) + \sum_{j=1}^m z_j^* g_j(u) \geq 0, \\ 0 \in \partial f(u) + t\partial(-h)(u) + \partial\left(\sum_{j=1}^m z_j^* g_j\right)(u) + N(S,u)}} t.$$

This dual is of *Wolfe type* and one of *Mond-Weir type* can be attached to (PQ^C) by this approach, too, namely

$$(DQ^C_{MW-JS}) \qquad \sup_{\substack{u \in S, z^* \in \mathbb{R}^m_+, z^* = (z_1^*, \ldots, z_m^*)^T, t \in \mathbb{R}_+, \\ f(u) - th(u) \geq 0, \sum_{j=1}^m z_j^* g_j(u) \geq 0, \\ 0 \in \partial f(u) + t\partial(-h)(u) + \partial\left(\sum_{j=1}^m z_j^* g_j\right)(u) + N(S,u)}} t.$$

From the way the feasible sets of these duals are constructed, it follows that whenever (u, z^*, t) is feasible to the Wolfe type dual one has $0 \leq t \leq (f(u) + \sum_{j=1}^m z_j^* g_j(u))/h(u)$, while when (u, z^*, t) is feasible to the Mond-Weir type dual there is $0 \leq t \leq f(u)/h(u)$. Moreover, the feasible set of (DQ^C_{MW-JS}) is contained in the one of (DQ^C_{W-JS}). Weak and strong duality statements for the problems just introduced follow.

Theorem 6.4.1. *One has* $v(DQ^C_{W-JS}) \leq v(PQ^C)$.

Proof. Let be $x = (x_1, \ldots, x_n)^T \in \mathcal{A}^Q$ and (u, z^*, t) be feasible to (DQ^C_{W-JS}), with $u = (u_1, \ldots, u_n)^T$ and $z^* = (z_1^*, \ldots, z_m^*)^T$. Then there are some $u^*, y^*, w^* \in \mathbb{R}^n$, $u^* = (u_1^*, \ldots, u_n^*)^T$, $y^* = (y_1^*, \ldots, y_n^*)^T$, $w^* = (w_1^*, \ldots, w_n^*)^T$, such that $u^* \in \partial f(u)$, $y^* \in \partial(\sum_{j=1}^m z_j^* g_j)(u)$, $w^* \in N(S, u)$ and $-u^* - y^* - w^* \in t\partial(-h)(u)$. We have $f(x) - f(u) \geq \sum_{i=1}^n u_i^*(x_i - u_i)$, $\sum_{j=1}^m z_j^* g_j(x) - \sum_{j=1}^m z_j^* g_j(u) \geq \sum_{i=1}^n y_i^*(x_i - u_i)$, $\delta_S(x) - \delta_S(u) \geq \sum_{i=1}^n w_i^*(x_i - u_i)$ and $t(-h(x) + h(u)) \geq -\sum_{i=1}^n (u_i^* + y_i^* + w_i^*)(x_i - u_i)$. Then $f(x) + \sum_{j=1}^m z_j^* g_j(x) - th(x) + \delta_S(x) \geq f(u) + \sum_{j=1}^m z_j^* g_j(u) - th(u) + \delta_S(u)$ and the term in the right-hand side is nonnegative according to the way the feasible set of the dual is defined. Taking into consideration that $x, u \in S$, we get

$f(x) + \sum_{j=1}^m z_j^* g_j(x) - th(x) \geq 0$, followed by $(f + \sum_{j=1}^m z_j^* g_j(x))/h(x) \geq t$. Since $\sum_{j=1}^m z_j^* g_j(x) \leq 0$, the latter inequality yields $f(x)/h(x) \geq t$. As the feasible points considered in the beginning of the proof were arbitrarily chosen, it follows $v(DQ^C_{W-JS}) \leq v(PQ^C)$. □

Theorem 6.4.2. *If* (PQ^C) *has an optimal solution* $\bar{x} \in \mathcal{A}^Q$ *and* $0 \in \operatorname{ri}(g(S) + \mathbb{R}^m_+)$, *then* $v(PQ^C) = v(DQ^C_{W-JS})$ *and there is some* $(\bar{z}^*, \bar{t}) \in \mathbb{R}^m_+ \times \mathbb{R}_+$ *such that* $(\bar{x}, \bar{z}^*, \bar{t})$ *is an optimal solution to the dual.*

Proof. From Dinkelbach's approach (cf. [59]) it is known that if $\bar{x} \in \mathcal{A}^Q$ solves (PQ^C), then it is an optimal solution to the convex minimization problem

$$\inf_{x \in \mathcal{A}^Q} \{f(x) - \bar{t}h(x)\},$$

where $\bar{t} := v(PQ^C) = f(\bar{x})/h(\bar{x}) \geq 0$, and the optimal objective value of the latter is 0. Using Theorem 6.1.2 for the above scalar optimization problem and its Wolfe dual, we get that there is some $\bar{z}^* \in \mathbb{R}^m_+$ such that $\sum_{j=1}^m \bar{z}_j^* g_j(\bar{x}) = 0$ and

$$0 \in \partial f(\bar{x}) + \bar{t}\partial(-h)(\bar{x}) + \partial\left(\sum_{j=1}^m \bar{z}_j^* g_j\right)(\bar{x}) + N(S, \bar{x}).$$

Consequently, $(\bar{x}, \bar{z}^*, \bar{t})$, where $\bar{z}^* = (\bar{z}_1^*, \ldots, \bar{z}_m^*)^T$, is feasible to (DQ^C_{W-JS}) and the objective functions of (PQ^C) and (DQ^C_{W-JS}) share a common value. By Theorem 6.4.1, $(\bar{x}, \bar{z}^*, \bar{t})$ is an optimal solution to (DQ^C_{W-JS}) and the strong duality is proven. □

Analogously, one can show the following assertions.

Theorem 6.4.3. *One has* $v(DQ^C_{MW-JS}) \leq v(PQ^C)$.

Theorem 6.4.4. *If* (PQ^C) *has an optimal solution* $\bar{x} \in \mathcal{A}^Q$ *and* $0 \in \operatorname{ri}(g(S) + \mathbb{R}^m_+)$, *then* $v(PQ^C) = v(DQ^C_{MW-JS})$ *and there is some* $(\bar{z}^*, \bar{t}) \in \mathbb{R}^m_+ \times \mathbb{R}$ *such that* $(\bar{x}, \bar{z}^*, \bar{t})$ *is an optimal solution to the dual.*

Another dual to (PQ^C) was introduced by Bector (cf. [13]) and its formulation requires the set $S \subseteq \mathbb{R}^n$ to be nonempty, convex and open, the functions $f : S \to \mathbb{R}$ and $g_j : S \to \mathbb{R}$, $j = 1, \ldots, m$, to be convex and Fréchet differentiable on S, and $h : S \to \mathbb{R}$ to be concave and Fréchet differentiable on S such that $S \cap g^{-1}(-C) \neq \emptyset$, where $g = (g_1, \ldots, g_m)^T$. We also assume that for all $x \in S$ $f(x) \geq 0$ and $h(x) > 0$. The so-called *Bector dual* problem *of Wolfe type* to (PQ^C) is

$$(DQ^C_{W-B}) \qquad \sup_{\substack{u \in S, z^* = (z_1^*, \ldots, z_m^*)^T \in \mathbb{R}^m_+, \\ \nabla\left(\frac{f + \sum_{j=1}^m z_j^* g_j}{h}\right)(u) = 0}} \frac{f(u) + \sum_{j=1}^m z_j^* g_j(u)}{h(u)}.$$

Analogously, a *Bector dual* problem *of Mond-Weir type* can be attached to (PQ^C), namely

$$(DQ^C_{MW-B}) \qquad \sup_{\substack{u\in S, z^*=(z_1^*,\ldots,z_m^*)^T\in\mathbb{R}^m_+,\\ \sum\limits_{j=1}^m z_j^* g_j(u)\geq 0,\\ \nabla\left(\frac{f+\sum\limits_{j=1}^m z_j^* g_j}{h}\right)(u)=0}} \frac{f(u)}{h(u)}.$$

Note that the feasible set of (DQ^C_{MW-B}) is contained in the one of (DQ^C_{W-B}). Weak and strong duality assertions for both these duals follow.

Theorem 6.4.5. *One has* $v(DQ^C_{W-B}) \leq v(PQ^C)$.

Proof. Let $x \in \mathcal{A}^Q$ and (u, z^*) be feasible to (DQ^C_{W-B}). Then one has

$$\nabla\left(\frac{f + \sum\limits_{j=1}^m z_j^* g_j}{h}\right)(u)^T(x-u) = 0,$$

which yields, via [126, Lemma 3.2.1],

$$\frac{f(x) + \sum\limits_{j=1}^m z_j^* g_j(x)}{h(x)} \geq \frac{f(u) + \sum\limits_{j=1}^m z_j^* g_j(u)}{h(u)}.$$

Because we also have $\sum_{j=1}^m z_j^* g_j(x) \leq 0$, it follows

$$\frac{f(x)}{h(x)} \geq \frac{f(u) + \sum\limits_{j=1}^m z_j^* g_j(u)}{h(u)}.$$

As x and (u, z^*) were arbitrarily chosen, we get $v(DQ^C_{W-B}) \leq v(PQ^C)$. □

Theorem 6.4.6. *If (PQ^C) has an optimal solution $\bar{x} \in \mathcal{A}^Q$ and the regularity condition $(RC^C_{KT})(\bar{x})$ is fulfilled, then $v(PQ^C) = v(DQ^C_{W-B})$ and there is a $\bar{z}^* \in \mathbb{R}^m_+$ such that $(\bar{x}, \bar{z}^*)$ is an optimal solution to the dual.*

Proof. The problem (PQ^C) has the same optimal solutions and optimal objective value as

$$\inf_{\substack{x\in S,\\ \frac{1}{h(x)}g(x)\in-\mathbb{R}^m_+}} \frac{f(x)}{h(x)}.$$

Thus $\bar{x}$ is an optimal solution to this optimization problem and applying Lemma 6.1.6 for it one obtains a $\bar{z}^* = (\bar{z}_1^*, \ldots, \bar{z}_m^*)^T \in \mathbb{R}^m_+$ for which

$$\nabla \left(\frac{f + \sum\limits_{j=1}^{m} \bar{z}_j^* g_j}{h} \right) (\bar{x}) = 0$$

and $\sum_{j=1}^{m} \bar{z}_j^* g_j(\bar{x}) = 0$. Thus $(\bar{x}, \bar{z}^*)$ is feasible to (DQ^C_{W-B}) and

$$\frac{f(\bar{x})}{h(\bar{x})} = \frac{f(\bar{x}) + \sum\limits_{j=1}^{m} \bar{z}_j^* g_j(\bar{x})}{h(\bar{x})}.$$

Using Theorem 6.4.5 it follows that $(\bar{x}, \bar{z}^*)$ is an optimal solution to the dual problem and we have strong duality. □

Analogously one can prove the following similar statements for the Mond-Weir type dual.

Theorem 6.4.7. *One has* $v(DQ^C_{MW-B}) \leq v(PQ^C)$.

Theorem 6.4.8. *If* (PQ^C) *has an optimal solution* $\bar{x} \in \mathcal{A}^Q$ *and the regularity condition* $(RC^C_{KT})(\bar{x})$ *is fulfilled, then* $v(PQ^C) = v(DQ^C_{MW-B})$ *and there is some* $\bar{z}^* \in \mathbb{R}^m_+$ *such that* $(\bar{x}, \bar{z}^*)$ *is an optimal solution to the dual.*

Remark 6.4.1. The regularity condition used in Theorem 6.4.6 and Theorem 6.4.8, can be replaced according to Remark 6.1.8 when the functions involved are Fréchet continuously differentiable on S.

Remark 6.4.2. In [100] the dual (DQ^C_{W-JS}) is considered to (PQ^C) when $\mathcal{A}^Q$ is assumed to be bounded and $S = \mathbb{R}^n$. In [126] the dual (DQ^C_{W-B}) with the additional constraint $f(u) + \sum_{j=1}^{m} z_j^* g_j(u) \geq 0$ is considered and also another similar dual where the constraint involving gradients is constructed analogously to $\left(DD^C_{\widetilde{W}}\right)$. A direct Mond-Weir dual to (PQ^C) is mentioned in [190]. Nevertheless, in [11] the dual (DQ^C_{W-B}) is considered, without the convexity assumptions on the functions involved, but for the weak and strong duality statements the objective function of the dual is assumed to be pseudoconvex. Note also that in [13] duals of both types are assigned to (PQ^C) which is considered only in the case when h is linear.

6.4.2 Wolfe and Mond-Weir duality in vector fractional programming

Several vector dual problems were proposed also for the primal vector fractional programming problem

$$(PVQ^C) \quad \operatorname*{Min}_{x \in \mathcal{A}^Q} f^Q(x),$$

where

$$\mathcal{A}^Q = \{x \in S : g(x) \in -\mathbb{R}^m_+\}$$

and

$$f^Q(x) = \begin{pmatrix} \frac{f_1(x)}{h_1(x)} \\ \vdots \\ \frac{f_k(x)}{h_k(x)} \end{pmatrix},$$

with $S \subseteq \mathbb{R}^n$ a nonempty convex set, the functions $f_i, g_j : \mathbb{R}^n \to \mathbb{R}$, convex and $h_i : \mathbb{R}^n \to \mathbb{R}$ concave for $i = 1, \dots, k$, $j = 1, \dots, m$, and $S \cap g^{-1}(-C) \neq \emptyset$, for $g = (g_1, \dots, g_m)^T$. We also assume that $f_i(x) \geq 0$ for $i = 1, ..., k$, and the following additional hypothesis

$$\exists a, b \in \mathbb{R}, 0 < a < b \text{ such that } h_i(x) \in [a, b], i = 1, \dots, k, \text{ for all } x \in S. \quad (6.6)$$

We recall that an element $\bar{x} \in \mathcal{A}^Q$ is said to be a properly efficient solution in the sense of Geoffrion to (PVQ^C) if $f^Q(\bar{x}) \in \operatorname{PMin}_{Ge}\left(f^Q(\mathcal{A}^Q), \mathbb{R}^k_+\right)$.

These vector duals were constructed starting from the scalar duals for the fractional programming problem presented in subsection 6.4.1. In the following we present some of these duals, considered with respect to properly efficient solutions in the sense of Geoffrion, alongside the corresponding duality statements.

We begin with extensions to vector duality of the scalar duals constructed by following the ideas of Jagannathan and Schaible. The *Wolfe type vector dual* introduced in [193] is

$$(DVQ^C_{W-JS}) \quad \operatorname*{Max}_{(\lambda, u, z^*, v) \in \mathcal{B}^Q_{W-JS}} h^Q_{W-JS}(\lambda, u, z^*, v),$$

where

$$\begin{aligned} \mathcal{B}^Q_{W-JS} = \Bigg\{ & (\lambda, u, z^*, v) \in \operatorname{int}(\mathbb{R}^k_+) \times S \times \mathbb{R}^m_+ \times \mathbb{R}^k_+ : \lambda = (\lambda_1, \dots, \lambda_k)^T, \\ & z^* = (z^*_1, \dots, z^*_m)^T, v = (v_1, \dots, v_k)^T, \\ & \sum_{i=1}^k \lambda_i (f_i(u) - v_i h_i(u)) + \sum_{j=1}^m z^*_j g_j(u) \geq 0, \\ & 0 \in \partial\Big(\sum_{i=1}^k \lambda_i f_i\Big)(u) + \sum_{i=1}^k v_i \partial(-\lambda_i h_i)(u) \\ & + \partial\Big(\sum_{j=1}^m z^*_j g_j\Big)(u) + N(S, u) \Bigg\} \end{aligned}$$

and

$$h^Q_{W-JS}(\lambda, u, z^*, v) = v,$$

while the one of *Mond-Weir type* is

$$(DVQ^C_{MW-JS}) \qquad \underset{(\lambda,uz^*,v)\in\mathcal{B}^Q_{MW-JS}}{\text{Max}} \quad h^Q_{MW-JS}(\lambda,u,z^*,v),$$

where

$$\mathcal{B}^Q_{MW-JS} = \Bigg\{(\lambda,u,z^*,v) \in \text{int}(\mathbb{R}^k_+) \times S \times \mathbb{R}^m_+ \times \mathbb{R}^k_+ : \lambda = (\lambda_1,\ldots,\lambda_k)^T, \\ z^* = (z_1^*,\ldots,z_m^*)^T, v = (v_1,\ldots,v_k)^T, \\ \sum_{i=1}^k \lambda_i(f_i(u) - v_i h_i(u)) \geq 0, \sum_{j=1}^m z_j^* g_j(u) \geq 0, \\ 0 \in \partial\bigg(\sum_{i=1}^k \lambda_i f_i\bigg)(u) + \sum_{i=1}^k v_i \partial(-\lambda_i h_i)(u) \\ + \partial\bigg(\sum_{j=1}^m z_j^* g_j\bigg)(u) + N(S,u)\Bigg\}$$

and

$$h^Q_{MW-JS}(\lambda,u,z^*,v) = v.$$

Note that $\mathcal{B}^Q_{MW-JS} \subseteq \mathcal{B}^Q_{W-JS}$. Weak and strong duality statements follow.

Theorem 6.4.9. *There is no $x \in \mathcal{A}^Q$ and no $(\lambda,u,z^*,v) \in \mathcal{B}^Q_{W-JS}$ such that $f_i^Q(x) \leq h^Q_{W-JSi}(\lambda,u,z^*,v)$ for $i = 1,\ldots,k$, and $f_j^Q(x) \leq h^Q_{W-JSj}(\lambda,u,z^*,v)$ for at least one $j \in \{1,\ldots,k\}$.*

Proof. Assume that there are some $x \in \mathcal{A}^Q$ and $(\lambda,u,z^*,v) \in \mathcal{B}^Q_{W-JS}$ such that $f_i^Q(x) \leq h^Q_{Wi-JS}(\lambda,u,z^*,v)$ for $i = 1,\ldots,k$, and $f_j^Q(x) < h^Q_{Wj-JS}(\lambda,u,z^*,v)$ for at least one $j \in \{1,\ldots,k\}$. This yields $f_i(x) \leq v_i h_i(x)$ for $i = 1,\ldots,k$, and $f_j(x) < v_j h_j(x)$ for at least one $j \in \{1,\ldots,k\}$, followed by $\sum_{i=1}^k \lambda_i f_i(x) < \sum_{i=1}^k \lambda_i v_i h_i(x)$, i.e.

$$\sum_{i=1}^k \lambda_i(f_i(x) - v_i h_i(x)) < 0. \tag{6.7}$$

On the other hand,

$$0 \in \partial\bigg(\sum_{i=1}^k \lambda_i f_i\bigg)(u) + \sum_{i=1}^k v_i \partial(-\lambda_i h_i)(u) + \partial\bigg(\sum_{j=1}^m z_j^* g_j\bigg)(u) + N(S,u)$$

means that there are $u^*, y^*, w_i^*, t^* \in \mathbb{R}^n$, $i = 1,\ldots,k$, such that $u^* \in \partial(\sum_{i=1}^k \lambda_i f_i)(u)$, $w_i^* \in \partial(-\lambda_i h_i)(u)$, $i = 1,\ldots,k$, $y^* \in \partial(\sum_{j=1}^m z_j^* g_j)(u)$, $t^* \in N(S,u)$ and $u^* + \sum_{i=1}^k v_i w_i^* + y^* + t^* = 0$. Then $\sum_{i=1}^k \lambda_i f_i(x) - \sum_{i=1}^k v_i \lambda_i h_i(x) + \sum_{j=1}^m z_j^* g_j(x) + \delta_S(x) \geq \sum_{i=1}^k \lambda_i f_i(u) - \sum_{i=1}^k v_i \lambda_i h_i(u) +$

$\sum_{j=1}^m z_j^* g_j(u) + \delta_S(u) \geq 0$ and, as $\sum_{j=1}^m z_j^* g_j(x) \leq 0$, we get $\sum_{i=1}^k \lambda_i(f_i(x) - v_i h_i(x)) \geq 0$, which contradicts (6.7). Consequently, the desired weak duality assertion holds. □

Theorem 6.4.10. *If $\bar{x} \in \mathcal{A}^Q$ is a properly efficient solution to (PVQ^C) in the sense of Geoffrion and $0 \in \operatorname{ri}(g(S) + \mathbb{R}_+^m)$, then there exists $(\bar{\lambda}, \bar{z}^*, \bar{v}) \in \operatorname{int}(\mathbb{R}_+^k) \times \mathbb{R}_+^m \times \mathbb{R}_+^k$ such that $(\bar{\lambda}, \bar{x}, \bar{z}^*, \bar{v})$ is an efficient solution to (DVQ_{W-JS}^C) and $f_i^Q(\bar{x}) = h_{W-JSi}^Q(\bar{\lambda}, \bar{x}, \bar{z}^*, \bar{v})$ for $i = 1, \dots, k$.*

Proof. In both [193, Theorem 3] and [112, Lemma 1] it is proven that $\bar{x}$ is properly efficient to (PVQ^C) in the sense of Geoffrion if and only if it is properly efficient in the sense of Geoffrion to the vector optimization problem

$$\operatorname*{Min}_{x \in \mathcal{A}^Q} \begin{pmatrix} f_1(x) - \frac{f_1(\bar{x})}{h_1(\bar{x})} h_1(x) \\ \vdots \\ f_k(x) - \frac{f_k(\bar{x})}{h_m(\bar{x})} h_m(x) \end{pmatrix}.$$

In fact this is the place where we need to assume (6.6). Unfortunately, this assumption is omitted in [112, 193], but we have doubts that without it the above-mentioned equivalence is valid. Proposition 2.4.18(*b*) ensures now that the properly minimal elements of the set

$$\left(f_1 - \frac{f_1(\bar{x})}{h_1(\bar{x})} h_1, \dots, f_k - \frac{f_k(\bar{x})}{h_k(\bar{x})} h_k \right)^T (\mathcal{A}^Q)$$

considered in the sense of Geoffrion coincide with its properly efficient solutions in the sense of linear scalarization. Thus thus there is some $\bar{\lambda} = (\bar{\lambda}_1, \dots, \bar{\lambda}_k)^T \in \operatorname{int}(\mathbb{R}_+^k)$ such that $\bar{x}$ minimizes the scalar convex optimization problem

$$\inf_{x \in \mathcal{A}^Q} \sum_{i=1}^k \bar{\lambda}_i \left(f_i(x) - \frac{f_i(\bar{x})}{h_i(\bar{x})} h_i(x) \right).$$

Applying now Theorem 6.1.2, there is strong duality for this scalar optimization problem and its Wolfe dual, thus there is some $\bar{z}^* \in \mathbb{R}_+^m$ for which $\bar{z}^{*T} g(\bar{x}) = 0$ and

$$0 \in \partial\bigg(\sum_{i=1}^k \bar{\lambda}_i f_i \bigg)(\bar{x}) + \sum_{i=1}^k \bar{v}_i \partial(-\bar{\lambda}_i h_i)(\bar{x}) + \partial\bigg(\sum_{j=1}^m \bar{z}_j^* g_j \bigg)(\bar{x}) + N(S, \bar{x}),$$

where $\bar{v}_i = f_i(\bar{x})/h_i(\bar{x})$, $i = 1, \dots, k$. Consequently, $(\bar{\lambda}, \bar{x}, \bar{z}^*, \bar{v}) \in \mathcal{B}_{W-JS}^Q$ and $f^Q(\bar{x}) = h_{W-JS}^Q(\bar{\lambda}, \bar{x}, \bar{z}^*, \bar{v})$. The efficiency of $(\bar{\lambda}, \bar{x}, \bar{z}^*, \bar{v})$ to the vector dual follows by Theorem 6.4.9. □

Analogously one can prove the following duality statements for the Mond-Weir type vector dual.

Theorem 6.4.11. *There is no* $x \in \mathcal{A}^Q$ *and no* $(\lambda, u, z^*, v) \in \mathcal{B}^Q_{MW-JS}$ *such that* $f_i^Q(x) \leq h^Q_{MW-JSi}(\lambda, u, z^*, v)$ *for* $i = 1, \dots, k$, *and* $f_j^Q(x) \leq h^Q_{MW-JSj}(\lambda, u, z^*, v)$ *for at least one* $j \in \{1, \dots, k\}$.

Theorem 6.4.12. *If* $\bar{x} \in \mathcal{A}^Q$ *is a properly efficient solution to* (PVQ^C) *in the sense of Geoffrion and* $0 \in \operatorname{ri}(g(S) + \mathbb{R}^m_+)$, *then there exists* $(\bar{\lambda}, \bar{z}^*, \bar{v}) \in \operatorname{int}(\mathbb{R}^k_+) \times \mathbb{R}^m_+ \times \mathbb{R}^k_+$ *such that* $(\bar{\lambda}, \bar{x}, \bar{z}^*, \bar{v})$ *is an efficient solution to* (DVQ^C_{MW-JS}) *and* $f_i^Q(\bar{x}) = h^Q_{MW-JSi}(\bar{\lambda}, \bar{x}, \bar{z}^*, \bar{v})$ *for* $i = 1, \dots, k$.

Now let us turn our attention to the vector fractional duals that can be constructed starting from the scalar Bector duals considered in section 6.1. For this we consider the primal vector fractional programming problem with respect to weakly efficient solutions

$$(PVQ^C_w) \quad \operatorname*{WMin}_{x \in \mathcal{A}^Q} f^Q(x),$$

where

$$\mathcal{A}^Q = \{x \in S : g(x) \in -\mathbb{R}^m_+\}$$

and

$$f^Q(x) = \begin{pmatrix} \frac{f_1(x)}{h_1(x)} \\ \vdots \\ \frac{f_k(x)}{h_k(x)} \end{pmatrix}.$$

We assume that $S \subseteq \mathbb{R}^n$ is a nonempty, convex and open set, the functions $f_i : S \to \mathbb{R}$, $g_j : S \to \mathbb{R}$, $i = 1, ..., k$, $j = 1, \dots, m$, are convex and Fréchet differentiable on S, and the functions $h_i : S \to \mathbb{R}$, $i = 1, ..., k$, are concave and Fréchet differentiable on S such that $S \cap g^{-1}(-C) \neq \emptyset$, where $g = (g_1, \dots, g_m)^T$. We also assume that $f_i(x) \geq 0$ and $h_i(x) > 0$ for $i = 1, ..., k$, and all $x \in S$. As we consider here a different approach to the one described above, we allow us to weaken the assumption (6.6) concerning the denominators of the components of the primal objective function. Following [194] we consider the following *Bector vector dual of Wolfe type* to (PVQ^C_w) with respect to weakly efficient solutions

$$(DVQ^C_{W-Bw}) \quad \operatorname*{WMax}_{(\lambda, u, z^*) \in \mathcal{B}^Q_{W-Bw}} h^Q_{W-Bw}(\lambda, u, z^*),$$

where

$$\mathcal{B}^Q_{W-Bw} = \Bigg\{ (\lambda, u, z^*) \in (\mathbb{R}^k_+ \setminus \{0\}) \times S \times \mathbb{R}^m_+ : \lambda = (\lambda_1, \ldots, \lambda_k)^T,$$

$$z^* = (z_1^*, \ldots, z_m^*)^T, \sum_{i=1}^k \lambda_i = 1,$$

$$\sum_{i=1}^k \lambda_i \nabla \left(\frac{f_i + \sum\limits_{j=1}^m z_j^* g_j}{h_i} \right)(u) = 0 \Bigg\}$$

and

$$h^Q_{W-Bw}(\lambda, u, z^*) = \begin{pmatrix} \frac{f_1(u) + \sum\limits_{j=1}^m z_j^* g_j(u)}{h_1(u)} \\ \vdots \\ \frac{f_k(u) + \sum\limits_{j=1}^m z_j^* g_j(u)}{h_k(u)} \end{pmatrix}.$$

Analogously, a *Bector vector dual of Mond-Weir type* with respect to weakly efficient solutions can be given

$$(DVQ^C_{MW-Bw}) \quad \underset{(\lambda,u,z^*) \in \mathcal{B}^Q_{MW-Bw}}{\text{WMax}} \quad h^Q_{MW-Bw}(\lambda, u, z^*),$$

where

$$\mathcal{B}^Q_{MW-Bw} = \Bigg\{ (\lambda, u, z^*) \in (\mathbb{R}^k_+ \setminus \{0\}) \times S \times \mathbb{R}^m_+ : \lambda = (\lambda_1, \ldots, \lambda_k)^T,$$

$$z^* = (z_1^*, \ldots, z_m^*)^T, \sum_{i=1}^k \lambda_i = 1, \sum_{j=1}^m z_j^* g_j(u) \geq 0,$$

$$\sum_{i=1}^k \lambda_i \nabla \left(\frac{f_i + \sum\limits_{j=1}^m z_j^* g_j}{h_i} \right)(u) = 0 \Bigg\}$$

and

$$h^Q_{MW-Bw}(\lambda, u, z^*) = \begin{pmatrix} \frac{f_1(u)}{h_1(u)} \\ \vdots \\ \frac{f_k(u)}{h_k(u)} \end{pmatrix}.$$

Note that $\mathcal{B}^Q_{MW-Bw} \subseteq \mathcal{B}^Q_{W-Bw}$ and $h^Q_{MW-Bw}(\mathcal{B}^Q_{MW-Bw}), h^Q_{W-Bw}(\mathcal{B}^Q_{W-Bw}) \subseteq \mathbb{R}^k$. Weak and strong duality for these vector dual problems follow.

Theorem 6.4.13. *There is no $x \in \mathcal{A}^Q$ and no $(\lambda, u, z^*) \in \mathcal{B}^Q_{W-Bw}$ such that $f_i^Q(x) < h^Q_{W-Bwi}(\lambda, u, z^*)$ for $i = 1, \ldots, k$.*

Proof. Take some arbitrary feasible elements $x \in \mathcal{A}^Q$ and $(\lambda, u, z^*) \in \mathcal{B}^Q_{W-Bw}$ for which $f_i^Q(x) < h^Q_{W-Bwi}(\lambda, u, z^*)$ for $i = 1, \dots, k$. Since $\lambda_i \geq 0$ and $h_i(x)/h_i(u) > 0$ for $i = 1, \dots, k$, it holds

$$\sum_{i=1}^{k} \lambda_i \frac{h_i(x)}{h_i(u)} \left(\frac{f_i(x)}{h_i(x)} - \frac{f_i(u) + \sum\limits_{j=1}^{m} z_j^* g_j(u)}{h_i(u)} \right) < 0.$$

Consider the function $\varphi : S \to \mathbb{R}$,

$$\varphi(x) = \sum_{i=1}^{k} \frac{\lambda_i}{h_i(u)^2} \left(h_i(u) \left(f_i(x) + \sum_{j=1}^{m} z_j^* g_j(x) \right) - h_i(x) \left(f_i(u) + \sum_{j=1}^{m} z_j^* g_j(u) \right) \right).$$

The convexity hypotheses imply that φ is convex on S. Moreover, $\varphi(u) = 0$ and $\nabla \varphi(u) = 0$. Consequently $\varphi(x) \geq 0$, which, taking into account that $\sum_{j=1}^{m} z_j^* g_j(x) \leq 0$, yields

$$\sum_{i=1}^{k} \lambda_i \frac{h_i(x)}{h_i(u)} \left(\frac{f_i(x)}{h_i(x)} - \frac{f_i(u) + \sum\limits_{j=1}^{m} z_j^* g_j(u)}{h_i(u)} \right) \geq 0.$$

This provides a contradiction and, consequently, the desired weak duality assertion holds. □

Theorem 6.4.14. *If $\bar{x} \in \mathcal{A}^Q$ is a weakly efficient solution to (PVQ^C_w) and the regularity condition $(RC^C_{KT})(\bar{x})$ is fulfilled, then there exists $(\bar{\lambda}, \bar{z}^*) \in (\mathbb{R}^k_+ \setminus \{0\}) \times \mathbb{R}^m_+$ such that $(\bar{\lambda}, \bar{x}, \bar{z}^*)$ is a weakly efficient solution to (DVQ^C_{W-Bw}) and $f_i^Q(\bar{x}) = h^Q_{W-Bwi}(\bar{\lambda}, \bar{x}, \bar{z}^*)$ for $i = 1, \dots, k$.*

Proof. The hypotheses of the theorem allow applying [54, Theorem 1]. Thus there exist $\bar{\lambda} = (\bar{\lambda}_1, ..., \bar{\lambda}_k)^T \in \mathbb{R}^k_+ \backslash \{0\}$ with $\sum_{i=1}^{k} \bar{\lambda}_i = 1$ and $\tilde{z}^* = (\tilde{z}_1^*, \dots, \tilde{z}_m^*)^T \in \mathbb{R}^m_+$ such that $\sum_{j=1}^{m} \tilde{z}_j^* g_j(\bar{x}) = 0$ and

$$\sum_{i=1}^{k} \bar{\lambda}_i \nabla \left(\frac{f_i}{h_i} \right)(\bar{x}) + \nabla \left(\sum_{j=1}^{m} \tilde{z}_j^* g_j \right)(\bar{x}) = 0.$$

Consider

$$\bar{z}^* := \frac{1}{\sum\limits_{i=1}^{k} \frac{\bar{\lambda}_i}{h_i(\bar{x})}} \tilde{z}^* \in \mathbb{R}^m_+.$$

Then $\sum_{j=1}^{m} \bar{z}_j^* g_j(\bar{x}) = 0$ and

$$\sum_{i=1}^{k} \bar{\lambda}_i \nabla \left(\frac{\sum\limits_{j=1}^{m} \bar{z}_j^* g_j}{h_i} \right) (\bar{x}) = \nabla\big(\bar{z}^{*^T} g\big)(\bar{x}).$$

Consequently,

$$\sum_{i=1}^{k} \bar{\lambda}_i \nabla \left(\frac{f_i + \sum\limits_{j=1}^{m} \bar{z}_j^* g_j}{h_i} \right) (\bar{x}) = 0,$$

thus $(\bar{\lambda}, \bar{x}, \bar{z}^*)$ is feasible to (DVQ^C_{W-Bw}) and $f_i^Q(\bar{x}) = h^Q_{W-Bwi}(\bar{\lambda}, \bar{x}, \bar{z}^*)$ for $i = 1, \dots, k$. The weak efficiency of $(\bar{\lambda}, \bar{x}, \bar{z}^*)$ to (DVQ^C_{W-Bw}) follows via Theorem 6.4.13. □

For the Mond-Weir type vector dual the proofs are analogous.

Theorem 6.4.15. *There is no $x \in \mathcal{A}^Q$ and no $(\lambda, u, z^*) \in \mathcal{B}^Q_{MW-Bw}$ such that $f_i^Q(x) < h^Q_{MW-Bwi}(\lambda, u, z^*)$ for $i = 1, \dots, k$.*

Theorem 6.4.16. *If $\bar{x} \in \mathcal{A}^Q$ is a weakly efficient solution to (PVQ^C_w) and the regularity condition $(RC^C_{KT})(\bar{x})$ is fulfilled, then there exists $(\bar{\lambda}, \bar{z}^*) \in (\mathbb{R}^k_+ \setminus \{0\}) \times \mathbb{R}^m_+$ such that $(\bar{\lambda}, \bar{x}, \bar{z}^*)$ is a weakly efficient solution to (DVQ^C_{MW-Bw}) and $f_i^Q(\bar{x}) = h^Q_{MW-Bwi}(\bar{\lambda}, \bar{x}, \bar{z}^*)$ for $i = 1, \dots, k$.*

Remark 6.4.3. The regularity condition used in Theorem 6.4.14 and Theorem 6.4.16, can be replaced according to Remark 6.1.8 when all the functions involved are Fréchet continuously differentiable on S.

Remark 6.4.4. Relaxing in their formulation the geometric constraint $\lambda \in \operatorname{int}(\mathbb{R}^k_+)$ to $\lambda \in \mathbb{R}^k_+ \setminus \{0\}$, the vector duals (DVQ^C_{W-JS}) and (DVQ^C_{MW-JS}) turn into ones for which weakly efficient solutions to the primal problem (PVQ^C) can be investigated, too. The weak and strong duality statements follow analogously and one does not have to impose (6.6) in this case, but only the positivity of the denominators on S.

Remark 6.4.5. In [193] instead of $\mathbb{R}^m_+$ an arbitrary convex closed cone in $\mathbb{R}^m$ is considered. Considering additional Fréchet differentiability hypotheses on the functions f, g and h, (DVQ^C_{W-JS}) turns into the vector dual given in [194]. The vector dual given in [12,112] is actually (DVQ^C_{MW-JS}) for the case when the functions involved are assumed Fréchet differentiable. Note also the vector Mond-Weir dual to (PVQ^C) from [190].

Remark 6.4.6. In papers like [112,190,193,194] it is claimed that properly efficient solutions to the vector duals considered there for (PVQ^C) are obtained via strong duality. We doubt that the proofs of those results are correct or at least complete.

Remark 6.4.7. When $k = 1$ the duals and the duality statements from the vector case collapse into the ones from the scalar case.

6.5 Generalized Wolfe and Mond-Weir duality: a perturbation approach

In the following we show that a perturbation theory similar to the one developed in the beginning of chapter 3 can be successfully employed to the Wolfe and Mond-Weir duality concepts.

6.5.1 Wolfe type and Mond-Weir type duals for general scalar optimization problems

Like in chapter 3, let X and Y be Hausdorff locally convex spaces and consider the proper function $F : X \to \overline{\mathbb{R}}$. Assume moreover that the topological dual spaces X^* and Y^* are endowed with the corresponding weak* topologies. Making use of a proper perturbation function $\Phi : X \times Y \to \overline{\mathbb{R}}$ fulfilling $\Phi(x, 0) = F(x)$ for all $x \in X$, to the general optimization problem

$$(PG) \quad \inf_{x \in X} F(x),$$

we attach, besides

$$(DG) \quad \sup_{y^* \in Y^*} \{-\Phi^*(0, y^*)\},$$

(see section 3.1), two more dual problems, namely a *Wolfe type* one

$$(DG_W) \quad \sup_{\substack{u \in X, y \in Y, y^* \in Y^*, \\ (0, y^*) \in \partial\Phi(u, y)}} \{-\Phi^*(0, y^*)\},$$

and a *Mond-Weir type* one

$$(DG_M) \quad \sup_{\substack{u \in X, y^* \in Y^*, \\ (0, y^*) \in \partial\Phi(u, 0)}} \{\Phi(u, 0)\}.$$

Next we show that weak duality holds for (PG) and its new duals, too, as a consequence of the way these dual problems are defined.

Theorem 6.5.1. *There is*

$$-\infty \leq v(DG_M) \leq v(DG_W) \leq v(DG) \leq v(PG) \leq +\infty.$$

Proof. Noting that (DG_M) can be obtained from (DG_W) by taking $y = 0$ it follows that $-\infty \leq v(DG_M) \leq v(DG_W)$. On the other hand, (DG_W) is actually the problem (DG) introduced in the third chapter, with an additional constraint. Consequently, $v(DG_W) \leq v(DG)$ and, using Theorem 3.1.1, we are done. □

The strong duality statement comes next. The regularity conditions it uses were introduced in section 3.2.

Theorem 6.5.2. *Let $\Phi : X \times Y \to \overline{\mathbb{R}}$ be a proper and convex function such that $0 \in \Pr_Y(\operatorname{dom} \Phi)$. If (PG) has an optimal solution $\bar{x} \in X$ and one of the regularity conditions (RC_i^{Φ}), $i \in \{1,2,3,4\}$, is fulfilled, then $v(PG) = v(DG_W) = v(DG_M)$ and there is some $\bar{y}^* \in Y^*$ such that $(\bar{x}, 0, \bar{y}^*)$ is an optimal solution to (DG_W) and $(\bar{x}, \bar{y}^*)$ an optimal solution to (DG_M).*

Proof. Theorem 3.3.1 guarantees that under the present hypotheses there is some $\bar{y}^* \in Y^*$ such that $(0, \bar{y}^*) \in \partial\Phi(\bar{x}, 0)$. Thus $(\bar{x}, 0, \bar{y}^*)$ and $(\bar{x}, \bar{y}^*)$ are feasible elements to (DG_W) and (DG_M), respectively. Moreover, $\Phi(\bar{x}, 0) = v(PG) = -\Phi^*(0, \bar{y}^*) \geq v(DG_W) \geq v(DG_M) \geq \Phi(\bar{x}, 0)$, which yields the strong duality for (PG) and both its duals (DG_W) and (DG_M). Then $(\bar{x}, 0, \bar{y}^*)$ turns out to be an optimal solution to (DG_W) and $(\bar{x}, \bar{y}^*)$ to (DG_M), respectively. □

Let us see now how do the dual problems arising from (DG_W) and (DG_M) look for some classes of primal problems considered in subsections 3.1.2 and 3.1.3.

6.5.2 Wolfe type and Mond-Weir type duals for different scalar optimization problems

Consider first the primal optimization problem

$$(P^A) \quad \inf_{x \in X} \{f(x) + g(Ax)\},$$

where $f : X \to \overline{\mathbb{R}}$ and $g : Y \to \overline{\mathbb{R}}$ are proper functions and $A \in \mathcal{L}(X, Y)$ fulfills $\operatorname{dom} f \cap A^{-1}(\operatorname{dom} g) \neq \emptyset$. Like in section 3.1.2, the perturbation function considered for assigning the Wolfe type and Mond-Weir type dual problems to (P^A) is $\Phi^A : X \times Y \to \overline{\mathbb{R}}$, $\Phi^A(x, y) = f(x) + g(Ax + y)$. After some calculations, these duals turn out to be

$$(D_W^A) \qquad \sup_{\substack{u \in X, y \in Y, y^* \in Y^*, \\ y^* \in (A^*)^{-1}(-\partial f(u)) \cap \partial g(Au + y)}} \{-f^*(-A^* y^*) - g^*(y^*)\},$$

and, respectively,

$$(D_M^A) \qquad \sup_{\substack{u \in X, \\ (A^*)^{-1}(-\partial f(u)) \cap \partial g(Au) \neq \emptyset}} \{f(u) + g(Au)\}.$$

By Theorem 6.5.1 we obtain $v(D_M^A) \leq v(D_W^A) \leq v(D^A) \leq v(P^A)$, i.e. weak duality for these dual problems to (P^A), while for strong duality one needs to assume that the functions f and g are convex, the existence of an optimal solution to (P^A) and the fulfillment of a regularity condition from (RC_i^A), $i \in \{1,2,3,4\}$, i.e. the hypotheses of Theorem 3.3.4(*a*).

Like in chapter 3 the problem (P^A) can be specialized, for different choices of the functions involved, to turn into different optimization problems, namely

(P^{id}), (P^{A_g}) and (P^{Σ}), for which the duals resulting from (D_M^A) and (D_W^A) can be derived, too. Because in the literature both Wolfe and Mond-Weir duality concepts apply only for constrained optimization problems we will not insist further on the unconstrained case.

Now let us turn our attention to the constrained optimization problems treated in section 3.1.3. Let Z be another Hausdorff locally convex space partially ordered by the convex cone $C \subseteq Z$. Consider the nonempty set $S \subseteq X$ and the proper functions $f : X \to \overline{\mathbb{R}}$ and $g : X \to \overline{Z}$, fulfilling $\operatorname{dom} f \cap S \cap g^{-1}(-C) \neq \emptyset$. The primal problem we treat further is

$$(P^C) \quad \inf_{x \in \mathcal{A}} f(x).$$
$$\mathcal{A} = \{x \in S : g(x) \in -C\}$$

Using the perturbation functions from subsection 3.1.3 we assign to (P^C) three pairs of duals arising from (DG_W) and (DG_M), respectively.

Taking the perturbation function Φ^{C_L}, we obtain from (DG_W) the following dual problem to (P^C)

$$(D_W^{C_L}) \quad \sup_{\substack{u \in S, z \in Z, z^* \in -C^*, \\ g(u) - z \in -C, \langle z^*, g(u) - z\rangle = 0, \\ 0 \in \partial(f + (-z^* g) + \delta_S)(u)}} \{f(u) - \langle z^*, z\rangle\},$$

which is nothing else than

$$(D_W^{C_L}) \quad \sup_{\substack{u \in S, z^* \in C^*, \\ 0 \in \partial(f + (z^* g) + \delta_S)(u)}} \{f(u) + \langle z^*, g(u)\rangle\}.$$

We call this the *Wolfe dual of Lagrange type* to (P^C), because it was obtained via the perturbation function used in the framework of chapter 3 to get the Lagrange dual to (P^C). We shall see that in the particular instance where the classical Wolfe duality was considered this dual turns into the well-known *Wolfe dual problem.*

Analogously we get a dual problem to (P^C) arising from (DG_M), i.e.

$$(D_M^{C_L}) \quad \sup_{\substack{u \in S, z^* \in C^*, \\ g(u) \in -C, \langle z^*, g(u)\rangle \geq 0, \\ 0 \in \partial(f + (z^* g) + \delta_S)(u)}} f(u),$$

further referred to as the *Mond-Weir dual of Lagrange type* to (P^C). We name it like this because it was obtained via the perturbation Φ^{C_L} and due to the fact that in the particular instances where the classical Mond-Weir duality was considered this dual turns into the Mond-Weir dual problem with a constraint more, namely $g(u) \in -C$.

This makes us consider also a dual problem to (P^C) which is obtained from $(D_M^{C_L})$ by removing the constraint $g(u) \in -C$, being

$$(D^{C_L}_{MW}) \qquad \sup_{\substack{u\in S, z^*\in C^*, \langle z^*, g(u)\rangle\geq 0,\\ 0\in\partial(f+(z^*g)+\delta_S)(u)}} f(u).$$

By construction it is clear that $v(D^{C_L}_M) \leq v(D^{C_L}_{MW})$. On the other hand, whenever (u, z^*) is feasible to $(D^{C_L}_{MW})$ it is feasible to $(D^{C_L}_W)$, too, and we moreover have $\langle z^*, g(u)\rangle \geq 0$. This yields $f(u) \leq f(u) + \langle z^*, g(u)\rangle \leq v(D^{C_L}_W)$. Considering the supremum over all the pairs (u, z^*) feasible to $(D^{C_L}_{MW})$ we obtain $v(D^{C_L}_{MW}) \leq v(D^{C_L}_W)$. Applying the weak duality statement we get $v(D^{C_L}_M) \leq v(D^{C_L}_{MW}) \leq v(D^{C_L}_W) \leq v(D^{C_L}) \leq v(P^C)$.

Scalar dual problems to (P^C) can be obtained from (DG_W) and (DG_M) with the perturbation function Φ^{C_F}, too. After some calculations, (DG_W) turns into

$$(D^{C_F}_W) \qquad \sup_{\substack{u\in S, y\in X, y^*\in X^*,\\ y^*\in\partial f(u+y)\cap(-N(\mathcal{A},(u)))}} \{\langle y^*, u\rangle - f^*(y^*)\},$$

further called the *Wolfe dual of Fenchel type* to (P^C) since it was obtained via the perturbation used earlier to assign the Fenchel dual to (P^C) and because (DG_W) was obtained by generalizing the classical Wolfe duality.

Similarly, the dual problem to (P^C) arising from (DG_M) is

$$(D^{C_F}_M) \qquad \sup_{\substack{u\in S,\\ 0\in\partial f(u)+N(\mathcal{A},(u))}} f(u),$$

called the *Mond-Weir dual of Fenchel type* to (P^C). From the weak duality statement we get $v(D^{C_F}_M) \leq v(D^{C_F}_W) \leq v(D^{C_F}) \leq v(P^C)$.

The last perturbation function considered here is $\Phi^{C_{FL}}$, which leads to the following dual to (P^C) we obtained from (DG_W)

$$(D^{C_{FL}}_W) \qquad \sup_{\substack{u\in S, y\in X, y^*\in X^*, z^*\in C^*,\\ y^*\in\partial f(u+y)\cap(-\partial((z^*g)+\delta_S)(u))}} \{\langle y^*, u\rangle + \langle z^*, g(u)\rangle - f^*(y^*)\},$$

further referred to as the *Wolfe dual of Fenchel-Lagrange type* to (P^C).

Analogously, the dual problem to (P^C) arising from (DG_M) is

$$(D^{C_{FL}}_M) \qquad \sup_{\substack{u\in S, z^*\in C^*,\\ \langle z^*, g(u)\rangle\geq 0, g(u)\in -C,\\ 0\in\partial f(u)+\partial((z^*g)+\delta_S)(u)}} f(u),$$

called the *Mond-Weir dual of Fenchel-Lagrange type* to (P^C).

Removing the constraint $g(u) \in -C$, we obtain from $(D^{C_{FL}}_M)$ the following dual problem to (P^C),

$$(D^{C_{FL}}_{MW}) \qquad \sup_{\substack{u\in S, z^*\in C^*, \langle z^*, g(u)\rangle\geq 0,\\ 0\in\partial f(u)+\partial((z^*g)+\delta_S)(u)}} f(u).$$

Applying the weak duality statement and using similar arguments to the ones used concerning $(D^{C_L}_{MW})$, we get $v(D^{C_{FL}}_M) \leq v(D^{C_{FL}}_{MW}) \leq v(D^{C_{FL}}_W) \leq v(D^{C_{FL}}) \leq v(P^C)$.

Analogously to the proofs of Proposition 3.1.5 and Proposition 3.1.6, one can show the following inequalities concerning the relations between the optimal objective values of the dual problems introduced in this section.

Proposition 6.5.3. *One has*

(i) $v(D^{C_{FL}}_M) \leq \begin{matrix} v(D^{C_L}_M) \\ v(D^{C_F}_M) \end{matrix} \leq v(P^C)$;

(ii) $v(D^{C_{FL}}_{MW}) \leq v(D^{C_L}_{MW}) \leq v(P^C)$.

Remark 6.5.1. By Theorem 3.5.8 one can give sufficient conditions that ensure that $(D^{C_L}_W)$ coincides with (D^C_W) and, respectively, $(D^{C_L}_{MW})$ with (D^C_{MW}), like the additional conditions (i) – (iv) from Theorem 6.1.2. Moreover, even conditions of closedness type can be considered. As the hypotheses of Remark 6.1.1, Remark 6.1.2 and subsection 6.1.2 guarantee the fulfillment of at least first two of these conditions, our claim from the beginning of the section that we have embedded the Wolfe and Mond-Weir duality concepts into the perturbational approach is sustained. Furthermore, by Theorem 3.5.6, Theorem 3.5.9 and Theorem 3.5.13 one can give sufficient conditions that allow us to have sums of subdifferentials instead of subdifferentials of sums of functions in the feasible sets of $(D^{C_L}_{MW})$ and $(D^{C_L}_W)$.

From the strong duality statement for (PG) and its duals (DG_W) and (DG_M), Theorem 6.5.2, one can obtain strong duality results for the duals considered in this subsection, by using the regularity conditions considered in section 3.2.

6.5.3 Wolfe type and Mond-Weir type duals for general vector optimization problems

Let X, Y and V be Hausdorff locally convex spaces and assume that V is partially ordered by the nontrivial pointed convex cone $K \subseteq V$. Let $F : X \to \overline{V}$ be a proper and K-convex function and consider the general vector optimization problem

$$(PVG) \quad \operatorname*{Min}_{x \in X} F(x).$$

Using the vector perturbation function $\Phi : X \times Y \to \overline{V}$, assumed proper and fulfilling $\Phi(x, 0) = F(x)$ for all $x \in X$, we attach to (PVG) the following vector dual problems with respect to properly efficient solutions in the sense of linear scalarization

$$(DVG_W) \quad \operatorname*{Max}_{(v^*,u,y,y^*,r) \in \mathcal{B}^G_W} h^G_W(v^*, u, y, y^*, r),$$

where

$$\mathcal{B}_W^G = \{(v^*, u, y, y^*, r) \in K^{*0} \times X \times Y \times Y^* \times (K\backslash\{0\}) : (0, y^*) \in \partial(v^*\Phi)(u, y)\}$$

and

$$h_W^G(v^*, u, y, y^*, r) = \Phi(u, y) - \frac{\langle y^*, y\rangle}{\langle v^*, r\rangle} r$$

and

$$(DVG_M) \quad \underset{(v^*, u, y^*) \in \mathcal{B}_M^G}{\text{Max}} h_M^G(v^*, u, y^*),$$

where

$$\mathcal{B}_M^G = \{(v^*, u, y^*) \in K^{*0} \times X \times Y^* : (0, y^*) \in \partial(v^*\Phi)(u, 0)\}$$

and

$$h_M^G(v^*, u, y^*) = \Phi(u, 0).$$

Note that $h_M^G(\mathcal{B}_M^G) \subseteq h_W^G(\mathcal{B}_W^G) \subseteq V$.

Remark 6.5.2. Considering the primal vector problem with respect to weakly efficient solutions

$$(PVG_w) \quad \underset{x \in X}{\text{WMin}} F(x),$$

these vector duals in this subsection can be modified to become ones with respect to weakly efficient solutions, too, by taking v^* in the larger set $K^*\backslash\{0\}$ and, concerning only (DVG_W), restricting the variable r to take values only in $\text{int}(K)$.

Remark 6.5.3. Fixing $r \in K\backslash\{0\}$, we can also construct, starting from the vector dual (DVG_W), a family of dual problems to (PVG), defined as

$$(DVG_W(r)) \quad \underset{(v^*, u, y, y^*) \in \mathcal{B}_{W_r}^G}{\text{Max}} h_{W_r}^G(v^*, u, y, y^*),$$

where

$$\mathcal{B}_{W_r}^G = \{(v^*, u, y, y^*) \in K^{*0} \times X \times Y \times Y^* : (0, y^*) \in \partial(v^*\Phi)(u, y), \langle v^*, r\rangle = 1\}$$

and

$$h_{W_r}^G(v^*, u, y, y^*) = \Phi(u, y) - \langle y^*, y\rangle r.$$

The way these vector dual problems to (PVG) are constructed is partially based on ideas from [44, 200] and on a vector duality approach which has been discussed in chapter 4. The duality statements are given without proofs, which can be made either directly, or by analogy to the corresponding results from chapter 4.

Theorem 6.5.4. *There is no $x \in X$ and no $(v^*, u, y, y^*, r) \in \mathcal{B}_W^G$ such that $F(x) \leq_K h_W^G(v^*, u, y, y^*, r)$.*

Theorem 6.5.5. *There is no $x \in X$ and no $(v^*, u, y^*) \in \mathcal{B}_M^G$ such that $F(x) \leq_K h_M^G(v^*, u, y^*)$.*

Theorem 6.5.6. *Assume the fulfillment of the following regularity condition*

$$(RCV^{\Phi}) \mid \exists x' \in X \text{ such that } (x', 0) \in \operatorname{dom} \Phi \text{ and } \Phi(x', \cdot) \text{ is continuous at } 0.$$

If $\bar{x} \in X$ is a properly efficient solution to (PVG), then there exists some $(\bar{v}^*, \bar{y}, \bar{y}^*, \bar{r}) \in K^{*0} \times Y \times Y^* \times (K \setminus \{0\})$ such that $(\bar{v}^*, \bar{x}, \bar{y}, \bar{y}^*, \bar{r})$ is an efficient solution to (DVG_W), $(\bar{v}^*, \bar{x}, \bar{y}^*)$ is an efficient solution to (DVG_M) and $F(\bar{x}) = h_W^G(\bar{v}^*, \bar{x}, \bar{y}, \bar{y}^*, \bar{r}) = h_M^G(\bar{v}^*, \bar{x}, \bar{y}^*)$.

Remark 6.5.4. As a regularity condition in Theorem 6.5.6 one can use any condition that ensures for $v^* \in K^{*0}$ the stability of the scalar optimization problem

$$\inf_{x \in X} (\bar{v}^* \Phi)(x, 0)$$

with respect to its general conjugate dual, like (RC_i^{Φ}), $i \in \{1, 2, 3\}$ from Theorem 3.2.1. A closedness type regularity condition (see (RC_4^{Φ}) in Theorem 3.2.3) could be taken into consideration, too, but it has as disadvantage the fact that $\bar{v}^*$ appears in its formulation.

Remark 6.5.5. In the particular instance when (PVG) is (PV^C) one can consider the perturbation functions introduced in subsection 4.3.2, obtaining thus new vector duals. The vector dual problems treated in subsection 6.2.1 turn out to be derivable from the duals introduced above considered when $V = \mathbb{R}^k$ and $K = \mathbb{R}_+^k$, the Wolfe vector dual after fixing $r = e$ and the Mond-Weir dual after removing the constraint $g(u) \in -C$, like in the scalar case, from (DVG_M).

Remark 6.5.6. In case $V = \mathbb{R}$ and $K = \mathbb{R}_+$ for $F : X \to \overline{\mathbb{R}}$ proper and convex the problem (PVG) becomes (PG), while the duals (DVG_W) and (DVG_M) turn into (DG_W) and (DG_M), respectively.

Remark 6.5.7. It is also possible to assign to (PVG) vector duals of Wolfe and Mond-Weir type by following the approach in section 4.3, by employing the scalar duals introduced in subsection 6.5.1.

Bibliographical notes

The Wolfe dual to a constrained scalar optimization problem was introduced by Wolfe in [202], while the Mond-Weir dual followed twenty years later, due to Mond and Weir [138]. In both cases the functions involved were considered differentiable and endowed with generalized convexity properties. Then these duality concepts evolved parallelly and one can distinguish two main directions for both of them. On one hand, considering the primal problem convex, the differentiability assumptions were dropped and the gradients that appear in the duals were replaced by subdifferentials. We mention here papers like [109, 166]. On the other hand, especially for Mond-Weir duality, the differentiability continued to play an important role and the convexity assumptions on the functions involved were weakened, in papers like [11, 126, 138]. Besides the classical Mond-Weir dual, different closely related Mond-Weir type duals were also proposed, as well as combinations of the Wolfe and Mond-Weir duals.

In the vector case, what we called here the classical Wolfe type and Mond Weir type duals were constructed by following ideas from the literature. Wolfe type vector duals were considered in [61,99,191,192,197,203], while for Mond-Weir type vector duals we refer to [6,61,62,195,197]. In these papers different duality instances were considered, with respect to properly efficient, efficient and weakly efficient solutions, respectively. In both scalar and vector cases there is also a rich literature on both Wolfe type and Mond Weir type duality for optimization problems with geometric and both inequality and equality constraints. Because they can be seen as special cases of the problems with geometric and cone constraints, such problems were not treated here.

As they can be employed also to non-convex optimization problems, the Wolfe and Mond-Weir duality concepts were considered in different applications, like duality without regularity condition (cf. [63, 109, 196, 198]), symmetric and self duality (cf. [111, 115, 116, 138, 171]), fractional programming (cf. [5, 12, 46, 112]) etc.

In the last section we used the perturbation theory from chapters 3 and 4 in relation to the Wolfe and Mond-Weir duality concepts.

7

Duality for set-valued optimization problems based on vector conjugacy

Since the early eighties of the last century there have been attempts to extend the perturbation approach for scalar duality (as developed in chapter 3 from scalar optimization problems) to vector duality in connection with different generalizations of the conjugacy concept. The idea of conjugate functions and subdifferentiability in scalar optimization is also fruitful in vector optimization. But, because of the different solution notions in vector optimization and the occurring partial orderings, there are several possibilities to define conjugate maps and vector-valued subgradients. In the current chapter we present some of these approaches extended to set-valued maps and, starting from these investigation, we develop a duality scheme for set-valued optimization problems.

7.1 Conjugate duality based on efficient solutions

The perturbation approach to vector duality based on efficiency has been developed in finite dimensional spaces by Tanino, Sawaragi and Nakayama in [163, 180]. We give here an extended approach in topological vector spaces, while the optimization problems we treat involve set-valued maps instead of only vector-valued functions.

7.1.1 Conjugate maps and the subdifferential of set-valued maps

First of all we need some preliminaries and definitions concerning minimality and maximality notions for sets in extended topological vector spaces. To this end we extend the definitions and results from subsection 2.4.1 to this general setting.

Unless otherwise mentioned, in the following we consider X and V to be topological vector spaces with X^* and V^* topological dual spaces, respectively. Moreover, let V be partially ordered by the nontrivial pointed convex cone $K \subseteq V$. Next we define minimal and maximal elements of a set $M \subseteq \overline{V} =$

R.I. Boţ et al., *Duality in Vector Optimization*, Vector Optimization,
DOI 10.1007/978-3-642-02886-1_7,

$V \cup \{\pm\infty_K\}$ with respect to the partial ordering induced by the cone K. As one can easily notice, the definition below extends Definition 2.4.1 to sets in $\overline{V}$.

Definition 7.1.1. *Let K be a nontrivial pointed convex cone in V and $M \subseteq \overline{V}$. In the case $M = \emptyset$, take by convention* $\mathrm{Min}(M,K) = \{+\infty_K\}$ *and $Max(M,K) = \{-\infty_K\}$. Otherwise, we say that $\bar{v} \in M$ is a minimal element of M if there is no $v \in M$ such that $v \leq_K \bar{v}$. The set of all minimal elements of M is denoted by* $\mathrm{Min}(M,K)$ *and it is called the minimal set of M. Accordingly, we say that $\bar{v} \in M$ is a maximal element of M if there is no $v \in M$ such that $v \geq_K \bar{v}$. The set of all maximal elements of M is denoted by* $\mathrm{Max}(M,K)$ *and it is called the maximal set of M.*

It is easy to see that for the notions introduces in Definition 7.1.1 we have $\mathrm{Max}(M,K) = \mathrm{Min}(M,-K) = -\mathrm{Min}(-M,K)$. The operations with sets in $\overline{V}$ are taken like for sets in V, as mentioned in section 2.1. We also notice that if $-\infty_K \in \mathrm{Max}(M,K)$, then $M = \{-\infty_K\}$ or $M = \emptyset$, while when $+\infty_K \in \mathrm{Min}(M,K)$ one has $M = \{+\infty_K\}$ or $M = \emptyset$.

Throughout this chapter we agree that if $K \subseteq V$ is a given ordering cone and there is no possibility of confusion, then instead of $\mathrm{Min}(M,K)$ and $\mathrm{Max}(M,K)$ the simplified notations $\mathrm{Min}\, M$ and $\mathrm{Max}\, M$, respectively, are used.

Next we define the conjugate map, the biconjugate map and the subdifferential of a set-valued map. For a set-valued map $F : X \rightrightarrows \overline{V}$ let its *graph* be $\mathrm{gph}\, F := \{(x,v) \in X \times \overline{V} : v \in F(x)\}$, its *domain* be $\mathrm{dom}\, F := \{x \in X : F(x) \neq \emptyset \text{ and } F(x) \neq \{+\infty_K\}\}$ and, for a set $U \subseteq X$, denote $F(U) := \cup_{x \in U} F(x)$.

Definition 7.1.2. *Let $F : X \rightrightarrows \overline{V}$ be a set-valued map.*

(a) The set-valued map

$$F^* : \mathcal{L}(X,V) \rightrightarrows \overline{V},\ F^*(T) = \mathrm{Max} \underset{x \in X}{\cup} [Tx - F(x)],$$

is called the conjugate map of F.

(b) The set-valued map

$$F^{**} : X \rightrightarrows \overline{V},\ F^{**}(x) = \mathrm{Max} \underset{T \in \mathcal{L}(X,V)}{\cup} [Tx - F^*(T)],$$

is called the biconjugate map of F.

(c) The operator $T \in \mathcal{L}(X,V)$ is said to be a subgradient of F at $(x,v) \in$ $\mathrm{gph}\, F \cap (X \times V)$ *if*

$$Tx - v \in \mathrm{Max} \underset{y \in X}{\cup} [Ty - F(y)].$$

The set of all subgradients of F at $(x,v) \in$ $\mathrm{gph}\, F \cap (X \times V)$ *is called the subdifferential of F at (x,v) and it is denoted by $\partial F(x;v)$. Further, for all $x \in X$ denote $\partial F(x) := \cup_{v \in F(x) \cap V} \partial F(x;v)$. If for all $v \in F(x) \cap V$ we have $\partial F(x;v) \neq \emptyset$ then F is said to be subdifferentiable at x.*

It is easy to see that the previous definitions are inspired by the definitions of the conjugate, biconjugate, subgradient and, respectively, subdifferential of a function with values in the extended real-valued space. By particularizing the above definition to the scalar case, we do not completely get the classical definition from chapter 2 as far as conjugacy is concerned, because we use "max" instead of "sup". There are also notions of supremum and infimum corresponding to maximality and minimality, but they have some computational disadvantages. In section 7.4 we deal with similar relations based on weak minimality and weak maximality. That approach has better properties and allows to develop a self-contained theory for vector conjugacy, subdifferentiability and duality based on those weak type notions.

Remark 7.1.1. The particularization of Definition 7.1.2 to vector-valued functions $f : X \to \overline{V}$ follows directly. One gets for all $T \in \mathcal{L}(X,V)$ for the conjugate of f the following formula $f^*(T) = \operatorname{Max}\{Tx - f(x) : x \in X\} = \operatorname{Max}\{Tx - f(x) : x \in \operatorname{dom} f\}$. If $T \in \mathcal{L}(X,V)$ is a subgradient of f at $(\bar{x}, f(\bar{x}))$, where $\bar{x} \in X$ and $f(\bar{x}) \in V$, we simply say that it is a *subgradient of f at $\bar{x}$*. Thus $T\bar{x} - f(\bar{x}) \in \operatorname{Max}\{Tx - f(x) : x \in X\}$, which is equivalent to the fact that there is no $x \in X$ such that $Tx - f(x) \geq_K T\bar{x} - f(\bar{x})$, or, in other words, for all $x \in X$ there is $f(x) - f(\bar{x}) \nleq_K T(x - \bar{x})$. This is further equivalent to $f(x) - f(\bar{x}) \nleq_K T(x - \bar{x})$ for all $x \in X$. The set of all subgradients of f at $\bar{x} \in X$, when $f(\bar{x}) \in V$, is said to be the *subdifferential of f at $\bar{x}$* and it is denoted by $\partial f(\bar{x})$. It is easy to see that for $V = \mathbb{R}$, $K = \mathbb{R}_+$, $f : X \to \overline{\mathbb{R}}$, $\bar{x} \in X$ with $f(\bar{x}) \in \mathbb{R}$ and $T \in \mathcal{L}(X,\mathbb{R}) = X^*$, the fact that $f(x) - f(\bar{x}) \nleq_K T(x - \bar{x})$ for all $x \in X$ is equivalent to $f(x) - f(\bar{x}) \geq \langle T, x - \bar{x}\rangle$ for all $x \in X$. In this way we rediscover the classical definition of the scalar subgradient as given in Definition 2.3.2.

The conjugate map of the set-valued map $F : X \rightrightarrows \overline{V}$ has some useful properties. Let $G : X \rightrightarrows \overline{V}$ be such that $G(x) = F(x - x_0)$, where $x_0 \in X$ is arbitrarily taken. Then it holds

$$G^*(T) = F^*(T) + Tx_0 \ \forall T \in \mathcal{L}(X,V) \tag{7.1}$$

and

$$G^{**}(x) = F^{**}(x - x_0) \ \forall x \in X. \tag{7.2}$$

Indeed, for $T \in \mathcal{L}(X,V)$ one has

$$G^*(T) = \operatorname{Max} \bigcup_{x\in X}[Tx - G(x)] = \operatorname{Max} \bigcup_{x\in X}[Tx - F(x - x_0)]$$

$$= \operatorname{Max} \bigcup_{y\in X}[T(y + x_0) - F(y)] = \operatorname{Max} \bigcup_{y\in X}[Ty - F(y)] + Tx_0 = F^*(T) + Tx_0,$$

while, when $x \in X$, it holds

$$G^{**}(x) = \operatorname{Max} \bigcup_{T\in\mathcal{L}(X,V)}[Tx - G^*(T)]$$

$$= \operatorname{Max} \bigcup_{T \in \mathcal{L}(X,V)} [T(x - x_0) - F^*(T)] = F^{**}(x - x_0).$$

Next we provide a result which can be seen as a generalization of the Young-Fenchel inequality (cf. Proposition 2.3.2(*a*)).

Proposition 7.1.1. *Let $F : X \rightrightarrows \overline{V}$ be a set-valued map, $x \in X$ and $T \in \mathcal{L}(X,V)$. Then for all $v \in F(x)$ and all $v^* \in F^*(T)$ it holds*

$$v + v^* \nleq_K Tx. \tag{7.3}$$

Proof. Let $x \in X$ and $T \in \mathcal{L}(X,V)$ be fixed and take $v \in F(x)$ and $v^* \in F^*(T)$.

Let us assume that $v \in V$. We prove first that it cannot hold $v^* = -\infty_K$. Otherwise, $-\infty_K \in \operatorname{Max} \bigcup_{x \in X}[Tx - F(x)]$ and one can easily deduce that either $F(x) = \{+\infty_K\}$ for all $x \in X$, or $F(x) = \emptyset$ for all $x \in X$. But this is not in accordance with the above assumption $v \in F(x) \cap V$. Thus $v* \in V \cup \{+\infty_K\}$. If $v^* \in V$, by taking into consideration the definition of maximality, one can easily demonstrate that it is binding to have $v^* \nleq_K Tx - v$, i.e. (7.3) is true. On the other hand, when $v^* = +\infty_K$, by taking into consideration the rules for the addition with $\pm\infty_K$, one can easily show that (7.3) is fulfilled in this case, too.

Assume now that $v = -\infty_K$. Then $+\infty_K \in \bigcup_{x \in X}[Tx - F(x)]$ and, by Definition 7.1.1 it yields $v^* = +\infty_K$. By (2.1) we have $v + v^* = +\infty_K$ and so (7.3) is fulfilled in this case, too.

In the third situation, namely when $v = +\infty_K$, one can notice that independently of the value of v^* relation (7.3) is always true. □

Remark 7.1.2. For $v \in F(x)$ and $v^* \in F^*(T^*)$ the generalized Young-Fenchel inequality can be equivalently written as $v + v^* - Tx \notin -K \setminus \{0\}$ or $Tx - (v + v^*) \notin K \setminus \{0\}$.

In the scalar case the inequality $f^{**} \leq f$ is always valid, as seen in Proposition 2.3.4. An analogous result can be proven for set-valued maps.

Proposition 7.1.2. *Let $F : X \rightrightarrows \overline{V}$ be a set-valued map and $x \in X$. For all $v \in F(x)$ and all $u \in F^{**}(x)$ we have $v \nleq_K u$.*

Proof. Let us first demonstrate the assertion for $x = 0$. According to Proposition 7.1.1, for each $v \in F(0)$ and each $\bar{v} \in -\bigcup_{T \in \mathcal{L}(X,V)} F^*(T)$ there is $v - \bar{v} \nleq_K T0 = 0$ or, equivalently, $\bar{v} \notin v + (K \setminus \{0\})$. As $F^{**}(0) = \operatorname{Max} \bigcup_{T \in \mathcal{L}(X,V)}[T0 - F^*(T)] = \operatorname{Max} \bigcup_{T \in \mathcal{L}(X,V)}[-F^*(T)]$, it holds $v \nleq_K u$ for all $u \in F^{**}(0)$. Observe that this relation is valid also if $\bigcup_{T \in \mathcal{L}(X,V)}[-F^*(T)] = \emptyset$, because in this situation $F^{**}(0) = \{-\infty_K\}$.

Now let be $x \neq 0$. We define the set-valued map $G : X \rightrightarrows \overline{V}$, $G(y) = F(y + x)$. Taking into consideration relation (7.2), one has $G^{**}(0) = F^{**}(x)$. Consequently, for all $v \in F(x) = G(0)$ and all $u \in F^{**}(x) = G^{**}(0)$ the assertion follows directly from the first part of the proof. □

Next, some elementary properties for conjugate maps and subgradients are gathered. One can easily see that these results are actually generalizations of the corresponding ones given for extended real-valued functions.

Proposition 7.1.3. *(a) Let $F : X \rightrightarrows \overline{V}$ be a set-valued map. Then for $(x, v) \in \operatorname{gph} F \cap (X \times V)$ it holds $T \in \partial F(x; v)$ if and only if $Tx - v \in F^*(T)$.*
(b) Let $f : X \to \overline{V}$ be a vector-valued function. Then for $x \in X$ with $f(x) \in V$ there is $T \in \partial f(x)$ if and only if $Tx - f(x) \in f^*(T)$.

Proof. The assertions are direct consequences of Definition 7.1.2. □

Proposition 7.1.4. *(a) Let $F : X \rightrightarrows \overline{V}$ be a set-valued map and $(x, v) \in \operatorname{gph} F \cap (X \times V)$. Then $\partial F(x; v) \neq \emptyset$ if and only if $v \in F^{**}(x)$. Consequently, F is subdifferentiable at x if and only if $F(x) \cap V \subseteq F^{**}(x)$.*
(b) Let $f : X \to \overline{V}$ be a vector-valued function. For any $x \in X$ with $f(x) \in V$ we have $\partial f(x) \neq \emptyset$ if and only if $f(x) \in f^{**}(x)$.

Proof. (a) Let $(x, v) \in \operatorname{gph} F \cap (X \times V)$. Suppose that $\partial F(x; v) \neq \emptyset$ and let $\overline{T} \in \partial F(x; v)$ be arbitrarily taken. According to Proposition 7.1.3 it holds $v \in \overline{T}x - F^*(\overline{T}) \subseteq \cup_{T \in \mathcal{L}(X,V)}[Tx - F^*(T)]$. Proposition 7.1.1 says that $v \nleq_K Tx - v^*$ for all $T \in \mathcal{L}(X, V)$ and all $v^* \in F^*(T)$. But this actually means that there is no $\tilde{v} \in \cup_{T \in \mathcal{L}(X,V)}[Tx - F^*(T)]$ such that $v \leq_K \tilde{v}$ and $v \in \operatorname{Max} \cup_{T \in \mathcal{L}(X,V)}[Tx - F^*(T)] = F^{**}(x)$.

Now assume that $v \in F^{**}(x)$. By definition it holds $v \in \overline{T}x - F^*(\overline{T})$ for some $\overline{T} \in \mathcal{L}(X, V)$. Proposition 7.1.3 yields $\overline{T} \in \partial F(x; v)$ and the first equivalence is proven.

The equivalent characterization of the subdifferentiability of F at x follows then directly from Definition 7.1.2.

(b) The assertion follows from (a) by taking into consideration Remark 7.1.1. □

Lemma 7.1.5. *Let $M_1, M_2 \subseteq \overline{V}$ be given sets. The following statements are true*

(a) if $\operatorname{Min} M_1 \neq \emptyset$ and $\operatorname{Min} M_2 \neq \emptyset$, then $\operatorname{Min}(M_1 + M_2) \subseteq \operatorname{Min} M_1 + \operatorname{Min} M_2$;
(b) if $\operatorname{Max} M_1 \neq \emptyset$ and $\operatorname{Max} M_2 \neq \emptyset$, then $\operatorname{Max}(M_1 + M_2) \subseteq \operatorname{Max} M_1 + \operatorname{Max} M_2$.

Proof. (a) Let $v \in \operatorname{Min}(M_1 + M_2)$ be arbitrarily taken. We treat three cases.

Suppose first that $v \in V$. Then there exist $v_1 \in M_1$ and $v_2 \in M_2$ such that $v = v_1 + v_2$. If $v_1 \notin \operatorname{Min} M_1$, there exists $\bar{v}_1 \in M_1$ such that $\bar{v}_1 \leq_K v_1$. Then $\bar{v}_1 + v_2 \in M_1 + M_2$ and $\bar{v}_1 + v_2 \leq_K v_1 + v_2 = v$, which contradicts the fact that v is a minimal element of $M_1 + M_2$. Thus $v_1 \in \operatorname{Min} M_1$. By the same argument one can prove that $v_2 \in \operatorname{Min} M_2$, too.

Assume that $v = +\infty_K$. Then we actually have $\operatorname{Min}(M_1 + M_2) = \{+\infty_K\}$ and this means that either $M_1 + M_2 = \{+\infty_K\}$ or $M_1 + M_2 = \emptyset$. If $M_1 + M_2 =$

$\{+\infty_K\}$, then the sets M_1 and M_2 are nonempty and $M_1 = \{+\infty_K\}$ or $M_2 = \{+\infty_K\}$. In both cases, because $\operatorname{Min} M_1 \neq \emptyset$ and $\operatorname{Min} M_2 \neq \emptyset$, the right hand side of the inclusion relation in (a) is equal to $\{+\infty_K\}$. If $M_1 + M_2 = \emptyset$, then $M_1 = \emptyset$ or $M_2 = \emptyset$, meaning again that (a) is fulfilled.

Finally, we assume that $v = -\infty_K$. Then $\{-\infty_K\} = \operatorname{Min}(M_1 + M_2)$ and $-\infty_K \in M_1 + M_2$. This also implies that $M_1 \neq \emptyset$ and $M_2 \neq \emptyset$. Without loss of generality we may assume that $-\infty_K \in M_1$. Taking into consideration the calculation rules when dealing with $\pm\infty_K$ one can see that $M_2 \neq \{+\infty_K\}$ and so $+\infty_K \notin \operatorname{Min} M_2$. Even more, since $\operatorname{Min} M_2$ is nonempty and $\operatorname{Min} M_1 = \{-\infty_K\}$, it holds $\operatorname{Min} M_1 + \operatorname{Min} M_2 = \{-\infty_K\}$.

(b) This part follows from (a) by taking account of the fact that for $M \subseteq \overline{V}$ one has $\operatorname{Max} M = -\operatorname{Min}(-M)$. $\square$

Corollary 7.1.6. *Let* $M_1, M_2 \subseteq V$ *be given. Then the inclusions* $\operatorname{Min}(M_1 + M_2) \subseteq \operatorname{Min} M_1 + \operatorname{Min} M_2$ *and* $\operatorname{Max}(M_1 + M_2) \subseteq \operatorname{Max} M_1 + \operatorname{Max} M_2$ *hold without additional assumptions.*

Remark 7.1.3. If in the situation considered in Corollary 7.1.6 $\operatorname{Min} M_1 = \emptyset$ or $\operatorname{Min} M_2 = \emptyset$, then $\operatorname{Min}(M_1 + M_2) = \emptyset$, too (see the first part of the proof of Lemma 7.1.5).

Definition 7.1.3. *Let* $M \subseteq \overline{V}$ *be a given set.*

(a) The set $\operatorname{Min} M$ *is said to be externally stable if* $M \setminus \{+\infty_K\} \subseteq \operatorname{Min} M + K$.
(b) The set $\operatorname{Max} M$ *is said to be externally stable if* $M \setminus \{-\infty_K\} \subseteq \operatorname{Max} M - K$.

Remark 7.1.4. External stability plays an important role in the following considerations. Assuming it fulfilled for $\operatorname{Min} M$, it means that each point of $M \setminus \{+\infty_K\}$ outside the set of minimal elements of M is dominated by a minimal element of M. That is why this property is sometimes called in the literature the *domination property* (first introduced in [181]). Although different criteria ensuring external stability can be found in the existent literature (see [125] for more details), we quote here only one result given in finite dimensional spaces (see [163, Theorem 3.2.9]).

Let K be a convex cone in $\mathbb{R}^m$. A set $M \subseteq \mathbb{R}^m$ is called *K-compact* if the set $(v - \operatorname{cl}(K)) \cap M$ is compact for any $v \in M$. If K is a pointed convex closed cone in $\mathbb{R}^m$ and $M \subseteq \mathbb{R}^m$ is a nonempty K-compact set, then $\operatorname{Min} M$ is externally stable.

Proposition 7.1.7. *Let* $F,\ G : X \rightrightarrows \overline{V}$ *be set-valued maps. Assume that for all* $x \in X$ *with* $+\infty_K \in F(x)$ *it holds* $G(x) \neq \emptyset$ *and* $\operatorname{Max} G(x) \neq \emptyset$. *Then*

$$\operatorname{Max} \bigcup_{x \in X} [F(x) + G(x)] \subseteq \operatorname{Max} \bigcup_{x \in X} [F(x) + \operatorname{Max} G(x)]. \tag{7.4}$$

If, additionally, $\operatorname{Max} G(x)$ *is externally stable for all* $x \in X$, *then the reverse inclusion holds, too.*

Proof. First we prove (7.4). Take an arbitrary $v \in \operatorname{Max} \cup_{x\in X}[F(x)+G(x)]$. Further we treat three cases.

Suppose first that $v \in V$. Then there exist $\bar{x} \in X$, $v_1 \in F(\bar{x}) \cap V$ and $v_2 \in G(\bar{x}) \cap V$ such that $v = v_1 + v_2$. If $v_2 \notin \operatorname{Max} G(\bar{x})$, by definition there exists $\bar{v}_2 \in G(\bar{x})$ such that $v_2 \leq_K \bar{v}_2$. Then $v = v_1 + v_2 \leq_K v_1 + \bar{v}_2 \in \cup_{x\in X}[F(x)+G(x)]$ and this contradicts the maximality of v. Therefore $v_2 \in \operatorname{Max} G(\bar{x})$ and so $v = v_1 + v_2 \in \cup_{x\in X}[F(x)+\operatorname{Max} G(x)]$. That v is a maximal element of the set $\cup_{x\in X}[F(x)+\operatorname{Max} G(x)]$ is a trivial consequence of the fact that in this case the inclusion $\cup_{x\in X}[F(x)+\operatorname{Max} G(x)] \subseteq \cup_{x\in X}[F(x)+G(x)]$ holds.

Assume that $v = +\infty_K$. Then $+\infty_K \in \cup_{x\in X}[F(x)+G(x)]$ and this secures the existence of some $\bar{x} \in X$ such that $+\infty_K \in F(\bar{x}) + G(\bar{x})$. Then two situations can occur: either $+\infty_K \in F(\bar{x})$ and $G(\bar{x}) \neq \emptyset$ or $F(\bar{x}) \neq \emptyset$ and $+\infty_K \in G(\bar{x})$. If the first situation is valid, then we cannot have $\operatorname{Max} G(\bar{x}) = \emptyset$ (the assumptions of the proposition impose this). Thus $+\infty_K \in F(\bar{x}) + \operatorname{Max} G(\bar{x})$ and from here we deduce that $+\infty_K \in \cup_{x\in X}[F(x)+\operatorname{Max} G(x)]$. Consequently, $\operatorname{Max} \cup_{x\in X}[F(x)+\operatorname{Max} G(x)] = \{+\infty_K\}$. Following a similar reasoning the same equality can be proven also if the second situation occurs. Finally, in this case it holds

$$\operatorname{Max} \bigcup_{x\in X}[F(x)+G(x)] = \operatorname{Max} \bigcup_{x\in X}[F(x)+\operatorname{Max} G(x)] = \{+\infty_K\}.$$

Let us assume that $v = -\infty_K$. Then, according to Definition 7.1.1 either $\cup_{x\in X}[F(x)+G(x)] = \{-\infty_K\}$, or $\cup_{x\in X}[F(x)+G(x)] = \emptyset$.

In the first situation for all $x \in X$ there is either $F(x)+G(x) = \{-\infty_K\}$, or $F(x)+G(x) = \emptyset$ and at least for one $y \in X$ there is $F(y)+G(y) = \{-\infty_K\}$. Let $x \in X$ be fixed. If $F(x)+G(x) = \{-\infty_K\}$ then $F(x) = \{-\infty_K\}$ and $+\infty_K \notin G(x)$, $G(x) \neq \emptyset$ or $G(x) = \{-\infty_K\}$ and $\{+\infty_K\} \notin F(x)$, $F(x) \neq \emptyset$. If $F(x)+G(x) = \emptyset$, then $F(x) = \emptyset$ or $G(x) = \emptyset$. If $G(x) = \emptyset$ then $+\infty_K \notin F(x)$ in accordance with our assumptions. In all these cases one can see easily that $F(x)+\operatorname{Max} G(x)$ is equal to $\{-\infty_K\}$ or $\emptyset$. Therefore, $\operatorname{Max} \cup_{x\in X}[F(x)+\operatorname{Max} G(x)] = \{-\infty_K\}$. In fact, we have

$$\operatorname{Max} \bigcup_{x\in X}[F(x)+G(x)] = \operatorname{Max} \bigcup_{x\in X}[F(x)+\operatorname{Max} G(x)] = \{-\infty_K\}. \quad (7.5)$$

It remains to consider the second situation in this third case, namely when the set $\cup_{x\in X}[F(x)+G(x)]$ is empty. Then $F(x)+G(x) = \emptyset$ for all $x \in X$. Let $x \in X$ be fixed. If $F(x) = \emptyset$, then $F(x)+\operatorname{Max} G(x) = \emptyset$. In the case $G(x) = \emptyset$, by the assumptions made we must have that $+\infty_K \notin F(x)$ and therefore $F(x)+\operatorname{Max} G(x) = F(x)+\{-\infty_K\}$ which is empty if $F(x) = \emptyset$ and $\{-\infty_K\}$, otherwise. Thus $\cup_{x\in X}[F(x)+\operatorname{Max} G(x)]$ is equal to $\{-\infty_K\}$ or $\emptyset$ and in both situations we have $\operatorname{Max} \cup_{x\in X}[F(x)+\operatorname{Max} G(x)] = \{-\infty_K\}$. Consequently, (7.5) is also in this case valid.

Now let us prove that the reverse inclusion holds when $\operatorname{Max} G(x)$ is externally stable for all $x \in X$.

Let be $v \in \operatorname{Max} \cup_{x \in X}[F(x) + \operatorname{Max} G(x)]$. Suppose first that $v \in V$. Then there exists $\bar{x} \in X$, $v_1 \in F(\bar{x}) \cap V$ and $v_2 \in \operatorname{Max} G(\bar{x}) \cap V$ such that $v = v_1 + v_2$. Thus $v_2 \in G(\bar{x}) \cap V$. Consequently, $v \in \cup_{x \in X}[F(x) + G(x)]$. If $v \notin \operatorname{Max} \cup_{x \in X}[F(x) + G(x)]$ then there exists $\tilde{v} \in \cup_{x \in X}[F(x) + G(x)]$ such that $\tilde{v} \geq_K v$. By (7.4) follows that $\tilde{v} \in V$. Let be $\tilde{x} \in X$, $\tilde{v}_1 \in F(\tilde{x}) \cap V$ and $\tilde{v}_2 \in G(\tilde{x}) \cap V$, with $\tilde{v} = \tilde{v}_1 + \tilde{v}_2$. Since $G(\tilde{x}) \setminus \{-\infty_K\} \subseteq \operatorname{Max} G(\tilde{x}) - K$, there exists $\tilde{v}_3 \in \operatorname{Max} G(\tilde{x})$ with $\tilde{v}_3 \geqq_K \tilde{v}_2$ and consequently $\tilde{v}_1 + \tilde{v}_3 \geq_K v$. As $\tilde{v}_1 + \tilde{v}_3 \in \cup_{x \in X}[F(x) + \operatorname{Max} G(x)]$, this leads to a contradiction.

Assume now that $v = +\infty_K$. Then there exists $\bar{x} \in X$ such that either $+\infty_K \in F(\bar{x})$ and $\operatorname{Max} G(\bar{x}) \neq \emptyset$, or $F(\bar{x}) \neq \emptyset$ and $+\infty_K \in \operatorname{Max} G(\bar{x})$. In the first situation, by the assumption we made, it holds that $G(\bar{x}) \neq \emptyset$ and so $+\infty_K \in F(\bar{x}) + G(\bar{x})$. In the second situation we have $+\infty_K \in G(\bar{x})$ and also in this case one has $+\infty_K \in F(\bar{x}) + G(\bar{x})$. In conclusion, one can easily see that $\operatorname{Max} \cup_{x \in X}[F(x) + G(x)] = \{+\infty_K\}$.

We come now to the case when $v = -\infty_K$. Then either $\cup_{x \in X}[F(x) + \operatorname{Max} G(x)] = \{-\infty_K\}$, or $\cup_{x \in X}[F(x) + \operatorname{Max} G(x)] = \emptyset$. In the first situation, for all $x \in X$ there is $F(x) + \operatorname{Max} G(x) = \{-\infty_K\}$ or $F(x) + \operatorname{Max} G(x) = \emptyset$ and $F(y) + \operatorname{Max} G(y) = \{-\infty_K\}$ at least for one $y \in X$. Let $x \in X$ be fixed. If $F(x) + \operatorname{Max} G(x) = \{-\infty_K\}$, then either $F(x) = \{-\infty_K\}$ and $+\infty_K \notin \operatorname{Max} G(x)$, $\operatorname{Max} G(x) \neq \emptyset$, or $\operatorname{Max} G(x) = \{-\infty_K\}$ and $+\infty_K \notin F(x)$, $F(x) \neq \emptyset$. In both cases $F(x) + G(x)$ is equal to $\{-\infty_K\}$ or $\emptyset$. Assume now that $F(x) + \operatorname{Max} G(x) = \emptyset$. Then $F(x) = \emptyset$, as we have that $\operatorname{Max} G(x) = \emptyset$ cannot occur because of the external stability of $\operatorname{Max} G(x)$. Thus $F(x) + G(x) = \emptyset$ and one can see that in all these situations $F(x) + G(x)$ is equal to $\{-\infty_K\}$ or $\emptyset$. Consequently, $\operatorname{Max} \cup_{x \in X}[F(x) + G(x)] = \{-\infty_K\}$.

Now we consider the second situation in this third case, namely when $F(x) + \operatorname{Max} G(x) = \emptyset$ for all $x \in X$. As seen above, this can be the case only if $F(x) = \emptyset$ for all $x \in X$. This means that $\cup_{x \in X}[F(x) + G(x)] = \emptyset$ and so $\operatorname{Max} \cup_{x \in X}[F(x) + G(x)] = \{-\infty_K\}$. One can easily conclude that also in this case $v \in \operatorname{Max} \cup_{x \in X}[F(x) + G(x)]$ and the reverse inclusion is proven. □

Corollary 7.1.8. *Let $F : X \rightrightarrows \overline{V}$ be a set-valued map. Then for any $T \in \mathcal{L}(X, V)$ it holds*

$$F^*(T) \subseteq \operatorname{Max} \bigcup_{x \in X} [Tx - \operatorname{Min} F(x)].$$

If $\operatorname{Min} F(x)$ is externally stable for all $x \in X$, then the converse inclusion holds, too.

Proof. Let us consider the maps $\widetilde{F} : X \rightrightarrows V$, $\widetilde{F}(x) = \{Tx\}$ and $\widetilde{G} : X \rightrightarrows \overline{V}$, $\widetilde{G}(x) = -F(x)$. Since for all $x \in X$ it is impossible to have $+\infty_K \in \widetilde{F}(x)$, the assumption of Proposition 7.1.7 is automatically fulfilled. Even more, as $\operatorname{Max} \widetilde{G}(x) = \operatorname{Max}(-F(x)) = -\operatorname{Min} F(x)$, the external stability of $\operatorname{Min} F(x)$ guarantees the external stability of $\operatorname{Max} \widetilde{G}(x)$. Proposition 7.1.7 applied to $\widetilde{F}$ and $\widetilde{G}$ leads to the desired inclusions. □

Corollary 7.1.9. *Let* $F : X \rightrightarrows \overline{V}$ *be a set-valued map and* $\operatorname{Max} F(x)$ *be externally stable for all* $x \in X$. *Then* $\operatorname{Max} \cup_{x \in X} F(x) = \operatorname{Max} \cup_{x \in X} \operatorname{Max} F(x)$.

Proof. Apply Proposition 7.1.7 to the maps $\widetilde{F} : X \rightrightarrows V$, $\widetilde{F}(x) = \{0\}$ and $\widetilde{G} : X \rightrightarrows \overline{V}$, $\widetilde{G}(x) = F(x)$. □

Remark 7.1.5. (*a*) The assertions of Proposition 7.1.7, Corollary 7.1.8 and Corollary 7.1.9 may be formulated also for minimal elements. For instance one has $\operatorname{Min} \cup_{x \in X} F(x) = \operatorname{Min} \cup_{x \in X} \operatorname{Min} F(x)$ if $\operatorname{Min} F(x)$ is externally stable for all $x \in X$. We would also like to underline the fact that in Proposition 7.1.7 and its corollaries the external stability cannot be omitted (for details see [163, Remark 6.1.1], where a counterexample has been given in finite dimensional spaces).

(*b*) It is worth mentioning that all the basic notions and results presented in this subsection concerning minimality and maximality remain true if considered in the framework of vector spaces X and V, the latter being partially ordered by a nontrivial pointed convex cone $K \subseteq V$.

7.1.2 The perturbation approach for conjugate duality

In this subsection the perturbation approach from section 3.1 for scalar optimization problems will be partially extended to set-valued optimization problems. Let $F : X \rightrightarrows V \cup \{+\infty_K\}$ be a set-valued map whose domain is a nonempty set. We consider the general set-valued optimization problem

$$(PSVG) \quad \operatorname*{Min}_{x \in X} F(x).$$

This actually means to find the minimal elements of the image set $F(X) \subseteq V \cup \{+\infty_K\}$ with respect to the partial ordering induced by the nontrivial pointed convex cone $K \subseteq V$ or, in other words, to look for an element $\bar{x} \in X$ such that there exists $\bar{v} \in F(\bar{x})$ with $\bar{v} \in \operatorname{Min} F(X)$. In this situation the element $\bar{x}$ is said to be an *efficient solution* to the problem $(PSVG)$ and $(\bar{x}, \bar{v})$ is called a *minimal pair* to the problem $(PSVG)$. A particular instance of the previous problem arises when the set-valued map F is replaced with the proper vector-valued function $f : X \to V \cup \{+\infty_K\}$. In this case the problem $(PSVG)$ becomes

$$(PVG) \quad \operatorname*{Min}_{x \in X} f(x)$$

and one looks for efficient solutions $\bar{x} \in X$ fulfilling $f(\bar{x}) \in \operatorname{Min} f(X)$. The efficient solutions to the problems $(PSVG)$ and (PVG) may be described via the subdifferential like in the scalar case, as follows from the way this is defined.

Proposition 7.1.10. *(a) An element* $\bar{x} \in X$ *is an efficient solution to* $(PSVG)$ *if and only if there exists* $\bar{v} \in F(\bar{x}) \cap V$ *such that* $0 \in \partial F(\bar{x}; \bar{v})$.

(b) *An element* $\bar{x} \in X$*, for which* $f(\bar{x}) \in V$*, is an efficient solution to* (PVG) *if and only if* $0 \in \partial f(\bar{x})$.

Our next aim is to attach a set-valued dual problem to $(PSVG)$. Notice that all what follows can be done also for the problem (PVG), since, as seen above, this problem can be treated as a special instance of $(PSVG)$. Similarly to section 3.1 we introduce a *set-valued perturbation map* $\Phi : X \times Y \rightrightarrows V \cup \{+\infty_K\}$ such that $\Phi(x,0) = F(x)$ for all $x \in X$. The topological vector space Y is called *perturbation space* and its topological dual space is denoted by Y^*. Then $(PSVG)$ is embedded into a family of perturbed problems

$$(PSVG_y) \ \operatorname*{Min}_{x \in X} \Phi(x,y),$$

where $y \in Y$ is the *perturbation variable*. Clearly, the problem $(PSVG_0)$ coincides with the problem $(PSVG)$. As in the scalar case the dual problem is defined by making use of the conjugate of the perturbation map

$$\Phi^* : \mathcal{L}(X,V) \times \mathcal{L}(Y,V) \rightrightarrows \overline{V}, \ \Phi^*(T,\Lambda) = \operatorname{Max} \bigcup_{x \in X, y \in Y} [Tx + \Lambda y - \Phi(x,y)].$$

Thus to the primal problem $(PSVG)$ one can attach the set-valued dual problem

$$(DSVG) \quad \operatorname*{Max}_{\Lambda \in \mathcal{L}(Y,V)} \{-\Phi^*(0,\Lambda)\}.$$

More precisely, we look for $\overline{\Lambda} \in \mathcal{L}(Y,V)$ such that there exists $\bar{v}^* \in -\Phi^*(0,\overline{\Lambda})$ fulfilling $\bar{v}^* \in \operatorname{Max} \cup_{\Lambda \in \mathcal{L}(Y,V)} \{-\Phi^*(0,\Lambda)\}$. In this case $\overline{\Lambda} \in \mathcal{L}(Y,V)$ is called *efficient solution* to $(DSVG)$ and $(\overline{\Lambda}, \bar{v}^*)$ is said to be a *maximal pair* to $(DSVG)$. As expected, having in mind the scalar case, *weak duality* holds in this very general setting.

Theorem 7.1.11. *For all* $x \in X$ *and all* $\Lambda \in \mathcal{L}(Y,V)$ *there is* $\Phi(x,0) \cap \{-\Phi^*(0,\Lambda) - (K \backslash \{0\})\} = \emptyset$.

Proof. Let $x \in X$ and $\Lambda \in \mathcal{L}(Y,V)$ be arbitrarily taken. Assume that there exists $v \in \Phi(x,0) \cap \{-\Phi^*(0,\Lambda) - (K \backslash \{0\})\}$. Then one can find a $v^* \in \Phi^*(0,\Lambda)$ such that $v + v^* \in -K \backslash \{0\}$. But by Proposition 7.1.1 it holds $v + v^* \nleq_K 0$ and this leads to a contradiction. □

One can give for the weak duality result from Theorem 7.1.11 the following equivalent formulation, followed by an immediate statement for the efficient solutions to $(PSVG)$ and $(DSVG)$.

Corollary 7.1.12. (a) *For all* $x \in X$ *and all* $\Lambda \in \mathcal{L}(Y,V)$*, it holds* $v \nleq_K v^*$ *whenever* $v \in F(x)$ *and* $v^* \in -\Phi^*(0,\Lambda)$.

(b) *Let be* $\bar{v} \in F(\bar{x}) \cap \{-\Phi^*(0,\overline{\Lambda})\}$ *for* $\bar{x} \in X$ *and* $\overline{\Lambda} \in \mathcal{L}(Y,V)$*. Then* $\bar{x}$ *is an efficient solution and* $(\bar{x},\bar{v})$ *is a minimal pair to* $(PSVG)$*, while* $\overline{\Lambda}$ *is an efficient solution and* $(\overline{\Lambda},\bar{v})$ *is a maximal pair to* $(DSVG)$.

Proof. (*a*) Obviously, the assertion is nothing but an equivalent formulation of the statement of Theorem 7.1.11.

(*b*) Assume that $(\bar{x}, \bar{v})$ is not a minimal pair to $(PSVG)$. Then there exist $x \in X$ and $v \in F(x) = \Phi(x, 0)$ such that $v \leq_K \bar{v}$, i.e. $v \in \Phi(x, 0) \cap \{-\Phi^*(0, \overline{\Lambda}) - (K \backslash \{0\})\}$, which contradicts Theorem 7.1.11. Assuming $(\overline{\Lambda}, \bar{v})$ not being a maximal pair to $(DSVG)$ allows to deduce a contradiction to Theorem 7.1.11 in a similar way. □

Like for the scalar optimization problem (PG) considered in section 3.1 we introduce in the set-valued case in an analogous way the *minimal value map* $H : Y \rightrightarrows V \cup \{+\infty_K\}$ defined by $H(y) = \operatorname{Min} \cup_{x \in X} \Phi(x, y) = \operatorname{Min} \Phi(X, y)$. It is easy to see that for all $y \in Y$ the set $H(y)$ is actually the set of minimal elements of the image set of the problem $(PSVG_y)$, while $H(0) = \operatorname{Min} \Phi(X, 0) = \operatorname{Min} F(X)$ is the set of minimal elements of the image set of $(PSVG)$. For some arbitrary $y \in Y$ we say that $H(y)$ is *externally stable* if $\operatorname{Min} \cup_{x \in X} \Phi(x, y)$ is externally stable in the sense of Definition 7.1.3.

Lemma 7.1.13. *For all $\Lambda \in \mathcal{L}(Y, V)$ it holds $\Phi^*(0, \Lambda) \subseteq H^*(\Lambda)$. If $H(y)$ is externally stable for all $y \in Y$, then $\Phi^*(0, \Lambda) = H^*(\Lambda)$.*

Proof. Let $\Lambda \in \mathcal{L}(X, V)$ be fixed. We have

$$\Phi^*(0, \Lambda) = \operatorname{Max} \bigcup_{x \in X, y \in Y} [\Lambda y - \Phi(x, y)] = \operatorname{Max} \bigcup_{y \in Y} \left[\Lambda y - \bigcup_{x \in X} \Phi(x, y)\right]$$

and

$$\begin{aligned} H^*(\Lambda) &= \operatorname{Max} \bigcup_{y \in Y} [\Lambda y - H(y)] = \operatorname{Max} \bigcup_{y \in Y} \left[\Lambda y - \operatorname{Min} \bigcup_{x \in X} \Phi(x, y)\right] \\ &= \operatorname{Max} \bigcup_{y \in Y} \left[\Lambda y + \operatorname{Max}\left(- \bigcup_{x \in X} \Phi(x, y)\right)\right]. \end{aligned}$$

Applying Proposition 7.1.7 to $F : Y \rightrightarrows V$, $F(y) = \Lambda y$, and $G : Y \rightrightarrows \overline{V}$, $G(y) = -\cup_{x \in X} \Phi(x, y)$, we get

$$\operatorname{Max} \bigcup_{y \in Y} \left[\Lambda y - \bigcup_{x \in X} \Phi(x, y)\right] \subseteq \operatorname{Max} \bigcup_{y \in Y} \left[\Lambda y + \operatorname{Max}\left(- \bigcup_{x \in X} \Phi(x, y)\right)\right]. \quad (7.6)$$

Therefore $\Phi^*(0, \Lambda) \subseteq H^*(\Lambda)$. If $\operatorname{Min} \cup_{x \in X} \Phi(x, y)$ is externally stable, then $\operatorname{Max}(-\cup_{x \in X} \Phi(x, y))$ is externally stable for all $y \in Y$, too. In this case (7.6) is fulfilled as equation, i.e. $\Phi^*(0, \Lambda) = H^*(\Lambda)$. □

For the subsequent considerations we assume that $H(y)$ is externally stable for all $y \in Y$. Then $H^*(\Lambda) = \Phi^*(0, \Lambda)$ and the dual problem $(DSVG)$ may be equivalently written in the form

$$(DSVG) \quad \operatorname{Max} \bigcup_{\Lambda \in \mathcal{L}(Y, V)} \{-H^*(\Lambda)\}.$$

Lemma 7.1.14. *It holds $\operatorname{Max} \cup_{\Lambda \in \mathcal{L}(Y, V)} \{-\Phi^*(0, \Lambda)\} = H^{**}(0)$.*

Proof. Using Lemma 7.1.13 it holds

$$H^{**}(0) = \operatorname{Max} \bigcup_{\Lambda \in \mathcal{L}(Y,V)} \{-H^*(\Lambda)\} = \operatorname{Max} \bigcup_{\Lambda \in \mathcal{L}(Y,V)} \{-\Phi^*(0, \Lambda)\}$$

and this provides the desired conclusion. □

Remark 7.1.6. It is not hard to see that Lemma 7.1.14 is a generalization of Proposition 3.1.2 given for the scalar primal-dual pair $(PG) - (DG)$.

Since $H(0)$ and $H^{**}(0)$ represent the sets of the minimal and maximal elements of the image sets of the primal and dual problem, respectively, the duality properties can be reflected by means of the relations between $H(0)$ and $H^{**}(0)$. *Strong duality* applies for $(PSVG) - (DSVG)$ if $H(0) = H^{**}(0)$, but also weakened versions like $H(0) \subseteq H^{**}(0)$ are of interest. In the first three sections of this chapter by strong duality we actually mean the latter formulation.

Definition 7.1.4. *The problem $(PSVG)$ is called stable with respect to the perturbation map Φ if the minimal value map H is subdifferentiable at* 0.

We notice that in our hypotheses $\pm\infty_K \notin H(0)$.

A quick look at Definition 7.1.4 shows the analogy with the definition of stability for scalar programming problems. Of course, this definition (see also [163] for finite dimensional spaces and vector-valued objective functions) is motivated by Proposition 7.1.4, expressing that H is subdifferentiable at 0 if and only if $H(0) \subseteq H^{**}(0)$. Thus stability equivalently characterizes strong duality as formulated in the following theorem.

Theorem 7.1.15. *The problem $(PSVG)$ is stable if and only if for each efficient solution $\bar{x} \in X$ to $(PSVG)$ and $\bar{v} \in F(\bar{x})$ such that $(\bar{x}, \bar{v})$ is a minimal pair to $(PSVG)$ there exists an efficient solution $\overline{\Lambda} \in \mathcal{L}(Y, V)$ to $(DSVG)$ such that $\bar{v} \in -\Phi^*(0, \overline{\Lambda})$ and $(\overline{\Lambda}, \bar{v})$ is a maximal pair to $(DSVG)$.*

Proof. Let $(PSVG)$ be stable. Then H is subdifferentiable at 0 and, according to Proposition 7.1.4, as $\pm\infty_K \notin H(0)$, this is equivalent to $H(0) \subseteq H^{**}(0)$. Take $\bar{x} \in X$ efficient to $(PSVG)$ with a corresponding $\bar{v} \in F(\bar{x})$ such that $\bar{v} \in H(0)$. Then Lemma 7.1.14 secures the existence of $\overline{\Lambda} \in \mathcal{L}(Y, V)$ such that $\bar{v} \in -\Phi^*(0, \overline{\Lambda})$. Further, Corollary 7.1.12 guarantees that $\bar{v}$ is an efficient solution and $(\overline{\Lambda}, \bar{v})$ is a maximal pair to $(DSVG)$.

The above considerations can be done in the opposite direction yielding $H(0) \subseteq H^{**}(0)$. But this means that H is subdifferentiable at 0 and thus the stability of $(PSVG)$ has been proven. □

Remark 7.1.7. Theorem 7.1.15 is a strong duality type assertion since it provides the existence of a common element $\bar{v}$ in the objective values of the primal and dual set-valued problems.

We emphasize once again that external stability of $H(y)$ for all $y \in Y$ is supposed. The importance of this fact also for Theorem 7.1.15 is underlined by the above proof. The strong duality claimed in Theorem 7.1.15 implies the following necessary optimality conditions of subdifferential type.

Theorem 7.1.16. *The minimal pair $(\bar{x}, \bar{v})$ to $(PSVG)$ and the corresponding maximal pair $(\overline{\Lambda}, \bar{v})$ to $(DSVG)$ from Theorem 7.1.15 satisfy the optimality condition $(0, \overline{\Lambda}) \in \partial\Phi(\bar{x}, 0; \bar{v})$ or, equivalently, $\overline{\Lambda} \in \partial H(0; \bar{v})$.*

Proof. According to Theorem 7.1.15, there is $\bar{v} \in F(\bar{x}) = \Phi(\bar{x}, 0)$ and $\bar{v} \in -\Phi^*(0, \overline{\Lambda})$. For $\overline{T} := 0 \in \mathcal{L}(X, V)$ this may be written as $\overline{T}\bar{x} + \overline{\Lambda}0 - \bar{v} \in \Phi^*(0, \overline{\Lambda})$. Proposition 7.1.3$(a)$ says that this relation is equivalent to $(0, \overline{\Lambda}) \in \partial\Phi(\bar{x}, 0; \bar{v})$. On the other hand, there holds $\bar{v} \in H(0)$ and $\bar{v} \in -H^*(\overline{\Lambda})$, which is nothing but $\overline{\Lambda}0 - \bar{v} \in H^*(\overline{\Lambda})$. Applying again Proposition 7.1.3(a) yields the equivalence to $\overline{\Lambda} \in \partial H(0; \bar{v})$. □

It turns out that the necessary subdifferential conditions from Theorem 7.1.16 are sufficient for strong duality and for the existence of the primal and dual efficient solutions, too.

Theorem 7.1.17. *Let $(\bar{x}, \bar{v}) \in \operatorname{gph} F \cap (X \times V)$ and $\overline{\Lambda} \in \mathcal{L}(Y, V)$ fulfill the condition $(0, \overline{\Lambda}) \in \partial\Phi(\bar{x}, 0; \bar{v})$. Then $\bar{x}$ is an efficient solution and $(\bar{x}, \bar{v})$ is a minimal pair to $(PSVG)$, while $\overline{\Lambda}$ is an efficient solution and $(\overline{\Lambda}, \bar{v})$ is a maximal pair to $(DSVG)$.*

Proof. Since $(0, \overline{\Lambda}) \in \partial\Phi(\bar{x}, 0; \bar{v})$, according to Proposition 7.1.3(a), this is equivalent to $-\bar{v} \in \Phi^*(0, \overline{\Lambda})$. As $\bar{v} \in F(\bar{x})$, we finally get $\bar{v} \in F(\bar{x}) \cap \{-\Phi^*(0, \overline{\Lambda})\}$. Corollary 7.1.12$(b)$ provides the claimed assertion concerning the efficiency of $\bar{x}$ and $\overline{\Lambda}$, as well as the minimality of $(\bar{x}, \bar{v})$ and the maximality of $(\overline{\Lambda}, \bar{v})$. □

Remark 7.1.8. One could notice that for the above result no external stability for $H(y)$, when $y \in Y$, is required. Assuming that $H(y)$ is externally stable for all $y \in Y$, in order to obtain the same conclusion one can equivalently ask instead of $(0, \overline{\Lambda}) \in \partial\Phi(\bar{x}, 0; \bar{v})$ that $\overline{\Lambda} \in \partial H(0, \bar{v})$.

Next we define the notions of *epigraph* and *cone-convexity* for a set-valued map with values in $V \cup \{+\infty_K\}$.

Definition 7.1.5. *Let $F : X \rightrightarrows V \cup \{+\infty_K\}$ be a set-valued map.*

(a) The set

$$\operatorname{epi}_K F = \{(x, v) \in X \times V : v \in F(x) + K\}$$

is called the K-epigraph of F.

(b) The map F is said to be K-convex if $\operatorname{epi}_K F$ is a convex subset of $X \times V$.

It is straightforward to prove that F is K-convex if and only if for all $x_1, x_2 \in X$ and all $\lambda \in [0,1]$ it holds

$$\lambda F(x_1) \cap V + (1-\lambda)F(x_2) \cap V \subseteq F(\lambda x_1 + (1-\lambda)x_2) + K.$$

Now we show the K-convexity of the minimal value map H under cone-convexity assumptions for the perturbation map Φ along with the external stability of $H(y)$ for all $y \in Y$.

Lemma 7.1.18. *Let $\Phi : X \times Y \rightrightarrows V \cup \{+\infty_K\}$ be a K-convex set-valued map and $H(y) = \operatorname{Min} \Phi(X, y)$ be externally stable for all $y \in Y$. Then H is a K-convex set-valued map.*

Proof. We prove that

$$\operatorname{epi}_K H = \Big\{(y,v) \in Y \times V : v \in \Phi(X,y) + K\Big\}. \tag{7.7}$$

Let be $(y,v) \in \operatorname{epi}_K H$. Then $v \in H(y) + K \subseteq \Phi(X,y) + K$ and the inclusion "$\subseteq$" is proven.

Take now $(y,v) \in Y \times V$ with $v \in \Phi(X,y) + K$. Then there exist $\bar{v} \in \Phi(X,y) \cap V$ and $\bar{k} \in K$ such that $v = \bar{v} + \bar{k}$. As $\bar{v} \subseteq H(y) + K$, one has $v \in H(y) + K$ and so $(y,v) \in \operatorname{epi}_K H$. Thus (7.7) is proven.

Now it is easy to see that $\operatorname{epi}_K H = \operatorname{Pr}_{Y \times V}(\operatorname{epi}_K \Phi)$. Since $\operatorname{epi}_K \Phi$ is a convex set, $\operatorname{epi}_K H$ is a convex set, too. □

The notion introduced in the following will play an important role in characterizing the subdifferentiability of a set-valued map. Let $F : X \rightrightarrows V \cup \{+\infty_K\}$ be a set-valued map and $T \in \mathcal{L}(X,V)$. We say that $F^*(T)$ can be *completely characterized by scalarization* when for a pair $(\bar{x}, \bar{v}) \in \operatorname{gph} F$ it holds $T\bar{x} - \bar{v} \in F^*(T)$ if and only if there exists $k^* \in K^* \backslash \{0\}$ such that

$$\langle k^*, T\bar{x} - \bar{v} \rangle = \max_{(x,v) \in \operatorname{gph} F} \langle k^*, Tx - v \rangle. \tag{7.8}$$

Remark 7.1.9. One can interpret this definition as follows. Assume that $F : X \rightrightarrows V$ is a set-valued map and that $\operatorname{int}(K) \neq \emptyset$. For $T \in \mathcal{L}(X,V)$ assume that each weakly maximal element of the set $\cup_{x \in X}[Tx - F(x)]$ is maximal and that the set $\cup_{x \in X}[Tx - F(x)] + K$ is convex. Then, via Corollary 2.4.26 for a pair $(\bar{x}, \bar{v}) \in \operatorname{gph} F$ one has that $(\bar{x}, \bar{v}) \in F^*(T)$ if and only if there exists $k^* \in K^* \backslash \{0\}$ satisfying (7.8). This means that under these conditions $F^*(T)$ can be completely characterized by scalarization.

Let us give now the announced sufficient condition for the subdifferentiability of a set-valued map. One can notice that in the proof of the following result we do not make use of the external stability of $H(y)$ for $y \in Y$.

Proposition 7.1.19. *Let $F : X \rightrightarrows V \cup \{+\infty_K\}$ be a K-convex set-valued map such that $\operatorname{Min} F(\bar{x}) = F(\bar{x})$ for some $\bar{x} \in \operatorname{int}(\operatorname{dom} F)$. If $\operatorname{int}(\operatorname{epi}_K F) \neq \emptyset$ and $F^*(T)$ can be completely characterized by scalarization for all $T \in \mathcal{L}(X,V)$, then F is subdifferentiable at $\bar{x}$.*

Proof. Without loss of generality we may assume that $\bar{x} = 0 \in \operatorname{int}(\operatorname{dom} F)$. One can notice that $\emptyset \neq F(0) \subseteq V$. Take an arbitrary $\bar{v} \in F(0) = \operatorname{Min} F(0)$. Then $(0, \bar{v}) \notin \operatorname{int}(\operatorname{epi}_K F)$ and, since $\operatorname{epi}_K F$ is convex, Theorem 2.1.2 can be applied. Hence there exists a pair $(x^*, k^*) \in X^* \times V^*$, $(x^*, k^*) \neq (0, 0)$, such that

$$\langle x^*, 0\rangle + \langle k^*, \bar{v}\rangle \leq \langle x^*, x\rangle + \langle k^*, v\rangle \text{ for all } (x, v) \in \operatorname{epi}_K F. \tag{7.9}$$

One can easily show that $k^* \in K^*$. Even more, it is not possible to have $k^* = 0$. Assuming the contrary, let U be a neighborhood of 0 in X with $U \subseteq \operatorname{dom} F$. Thus $\langle x^*, u\rangle \geq 0$ for all $u \in U$ and this has as consequence the fact that $x^* = 0$. This is a contradiction, therefore $k^* \neq 0$ and one can take $\tilde{v} \in V$ such that $\langle k^*, \tilde{v}\rangle = 1$. Define $T \in \mathcal{L}(X, V)$ by $Tx := -\langle x^*, x\rangle\tilde{v}$ for $x \in X$. Using (7.9) we deduce that for all $(x, v) \in \operatorname{epi}_K F$

$$\langle k^*, v - Tx\rangle = \langle k^*, v\rangle + \langle x^*, x\rangle \geq \langle k^*, \bar{v}\rangle = \langle k^*, \bar{v} - T0\rangle.$$

But this relation holds for all $(x, v) \in \operatorname{gph} F$, even if $v = +\infty_K$ since $\langle k^*, +\infty_K\rangle = +\infty$. By the assumption we made, $-\bar{v} \in F^*(T)$ and according to Proposition 7.1.3 we have $T \in \partial F(0; \bar{v})$. As $\bar{v} \in F(0)$ has been arbitrarily chosen, the subdifferentiability of F at 0 follows. □

In an analogous manner one can prove the following stability result.

Theorem 7.1.20. *Let the perturbation map $\Phi : X \times Y \rightrightarrows V \cup \{+\infty_K\}$ be K-convex and $H : Y \rightrightarrows V \cup \{+\infty_K\}$, $H(y) = \operatorname{Min} \Phi(X, y)$, be externally stable for all $y \in Y$. Assume that $\operatorname{int}(\operatorname{epi}_K H) \neq \emptyset$ and that for the set-valued map $\Psi : Y \rightrightarrows V \cup \{+\infty_K\}$, $\Psi(y) = \Phi(X, y)$, there is $0 \in \operatorname{int}(\operatorname{dom} \Psi)$. If $H^*(\Lambda)$ can be completely characterized by scalarization for all $\Lambda \in \mathcal{L}(Y, V)$, then the primal problem $(PSVG)$ is stable.*

Proof. One can easily see that the set-valued map Ψ is K-convex. Since $\Psi(y) \backslash \{+\infty_K\} \subseteq H(y) + K$ for all $y \in Y$, there is $\operatorname{dom} \Psi \subseteq \operatorname{dom} H$ and so $0 \in \operatorname{int}(\operatorname{dom} H)$. Moreover, $H(y) \neq \emptyset$ and so $\operatorname{Min} H(y) = H(y) = \operatorname{Min} \Phi(X, y)$ for all $y \in Y$. Further, by Lemma 7.1.18, H is K-convex and thus the subdifferentiability of H at 0 is a direct consequence of Proposition 7.1.19. Consequently, $(PSVG)$ is stable. □

Remark 7.1.10. Proposition 7.1.19 and Theorem 7.1.20 generalize corresponding results of [163] obtained there in finite dimensional spaces while the objective function of the primal problem is assumed to be vector-valued (see [163, Proposition 6.1.7 and Proposition 6.1.13]).

In Proposition 7.1.19 and Theorem 7.1.20 we have seen that $\operatorname{int}(\operatorname{epi}_K F)$ and, respectively, $\operatorname{int}(\operatorname{epi}_K H)$ must be nonempty in order to apply a standard separation theorem. Thus it is important to have simple criteria ensuring this. For introducing such criteria, the following definition is useful.

Definition 7.1.6. *A set-valued map $F : X \rightrightarrows V \cup \{+\infty_K\}$ is said to be weakly K-upper bounded on a set $S \subseteq X$ if there exists an element $b \in V$ such that $(x, b) \in \operatorname{epi}_K F$ for all $x \in S$.*

It is straightforward to see that F is weakly K-upper bounded on S if and only if there exists $b \in V$ such that $F(x) \cap (b - K) \neq \emptyset$ for all $x \in S$. To establish a connection between the weakly K-upper boundedness and the assumption $\operatorname{epi}_K F \neq \emptyset$ we have to suppose that $\operatorname{int}(K) \neq \emptyset$. For this reason in the following we assume it to be fulfilled.

Lemma 7.1.21. *Let be $\operatorname{int}(K) \neq \emptyset$ and $F : X \rightrightarrows V \cup \{+\infty_K\}$ be a set-valued map. Then the following statements are equivalent:*

(i) $\operatorname{int}(\operatorname{epi}_K F) \neq \emptyset$;
(ii) there exists $x' \in \operatorname{dom} F$ such that F is weakly K-upper bounded on some neighborhood of x' in X.

The proof is straightforward. For details see [170, Theorem 6.1].

Remark 7.1.11. One can also imagine that appropriate topological properties of a set-valued map F, like in scalar programming, imply the nonemptiness of the interior of the K-epigraph of F. To this end one can assume that the set-valued map $F : X \rightrightarrows V \cup \{+\infty_K\}$ is *K-Hausdorff lower continuous* on $\operatorname{int}(\operatorname{dom} F)$. According to [147], F is *K-Hausdorff lower continuous* at $x' \in X$ if for every neighborhood W of 0 in V there exists a neighborhood U of x' in X such that $F(x') \subseteq F(x) + W + K$ for all $x \in \operatorname{dom} F \cap U$.

Indeed, assume that for $x' \in \operatorname{dom} F$ the map F is K-Hausdorff lower continuous at x'. Let be $k' \in \operatorname{int}(K)$ and W a neighborhood of 0 in V such that $k' + W \subseteq K$. Then there exists a neighborhood U of x' in X with $U \subseteq \operatorname{dom} F$ and $F(x') \subseteq F(x) + W + K$ for all $x \in \operatorname{dom} F \cap U$. Let be $b \in F(x') \cap V$. Then for all $x \in U$ one has $b + k' \in F(x') + k' \subseteq F(x) + k' + W + K \subseteq F(x) + K$, and thus statement *(ii)* in Lemma 7.1.21 is fulfilled.

Lemma 7.1.21 allows to substitute in both Proposition 7.1.19 and Theorem 7.1.20 the assumptions concerning the nonemptiness of the K-epigraph with the weakly K-upper boundedness condition. We come now to another assertion which is useful for characterizing the stability of the set-valued optimization problem $(PSVG)$.

Proposition 7.1.22. *Assume that $\Psi : Y \rightrightarrows V \cup \{+\infty_K\}$, $\Psi(y) = \Phi(X, y)$, is K-convex, $H : Y \rightrightarrows V \cup \{+\infty_K\}$, $H(y) = \operatorname{Min} \Psi(y)$, is externally stable for all $y \in Y$ and there exists $y' \in \operatorname{dom} \Psi$ such that Ψ is weakly K-upper bounded on some neighborhood of y'. Then H is K-convex and $\operatorname{int}(\operatorname{epi}_K H) \neq \emptyset$.*

Proof. As follows from 7.7, it holds $\operatorname{epi}_K H = \operatorname{epi}_K \Psi$ and, since Ψ is K-convex, H is K-convex, too. The fact that $\operatorname{int}(\operatorname{epi}_K H) \neq \emptyset$ follows by Lemma 7.1.21. □

Proposition 7.1.22 allows to reformulate Theorem 7.1.20, by replacing the hypothesis $\operatorname{int}(\operatorname{epi}_K H) \neq \emptyset$ with the assumptions concerning Ψ considered in this las result.

Next we intend to expand the subdifferentiability and stability criteria for set-valued maps by assuming convexity assumptions which allow to skip imposing that the conjugate map can be completely characterized by scalarization.

Definition 7.1.7. *Let $F : X \rightrightarrows V \cup \{+\infty_K\}$ be a set-valued map.*

(a) F is said to be strictly K-convex if it is K-convex and if for all $x_1, x_2 \in X$, $x_1 \neq x_2$, and all $\lambda \in (0,1)$ there is

$$\lambda F(x_1) \cap V + (1-\lambda)F(x_2) \cap V \subseteq F(\lambda x_1 + (1-\lambda)x_2) + \operatorname{int}(K).$$

(b) F is said to be K-convexlike on the nonempty set $S \subseteq X$ if for all $v_1, v_2 \in F(S) \cap V$ and all $\lambda \in [0,1]$ there exists $\bar{x} \in S$ such that

$$\lambda v_1 + (1-\lambda)v_2 \in F(\bar{x}) + K.$$

(c) F is said to be strictly K-convexlike on the nonempty set $S \subseteq X$ if for all $v_1, v_2 \in F(S) \cap V$, $v_1 \neq v_2$, and all $\lambda \in (0,1)$ there exists $\bar{x} \in S$ such that

$$\lambda v_1 + (1-\lambda)v_2 \in F(\bar{x}) + \operatorname{int}(K).$$

(d) When in (b) and (c) the set S coincides with the whole space, then we call the set-valued map K-convexlike and strictly K-convexlike, respectively.

In the case $F : X \to V \cup \{+\infty_K\}$ is a vector-valued function, then F is K-convexlike on S if and only if $(F(S) \cap V) + K$ is convex. On the other hand, a vector-valued function $F : X \to V$ is strictly K-convex if and only if for all $x_1, x_2 \in X$, $x_1 \neq x_2$, and all $\lambda \in (0,1)$ the inequality

$$F(\lambda x_1 + (1-\lambda)x_2) <_K \lambda F(x_1) + (1-\lambda)F(x_2)$$

holds.

We notice that, different to the situation when vector-valued functions are considered, only assuming the inclusion relation in Definition 7.1.7(a) to be fulfilled does not imply that the set-valued map F is K-convex. On the other hand, let us notice that if $S \subseteq X$ is a nonempty convex set and $F : X \rightrightarrows V \cup \{+\infty_K\}$ is a K-convex set-valued map, then F is K-convexlike on S, while if F is strictly K-convex, then it is strictly K-convexlike on S, too.

Definition 7.1.8. *Let $\Phi : X \times Y \rightrightarrows V \cup \{+\infty_K\}$ be a set-valued map. The map Φ is called*

(a) K-convexlike-convex if for all $y_i \in Y$, $v_i \in \Phi(X, y_i) \cap V$, $i = 1, 2$, and all $\lambda \in [0,1]$ there exists $\bar{x} \in X$ such that

$$\lambda v_1 + (1-\lambda)v_2 \in \Phi(\bar{x}, \lambda y_1 + (1-\lambda)y_2) + K; \tag{7.10}$$

(b) *strictly-K-convexlike-convex if it is K-convexlike-convex and if for all $y_i \in Y$, $v_i \in \Phi(X, y_i) \cap V$, $i = 1, 2$, with $y_1 \neq y_2$, and all $\lambda \in (0,1)$ there exists $\bar{x} \in X$ such that*

$$\lambda v_1 + (1-\lambda)v_2 \in \Phi(\bar{x}, \lambda y_1 + (1-\lambda)y_2) + \operatorname{int}(K). \tag{7.11}$$

If Φ is K-convex then it is K-convexlike-convex, too, while when Φ is strictly K-convex, it is strictly-K-convexlike-convex, too.

As above we consider $\Psi : Y \rightrightarrows V \cup \{+\infty_K\}$, $\Psi(y) = \Phi(X, y)$, and $H : Y \rightrightarrows V \cup \{+\infty_K\}$, $H(y) = \operatorname{Min} \Psi(y)$. The proof of the following statement is straightforward.

Lemma 7.1.23. *The map Φ is K-convexlike-convex if and only if the map Ψ is K-convex, while Φ is strictly-K-convexlike-convex if and only if Ψ is strictly K-convex.*

Lemma 7.1.24. *Let Ψ be strictly K-convex and $H(y)$ externally stable for all $y \in Y$. Then H is strictly K-convex.*

Proof. As we have noticed in Proposition 7.1.22, under the imposed assumptions it holds $\operatorname{epi}_K H = \operatorname{epi}_K \Psi$ and this guarantees the K-convexity of H. Consider now $y_1, y_2 \in Y$, $y_1 \neq y_2$, and $\lambda \in (0,1)$. Therefore

$$\lambda H(y_1) \cap V + (1-\lambda)H(y_2) \cap V \subseteq \lambda \Psi(y_1) \cap V + (1-\lambda)\Psi(y_2) \cap V$$

$$\subseteq \Psi(\lambda y_1 + (1-\lambda)y_2) \cap V + \operatorname{int}(K) \subseteq H(\lambda y_1 + (1-\lambda)y_2) + K + \operatorname{int}(K)$$
$$= H(\lambda y_1 + (1-\lambda)y_2) + \operatorname{int}(K).$$

This completes the proof. □

By combining Lemma 7.1.23 and Lemma 7.1.24 one obtains the following statement, which generalizes Lemma 7.1.18.

Lemma 7.1.25. (a) *Let Φ be K-convexlike-convex and $H(y)$ externally stable for all $y \in Y$. Then H is K-convex.*

(b) *Let Φ be strictly-K-convexlike-convex and $H(y)$ externally stable for all $y \in Y$. Then H is strictly K-convex.*

Now let us formulate a subdifferentiability result for set-valued maps by assuming strict K-convexity. Comparing it to Proposition 7.1.19, one can notice that one needs here a stronger convexity assumption, but the restrictive property that $F^*(T)$ can be completely characterized by scalarization for all $T \in \mathcal{L}(X, V)$ can be avoided. In the proof of the following result we do not make use of the external stability of $H(y)$ for $y \in Y$.

Proposition 7.1.26. *Let the set-valued map $F : X \rightrightarrows V \cup \{+\infty_K\}$ be strictly K-convex and weakly K-upper bounded on some neighborhood of $x' \in \operatorname{dom} F$ (or, equivalently, $\operatorname{int}(\operatorname{epi}_K F) \neq \emptyset$). If $\bar{x} \in \operatorname{int}(\operatorname{dom} F)$ and $\operatorname{Min} F(\bar{x}) = F(\bar{x})$, then F is subdifferentiable at $\bar{x}$.*

Proof. Without loss of generality let $\bar{x} = 0$. Obviously, $\emptyset \neq F(0) \subseteq V$. Take an arbitrary $\bar{v} \in F(0) = \operatorname{Min} F(0)$. Then $(0, \bar{v}) \notin \operatorname{int}(\operatorname{epi}_K F)$, which is a nonempty set. Further, $\operatorname{epi}_K F$ is convex and this allows to use Theorem 2.1.2 like in the proof of Proposition 7.1.19. Hence there are $x^* \in X^*$ and $k^* \in K^* \backslash \{0\}$ such that

$$\langle k^*, \bar{v} \rangle \leq \langle x^*, x \rangle + \langle k^*, v \rangle \text{ for all } (x, v) \in \operatorname{epi}_K F. \tag{7.12}$$

Now, the strict convexity of F allows to verify even the strict inequality

$$\langle k^*, \bar{v} \rangle < \langle x^*, x \rangle + \langle k^*, v \rangle \text{ for all } (x, v) \in \operatorname{epi}_K F \text{ with } x \neq 0. \tag{7.13}$$

Indeed, were (7.13) not fulfilled, then there would exist $(x_1, v_1) \in \operatorname{epi}_K F$ with $x_1 \neq 0$, such that $\langle k^*, \bar{v} \rangle = \langle x^*, x_1 \rangle + \langle k^*, v_1 \rangle$. As

$$\frac{1}{2}v_1 + \frac{1}{2}\bar{v} \in \frac{1}{2}\left((F(x_1) \cap V) + K\right) + \frac{1}{2}(F(0) \cap V) = \frac{1}{2}F(x_1) \cap V +$$

$$\frac{1}{2}F(0) \cap V + K \subseteq F\left(\frac{1}{2}x_1\right) + \operatorname{int}(K) + K = F\left(\frac{1}{2}x_1\right) + \operatorname{int}(K),$$

there exists in this situation $k_1 \in \operatorname{int}(K)$ such that $(1/2)v_1 + (1/2)\bar{v} - k_1 \in F((1/2)x_1)$. Thus we have

$$\left(\frac{1}{2}x_1, \frac{1}{2}v_1 + \frac{1}{2}\bar{v} - k_1\right) \in \operatorname{epi}_K F$$

and using (7.12) yields

$$\langle k^*, \bar{v} \rangle \leq \left\langle x^*, \frac{1}{2}x_1 \right\rangle + \left\langle k^*, \frac{1}{2}v_1 + \frac{1}{2}\bar{v} - k_1 \right\rangle$$

or, equivalently,

$$\frac{1}{2}\langle k^*, \bar{v} \rangle \leq \frac{1}{2}\langle x^*, x_1 \rangle + \frac{1}{2}\langle k^*, v_1 \rangle - \langle k^*, k_1 \rangle.$$

Since $\langle k^*, k_1 \rangle > 0$, one has $\langle k^*, \bar{v} \rangle < \langle x^*, x_1 \rangle + \langle k^*, v_1 \rangle$ and this contradicts the equality from above. Thus (7.13) is valid. Let be $\tilde{v} \in V$ with $\langle k^*, \tilde{v} \rangle = 1$ and $T \in \mathcal{L}(X, V)$ defined by $Tx := -\langle x^*, x \rangle \tilde{v}$ for $x \in X$. Then from (7.13) we get

$$\langle k^*, \bar{v} \rangle < \langle x^*, x \rangle + \langle k^*, v \rangle = \langle k^*, -Tx \rangle + \langle k^*, v \rangle = \langle k^*, v - Tx \rangle \tag{7.14}$$

for all $(x, v) \in \operatorname{epi}_K F$ with $x \neq 0$. Notice that this result is true for arbitrary $x \in X$ whenever $v = +\infty_K$. We prove that $\bar{v} \in \operatorname{Min} \cup_{x \in X} [F(x) - Tx]$. Assuming the contrary, there exist $\tilde{x} \in X \backslash \{0\}$, and $\tilde{v} \in F(\tilde{x})$ such that $\tilde{v} - T\tilde{x} \leq_K \bar{v}$. This means that

$$\langle k^*, \tilde{v} - T\tilde{x} \rangle \leq \langle k^*, \bar{v} \rangle.$$

On the other hand, (7.14) implies $\langle k^*, \bar{v} \rangle < \langle k^*, \tilde{v} - T\tilde{x} \rangle$, and this leads to a contradiction. Consequently, $\bar{v} \in \operatorname{Min} \cup_{x \in X} [F(x) - Tx]$, or, equivalently, $-\bar{v} \in \operatorname{Max} \cup_{x \in X} [Tx - F(x)]$, which means $T \in \partial F(0; \bar{v})$. Since $\bar{v} \in F(0) \cap V$ was arbitrarily chosen, F is subdifferentiable at 0. □

Proposition 7.1.26 allows to formulate a stability criterion in analogy to Theorem 7.1.20, but which does not require that $H^*(\Lambda)$ can be completely characterized by scalarization for $\Lambda \in \mathcal{L}(Y,V)$.

Theorem 7.1.27. *Let the perturbation map $\Phi : X \times Y \rightrightarrows V \cup \{+\infty_K\}$ be strictly-K-convexlike-convex (or, equivalently, $\Psi : Y \rightrightarrows V \cup \{+\infty_K\}$, $\Psi(y) = \Phi(X,y)$, be strictly K-convex) and $H : Y \rightrightarrows V \cup \{+\infty_K\}$, $H(y) = \operatorname{Min} \Phi(X,y)$, be externally stable for all $y \in Y$. Assume there exists $y' \in \operatorname{dom} \Psi$ such that Ψ is weakly K-upper bounded on some neighborhood of y'. If $0 \in \operatorname{int}(\operatorname{dom} \Psi)$, then the primal problem $(PSVG)$ is stable.*

Proof. Lemma 7.1.25 ensures that H is strictly K-convex and via Proposition 7.1.22 follows that $\operatorname{int}(\operatorname{epi}_K H) \neq \emptyset$. Because $H(0) = \operatorname{Min} \Phi(X,0)$ it holds $\operatorname{Min} H(0) = H(0)$ and, moreover, as shown in the proof of Theorem 7.1.20, one has that $\operatorname{dom} \Psi \subseteq \operatorname{dom} H$. Consequently, $0 \in \operatorname{int}(\operatorname{dom} H)$. Proposition 7.1.26 ensures the subdifferentiability of H at 0 or, equivalently, the stability of $(PSVG)$. □

7.1.3 A special approach - vector k-conjugacy and duality

The considerations made in the previous subsections of this chapter extend the ones from [163] made when $X = \mathbb{R}^n$, $V = \mathbb{R}^k$, $K = \mathbb{R}^k_+$ and the primal objective function is vector-valued. When the objective space V is finite dimensional one can find in [180] another approach for defining conjugate maps and subgradients, by replacing $T \in \mathcal{L}(X,V)$ (or in the finite dimensional case the corresponding matrix $T \in \mathbb{R}^{k \times n}$) with a linear continuous functional mapping from X into $\mathbb{R}$. In the following we extend that approach to infinite dimensional spaces and set-valued maps.

Throughout this subsection let X, Y and V be topological vector spaces with topological dual spaces X^*, Y^* and V^*, respectively. Let V be partially ordered by the nontrivial pointed convex cone $K \subseteq V$. Concerning the basic notations and conventions with respect to minimal and maximal elements of sets in $\overline{V}$ we refer to subsection 7.1.1.

Definition 7.1.9. *Let $F : X \rightrightarrows \overline{V}$ be a set-valued map and $k \in V \setminus \{0\}$ a given element.*

(a) The set-valued map

$$F_k^* : X^* \rightrightarrows \overline{V},\ F_k^*(x^*) = \operatorname{Max} \bigcup_{x \in X} [\langle x^*, x\rangle k - F(x)],$$

is called the k-conjugate map of F.

(b) The set-valued map

$$F_k^{**} : X \rightrightarrows \overline{V},\ F_k^{**}(x) = \operatorname{Max} \bigcup_{x^* \in X^*} [\langle x^*, x\rangle k - F_k^*(x^*)],$$

is called the $k-$biconjugate map of F.

(c) The element $x^* \in X^*$ is said to be a k-subgradient of F at (x, v), where $(x, v) \in \operatorname{gph} F \cap (X \times V)$, if

$$\langle x^*, x\rangle k - v \in \operatorname{Max} \bigcup_{y \in X} [\langle x^*, y\rangle k - F(y)].$$

The set of all k-subgradients of F at $(x, v) \in \operatorname{gph} F \cap (X \times V)$ is called the k-subdifferential of F at (x, v) and is denoted by $\partial_k F(x; v)$. Further, for all $x \in X$ denote $\partial_k F(x) := \cup_{v \in F(x)} \partial F(x; v)$. If for all $v \in F(x) \cap V$ we have $\partial_k F(x; v) \neq \emptyset$, then F is said to be k-subdifferentiable at x.

Remark 7.1.12. We refer also to Remark 7.1.1 concerning the vector-valued case. Now, for a vector-valued function $f : X \to \overline{V}$, we use the notations f_k^*, f_k^{**} and $\partial_k f$ when particularizing Definition 7.1.9. If we consider $V = \mathbb{R}^k$, $K = \mathbb{R}^k_+$, the vector $k := (1, ..., 1)^T$ and the function $F : X \to \mathbb{R}^k$ vector-valued, one rediscovers the original approach of this kind of conjugacy as established in [180]. Even more, most of the results presented there can be derived as particular instances of the results we give below.

In Definition 7.1.9, different from Definition 7.1.2, $T \in \mathcal{L}(X, V)$ has the special formulation $Tx = \langle x^*, x\rangle k$ for $x \in X$. The generalized Young-Fenchel inequality from Proposition 7.1.1 holds also for $F_k^*(x^*)$, as a particular case. The other results and properties obtained in subsection 7.1.1 can be proven in a similar way. Nevertheless, in particular, assertions containing F_k^{**} require new considerations since the k-biconjugate functions cannot be seen as a direct particularization of the biconjugate from the previous subsections. Although some modifications are necessary, the transfer of the proofs is straightforward. Therefore Proposition 7.1.2, Proposition 7.1.3, Proposition 7.1.4 and Corollary 7.1.8 remain true for corresponding notions and assertions based on Definition 7.1.9.

Concerning the subgradient, it is easy to see that for $x^* \in X^*$, if $T \in \mathcal{L}(X, V)$, defined by $Tx = \langle x^*, x\rangle k$ for $x \in X$, is a subgradient of F in the sense of Definition 7.1.2, then x^* is also a $k-$subgradient of F and vice versa. But, in general one can have subgradients of F in the sense of Definition 7.1.2 that are not $k-$subgradients of F. Therefore the k-subdifferential of F at $(x, v) \in \operatorname{gph} F \cap (X \times V)$ is a subset of the subdifferential of F at the same point in the sense of Definition 7.1.2. Thus, if F is k-subdifferentiable at $x \in X$, then F is subdifferentiable at $x \in X$, too. The opposite statement might fail.

Similar to the investigations made in the previous section, one can attach to the primal problem $(PSVG)$ a set-valued dual problem, this time based on k-conjugate functions. Weak and strong duality assertions, like in subsection 7.1.2, can be proven in this case, too. Even more, Proposition 7.1.10 remains true if $0 \in \partial F(\bar{x}; \bar{v})$ is replaced by $0 \in \partial_k F(\bar{x}; \bar{v})$ in the statement (a), while $0 \in \partial f(\bar{x})$ is replaced by $0 \in \partial_k f(\bar{x})$ in the statement (b).

Let us turn our attention to the set-valued optimization problem

$$(PSVG) \quad \operatorname*{Min}_{x \in X} F(x)$$

treated in the previous subsections. As there we take as perturbation map $\Phi : X \times Y \rightrightarrows V \cup \{+\infty_K\}$ such that $\Phi(x, 0) = F(x)$ for all $x \in X$. We consider a fixed $k \in V \backslash \{0\}$. Then $\Phi_k^* : X^* \times Y^* \rightrightarrows \overline{V}$ is defined by $\Phi_k^*(x^*, y^*) = \operatorname{Max} \cup_{x \in X, y \in Y} [(\langle x^*, x\rangle + \langle y^*, y\rangle)k - \Phi(x, y)]$ and to $(PSVG)$ we attach in this way the set-valued dual problem

$$(DSVG_k) \quad \operatorname*{Max}_{y^* \in Y^*} \{-\Phi_k^*(0, y^*)\}.$$

By solving $(DSVG_k)$ we mean finding those elements $\bar{y}^* \in Y^*$ such that there exists $\bar{v}^* \in -\Phi_k^*(0, \bar{y}^*)$ fulfilling $\bar{v}^* \in \operatorname{Max} \cup_{y^* \in Y^*} \{-\Phi_k^*(0, y^*)\}$. In this situation $\bar{y}^*$ is said to be an *efficient solution* to $(PSVG)$, while the tuple $(\bar{y}^*, \bar{v}^*)$ is said to be a *maximal pair* to $(DSVG_k)$.

Further we give some essential results without proofs, as these can be done in a similar manner as in the previous subsection. First, let us consider the weak duality assertion.

Theorem 7.1.28. *For all $x \in X$ and all $y^* \in Y^*$ there is $\Phi(x, 0) \cap \{-\Phi_k^*(0, y^*) - (K \backslash \{0\})\} = \emptyset$.*

Corollary 7.1.29. *(a) For all $x \in X$ and all $y^* \in Y^*$ it holds $v \nleq_K v^*$, whenever $v \in F(x)$ and $v^* \in -\Phi_k^*(0, y^*)$.*
(b) Let be $\bar{v} \in F(\bar{x}) \cap \{-\Phi_k^*(0, \bar{y}^*)\}$ for $\bar{x} \in X$ and $\bar{y}^* \in Y^*$. Then $\bar{x}$ is an efficient solution and $(\bar{x}, \bar{v})$ a minimal pair to $(PSVG)$, while $\bar{y}^*$ is an efficient solution and $(\bar{y}^*, \bar{v})$ is a maximal pair to $(DSVG_k)$.

As in subsection 7.1.2 the *minimal value map* of Φ is taken to be the set-valued map $H : Y \rightrightarrows V \cup \{+\infty_K\}$, $H(y) = \Phi(X, y)$.

Lemma 7.1.30. *The inclusion $\Phi_k^*(0, y^*) \subseteq H_k^*(y^*)$ holds for all $y^* \in Y^*$. Moreover, if $H(y)$ is externally stable for all $y \in Y$, then $\Phi_k^*(0, y^*) = H_k^*(y^*)$ for all $y^* \in Y^*$.*

In the following we suppose that $H(y)$ is externally stable for all $y \in Y$. Then the dual problem $(DSVG_k)$ may be formally written as being

$$(DSVG_k) \quad \operatorname{Max} \bigcup_{y^* \in Y^*} \{-H_k^*(y^*)\}.$$

Lemma 7.1.31. *It holds* $\operatorname{Max} \cup_{y^* \in Y^*} \{-\Phi_k^*(0, y^*)\} = H_k^{**}(0)$.

As far as stability is concerned, we refer to Definition 7.1.4 and call $(PSVG)$ *k-stable* (with respect to the perturbation map Φ) if the minimal value map H is k-subdifferentiable at 0. The next result concerns strong duality.

Theorem 7.1.32. *The problem $(PSVG)$ is k-stable if and only if for each efficient solution $\bar{x} \in X$ to $(PSVG)$ and $\bar{v} \in F(\bar{x})$ such that $(\bar{x}, \bar{v})$ is a minimal pair to $(PSVG)$ there exist an efficient solution $\bar{y}^* \in Y^*$ such that $\bar{v} \in -\Phi_k^*(0, \bar{y}^*)$ and $(\bar{y}^*, \bar{v})$ is a maximal pair to $(DSVG_k)$.*

Obviously, this result corresponds to Theorem 7.1.15 and the external stability of H is an essential feature in the proof.

Like in Theorem 7.1.16 and Theorem 7.1.17 one can state the following optimality conditions.

Theorem 7.1.33. *The minimal pair $(\bar{x}, \bar{v})$ to $(PSVG)$ and the corresponding maximal pair $(\bar{y}^*, \bar{v})$ to $(DSVG_k)$ from Theorem 7.1.32 satisfy the optimality condition $(0, \bar{y}^*) \in \partial_k \Phi(\bar{x}, 0; \bar{v})$ or, equivalently, $\bar{y}^* \in \partial_k H(0; \bar{v})$.*

Theorem 7.1.34. *Let $(\bar{x}, \bar{v}) \in \operatorname{gph} F \cap (X \times V)$ and $\bar{y}^* \in Y^*$ fulfill the condition $(0, \bar{y}^*) \in \partial_k \Phi(\bar{x}, 0; \bar{v})$. Then $\bar{x}$ is an efficient solution and $(\bar{x}, \bar{v})$ is a minimal pair to $(PSVG)$, while $\bar{y}^* \in Y^*$ is an efficient solution and $(\bar{y}^*, \bar{v})$ is a maximal pair to $(DSVG_k)$.*

Remark 7.1.13. One can notice that in Theorem 7.1.34 no external stability for $H(y)$, when $y \in Y$, is asked. Assuming this additional hypothesis for all $y \in Y$, in order to get the same conclusion in the theorem above, instead of $(0, \bar{y}^*) \in \partial_k \Phi(\bar{x}, 0; \bar{v})$ one can equivalently ask that $\bar{y}^* \in \partial_k H(0; \bar{v})$.

Let $x^* \in X^*$ be fixed. We say that $F_k^*(x^*)$ can be *completely characterized by scalarization* when for a pair $(\bar{x}, \bar{v}) \in \operatorname{gph} F$ one has $\langle x^*, \bar{x} \rangle k - \bar{v} \in F_k^*(x^*)$ if and only if there exists $k^* \in K^* \backslash \{0\}$ such that

$$\langle k^*, \langle x^*, \bar{x} \rangle k - \bar{v} \rangle = \max_{(x,v) \in \operatorname{gph} F} \langle k^*, \langle x^*, x \rangle k - v \rangle.$$

The next result corresponds to the assertion in Proposition 7.1.19. Nevertheless, we prefer to give its proof, since the construction of the k-subgradient is a bit more different. Moreover, for the next result we need to assume that the interior of K is not empty. The external stability for $H(y)$, when $y \in Y$, is for the following result not assumed.

Proposition 7.1.35. *Let $F : X \rightrightarrows V \cup \{+\infty_K\}$ be a K-convex set-valued map such that $\operatorname{Min} F(\bar{x}) = F(\bar{x})$ for some $\bar{x} \in \operatorname{int}(\operatorname{dom} F)$ and assume that $k \in \operatorname{qi}(K) \cup (-\operatorname{qi}(K))$. If $\operatorname{int}(\operatorname{epi}_K F) \neq \emptyset$ and $F_k^*(x^*)$ can be completely characterized by scalarization for all $x^* \in X^*$, then F is k-subdifferentiable at $\bar{x}$.*

Proof. Without loss of generality we assume that $\bar{x} = 0$. Obviously one has $\emptyset \neq F(0) \subseteq V$. Following the same idea as in the proof of Proposition 7.1.19, one can prove that for an arbitrary $\bar{v} \in F(0) = \operatorname{Min} F(0)$ there exists $(x^*, k^*) \in X^* \times V^*$, $k^* \in K^* \backslash \{0\}$, such that the inequality

$$\langle k^*, \bar{v} \rangle \leq \langle x^*, x \rangle + \langle k^*, v \rangle \tag{7.15}$$

holds for all $(x, v) \in \operatorname{epi}_K F$. Since $k \in \operatorname{qi}(K) \cup (-\operatorname{qi}(K))$, one can notice that $k \neq 0$. Assuming the contrary, one would have that $K^* = \{0\}$, which would lead to a contradiction. Thus $k \neq 0$ and therefore $\langle k^*, k\rangle \neq 0$. Consequently, we can define $z^* := -(1/\langle k^*, k\rangle)x^* \in X^*$. Then, from (7.15), for all $(x, v) \in \operatorname{epi}_K F$ we acquire

$$\langle k^*, \bar{v}\rangle \leq -\langle k^*, k\rangle\langle z^*, x\rangle + \langle k^*, v\rangle = \langle k^*, v - \langle z^*, x\rangle k\rangle. \tag{7.16}$$

Since $\langle k^*, +\infty_K\rangle = +\infty_K$ for $k^* \in K^*\backslash\{0\}$, one can easily conclude that (7.16) holds for all $x \in X$ and all $v \in F(x)$, even if $v = +\infty_K$. Thus $-\bar{v} \in F_k^*(z^*)$ and so $z^* \in \partial_k F(0; \bar{v})$. As $\bar{v}$ was arbitrarily taken in $F(0)$, we finally get that F is k-subdifferentiable at 0. □

With these preparations the subsequent stability criterion can be delivered (see Theorem 7.1.20 and its proof).

Theorem 7.1.36. *Let the perturbation map $\Phi : X \times Y \rightrightarrows V \cup \{+\infty_K\}$ be K-convex and $H : Y \rightrightarrows V \cup \{+\infty_K\}$, $H(y) = \operatorname{Min} \Phi(X, y)$, be externally stable for all $y \in Y$. Assume that $k \in \operatorname{qi}(K) \cup (-\operatorname{qi}(K))$, $\operatorname{int}(\operatorname{epi}_K H) \neq \emptyset$ and that for the set-valued map $\Psi : Y \rightrightarrows V \cup \{+\infty_K\}$, $\Psi(y) = \Phi(X, y)$, there is $0 \in \operatorname{int}(\operatorname{dom} \Psi)$. If $H_k^*(y^*)$ can be completely characterized by scalarization for each $y^* \in Y^*$, then the primal problem $(PSVG_k)$ is k-stable.*

As we know from Theorem 7.1.32 this criterion ensures strong duality.

Finally, to round the things out we notice that as in subsection 7.1.2 the last two results concerning the k-subdifferentiability and, respectively, the k-stability of $(PSVG)$ may be reformulated by taking into account Lemma 7.1.21 and Proposition 7.1.22. In other words, one can replace in these statements the nonemptiness of the interiors of $\operatorname{epi}_K F$ and $\operatorname{epi}_K H$ with the weakly K-upper boundedness of F and H, respectively.

Furthermore, it is possible to refine these results via strict K-convexity notions as in subsection 7.1.2. In particular, Proposition 7.1.26 and Theorem 7.1.27 can be reformulated regarding k-subdifferentiability and k-stability. In this way we get some extensions of Proposition 7.1.35 and Theorem 7.1.36. The assumptions are identical allowing to present here an abridged form of the facts.

Corollary 7.1.37. *(a) Assume that for $\bar{x} \in \operatorname{int}(\operatorname{dom} F)$ the hypotheses of Proposition 7.1.26 are fulfilled and that $k \in \operatorname{int}(K) \cup (-\operatorname{int}(K))$. Then F is k-subdifferentiable at $\bar{x}$.*

(b) Assume that the hypotheses of Theorem 7.1.27 are fulfilled and that $k \in \operatorname{int}(K) \cup (-\operatorname{int}(K))$. Then the primal problem $(PSVG_k)$ is k-stable.

7.2 The set-valued optimization problem with constraints

In the second section of this chapter we apply the perturbation approach developed in the previous one to the general set-valued optimization problem with constraints. We consider different perturbations of the primal problem, each of them providing a different set-valued dual problem. As primal set-valued optimization problem we consider

$$(PSV^C) \quad \underset{x \in \mathcal{A}}{\text{Min}}\, F(x),$$
$$\mathcal{A} = \{x \in S : G(x) \cap (-C) \neq \emptyset\}$$

where X, Z and V are topological vector spaces with Z partially ordered by the convex cone $C \subseteq Z$ and V partially ordered by the nontrivial pointed convex cone $K \subseteq V$, $S \subseteq X$ is a nonempty set, while $F : X \rightrightarrows V \cup \{+\infty_K\}$ and $G : X \rightrightarrows Z$ are set-valued maps such that $\operatorname{dom} F \cap S \cap G^{-1}(-C) \neq \emptyset$, with $G^{-1}(-C) = \{x \in X : G(x) \cap (-C) \neq \emptyset\}$. By X^*, Z^* and V^* we denote the corresponding topological dual spaces of X, Z and V, respectively.

We deal with the minimal elements of the image set $F(\mathcal{A})$ with respect to the partial ordering induced by K. An element $\bar{x} \in \mathcal{A}$ is said to be an *efficient solution* and $(\bar{x}, \bar{v})$ is said to be a *minimal pair* to (PSV^C) if $\bar{v} \in F(\bar{x})$ and $\bar{v} \in \operatorname{Min} F(\mathcal{A})$. Obviously, (PSV^C) may be considered as a particular case of the problem $(PSVG)$ we dealt with in section 7.1. We observe that the constraint $G(x) \cap (-C) \neq \emptyset$ is nothing else than the natural generalization of the cone constraint $g(x) \in -C$ for a vector-valued function $g : X \to Z$. If G is single-valued, then $G(x) \cap (-C) \neq \emptyset$ reduces to the mentioned cone constraint.

7.2.1 Duality based on general vector conjugacy

Let us define several set-valued perturbation maps which give, via their conjugate maps, different dual problems to (PSV^C): the *Lagrange perturbation map* $\Phi^{C_L} : X \times Z \rightrightarrows V \cup \{+\infty_K\}$,

$$\Phi^{C_L}(x, z) = \begin{cases} F(x), & \text{if } x \in S, G(x) \cap (z - C) \neq \emptyset, \\ \{+\infty_K\}, & \text{otherwise,} \end{cases}$$

the *Fenchel perturbation map*

$$\Phi^{C_F} : X \times X \rightrightarrows V \cup \{+\infty_K\},\ \Phi^{C_F}(x, y) = \begin{cases} F(x + y), & \text{if } x \in \mathcal{A}, \\ \{+\infty_K\}, & \text{otherwise,} \end{cases}$$

and, respectively, the *Fenchel-Lagrange perturbation map* $\Phi^{C_{FL}} : X \times X \times Z \rightrightarrows V \cup \{+\infty_K\}$,

$$\Phi^{C_{FL}}(x, y, z) = \begin{cases} F(x + y), & \text{if } x \in S, G(x) \cap (z - C) \neq \emptyset, \\ \{+\infty_K\}, & \text{otherwise.} \end{cases}$$

As in subsection 7.1.2 we introduce for the different perturbation maps the corresponding minimal value maps

$$H^{C_L} : Z \rightrightarrows V \cup \{+\infty_K\},\ H^{C_L}(z) = \operatorname{Min} \bigcup_{x \in X} \Phi^{C_L}(x, z),\ z \in Z,$$

$$H^{C_F} : X \rightrightarrows V \cup \{+\infty_K\},\ H^{C_F}(y) = \operatorname{Min} \bigcup_{x\in X} \Phi^{C_F}(x,y),\ y \in X,$$

and, respectively, $H^{C_{FL}} : X \times Z \rightrightarrows V \cup \{+\infty_K\}$,

$$H^{C_{FL}}(y,z) = \operatorname{Min} \bigcup_{x\in X} \Phi^{C_{FL}}(x,y,z),\ y \in X,\ z \in Z.$$

By Lemma 7.1.13 it follows that in case $H^{C_L}(z)$ is externally stable for all $z \in Z$, then $(H^{C_L})^*(\Lambda) = (\Phi^{C_L})^*(0,\Lambda)$ for all $\Lambda \in \mathcal{L}(Z,V)$, while if $H^{C_F}(y)$ is externally stable for all $y \in X$, then $(H^{C_F})^*(\Gamma) = (\Phi^{C_F})^*(0,\Gamma)$ for all $\Gamma \in \mathcal{L}(X,V)$. Assuming that $H^{C_{FL}}(y,z)$ is externally stable for all $(y,z) \in X \times Z$, then $(H^{C_{FL}})^*(\Gamma,\Lambda) = (\Phi^{C_{FL}})^*(0,\Gamma,\Lambda)$ for all $\Gamma \in \mathcal{L}(X,V)$ and all $\Lambda \in \mathcal{L}(Z,V)$.

For formulating the dual problems we have to calculate the corresponding conjugate maps to these perturbation maps. To this end we use the general conjugacy concept as developed in subsections 7.1.1 and 7.1.2. For an arbitrary $\Lambda \in \mathcal{L}(Z,V)$ it holds

$$(\Phi^{C_L})^*(0,\Lambda) = \operatorname{Max} \bigcup_{x\in X, z\in Z} [\Lambda z - \Phi^{C_L}(x,z)] = \operatorname{Max} \bigcup_{x\in S, G(x)\cap(z-C)\neq\emptyset} [\Lambda z - F(x)].$$

Since $G(x) \cap (z - C) \neq \emptyset$ is equivalent to $z \in G(x) + C$, we obtain

$$(\Phi^{C_L})^*(0,\Lambda) = \operatorname{Max} \bigcup_{x\in S, z\in G(x)+C} [\Lambda z - F(x)] = \operatorname{Max} \bigcup_{x\in S} [\Lambda(G(x)) + \Lambda(C) - F(x)]. \tag{7.17}$$

Even more, if the set $\operatorname{Max} \Lambda(C)$ is externally stable, then Proposition 7.1.7 secures the equality

$$(\Phi^{C_L})^*(0,\Lambda) = \operatorname{Max} \Big\{ \bigcup_{x\in S} [\Lambda(G(x)) - F(x)] + \operatorname{Max} \Lambda(C) \Big\}.$$

Pursuing the approach given in subsection 7.1.2 we associate to (PSV^C) the following set-valued dual problem

$$(DSV^{C_L}) \quad \operatorname{Max} \bigcup_{\Lambda\in\mathcal{L}(Z,V)} \operatorname{Min} \bigcup_{x\in S} [F(x) - \Lambda(G(x) + C)].$$

In case $C \neq \{0\}$ a straightforward consideration shows that the existence of an element $c' \in C \setminus \{0\}$ with the property $\Lambda c' \in K \setminus \{0\}$ implies

$$\operatorname{Min} \bigcup_{x\in S} [F(x) - \Lambda(G(x) + C)] = \emptyset.$$

Therefore the mappings $\Lambda \in \mathcal{L}(Z,V)$ fulfilling $\Lambda(C) \cap K \neq \{0\}$ have no influence on the set of maximal elements of the dual problem (DSV^{C_L}). Hence the final form of the dual problem, which is obviously valid also when $C = \{0\}$, is

$$(DSV^{C_L}) \quad \operatorname{Max} \bigcup_{\substack{\Lambda\in\mathcal{L}(Z,V),\\ \Lambda(C)\cap(-K)=\{0\}}} \operatorname{Min} \bigcup_{x\in S} [F(x) + \Lambda(G(x) + C)].$$

We call (DSV^{C_L}) the *Lagrange set-valued dual problem* to (PSV^C).

Theorem 7.1.11 implies that for (PSV^C) and its Lagrange set-valued dual problem (DSV^{C_L}) weak duality holds in the most general situation, even without any assumptions like convexity or external stability. Under the stability assumption of (PSV^C) with respect to the perturbation map Φ^{C_L}, the strong duality for (PSV^C) and (DSV^{C_L}) arises as a consequence of Theorem 7.1.15. We skip the detailed formulation of this result.

Next, we derive the Fenchel set-valued dual problem to (PSV^C) by means of the perturbation map Φ^{C_F}. For an arbitrary $\Gamma \in \mathcal{L}(X,V)$ it holds

$$(\Phi^{C_F})^*(0,\Gamma) = \operatorname{Max} \bigcup_{x,y\in X} [\Gamma y - \Phi^{C_F}(x,y)] = \operatorname{Max} \bigcup_{x\in\mathcal{A},y\in X} [\Gamma y - F(x+y)].$$

Then we get

$$\begin{aligned}
(\Phi^{C_F})^*(0,\Gamma) &= \operatorname{Max} \bigcup_{x\in\mathcal{A},y\in X} [\Gamma y - F(y) - \Gamma x] \\
&= \operatorname{Max} \bigcup_{x,y\in X} \Big\{[\Gamma y - F(y)] + [-\Gamma x - \delta^V_{\mathcal{A}}(x)]\Big\} \\
&= \operatorname{Max} \Big\{ \bigcup_{y\in X}[\Gamma y - F(y)] + \bigcup_{x\in X}[-\Gamma x - \delta^V_{\mathcal{A}}(x)]\Big\} \\
&= \operatorname{Max} \Big\{ \bigcup_{y\in X}[\Gamma y - F(y)] - \Gamma(\mathcal{A})\Big\}.
\end{aligned}$$

If $F^*(\Gamma) = \operatorname{Max} \cup_{y\in X}[\Gamma y - F(y)]$ is externally stable, then the dual problem can be written in a more compact form. To this end one has to apply Proposition 7.1.7 for the set-valued maps $\widetilde{F} : X \rightrightarrows \overline{V}$, $\widetilde{F}(x) = [-\Gamma x - \delta^V_{\mathcal{A}}(x)]$, and $\widetilde{G} : X \rightrightarrows \overline{V}$, $\widetilde{G}(x) = \cup_{y\in X}[\Gamma y - F(y)]$). Thus we obtain

$$\begin{aligned}
(\Phi^{C_F})^*(0,\Gamma) &= \operatorname{Max} \bigcup_{x\in X} \Big\{[-\Gamma x - \delta^V_{\mathcal{A}}(x)] + \bigcup_{y\in X}[\Gamma y - F(y)]\Big\} \\
&= \operatorname{Max} \bigcup_{x\in X} \Big\{[-\Gamma x - \delta^V_{\mathcal{A}}(x)] + \operatorname{Max} \bigcup_{y\in X}[\Gamma y - F(y)]\Big\} \\
&= \operatorname{Max} \bigcup_{x\in X} \Big\{[-\Gamma x - \delta^V_{\mathcal{A}}(x)] + F^*(\Gamma)\Big\} = \operatorname{Max}\{F^*(\Gamma) - \Gamma(\mathcal{A})\}.
\end{aligned}$$

Assuming additionally that $(\delta^V_{\mathcal{A}})^*(-\Gamma) = \operatorname{Max} \cup_{x\in X}[-\Gamma x - \delta^V_{\mathcal{A}}(x)]$ is externally stable, which is the case when $\Gamma(\mathcal{A}) \subseteq \operatorname{Min} \Gamma(\mathcal{A}) + K$ or, equivalently, when $\operatorname{Min} \Gamma(A)$ is externally stable, one obtains again via Proposition 7.1.7 that

$$\begin{aligned}
(\Phi^{C_F})^*(0,\Gamma) &= \operatorname{Max}[F^*(\Gamma) + \operatorname{Max}(-\Gamma(\mathcal{A}))] \\
&= \operatorname{Max}[F^*(\Gamma) - \operatorname{Min} \Gamma(\mathcal{A})] = \operatorname{Max}\Big\{F^*(\Gamma) + (\delta^V_{\mathcal{A}})^*(\Gamma)\Big\}.
\end{aligned}$$

In the general case, without any external stability assumption, the *Fenchel set-valued dual problem* to (PSV^C) is

$$(DSV^{C_F}) \ \operatorname{Max} \bigcup_{\Gamma \in \mathcal{L}(X,V)} \operatorname{Min} \Big\{ \bigcup_{y \in X} [F(y) - \Gamma y] + \Gamma(\mathcal{A}) \Big\},$$

while when $F^*(\Gamma)$ and $\operatorname{Min} \Gamma(A)$ are supposed externally stable for all $\Gamma \in \mathcal{L}(X,V)$, the Fenchel set-valued dual becomes

$$(DSV^{C_F}) \ \operatorname{Max} \bigcup_{\Gamma \in \mathcal{L}(X,V)} \operatorname{Min}[-F^*(\Gamma) + \operatorname{Min} \Gamma(\mathcal{A})].$$

Dealing further with the third perturbation map $\Phi^{C_{FL}}$ we obtain for its conjugate map, for $\Gamma \in \mathcal{L}(X,V)$ and $\Lambda \in \mathcal{L}(Z,V)$, the following formulation

$$(\Phi^{C_{FL}})^*(0,\Gamma,\Lambda) = \operatorname{Max} \bigcup_{x,y \in X, z \in Z} [\Gamma y + \Lambda z - \Phi^{C_{FL}}(x,y,z)]$$

$$= \operatorname{Max} \bigcup_{\substack{x \in S, y \in X, z \in Z, \\ G(x) \cap (z - C) \neq \emptyset}} [\Gamma y + \Lambda z - F(x+y)] = \operatorname{Max} \bigcup_{\substack{x \in S, y \in X, \\ z \in G(x) + C}} [\Gamma y - F(y) - \Gamma x + \Lambda z]$$

$$= \operatorname{Max} \bigcup_{x \in S, y \in X} \{[\Gamma y - F(y)] + [-\Gamma x + \Lambda(G(x))] + \Lambda(C)\}$$

$$= \operatorname{Max} \Big\{ \bigcup_{y \in X} [\Gamma y - F(y)] + \bigcup_{x \in S} [-\Gamma x + \Lambda(G(x))] + \Lambda(C) \Big\}.$$

In this general situation one can attach to (PSV^C) the so-called *Fenchel-Lagrange set-valued dual problem*

$$(DSV^{C_{FL}}) \ \operatorname{Max} \bigcup_{\substack{\Gamma \in \mathcal{L}(X,V), \\ \Lambda \in \mathcal{L}(Z,V), \\ \Lambda(C) \cap (-K) = \{0\}}} \operatorname{Min} \Big\{ \bigcup_{y \in X} [F(y) - \Gamma y] + \bigcup_{x \in S} [\Gamma x + \Lambda(G(x))] + \Lambda(C) \Big\}.$$

Also here the dual problem can be written in a more compact form if external stability conditions are imposed. If $F^*(\Gamma) = \operatorname{Max} \cup_{y \in X} [\Gamma y - F(y)]$ is externally stable for all $\Gamma \in \mathcal{L}(X,V)$, then the dual problem becomes

$$(DSV^{C_{FL}}) \ \operatorname{Max} \bigcup_{\substack{\Gamma \in \mathcal{L}(X,V), \\ \Lambda \in \mathcal{L}(Z,V), \\ \Lambda(C) \cap (-K) = \{0\}}} \operatorname{Min} \Big\{ -F^*(\Gamma) + \bigcup_{x \in S} [\Gamma x + \Lambda(G(x))] + \Lambda(C) \Big\}.$$

If, additionally, also $(\Lambda G)_S^*(-\Gamma) = \operatorname{Max} \cup_{x \in S} [-\Gamma x - \Lambda(G(x))]$ is externally stable for all $\Lambda \in \mathcal{L}(Z,V)$ and all $\Gamma \in \mathcal{L}(X,V)$, then we acquire the following formulation for the dual problem

$$(DSV^{C_{FL}}) \ \operatorname{Max} \bigcup_{\substack{\Gamma \in \mathcal{L}(X,V), \Lambda \in \mathcal{L}(Z,V), \\ \Lambda(C) \cap (-K) = \{0\}}} \operatorname{Min}[-F^*(\Gamma) - (\Lambda G)_S^*(-\Gamma) + \Lambda(C)].$$

Remark 7.2.1. The scalar optimization problem with geometric and cone constraints (P^C) studied in chapter 3 can be seen as a particular case of (PSV^C). As one can notice in the following, by particularizing the set-valued duals to the scalar setting we rediscover three dual problems similar to (D^{C_L}), (D^{C_F}) and $(D^{C_{FL}})$, respectively, that have been formulated in section 3.1 in connection to (P^C).

Not only for the Lagrange set-valued dual problem, but also for the Fenchel and Fenchel-Lagrange set-valued dual problems weak duality holds without supposing any particular hypotheses. This is an immediate consequence of Theorem 7.1.11. According to Theorem 7.1.15 strong duality is ensured under stability assumptions for (PSV^C) regarding the perturbation maps Φ^{C_F} and $\Phi^{C_{FL}}$, respectively. Using the strong duality and Remark 7.1.7 optimality conditions can be easily derived. Even more, taking into consideration Theorem 7.1.16 and Theorem 7.1.17 these optimality conditions may be represented in subdifferential form as necessary and sufficient conditions.

For the following we assume that the corresponding minimal value maps of the three perturbation maps are externally stable. First we present optimality conditions and strong duality by employing the Lagrange set-valued dual to (PSV^C)

$$(DSV^{C_L}) \quad \text{Max} \bigcup_{\substack{\Lambda \in \mathcal{L}(Z,V), \\ \Lambda(C)\cap(-K)=\{0\}}} \text{Min} \bigcup_{x \in S} [F(x) + \Lambda(G(x) + C)].$$

Theorem 7.2.1. *(a) Suppose that the problem (PSV^C) is stable with respect to the perturbation map Φ^{C_L}. Let $\bar{x} \in \mathcal{A}$ be an efficient solution to (PSV^C) and $\bar{v} \in F(\bar{x})$ such that $(\bar{x}, \bar{v})$ is a minimal pair to (PSV^C). Then there exists $\overline{\Lambda} \in \mathcal{L}(Z, V)$, an efficient solution to (DSV^{C_L}), with $(\overline{\Lambda}, \bar{v})$ corresponding maximal pair such that strong duality holds and the following conditions are fulfilled*

(i) $\bar{v} \in \text{Min} \cup_{x\in S}[F(x) + \overline{\Lambda}(G(x) + C)]$;

(ii) $\overline{\Lambda}(C) \cap (-K) = \{0\}$;

(iii) $\overline{\Lambda}(G(\bar{x}) + C) \cap (-K) = \{0\}$.

(b) Assume that for $\bar{x} \in \mathcal{A}$, $\bar{v} \in F(\bar{x}) \cap V$ and $\overline{\Lambda} \in \mathcal{L}(Z, V)$ the conditions $(i) - (iii)$ are fulfilled. Then $\bar{x}$ is an efficient solution and $(\bar{x}, \bar{v})$ is a minimal pair to (PSV^C), while $\overline{\Lambda}$ is an efficient solution and $(\overline{\Lambda}, \bar{v})$ is a maximal pair to (DSV^{C_L}).

Proof. (a) By Theorem 7.1.15 there exists an efficient solution $\overline{\Lambda} \in \mathcal{L}(Z, V)$ to (DSV^{C_L}) with corresponding maximal pair $(\overline{\Lambda}, \bar{v})$ fulfilling $\bar{v} \in \text{Min} \cup_{x\in S}[F(x) + \overline{\Lambda}(G(x) + C)]$ and so (i) is verified. The condition (ii) follows from the construction of the Lagrange set-valued dual problem. To prove (iii) notice first that from $\bar{x} \in \mathcal{A}$ follows $G(\bar{x}) \cap (-C) \neq \emptyset$, consequently $0 \in \overline{\Lambda}(G(\bar{x}) + C) \cap (-K)$. Assume now that there exists $\tilde{v} \in \overline{\Lambda}(G(\bar{x}) + C) \cap ((-K)\setminus\{0\})$. Then

$$\bar{v} + \tilde{v} \in F(\bar{x}) + \overline{\Lambda}(G(\bar{x}) + C) \subseteq \bigcup_{x\in S} [F(x) + \overline{\Lambda}(G(x) + C)]$$

and $\bar{v} + \tilde{v} \leqq_K \bar{v}$, contradicting (i). Altogether, (iii) is true.

(b) The proof of this result is a direct consequence of Theorem 7.1.17. □

Remark 7.2.2. Let us compare the optimality conditions $(i)-(iii)$ in Theorem 7.2.1 with the optimality conditions given for the primal dual pair $(P^C) - (D^{C_L})$ in Theorem 3.3.16. Consider as image space $V = \mathbb{R}$ with the ordering cone $K = \mathbb{R}_+$. Then $\mathcal{L}(Z, \mathbb{R}) = Z^*$ and an element $\Lambda \in Z^*$ fulfilling $\Lambda(C) \geq 0$ is in fact belonging to the dual cone C^*. Further assume that $F : X \to \mathbb{R} \cup \{+\infty\}$ and $G : X \to Z$ are single-valued maps. The Lagrange set-valued dual problem (DSV^{C_L}) takes in this case the form

$$\max_{\Lambda \in C^*} \min_{x \in S} [F(x) + \Lambda(G(x))]$$

and now it is easy to recognize the similar formulation to the one of the scalar Lagrange dual problem (D^{C_L}).

Concerning the optimality conditions, one can easily see that assertion (i) of Theorem 7.2.1 becomes in this case $F(\bar{x}) = \min_{x \in S}[F(x) + \overline{\Lambda}(G(x))]$, while condition (ii) states that $\overline{\Lambda} \in C^*$. Coming now to condition (iii), this looks now like $\overline{\Lambda}(G(\bar{x}) + C) \geq 0$, which is equivalent to $\overline{\Lambda}(G(\bar{x})) = 0$. In this way we rediscover as particular instance the three optimality conditions for $(P^C) - (D^{C_L})$ given in Theorem 3.3.16.

Next we provide optimality conditions for (PSV^C) and its Fenchel set-valued dual problem

$$(DSV^{C_F}) \quad \operatorname{Max} \bigcup_{\Gamma \in \mathcal{L}(X,V)} \operatorname{Min}[-F^*(\Gamma) + \operatorname{Min} \Gamma(\mathcal{A})],$$

under the additional assumptions that $F^*(\Gamma)$ and $\operatorname{Min} \Gamma(\mathcal{A})$ are externally stable for all $\Gamma \in \mathcal{L}(X, V)$. The proof of the following result is a consequence of Theorem 7.1.15, Theorem 7.1.16 and Theorem 7.1.17.

Theorem 7.2.2. *Let $F^*(\Gamma)$ and $\operatorname{Min} \Gamma(\mathcal{A})$ be externally stable for all $\Gamma \in \mathcal{L}(X, V)$.*

(a) Suppose that the problem (PSV^C) is stable with respect to the perturbation map Φ^{C_F}. Let $\bar{x} \in \mathcal{A}$ be an efficient solution to (PSV^C) and $\bar{v} \in F(\bar{x})$ such that $(\bar{x}, \bar{v})$ is a minimal pair to (PSV^C). Then there exists $\overline{\Gamma} \in \mathcal{L}(X, V)$, an efficient solution to (DSV^{C_F}), with $(\overline{\Gamma}, \bar{v})$ corresponding maximal pair such that strong duality holds and

$$\bar{v} \in \operatorname{Min}[-F^*(\overline{\Gamma}) + \operatorname{Min} \overline{\Gamma}(\mathcal{A})].$$

(b) Assume that for $\bar{x} \in \mathcal{A}$, $\bar{v} \in F(\bar{x}) \cap V$ and $\overline{\Gamma} \in \mathcal{L}(X, V)$ one has $\bar{v} \in \operatorname{Min}[-F^*(\overline{\Gamma}) + \operatorname{Min} \overline{\Gamma}(\mathcal{A})]$. Then $\bar{x}$ is an efficient solution and $(\bar{x}, \bar{v})$ a minimal pair to (PSV^C), while $\overline{\Gamma}$ is an efficient solution and $(\overline{\Gamma}, \bar{v})$ a maximal pair to (DSV^{C_F}).

Remark 7.2.3. Considering the same setting as in Remark 7.2.2, we have that $\mathcal{L}(X, \mathbb{R}) = X^*$. Assuming that for all $\Gamma \in X^*$ the supremum $F^*(\Gamma) = \sup_{x \in X}\{\Gamma x - F(x)\}$ and the infimum $\inf_{x \in \mathcal{A}}(\Gamma x)$ are attained, means that both sets $F^*(\Gamma)$ and $\operatorname{Min} \Gamma(\mathcal{A})$ are externally stable for all $\Gamma \in X^*$. The Fenchel set-valued dual problem looks then like

$$\max_{\Gamma\in X^*}\left[-F^*(\Gamma)+\min_{x\in\mathcal{A}}(\Gamma x)\right],$$

while the optimality condition in Theorem 7.2.2 can be written for $\bar{x}\in\mathcal{A}$ and $\overline{\Gamma}\in X^*$ as

$$F(\bar{x})=-F^*(\overline{\Gamma})+\min_{x\in\mathcal{A}}(\overline{\Gamma}x),$$

or, equivalently,

$$\min_{x\in\mathcal{A}}\langle\overline{\Gamma},x\rangle=\overline{\Gamma}\bar{x} \text{ and } F(\bar{x})+F^*(\overline{\Gamma})=\overline{\Gamma}\bar{x}.$$

The reader can notice the analogy of the optimality conditions offered above to the ones in Theorem 3.3.19.

Next we give strong duality and corresponding optimality conditions for (PSV^C) and the Fenchel-Lagrange set-valued dual problem

$$(DSV^{C_{FL}})\quad \text{Max}\bigcup_{\substack{\Gamma\in\mathcal{L}(X,V),\Lambda\in\mathcal{L}(Z,V),\\ \Lambda(C)\cap(-K)=\{0\}}}\text{Min}[-F^*(\Gamma)-(\Lambda G)^*_S(-\Gamma)+\Lambda(C)],$$

in case $F^*(\Gamma)$ and $(\Lambda G)^*_S(-\Gamma)$ are externally stable for all $\Gamma\in\mathcal{L}(X,V)$ and all $\Lambda\in\mathcal{L}(Z,V)$.

Theorem 7.2.3. *Let $F^*(\Gamma)$ and $(\Lambda G)^*_S(-\Gamma)$ be externally stable for all $\Gamma\in\mathcal{L}(X,V)$ and all $\Lambda\in\mathcal{L}(Z,V)$.*

(a) Suppose that the problem (PSV^C) is stable with respect to the perturbation map $\Phi^{C_{FL}}$. Let $\bar{x}\in\mathcal{A}$ be an efficient solution to (PSV^C) and $\bar{v}\in F(\bar{x})$ such that $(\bar{x},\bar{v})$ is a minimal pair to (PSV^C). Then there exists $(\overline{\Gamma},\overline{\Lambda})\in\mathcal{L}(X,V)\times\mathcal{L}(Z,V)$, an efficient solution to $(DSV^{C_{FL}})$, with $(\overline{\Gamma},\overline{\Lambda},\bar{v})$ corresponding maximal pair such that strong duality holds and the following conditions are fulfilled
(i) $\bar{v}\in\text{Min}[-F^*(\overline{\Gamma})-(\overline{\Lambda}G)^*_S(-\overline{\Gamma})+\overline{\Lambda}(C)]$;
(ii) $\overline{\Lambda}(C)\cap(-K)=\{0\}$;
(iii) $\overline{\Lambda}(G(\bar{x})+C)\cap(-K)=\{0\}$.

(b) Assume that for $\bar{x}\in\mathcal{A}$, $\bar{v}\in F(\bar{x})\cap V$ and $(\overline{\Gamma},\overline{\Lambda})\in\mathcal{L}(X,V)\times\mathcal{L}(Z,V)$ the conditions $(i)-(iii)$ are fulfilled. Then $\bar{x}$ is an efficient solution and $(\bar{x},\bar{v})$ a minimal pair to (PSV^C), while $(\overline{\Gamma},\overline{\Lambda})$ is an efficient solution and $(\overline{\Gamma},\overline{\Lambda},\bar{v})$ a maximal pair to $(DSV^{C_{FL}})$.

Proof. (*a*) By Theorem 7.1.15 there exists an efficient solution $(\overline{\Gamma},\overline{\Lambda})\in\mathcal{L}(X,V)\times\mathcal{L}(Z,V)$ to $(DSV^{C_{FL}})$ with corresponding maximal pair $(\overline{\Gamma},\overline{\Lambda},\bar{v})$ fulfilling $\bar{v}\in\text{Min}[-F^*(\overline{\Gamma})-(\overline{\Lambda}G)^*_S(-\overline{\Gamma})+\overline{\Lambda}(C)]$, which is in fact (*i*). The condition (*ii*) follows from the construction of the Fenchel-Lagrange set-valued dual problem. To prove (*iii*) notice first that from $\bar{x}\in\mathcal{A}$ follows $G(\bar{x})\cap(-C)\neq\emptyset$, consequently $0\in\overline{\Lambda}(G(\bar{x})+C)\cap(-K)$. Assume now that there exists $\tilde{v}\in\overline{\Lambda}(G(\bar{x})+C)\cap((-K)\backslash\{0\})$. Then

$$\bar{v}+\tilde{v}\in F(\bar{x})+\overline{\Lambda}(G(\bar{x})+C)=F(\bar{x})-\overline{\Gamma}(\bar{x})+\overline{\Gamma}(\bar{x})+\overline{\Lambda}(G(\bar{x}))+\overline{\Lambda}(C)$$
$$\subseteq \bigcup_{y\in X}[F(y)-\overline{\Gamma}(y)]+\bigcup_{x\in S}[\overline{\Gamma}(x)+\overline{\Lambda}(G(x))]+\overline{\Lambda}(C).$$

On the other hand, since $F^*(\overline{\Gamma})$ and $(\overline{\Lambda}G)^*_S(-\overline{\Gamma})$ are externally stable, from (i) it follows $\bar{v}\in \operatorname{Min}\left\{\cup_{y\in X}[F(y)-\overline{\Gamma}(y)]+\cup_{x\in S}[\overline{\Gamma}(x)+\overline{\Lambda}(G(x))]+\overline{\Lambda}(C)\right\}$. Since $\bar{v}+\tilde{v}\leq_K \bar{v}$, this leads to a contradiction and so (iii) is proven.

(b) The proof of this result is a direct consequence of Theorem 7.1.17. □

Remark 7.2.4. Considering again the setting from Remark 7.2.2 and Remark 7.2.3, we assume that for all $\Gamma\in X^*$ and all $\Lambda\in Z^*$ the suprema $F^*(\Gamma):=\sup_{x\in X}\{\langle\Gamma,x\rangle - F(x)\}$ and $(\Lambda G)^*_S(-\Gamma):=\sup_{x\in S}\{-\langle\Gamma,x\rangle-(\Lambda G)(x)\}$ are attained. Then the Fenchel-Lagrange set-valued dual problem $(DSV^{C_{FL}})$ is nothing else than

$$\max_{\Gamma\in X^*,\Lambda\in C^*}[-F^*(\Gamma)-(\Lambda G)^*_S(-\Gamma)],$$

while the optimality conditions $(i)-(iii)$ in the previous result can be formulated as $F(\bar{x})=-F^*(\overline{\Gamma})-(\overline{\Lambda}G)^*_S(-\overline{\Gamma})$, $\overline{\Lambda}\in C^*$ and $\overline{\Lambda}(G(\bar{x}))=0$, respectively. The first relation can be in this situation equivalently written as $F(\bar{x})+F^*(\overline{\Gamma})=\overline{\Gamma}\bar{x}$ and $\overline{\Lambda}(G(\bar{x}))+(\overline{\Lambda}G)^*_S(-\overline{\Gamma})=-\overline{\Gamma}\bar{x}$. The analogy to the optimality conditions stated for the primal-dual pair $(P^C)-(D^{C_{FL}})$ in Theorem 3.3.22 is also in this case remarkable.

7.2.2 Duality based on vector k-conjugacy

Within the current subsection we deal with duality for the primal set-valued optimization problem (PSV^C) based on the vector k-conjugacy approach introduced in subsection 7.1.3 and for this purpose we use the same perturbation maps Φ^{C_L}, Φ^{C_F} and $\Phi^{C_{FL}}$ which have been introduced at the beginning of subsection 7.2.1.

Let $k\in V\backslash\{0\}$ be fixed. We begin with the calculation of the map $(\Phi^{C_L})^*_k$. For an arbitrary $z^*\in Z^*$ it holds

$$(\Phi^{C_L})^*_k(0,z^*)=\operatorname{Max}\bigcup_{\substack{x\in X,\\ z\in Z}}[\langle z^*,z\rangle k-\Phi^{C_L}(x,z)]$$
$$=\operatorname{Max}\bigcup_{\substack{x\in S,z\in Z,\\ z\in G(x)+C}}[\langle z^*,z\rangle k-F(x)]=\operatorname{Max}\bigcup_{x\in S}[(z^*G)(x)k+z^*(C)k-F(x)],$$

where we denote $(z^*G)(x):=\{\langle z^*,z\rangle : z\in G(x)\}$. Therefore to the primal problem (PSV^C) we attach the *Lagrange set-valued dual problem*

$$(DSV_k^{C_L})\quad \operatorname{Max}\bigcup_{z^*\in Z^*}\operatorname{Min}\bigcup_{x\in S}[F(x)-(z^*G)(x)k-z^*(C)k].$$

Let us additionally assume in the following that $k\in K$. If $z^*\in -C^*$ then for all $c\in C$ it holds $-\langle z^*,c\rangle\geq 0$. Thus $-\langle z^*,c\rangle k\in K$ and a straightforward consideration shows that in this case the relation

$$\text{Min} \bigcup_{x \in S} [F(x) - (z^*G)(x)k - z^*(C)k] = \text{Min} \bigcup_{x \in S} [F(x) - (z^*G)(x)k]$$

holds. On the other hand, if $z^* \notin -C^*$ then there exists some $\bar{c} \in C$ such that $-\langle z^*, \bar{c} \rangle < 0$. Using the fact that for all $\alpha > 0$ it holds $\alpha \bar{c} \in C$ one can easily prove that

$$\text{Min} \bigcup_{x \in S} [F(x) - \langle z^*, G(x) \rangle k - z^*(C)k] = \emptyset.$$

Remark 7.2.5. Following a similar reasoning one can show that if $k \in -K$, then the previous comments regarding

$$\text{Min} \bigcup_{x \in S} [F(x) - (z^*G)(x)k - z^*(C)k]$$

remain valid if $z^* \in C^*$ and $z^* \notin C^*$, respectively. If, finally, $k \notin K \cup (-K)$, then in general the set $-z^*(C)k$ cannot be dropped in the formulation of the dual problem.

The Lagrange set-valued dual problem to (PSV^C) may be rewritten as

$$(DSV_k^{C_L}) \quad \text{Max} \bigcup_{z^* \in C^*} \text{Min} \bigcup_{x \in S} [F(x) + (z^*G)(x)k].$$

For $V = \mathbb{R}$ and $K = \mathbb{R}_+$ and when F and G are single-valued one can easily notice the analogy of $(DSV_k^{C_L})$ to the classical Lagrange dual problem (D^{C_L}) from subsection 3.1.3, the only difference being that in its formulation we have maximum and minimum instead of supremum and infimum, respectively.

Now let us determine the Fenchel set-valued dual problem to (PSV^C). To this end it is necessary to calculate the k-conjugate of the perturbation map Φ^{C_F}. For an arbitrary $y^* \in X^*$ it holds

$$(\Phi^{C_F})_k^*(0, y^*) = \text{Max} \bigcup_{x,y \in X} [\langle y^*, y \rangle k - \Phi^{C_F}(x, y)]$$

$$= \text{Max} \bigcup_{x \in \mathcal{A}, y \in X} [\langle y^*, y \rangle k - F(y) - \langle y^*, x \rangle k].$$

Because $k \in K$, there holds

$$(\Phi^{C_F})_k^*(0, y^*) = \text{Max} \left\{ \bigcup_{y \in X} [\langle y^*, y \rangle k - F(y)] - y^*(\mathcal{A})k \right\} = F_k^*(y^*) + \max_{x \in \mathcal{A}} \langle -y^*, x \rangle k.$$

Notice that we encountered in the above formula also the situation when $\min_{x \in \mathcal{A}} \langle y^*, x \rangle$ is not attained. In this case $(\Phi^{C_F})_k^*(0, y^*) = \emptyset$. This leads to the *Fenchel set-valued dual problem* to (PSV^C)

$$(DSV_k^{C_F}) \quad \text{Max} \bigcup_{y^* \in X^*} \left\{ -F_k^*(y^*) + \left[\min_{x \in \mathcal{A}} \langle y^*, x \rangle \right] k \right\}.$$

The analogy of $(DSV_k^{C_F})$ to the dual problem (D^{C_F}) stated for the scalar primal problem with geometric and cone constraints (P^C) in subsection 3.1.3 can be easily recognized.

Next we derive the Fenchel-Lagrange set-valued dual problem to (PSV^C) via vector k-conjugacy. For $y^* \in X^*$ and $z^* \in Z^*$ arbitrarily taken we have

$$(\Phi^{C_{FL}})^*_k(0, y^*, z^*) = \operatorname{Max} \bigcup_{x,y \in X, z \in Z} [(\langle y^*, y\rangle + \langle z^*, z\rangle)k - \Phi^{C_{FL}}(x, y, z)]$$

$$= \operatorname{Max} \bigcup_{\substack{x \in S, y \in X, \\ z \in G(x)+C}} [\langle y^*, y\rangle k - F(y) - \langle y^*, x\rangle k + \langle z^*, z\rangle k]$$

$$= \operatorname{Max} \bigcup_{x \in S, y \in X} \{[\langle y^*, y\rangle k - F(y)] + [-\langle y^*, x\rangle k + (z^*G)(x)k] + z^*(C)k\}.$$

Since we supposed that $k \in K$, we further have

$$(\Phi^{C_{FL}})^*_k(0, y^*, z^*) = \operatorname{Max} \bigcup_{y \in X} [\langle y^*, y\rangle k - F(y)] + \Big[\max_{x \in S} \big(-\langle y^*, x\rangle + (z^*G)(x)\big)\Big] k$$

$$= F^*_k(y^*) + \big[\max_{x \in S} \big(-\langle y^*, x\rangle + (z^*G)(x)\big)\big] k$$

if $z^* \in -C^*$ and $(\Phi^{C_{FL}})^*_k(0, y^*, z^*) = \emptyset$ if $z^* \notin -C^*$. Thus the *Fenchel-Lagrange set-valued dual problem* to (PSV^C) turns out to be

$$(DSV^{C_{FL}}_k) \quad \operatorname{Max} \bigcup_{\substack{y^* \in X^*, \\ z^* \in C^*}} \Big\{ -F^*_k(y^*) + \Big[\min_{x \in S} \big(\langle y^*, x\rangle + (z^*G)(x)\big)\Big] k\Big\}.$$

Also in this case it is easy to recognize the analogy to the formulation of the scalar Fenchel-Lagrange dual problem $(D^{C_{FL}})$ stated to the primal scalar problem (P^C) in subsection 3.1.3.

It follows from Theorem 7.1.28 that weak duality holds for (PSV^C) and the three duals $(DSV^{C_L}_k)$, $(DSV^{C_F}_k)$ and $(DSV^{C_{FL}}_k)$, respectively. While weak duality applies without any further assumptions, for having strong duality we require external stability for the corresponding minimal value maps of the three perturbation maps considered above. Indeed, Theorem 7.1.32 guarantees strong duality under the stability of (PSV^C) with respect to the perturbation maps Φ^{C_L} or Φ^{C_F} or $\Phi^{C_{FL}}$ and using k-subdifferentiability as considered in Definition 7.1.4. It is straightforward to transfer the formulation of Theorem 7.1.32 and Theorem 7.1.34 to the three dual problems derived in this subsection.

Theorem 7.2.4. *Let be $k \in K \backslash \{0\}$.*

(a) Suppose that the problem (PSV^C) is k-stable with respect to the perturbation map Φ^{C_L}. Let $\bar{x} \in \mathcal{A}$ be an efficient solution to (PSV^C) and $\bar{v} \in F(\bar{x})$ such that $(\bar{x}, \bar{v})$ is a minimal pair to (PSV^C). Then there exists $\bar{z}^* \in C^*$, an efficient solution to $(DSV^{C_L}_k)$, with $(\bar{z}^*, \bar{v})$ corresponding maximal pair such that strong duality holds and the following conditions are fulfilled

(i) $\bar{v} \in \operatorname{Min} \cup_{x \in S} [F(x) + (\bar{z}^*G)(x)k]$;

(ii) $\langle \bar{z}^*, z\rangle = 0$ for all $z \in G(\bar{x}) \cap (-C)$.

(b) *Assume that for* $\bar{x} \in \mathcal{A}$, $\bar{v} \in F(\bar{x}) \cap V$ *and* $\bar{z}^* \in C^*$ *the conditions* $(i)-(ii)$ *are fulfilled. Then* $\bar{x}$ *is an efficient solution and* $(\bar{x}, \bar{v})$ *a minimal pair to* (PSV^C)*, while* $\bar{z}^*$ *is an efficient solution and* $(\bar{z}^*, \bar{v})$ *a maximal pair to* $(DSV_k^{C_L})$.

Proof. (a) Let $\bar{x} \in \mathcal{A}$ and $\bar{v} \in F(\bar{x})$ be as supposed in the hypothesis. By Theorem 7.1.32 there exists $\bar{z}^* \in C^*$, an efficient solution to $(DSV_k^{C_L})$, with $(\bar{z}^*, \bar{v})$ corresponding maximal pair such that $\bar{v} \in \operatorname{Min} \cup_{x \in S}[F(x) + (\bar{z}^* G)(x)k]$. Next, we show that (ii) also holds. Indeed, assume to the contrary that there exists $\bar{z} \in G(\bar{x}) \cap (-C)$ such that $\langle \bar{z}^*, \bar{z} \rangle < 0$. But this contradicts (i), because $\bar{v} + \langle \bar{z}^*, \bar{z} \rangle k \in F(\bar{x}) + (\bar{z}^* G)(\bar{x})k \subseteq \cup_{x \in S}[F(x) + (\bar{z}^* G)(x)k]$ and $\bar{v} + \langle \bar{z}^*, \bar{z} \rangle k \in \bar{v} - \{K \backslash \{0\}\}$.

(b) The statement follows via Theorem 7.1.34. □

The next result concerns the pair of set-valued optimization problems $(PSV^C) - (DSV_k^{C_F})$.

Theorem 7.2.5. *Let be* $k \in K \backslash \{0\}$.

(a) *Suppose that the problem* (PSV^C) *is* k*-stable with respect to the perturbation map* Φ^{C_F}. *Let* $\bar{x} \in \mathcal{A}$ *be an efficient solution to* (PSV^C) *and* $\bar{v} \in F(\bar{x})$ *such that* $(\bar{x}, \bar{v})$ *is a minimal pair to* (PSV^C). *Then there exists* $\bar{y}^* \in X^*$, *an efficient solution to* $(DSV_k^{C_F})$, *with* $(\bar{y}^*, \bar{v})$ *corresponding maximal pair such that strong duality holds and the following conditions are fulfilled*
   (i) $\bar{v} \in -F_k^*(\bar{y}^*) + \langle \bar{y}^*, \bar{x} \rangle k$;
   (ii) $\langle \bar{y}^*, \bar{x} \rangle = \min_{x \in \mathcal{A}} \langle \bar{y}^*, x \rangle$.
(b) *Assume that for* $\bar{x} \in \mathcal{A}$, $\bar{v} \in F(\bar{x}) \cap V$ *and* $\bar{y}^* \in X^*$ *the conditions* $(i)-(ii)$ *are fulfilled. Then* $\bar{x}$ *is an efficient solution and* $(\bar{x}, \bar{v})$ *a minimal pair to* (PSV^C), *while* $\bar{y}^*$ *is an efficient solution and* $(\bar{y}^*, \bar{v})$ *a maximal pair to* $(DSV_k^{C_F})$.

Proof. (a) Let $\bar{x} \in \mathcal{A}$ and $\bar{v} \in F(\bar{x})$ be as assumed in the hypothesis. Then by Theorem 7.1.32 follows that there exists $\bar{y}^* \in X^*$, an efficient solution to $(DSV_k^{C_F})$, with $(\bar{y}^*, \bar{v})$ corresponding maximal pair such that $\bar{v} \in -F_k^*(\bar{y}^*) + [\min_{x \in \mathcal{A}} \langle \bar{y}^*, x \rangle]k$. Hence, $\bar{v} = -v^* + [\min_{x \in \mathcal{A}} \langle \bar{y}^*, x \rangle]k$ for some $v^* \in F_k^*(\bar{y}^*)$. We show now that $\langle \bar{y}^*, \bar{x} \rangle = \min_{x \in \mathcal{A}} \langle \bar{y}^*, x \rangle$. Assume the opposite, namely that $\min_{x \in \mathcal{A}} \langle \bar{y}^*, x \rangle < \langle \bar{y}^*, \bar{x} \rangle$. As $\bar{v} \in F(\bar{x})$ and $v^* \in F_k^*(\bar{y}^*)$, by Proposition 7.1.1 follows that $\bar{v} + v^* \nleq_K \langle \bar{y}^*, \bar{x} \rangle k$. But $\bar{v} + v^* = [\min_{x \in \mathcal{A}} \langle \bar{y}^*, x \rangle]k \leq_K \langle \bar{y}^*, \bar{x} \rangle k$ generates a contradiction. Thus (ii) holds and from here it follows that $\bar{v} \in -F_k^*(\bar{y}^*) + [\min_{x \in \mathcal{A}} \langle \bar{y}^*, x \rangle]k = -F_k^*(\bar{y}^*) + \langle \bar{y}^*, \bar{x} \rangle k$, i.e. (i) is also proven.

(b) The statement follows via Theorem 7.1.34. □

Finally, it remains to state the strong duality statement along with the optimality conditions for the Fenchel-Lagrange set-valued dual problem $(DSV_k^{C_{FL}})$.

Theorem 7.2.6. *Let be* $k \in K \backslash \{0\}$.

(a) Suppose that the problem (PSV^C) *is* k*-stable with respect to the perturbation map* $\Phi^{C_{FL}}$*. Let* $\bar{x} \in \mathcal{A}$ *be an efficient solution to* (PSV^C) *and* $\bar{v} \in F(\bar{x})$ *such that* $(\bar{x}, \bar{v})$ *is a minimal pair to* (PSV^C)*. Then there exists* $(\bar{y}^*, \bar{z}^*) \in X^* \times C^*$*, an efficient solution to* $(DSV_k^{C_{FL}})$*, with* $(\bar{y}^*, \bar{z}^*, \bar{v})$ *corresponding maximal pair such that strong duality holds and the following conditions are fulfilled*

(i) $\bar{v} \in -F_k^*(\bar{y}^*) + \langle \bar{y}^*, \bar{x} \rangle k$;

(ii) $\langle \bar{y}^*, \bar{x} \rangle = \min_{x \in S} \{\langle \bar{y}^*, x \rangle + (\bar{z}^* G)(x)\}$;

(iii) $\langle \bar{z}^*, z \rangle = 0$ *for all* $z \in G(\bar{x}) \cap (-C)$.

(b) Assume that for $\bar{x} \in \mathcal{A}$, $\bar{v} \in F(\bar{x}) \cap V$ *and* $(\bar{y}^*, \bar{z}^*) \in X^* \times C^*$ *the conditions* $(i)-(iii)$ *are fulfilled. Then* $\bar{x}$ *is an efficient solution and* $(\bar{x}, \bar{v})$ *a minimal pair to* (PSV^C)*, while* $(\bar{y}^*, \bar{z}^*)$ *is an efficient solution and* $(\bar{y}^*, \bar{z}^*, \bar{v})$ *a maximal pair to* $(DSV_k^{C_{FL}})$.

Proof. (a) Theorem 7.1.32 provides the existence of an efficient solution $(\bar{y}^*, \bar{z}^*) \in X^* \times C^*$ to $(DSV^{C_{FL}})$ with $(\bar{y}^*, \bar{z}^*, \bar{v})$ a corresponding maximal pair such that $\bar{v} \in -F_k^*(\bar{y}^*) + [\min_{x \in S}\{\langle \bar{y}^*, x \rangle + (\bar{z}^* G)(x)\}]k$. The remaining part of the proof can be done combining some similar considerations as within the proofs of Theorem 7.2.4 and Theorem 7.2.5.

(b) The statement follows via Theorem 7.1.34. □

Remark 7.2.6. (a) Let us note that the sufficiency of the optimality conditions in Theorem 7.2.4, Theorem 7.2.5 and Theorem 7.2.6 is guaranteed even without the external stability of the corresponding minimal value maps.

(b) Like in the previous subsection one can notice an analogy between the optimality conditions in Theorem 7.2.4, Theorem 7.2.5 and Theorem 7.2.6 and the optimality conditions stated in section 3.3 for the primal-dual pairs $(P^C) - (D^{C_L})$, $(P^C) - (D^{C_F})$ and $(P^C) - (D^{C_{FL}})$, respectively.

7.2.3 Stability criteria

This subsection is devoted to some results concerning the stability of the primal problem (PSV^C) with respect to the different perturbation maps we have considered in this chapter. Such results are very beneficial as they permit to get strong duality for (PSV^C) and its Lagrange, Fenchel and Fenchel-Lagrange set-valued dual problems, respectively. Stability of (PSV^C) is closely related to the underlying perturbation map. Let us start with Φ^{C_L}, which was used for deriving the Lagrange set-valued dual problems (DSV^{C_L}) and $(DSV_k^{C_L})$, respectively.

As in the previous subsections all considered spaces are supposed to be topological vector spaces, while the ordering cone $K \subseteq V$ is assumed to be nontrivial pointed convex with $\operatorname{int}(K) \neq \emptyset$.

We start by providing a stability criterion for Lagrange set-valued duality.

Theorem 7.2.7. *Let* Φ^{C_L} *be strictly-*K*-convexlike-convex and* $H^{C_L}(z) = \operatorname{Min} \Phi^{C_L}(X, z)$ *be externally stable for all* $z \in Z$*. If* F *is weakly* K*-upper* *bounded on* $\operatorname{dom} F \cap S$ *and* $0 \in \operatorname{int}[G(\operatorname{dom} F \cap S) + C]$, *then* (PSV^C) *is stable with respect to the perturbation map* Φ^{C_L}*; hence strong duality holds for* (PSV^C) *and its Lagrange set-valued dual problem* (DSV^{C_L}).

Proof. For all $z \in Z$ it holds $\Psi^{C_L}(z) \cap V = \Phi^{C_L}(X, z) \cap V = F(\operatorname{dom} F \cap S \cap G^{-1}(z - C)) \cap V$. The consequence of the regularity condition $0 \in \operatorname{int}[G(\operatorname{dom} F \cap S) + C]$ is the existence of a neighborhood W of 0 in Z with $W \subseteq G(\operatorname{dom} F \cap S) + C$. On the other hand, we have that $z \in G(\operatorname{dom} F \cap S) + C$ if and only if

$$\operatorname{dom} F \cap S \cap G^{-1}(z - C) \neq \emptyset$$

and this is further equivalent to

$$F(\operatorname{dom} F \cap S \cap G^{-1}(z - C)) \cap V \neq \emptyset.$$

Consequently, one has that $W \subseteq \operatorname{dom} \Psi^{C_L}$ and from here follows $0 \in \operatorname{int}(\operatorname{dom} \Psi^{C_L})$. Obviously Ψ^{C_L} is weakly K-upper bounded on W, since $\Psi^{C_L}(z) \cap V = F(\operatorname{dom} F \cap S \cap G^{-1}(z - C)) \cap V \neq \emptyset$ for all $z \in W$ and F is assumed to be weakly K-upper bounded on $\operatorname{dom} F \cap S$. Therefore all the hypotheses of Theorem 7.1.27 are fulfilled, which means that (PSV^C) is stable with respect to Φ^{C_L}. Finally, Theorem 7.2.1 enforces the claimed strong duality. □

The condition $0 \in \operatorname{int}[G(\operatorname{dom} F \cap S) + C]$ is fulfilled in particular if $G(\operatorname{dom} F \cap S) \cap (-\operatorname{int}(C)) \neq \emptyset$, which can be seen as a *generalized Slater regularity condition* for set-valued optimization problems. As we prove next, imposing this stronger condition, we can drop the assumption of F being weakly K-upper bounded on $\operatorname{dom} F \cap S$.

Theorem 7.2.8. *Let* Φ^{C_L} *be strictly-*K*-convexlike-convex and* $H^{C_L}(z) = \operatorname{Min} \Phi^{C_L}(X, z)$ *be externally stable for all* $z \in Z$. *If* $G(\operatorname{dom} F \cap S) \cap (-\operatorname{int}(C)) \neq \emptyset$, *then* (PSV^C) *is stable with respect to the perturbation map* Φ^{C_L}*; hence strong duality holds for* (PSV^C) *and its Lagrange set-valued dual problem* (DSV^{C_L}).

Proof. Let $x' \in \operatorname{dom} F \cap S$ be such that $G(x') \cap (-\operatorname{int}(C)) \neq \emptyset$. Thus $0 \in G(x') + \operatorname{int}(C) \subseteq \operatorname{int}(G(x') + C)$. Then there exists W, a neighborhood of 0 in Z, such that $W \subseteq G(x') + C \subseteq G(\operatorname{dom} F \cap S) + C$. As seen in the proof of Theorem 7.2.7, $W \subseteq \operatorname{dom} \Psi^{C_L}$ and so $0 \in \operatorname{int}(\operatorname{dom} \Psi^{C_L})$.

On the other hand, since $x' \in \operatorname{dom} F \cap S$ one can chose $b \in F(x') \cap V$. For all $z \in W$ we have that $x' \in G^{-1}(z - C)$ and further $b \in F(\operatorname{dom} F \cap S \cap G^{-1}(z - C)) \cap V = \Psi^{C_L}(z) \cap V \subseteq \Psi^{C_L}(z) + K$. Thus Ψ^{C_L} is weakly K-upper bounded on W and the conclusion follows via Theorem 7.1.27. □

Remark 7.2.7. Consider the set-valued map $F \times G : X \rightrightarrows (V \cup \{+\infty_K\}) \times Z$ defined by $(F \times G)(x) = F(x) \times G(x)$. We say that the set-valued map $F \times G$ is *strictly-K-convexlike-C-convexlike* on a nonempty set $S \subseteq X$ if it is $(K \times C)$-convexlike on S and if for $(v_i, z_i) \in (F \times G)(S) \cap (V \times Z)$, $i = 1, 2$, with $v_1 \neq v_2$, and $\lambda \in (0, 1)$ there exists $\bar{x} \in S$ such that

$$\lambda(v_1, z_1) + (1 - \lambda)(v_2, z_2) \in (F(\bar{x}), G(\bar{x})) + (\mathrm{int}(K) \times C). \tag{7.18}$$

We emphasize that Φ^{C_L} is strictly-K-convexlike-convex if and only if $F \times G$ is strictly-K-convexlike-C-convexlike on S. Indeed, we begin assuming that $F \times G$ is strictly-K-convexlike-C-convexlike on S. We prove first that Φ^{C_L} is K-convexlike-convex and consider to this purpose $z_i \in Z$, $v_i \in \Phi^{C_L}(X, z_i) \cap V$, $i = 1, 2$, and $\lambda \in [0, 1]$. Then there exist $x_i \in S$ such that $v_i \in \Phi^{C_L}(x_i, z_i) \cap V$ and, consequently, $z_i \in G(x_i) + C$ and $v_i \in F(x_i) \cap V$ for $i = 1, 2$. Further there exist $c_i \in C$ such that $(v_i, z_i - c_i) \in (F(x_i), G(x_i)) \cap (V \times Z)$ for $i = 1, 2$. As $F \times G$ is $(K \times C)$-convexlike on S, there is an $\bar{x} \in S$ such that

$$\lambda(v_1, z_1 - c_1) + (1 - \lambda)(v_2, z_2 - c_2) \in (F(\bar{x}), G(\bar{x})) + (K \times C).$$

This means that $\lambda v_1 + (1 - \lambda) v_2 \in F(\bar{x}) + K$ and

$$\lambda z_1 + (1 - \lambda) z_2 \in G(\bar{x}) + \lambda c_1 + (1 - \lambda) c_2 + C \subseteq G(\bar{x}) + C.$$

Thus $\Phi^{C_L}(\bar{x}, \lambda z_1 + (1 - \lambda) z_2) = F(\bar{x})$ and therefore $\lambda v_1 + (1 - \lambda) v_2 \in \Phi^{C_L}(\bar{x}, \lambda z_1 + (1 - \lambda) z_2) + K$. This verifies that Φ^{C_L} is K-convexlike-convex. In a similar manner, using this time (7.18), one can prove that for Φ^{C_L} relation (7.11) in the definition of the strictly-K-convexlike-convexity is fulfilled, guaranteeing that this perturbation function is a member of this class.

Vice versa, let us assume that Φ^{C_L} is strictly-K-convexlike-convex and prove that $F \times G$ is strictly-K-convexlike-C-convexlike on S. Actually we prove only that $F \times G$ is $(K \times C)$-convexlike on S, the demonstration of the second property in the definition of a strictly-K-convexlike-C-convexlike map follows by using similar arguments. So let be $(v_i, z_i) \in (F \times G)(S) \cap (V \times Z), i = 1, 2$, and $\lambda \in [0, 1]$. Then there exists $x_i \in S$ such that $z_i \in G(x_i)$ and $v_i \in F(x_i) \cap V$ and from here $v_i \in \Phi^{C_L}(x_i, z_i) \subseteq \Phi^{C_L}(X, z_i)$ for $i = 1, 2$. As Φ^{C_L} is K-convexlike-convex, there exists $\bar{x} \in X$ such that $\lambda v_1 + (1 - \lambda) v_2 \in \Phi^{C_L}(\bar{x}, \lambda z_1 + (1 - \lambda) z_2) + K$ and one has $\bar{x} \in S$, $\lambda z_1 + (1 - \lambda) z_2 \in G(\bar{x}) + C$ and $\lambda v_1 + (1 - \lambda) v_2 \in F(\bar{x}) + K$. This provides the $(K \times C)$-convexlikeness on S of $F \times G$.

Consequently, instead of assuming that Φ^{C_L} is strictly-K-convexlike-convex, one can require in the hypotheses of Theorem 7.2.7 and Theorem 7.2.8 conditions which are sufficient for having that $F \times G$ is strictly-K-convexlike-C-convexlike on S. For instance, if S is convex, F is strictly K-convex and G is C-convex this fact is guaranteed.

It is also worth noticing that, via Corollary 7.1.37(b) a completely analogous result can be given for the primal-dual pair $(PSV^C) - (DSV_k^{C_L})$.

Corollary 7.2.9. *Let be $k \in \operatorname{int}(K)$. Under the hypotheses of Theorem 7.2.7 or Theorem 7.2.8 the problem (PSV^C) is k-stable with respect to the perturbation map Φ^{C_L}; hence strong duality holds for (PSV^C) and its Lagrange set-valued dual problem $(DSV_k^{C_L})$.*

Remark 7.2.8. In [163] the multiobjective programming problem

$$\begin{aligned}&\operatorname*{Min}_{x \in \mathcal{A}} f(x),\\ &\mathcal{A} = \{x \in \mathbb{R}^n : g(x) \leqq 0\}\end{aligned}$$

where $f : \mathbb{R}^n \to \mathbb{R}^k$ and $g : \mathbb{R}^n \to \mathbb{R}^m$ are single-valued functions has been investigated. The ordering cone in the image space $V = \mathbb{R}^k$ is $K = \mathbb{R}^k_+$. This problem is a special case of (PSV^C). The associated Lagrange set-valued dual problem looks like

$$\operatorname{Max} \bigcup_{z^* \geqq 0} \operatorname{Min} \bigcup_{x \in \mathbb{R}^n} [f(x) + \langle z^*, g(x)\rangle e].$$

Obviously, this is a particular case of the general Lagrange set-valued dual problem $(DSV_k^{C_L})$ when $k = e$. Corollary 7.2.9 can be considered as a generalization of the stability result claimed in [163, Proposition 6.1.16], while [163, Theorem 6.1.4] turns out to be a special case of Theorem 7.2.4.

Next, we turn our attention to the Fenchel duality.

Theorem 7.2.10. *Let Φ^{C_F} be strictly-K-convexlike-convex and $H^{C_F}(y) = \operatorname{Min} \Phi^{C_F}(X, y)$ be externally stable for all $y \in X$. If there exists $x' \in \operatorname{dom} F \cap \mathcal{A}$ such that F is weakly K-upper bounded on some neighborhood of x', then (PSV^C) is stable with respect to the perturbation map Φ^{C_F}; hence strong duality holds for (PSV^C) and its Fenchel set-valued dual problem (DSV^{C_F}).*

Proof. For all $y \in Y$ we have $\Psi^{C_F}(y) \cap V = \Phi^{C_F}(X, y) \cap V = F(\mathcal{A} + y) \cap V$. From the supposed weak K-upper boundedness follows the existence of a neighborhood U of 0 in X and of an element $b \in V$ such that $F(x' + y) \cap (b - K) \neq \emptyset$ for all $y \in U$. In other words, $b \in F(\mathcal{A}+y) \cap V + K \subseteq \Psi^{C_F}(y) + K$. Thus Ψ^{C_F} is weakly K-upper bounded on U and $U \subseteq \operatorname{dom} \Psi^{C_F}$, which yields $0 \in \operatorname{int}(\operatorname{dom} \Psi^{C_F})$. Now strong duality follows via Theorem 7.1.27 and Theorem 7.2.2. □

Concerning the strong duality for (PSV^C) and its Fenchel set-valued dual problem (DSV^{C_F}) one can give an alternative result. This is based on a different perturbation map of the primal problem. If we define this as being $\widetilde{\Phi}^{C_F} : X \times X \rightrightarrows V \cup \{+\infty_K\}$,

$$\widetilde{\Phi}^{C_F}(x, y) = \begin{cases} F(x), & \text{if } x \in \mathcal{A} + y, \\ \{+\infty_K\}, & \text{otherwise} \end{cases} = F(x) + \delta^V_{\mathcal{A}}(x - y),$$

then a straightforward calculation shows that for $\Gamma \in \mathcal{L}(X, V)$

$$(\widetilde{\Phi}^{C_F})^*(0,\Gamma) = \operatorname{Max}\Big\{ \bigcup_{y\in X}[\Gamma y - F(y)] - \Gamma(\mathcal{A})\Big\} = (\Phi^{C_F})^*(0,\Gamma).$$

Thus both perturbation maps generate the same dual problem (DSV^{C_F}). The minimal value map is now $\widetilde{H}^{C_F} : X \rightrightarrows V \cup \{+\infty_K\}$, $\widetilde{H}^{C_F}(y) = \operatorname{Min}\widetilde{\Phi}^{C_F}(X,y)$ for $y \in X$. Notice that for all $y \in Y$ one has

$$\widetilde{\Psi}^{C_F}(y) = \widetilde{\Phi}^{C_F}(X,y) = \Phi^{C_F}(X,y) = \Psi^{C_F}(y),$$

and thus the corresponding minimal value map $\widetilde{H}^{C_F}$ coincides with H^{C_F}.

Theorem 7.2.11. *Let Φ^{C_F} be strictly-K-convexlike-convex and $H^{C_F}(y) = \operatorname{Min}\Phi^{C_F}(X,y)$ be externally stable for all $y \in X$. If $\operatorname{dom} F \cap \operatorname{int}(\mathcal{A}) \neq \emptyset$, then (PSV^C) is stable with respect to the perturbation map Φ^{C_F}; hence strong duality holds for (PSV^C) and its Fenchel set-valued dual problem (DSV^{C_F}).*

Proof. Let be $x' \in \operatorname{dom} F \cap \operatorname{int}(\mathcal{A})$ and $b \in F(x') \cap V$. Then there exists U, a neighborhood of 0 in X, such that $x' - y \in \mathcal{A}$ for all $y \in U$. Thus $U \subseteq \operatorname{dom}\Psi^{C_F}$ and, consequently, $0 \in \operatorname{int}(\operatorname{dom}\Psi^{C_F})$.

On the other hand, for all $y \in U$ one has $b \in F(x') + \delta^V_{\mathcal{A}}(x'-y) \subseteq \Psi^{C_F}(y) + K$ and this implies that Ψ^{C_F} is weakly K-upper bounded on U. The conclusion follows via Theorem 7.1.27. □

Remark 7.2.9. One can see that $\mathcal{A}$ is convex if, in particular, $S \subseteq X$ is convex and $G : X \rightrightarrows Z$ is a C-convex set-valued map. Thus, assuming additionally that F is K-convex, one gets that Φ^{C_F} is K-convex, too. Nevertheless, imposing that F is strictly K-convex, it is not enough to guarantee in general that Φ^{C_F} is strictly-K-convexlike-convex.

One can formulate completely analogous results for the problem (PSV^C) and its Fenchel set-valued dual problem $(DSV^{C_F}_k)$.

Corollary 7.2.12. *Let be $k \in \operatorname{int}(K)$. Under the hypotheses of Theorem 7.2.10 or Theorem 7.2.11 the problem (PSV^C) is k-stable with respect to the perturbation map Φ^{C_F}; hence strong duality holds for (PSV^C) and its Fenchel set-valued dual problem $(DSV^{C_F}_k)$.*

It remains to deal with the Fenchel-Lagrange set-valued dual problem concerning stability and strong duality.

Theorem 7.2.13. *Let $\Phi^{C_{FL}}$ be strictly-K-convexlike-convex and $H^{C_{FL}}(y,z) = \operatorname{Min}\Phi^{C_{FL}}(X,y,z)$ be externally stable for all $(y,z) \in X \times Z$. If F is weakly K-upper bounded on $\operatorname{dom} F$ and $U \times W$ is a neighborhood of $(0,0)$ in $X \times Z$ such that $W \subseteq \cap_{y\in U} G((\operatorname{dom} F - y) \cap S) + C$, then (PSV^C) is stable with respect to the perturbation map $\Phi^{C_{FL}}$; hence strong duality holds for (PSV^C) and its Fenchel-Lagrange set-valued dual problem $(DSV^{C_{FL}})$.*

Proof. Notice first that for all $(y, z) \in X \times Z$ one has

$$\Psi^{C_{FL}}(y,z) \cap V = \Phi^{C_{FL}}(X,y,z) \cap V = F\Big(S \cap G^{-1}(z-C) + y\Big) \cap V.$$

Let $U \times W$ be a neighborhood of $(0,0)$ in $X \times Z$ fulfilling $W \subseteq \cap_{y \in U} G((\operatorname{dom} F - y) \cap S) + C$. From here follows that for all $y \in U$ and $z \in W$ it holds $F\big(S \cap G^{-1}(z-C)+y\big) \cap V \neq \emptyset$ and this yields $\Psi^{C_{FL}}(y,z) \cap V \neq \emptyset$. Thus $U \times W \subseteq \operatorname{dom} \Psi^{C_{FL}}$ and so $(0,0) \in \operatorname{int}(\operatorname{dom} \Psi^{C_{FL}})$. Further, as F is assumed to be weakly K-upper bounded on $\operatorname{dom} F$, $\Psi^{C_{FL}}$ is weakly K-upper bounded on $U \times W$ and Theorem 7.1.27 leads to the desired conclusion. □

Remark 7.2.10. If $\operatorname{dom} F = X$ one can assume as regularity condition in the result above the existence of a neighborhood W of 0 in Z such that $W \subseteq G(S) + C$, in other words that $0 \in \operatorname{int}[G(S) + C]$, which is nothing else but the condition assumed in Theorem 7.2.7.

Another stability result for the Fenchel-Lagrange set-valued dual pair given by employing the weakly K-upper boundedness of F follows (cf. [23]).

Theorem 7.2.14. *Let $\Phi^{C_{FL}}$ be strictly-K-convexlike-convex and $H^{C_{FL}}(y,z) = \operatorname{Min} \Phi^{C_{FL}}(X,y,z)$ be externally stable for all $(y,z) \in X \times Z$. If there exist $x' \in S$ such that $0 \in \operatorname{int}(G(x') + C)$ and a neighborhood U of x' in X such that F is weakly K-upper bounded on U, then (PSV^C) is stable with respect to the perturbation map $\Phi^{C_{FL}}$; hence strong duality holds for (PSV^C) and its Fenchel-Lagrange set-valued dual problem $(DSV^{C_{FL}})$.*

Proof. Because F is weakly K-upper bounded on U there exists an element $b \in V$ such that $b \in F(x) + K$ for all $x \in U$. Then for all $y \in U - x'$, which is a neighborhood of 0 in X, it holds $b \in F(x' + y) + K$. Since we have $0 \in \operatorname{int}(G(x') + C)$, there exists a neighborhood W of 0 in Z such that $x' \in S \cap G^{-1}(z - C)$ for all $z \in W$. Consequently, for all $(y,z) \in (U - x') \times W$ it holds

$$b \in F(x' + y) + K \subseteq F(S \cap G^{-1}(z-C) + y) \cap V + K \subseteq \Psi^{C_{FL}}(y,z) + K.$$

Thus $(0,0) \in \operatorname{int}(\operatorname{dom} \Psi^{C_{FL}})$ and $\Psi^{C_{FL}}$ is weakly K-upper bounded on $(U - x') \times W$. The conclusion follows via Theorem 7.1.27. □

Remark 7.2.11. (*a*) One can easily check that $\Phi^{C_{FL}}$ is K-convex if S is convex, F is a K-convex and G is a C-convex set-valued map. Nevertheless, additionally assuming strict K-convexity for F is not sufficient in general to have that $\Phi^{C_{FL}}$ is strictly-K-convexlike-convex.

(*b*) Taking a close look at the proof of Theorem 7.2.14 one can easily see that for all $y \in U - x'$ and $z \in W$ it holds $z \in G((\operatorname{dom} F - y) \cap S) + C$. Thus $W \subseteq \cap_{y \in U - x'} G((\operatorname{dom} F - y) \cap S) + C$ and the neighborhood $(U - x') \times W$ of $(0,0)$ in $X \times Z$ makes the condition assumed in Theorem 7.2.13 valid. Nevertheless, the mentioned condition alone is not enough for having strong duality. As follows from the proof of Theorem 7.2.13 one cannot in general omit that F is weakly K-upper bounded on $\operatorname{dom} F$.

Again we state that an analogous result holds for (PSV^C) and its Fenchel-Lagrange set-valued dual problem $(DSV_k^{C_{FL}})$.

Corollary 7.2.15. *Let be $k \in \text{int}(K)$. Under the assumptions of Theorem 7.2.13 or Theorem 7.2.14, the problem (PSV^C) is k-stable with respect to the perturbation map $\Phi^{C_{FL}}$; hence strong duality holds for (PSV^C) and its Fenchel-Lagrange set-valued dual problem $(DSV_k^{C_{FL}})$.*

7.3 The set-valued optimization problem having the composition with a linear continuous mapping in the objective function

In this section we investigate the set-valued analog of the scalar problem (P^A) from subsection 3.1.2 with respect to duality and optimality conditions.

Consider X, Y and V topological vector spaces, V partially ordered by the nontrivial pointed convex cone $K \subseteq V$, X^*, Y^* and V^* their corresponding topological dual spaces, respectively, $F : X \rightrightarrows V \cup \{+\infty_K\}$ and $G : Y \rightrightarrows V \cup \{+\infty_K\}$ set-valued maps and $A \in \mathcal{L}(X, Y)$. The set-valued optimization problem under discussion is

$$(PSV^A) \quad \operatorname*{Min}_{x \in X} \{F(x) + G(Ax)\}.$$

Additionally, we impose the feasibility condition $\operatorname{dom} F \cap A^{-1}(\operatorname{dom} G) \neq \emptyset$. We refer the reader also to section 4.1 where we investigate the same problem in case F and G are vector-valued functions and to section 5.1 where we additionally assume that $V = \mathbb{R}^k$ and $K = \mathbb{R}^k_+$.

7.3.1 Fenchel set-valued duality

Following the general perturbation approach described in section 7.1 we begin by considering the set-valued perturbation map

$$\Phi^A : X \times Y \rightrightarrows V \cup \{+\infty_K\}, \ \Phi^A(x, y) = F(x) + G(Ax + y).$$

For all $x \in X$ it holds $\Phi^A(x, 0) = F(x) + G(Ax)$. The corresponding minimal value map $H^A : Y \rightrightarrows V \cup \{+\infty_K\}$ is defined by $H^A(y) = \operatorname{Min} \Phi^A(X, y)$. The problem (PSV^A) means to deal with minimal elements of the image set $\cup_{x \in X}\{F(x)+G(Ax)\}$, i.e. we look for an efficient solutions $\bar{x} \in X$ and a corresponding minimal pair $(\bar{x}, \bar{v})$ to (PSV^A), which means that $\bar{v} \in F(\bar{x})+G(A\bar{x})$ and $\bar{v} \in \operatorname{Min} \cup_{x \in X}\{F(x) + G(Ax)\}$. For the formulation of the dual problem we have to calculate the conjugate map to Φ^A. First we use the general conjugacy concept from the subsections 7.1.1 and 7.1.2. Afterwards we apply also the k-conjugation as developed in subsection 7.1.3.

For $\Gamma \in \mathcal{L}(Y, V)$ there is

$$\left(\Phi^A\right)^*(0,\Gamma) = \operatorname{Max} \bigcup_{\substack{x\in X,\\ y\in Y}} [\Gamma y - \Phi^A(x,y)] = \operatorname{Max} \bigcup_{\substack{x\in X,\\ y\in Y}} [\Gamma y - F(x) - G(Ax+y)]$$

$$= \operatorname{Max} \bigcup_{x\in X, y\in Y} [\Gamma y - \Gamma(Ax) - F(x) - G(y)]$$

$$= \operatorname{Max} \left\{ \bigcup_{y\in Y} [\Gamma y - G(y)] + \bigcup_{x\in X} [-\Gamma(Ax) - F(x)] \right\}.$$

Via the general duality approach one can associate to (PSV^A) as set-valued dual problem

$$(DSV^A) \quad \operatorname{Max} \bigcup_{\Gamma\in\mathcal{L}(Y,V)} \operatorname{Min} \left\{ \bigcup_{y\in Y} [G(y) - \Gamma y] + \bigcup_{x\in X} [\Gamma(Ax) + F(x)] \right\}.$$

We call (DSV^A) the *Fenchel set-valued dual problem* to (PSV^A). The weak duality result for (PSV^A) and (DSV^A) is a consequence of Theorem 7.1.11. From now on $H^A(y)$ is assumed to be externally stable for all $y \in Y$. Then Theorem 7.1.15 guarantees strong duality if stability for (PSV^A) with respect to Φ^A is ensured. For all $\Gamma \in \mathcal{L}(Y,V)$ we additionally assume that

$$F^*(-\Gamma \circ A) = \operatorname{Max} \bigcup_{x\in X} [-\Gamma(Ax) - F(x)]$$

and

$$G^*(\Gamma) = \operatorname{Max} \bigcup_{y\in Y} [\Gamma y - G(y)]$$

are externally stable. Then for all $\Gamma \in \mathcal{L}(Y,V)$ we obtain via Proposition 7.1.7

$$\left(\Phi^A\right)^*(0,\Gamma) = \operatorname{Max} \bigcup_{x\in X} \left\{ \bigcup_{y\in Y} [\Gamma y - G(y)] + [-\Gamma(Ax) - F(x)] \right\}$$

$$= \operatorname{Max} \bigcup_{x\in X} \left\{ \operatorname{Max} \bigcup_{y\in Y} [\Gamma y - G(y)] + [-\Gamma(Ax) - F(x)] \right\}$$

$$= \operatorname{Max} \left\{ G^*(\Gamma) + \bigcup_{x\in X} [-\Gamma(Ax) - F(x)] \right\}$$

$$= \operatorname{Max} \left\{ G^*(\Gamma) + \operatorname{Max} \bigcup_{x\in X} [-\Gamma(Ax) - F(x)] \right\} = \operatorname{Max}[G^*(\Gamma) + F^*(-\Gamma \circ A)]$$

and the Fenchel set-valued dual problem becomes

$$(DSV^A) \quad \operatorname{Max} \bigcup_{\Gamma\in\mathcal{L}(Y,V)} \operatorname{Min}[-F^*(-\Gamma \circ A) - G^*(\Gamma)].$$

The following result is a consequence of Theorem 7.1.15, Theorem 7.1.16 and Theorem 7.1.17

Theorem 7.3.1. *Let $F^*(-\Gamma \circ A)$ and $G^*(\Gamma)$ be externally stable for all $\Gamma \in \mathcal{L}(Y,V)$.*

(a) *Suppose that the problem* (PSV^A) *is stable with respect to the perturbation map* Φ^A. *Let* $\bar{x} \in X$ *be an efficient solution to* (PSV^A) *and* $\bar{v} \in F(\bar{x}) + G(A\bar{x})$ *such that* $(\bar{x}, \bar{v})$ *is a minimal pair to* (PSV^A). *Then there exists* $\overline{\Gamma} \in \mathcal{L}(Y, V)$, *an efficient solution to* (DSV^A), *with* $(\overline{\Gamma}, \bar{v})$ *corresponding maximal pair such that strong duality holds and*

$$\bar{v} \in \operatorname{Min}[-F^*(-\overline{\Gamma} \circ A) - G^*(\overline{\Gamma})].$$

(b) *Assume that for* $\bar{x} \in X$, $\bar{v} \in [F(\bar{x}) + G(A\bar{x})] \cap V$ *and* $\overline{\Gamma} \in \mathcal{L}(Y, V)$ *one has* $\bar{v} \in \operatorname{Min}[-F^*(-\overline{\Gamma} \circ A) - G^*(\overline{\Gamma})]$. *Then* $\bar{x}$ *is an efficient solution and* $(\bar{x}, \bar{v})$ *a minimal pair to* (PSV^A), *while* $\overline{\Gamma}$ *is an efficient solution and* $(\overline{\Gamma}, \bar{v})$ *a maximal pair to* (DSV^A).

Next we derive a Fenchel set-valued dual problem to (PSV^A) based on k-conjugation by using the perturbation map Φ^A. Let $k \in V \setminus \{0\}$ be fixed. For $y^* \in Y^*$ we have

$$(\Phi_k^A)^*(0, y^*) = \operatorname{Max} \bigcup_{x \in X, y \in Y} [\langle y^*, y \rangle k - \Phi^A(x, y)]$$

$$= \operatorname{Max} \left\{ \bigcup_{y \in Y} [\langle y^*, y \rangle k - G(y)] + \bigcup_{x \in X} [\langle -A^* y^*, x \rangle k - F(x)] \right\}.$$

Assuming the external stability of

$$F_k^*(-A^* y^*) = \operatorname{Max} \bigcup_{x \in X} [\langle -A^* y^*, x \rangle k - F(x)]$$

and of

$$G_k^*(y^*) = \operatorname{Max} \bigcup_{y \in Y} [\langle y^*, y \rangle k - G(y)],$$

from Proposition 7.1.7 we conclude that

$$(\Phi_k^A)^*(0, y^*) = \operatorname{Max}[G_k^*(y^*) + F_k^*(-A^* y^*)].$$

Thus, the following Fenchel set-valued dual problem to (PSV^A) arises from the general theory in subsection 7.1.3

$$(DSV_k^A) \quad \operatorname{Max} \bigcup_{y^* \in Y^*} \operatorname{Min} [-F_k^*(-A^* y^*) - G_k^*(y^*)].$$

For (PSV^A) and (DSV_k^A) weak duality is satisfied, due to Theorem 7.1.28. One can formulate by means of Theorem 7.1.32, Theorem 7.1.33 and Theorem 7.1.34 the analog of Theorem 7.3.1 for this kind of conjugacy.

Theorem 7.3.2. *Let* $F_k^*(-A^* y^*)$ *and* $G^*(y^*)$ *be externally stable for all* $y^* \in Y^*$.

(a) *Suppose that the problem* (PSV^A) *is* k*-stable with respect to the perturbation map* Φ^A. *Let* $\bar{x} \in X$ *be an efficient solution to* (PSV^A) *and* $\bar{v} \in F(\bar{x}) + G(A\bar{x})$ *such that* $(\bar{x}, \bar{v})$ *is a minimal pair to* (PSV^A). *Then there exists* $\bar{y}^* \in Y^*$, *an efficient solution to* (DSV_k^A), *with* $(\bar{y}^*, \bar{v})$ *corresponding maximal pair such that strong duality holds and*

$$\bar{v} \in \mathrm{Min}[-F_k^*(-A^*\bar{y}^*) - G_k^*(\bar{y}^*)].$$

(b) *Assume that for* $\bar{x} \in X$, $\bar{v} \in [F(\bar{x}) + G(A\bar{x})] \cap V$ *and* $\bar{y}^* \in Y^*$ *one has* $\bar{v} \in \mathrm{Min}[-F_k^*(-A^*\bar{y}^*) - G_k^*(\bar{y}^*)]$. *Then* $\bar{x}$ *is an efficient solution and* $(\bar{x}, \bar{v})$ *a minimal pair to* (PSV^A), *while* $\bar{y}^*$ *is an efficient solution and* $(\bar{y}^*, \bar{v})$ *a maximal pair to* (DSV_k^A).

We conclude this first subsection by giving some stability criteria implying strong duality for (PSV^A) and its both Fenchel set-valued dual problems (DSV^A) and (DSV_k^A), respectively. To this end we additionally assume that $\mathrm{int}(K) \neq \emptyset$.

Theorem 7.3.3. *Let* Φ^A *be strictly-*K*-convexlike-convex and* $H^A(y) = \mathrm{Min}\, \Phi^A(X, y)$ *be externally stable for all* $y \in Y$. *If there exists* $x' \in \mathrm{dom}\, F \cap A^{-1}(\mathrm{dom}\, G)$ *such that the map* G *is weakly* K*-upper bounded on some neighborhood of* Ax', *then* (PSV^A) *is stable with respect to the perturbation map* Φ^A; *hence strong duality holds for* (PSV^A) *and its Fenchel set-valued dual problem* (DSV^A).

Proof. The stated weak K-upper boundedness of G leads to the existence of a neighborhood U of 0 in Y and of some $b \in V$ such that $G(Ax'+y) \cap (b-K) \neq \emptyset$ for all $y \in U$. Let be $v' \in F(x') \cap V$. Then $v' + b \in F(x') + G(Ax' + y) + K$ and therefore finally $v' + b \in \Psi^A(y) + K$ for all $y \in U$. Hence, Ψ^A is weakly K-upper bounded on U and $U \subseteq \mathrm{dom}\, \Psi^A$. Thus $0 \in \mathrm{int}(\mathrm{dom}\, \Psi^A)$ and the conclusion follows from Theorem 7.1.27. □

Remark 7.3.1. In order to guarantee that Φ^A is strictly-K-convexlike-convex one has to assume that F and G are both strictly K-convex. Assuming this property for only one of the two maps along with the K-convexity of the other one is in general not sufficient for guaranteeing that Φ^A is strictly-K-convexlike-convex.

An analogous result applies via Corollary 7.1.37 to the primal-dual pair $(PSV^A) - (DSV_k^A)$.

Corollary 7.3.4. *Let be* $k \in \mathrm{int}(K) \cup (-\mathrm{int}(K))$. *Under the assumptions of Theorem 7.3.3, the problem* (PSV^A) *is* k*-stable with respect to the perturbation map* Φ^A; *hence strong duality holds for* (PSV^A) *and its Fenchel set-valued dual problem* (DSV_k^A).

7.3.2 Set-valued gap maps for vector variational inequalities

So-called *gap functions* have been introduced first for scalar variational inequalities (cf. [8,77]) and they represent a tool for characterizing the solutions of this class of problems.

Let us give first a brief introduction on this topic supporting on the classical case of a variational inequality in finite dimensional spaces. Let $F : \mathbb{R}^n \to \mathbb{R}^n$ be a vector-valued function and $\mathcal{A} \subseteq \mathbb{R}^n$ a nonempty set. The variational inequality problem consists in finding an element $x \in \mathcal{A}$ such that

$$(VI) \;\; F(x)^T(y-x) \geq 0 \; \forall y \in \mathcal{A}.$$

Variational inequality problems play an important role in different areas of mathematics and its applications, in particular when considering partial differential operators (cf. [117,208]) and optimization problems (cf. [86,149]).

A function $\gamma : \mathcal{A} \to \mathbb{R} \cup \{+\infty\}$ is said to be a *gap function* for the variational inequality problem (VI) if it satisfies the following conditions

(a) $\gamma(y) \geq 0$ for all $y \in \mathcal{A}$;
(b) $\gamma(x) = 0$ if and only if x solves (VI).

Auslender (cf. [8]) introduced the following gap function

$$\gamma_A^{VI} : \mathcal{A} \to \mathbb{R} \cup \{+\infty\}, \; \gamma_A^{VI}(x) = \sup_{y \in \mathcal{A}} F(x)^T(x-y),$$

and a closer look at its formulation makes clear that gap functions are closely related to the duality theory in optimization (cf. [2]). A natural generalization of (VI) are the vector variational inequalities (VVI) considered first by Giannessi (cf. [76]), while for the extension of the concept of gap function to (VVI) one can consult [48]. In this subsection we use the set-valued duality developed within this section to construct some set-valued gap maps for a general vector variational inequality.

First we introduce the *vector variational inequality* and the so-called *set-valued variational inequality* we want further to deal with. Let X and V be topological vector spaces, with V being partially ordered by the nontrivial pointed convex cone K, $F : X \to \mathcal{L}(X, V)$ be a given function and $\mathcal{A} \subseteq X$ a nonempty set. The vector variational inequality problem consists in finding $x \in \mathcal{A}$ such that

$$(VVI) \;\; F(x)(y-x) \nleq_K 0 \text{ for all } y \in \mathcal{A}.$$

The problem (VVI) may be seen as a particular case of the set-valued variational inequality problem $(SVVI)$ introduced below. Here, additionally one has a set-valued map $G : X \rightrightarrows V \cup \{+\infty_K\}$ with $G(x) \neq \emptyset$ for all $x \in X$ and $\operatorname{dom} G \neq \emptyset$. The problem to solve is to find an $x \in X$ such that

$$(SVVI) \;\; F(x)(y-x) \notin G(x) - G(y) - (K \setminus \{0\}) \text{ for all } y \in X.$$

More precisely, $x \in X$ solves $(SVVI)$ means that for all $y \in X$ there is no $v \in G(x) - G(y)$ such that $F(x)(y-x) \leq_K v$.

If $G : X \to V \cup \{+\infty_K\}, G = \delta_{\mathcal{A}}^V$, then $(SVVI)$ coincides with (VVI). Next we introduce the notion of a *set-valued gap map* for $(SVVI)$.

Definition 7.3.1. *A set-valued map $\gamma : X \rightrightarrows V \cup \{+\infty_K\}$ is said to be a gap map for $(SVVI)$ if it satisfies the following conditions*

(a) $\gamma(y) \cap (-K \setminus \{0\}) = \emptyset$ for all $y \in X$;
(b) $0 \in \gamma(x)$ if and only if x solves $(SVVI)$.

Remark 7.3.2. A set-valued map $\gamma : \mathcal{A} \rightrightarrows V \cup \{+\infty_K\}$ is said to be a gap map for (VVI) if it satisfies the following conditions

(a) $\gamma(y) \cap (-K \setminus \{0\}) = \emptyset$ for all $y \in \mathcal{A}$;
(b) $0 \in \gamma(x)$ if and only if x solves (VVI).

Obviously, the real-valued gap function defined above for (VI) results as a special case of the latter.

We observe that $x \in X$ is a solution to $(SVVI)$ if and only if $x \in \operatorname{dom} G$, x is an efficient solution and $(x, 0)$ a corresponding minimal pair to the set-valued optimization problem

$$(PV^{SVVI}; x) \quad \operatorname*{Min}_{y \in X} \{G(y) - G(x) + F(x)(y-x)\}.$$

Consider a fixed $x \in \operatorname{dom} G$. For $\widetilde{F} : X \rightrightarrows V \cup \{+\infty_K\}$, $\widetilde{F}(y) = G(y) - G(x)$, and $\widetilde{G} : X \rightrightarrows V \cup \{+\infty_K\}$, $\widetilde{G}(y) = F(x)(y-x)$, one has $\operatorname{dom} \widetilde{F} \cap \operatorname{dom} \widetilde{G} = \operatorname{dom} G \neq \emptyset$ and $(PV^{SVVI}; x)$ can be equivalently written as

$$\operatorname{Min} \bigcup_{y \in X} \{\widetilde{F}(y) + \widetilde{G}(y)\}.$$

The latter is a particular case of the vector-valued problem (PSV^A) in the particular situation when $X = Y$ and $A = \mathrm{id}_X$. Using as perturbation map

$$\Phi_x^{\mathrm{id}} : X \times X \rightrightarrows V \cup \{+\infty_K\}, \; \Phi_x^{\mathrm{id}}(y, z) = G(y) - G(x) + F(x)(y + z - x)$$

with the corresponding minimal value map $H_x^{\mathrm{id}} : X \rightrightarrows V \cup \{+\infty_K\}$, $H_x^{\mathrm{id}}(z) = \operatorname{Min} \Phi_x^{\mathrm{id}}(X, z)$, we obtain like in the previous subsection the following Fenchel set-valued dual problem to $(PV^{SVVI}; x)$

$$(DV^{SVVI}; x) \; \operatorname*{Max} \bigcup_{\Gamma \in \mathcal{L}(X,V)} \operatorname{Min} \Big[(F(x) - \Gamma)(X) + \bigcup_{y \in X} (\Gamma y + G(y)) - F(x)x - G(x)\Big].$$

Further we show that when for $\Gamma \in \mathcal{L}(X, V)$ one has $(F(x) - \Gamma)(X) \cap (K \backslash \{0\}) \neq \emptyset$, then

$$\operatorname{Min} \Big[(F(x) - \Gamma)(X) + \bigcup_{y \in X} (\Gamma y + G(y)) - F(x)x - G(x)\Big] = \emptyset.$$

If this were not true, there would exist $\tilde{v} \in V$ belonging to this minimum. Therefore one would have that there exist $\tilde{x} \in X$ and $\tilde{y} \in X$ such that

$$\tilde{v} \in (F(x) - \Gamma)\tilde{x} + (\Gamma\tilde{y} + G(\tilde{y})) - F(x)x - G(x).$$

Take any $\tilde{w} \in (F(x)-\Gamma)(X) \cap (K\backslash\{0\})$. Then $-\tilde{w} \in (F(x)-\Gamma)(X) \cap (-K\backslash\{0\})$ and $\tilde{v} - \tilde{w} \leqq_K \tilde{v}$ as well as

$$\tilde{v} - \tilde{w} \in (F(x) - \Gamma)(X) + \bigcup_{y \in X}(\Gamma y + G(y)) - F(x)x - G(x),$$

which generates a contradiction to the minimality of $\tilde{v}$. Thus the dual $(DV^{SVVI}; x)$ can be equivalently written as

$$(DV^{SVVI}; x) \ \operatorname{Max} \bigcup_{\substack{\Gamma \in \mathcal{L}(X,V),\\ (F(x)-\Gamma)(X) \cap K = \{0\}}} \operatorname{Min} \Big\{(F(x)-\Gamma)(X) + \bigcup_{y \in X}(\Gamma y + G(y)) - F(x)x - G(x)\Big\}.$$

In order to study the stability of $(PV^{SVVI}; x)$ with respect to the perturbation map Φ_x^{id}, we assume in the following that the minimal value map $H_x^{\mathrm{id}}(z) = \operatorname{Min} \Phi_x^{\mathrm{id}}(X, z)$ is externally stable for all $z \in X$.

Proposition 7.3.5. *The problem $(PV^{SVVI}; x)$ is stable with respect to the perturbation map Φ_x^{id}.*

Proof. We have to prove the subdifferentiability of H_x^{id} at 0. To this end we consider $v \in V$ with $v \in H_x^{\mathrm{id}}(0) = \operatorname{Min} \cup_{y \in X}[G(y) - G(x) + F(x)(y-x)]$. Next we calculate the conjugate of H_x^{id} at some arbitrary $T \in \mathcal{L}(X, V)$. It holds

$$(H_x^{\mathrm{id}})^*(T) = \operatorname{Max} \bigcup_{z \in X} \{Tz - H_x^{\mathrm{id}}(z)\}$$

$$= \operatorname{Max} \bigcup_{z \in X} \{Tz - \operatorname{Min} \bigcup_{y \in X}[G(y) - G(x) + F(x)(y + z - x)]\}$$

$$= \operatorname{Max} \bigcup_{z \in X} \{(T - F(x))z - \operatorname{Min} \bigcup_{y \in X}[G(y) - G(x) + F(x)(y - x)]\}.$$

Setting $T = F(x)$ we get

$$(H_x^{\mathrm{id}})^*(F(x)) = \operatorname{Max} \bigcup_{y \in X}[G(x) - G(y) + F(x)(x - y)] = -H_x^{\mathrm{id}}(0).$$

Hence $v \in -(H_x^{\mathrm{id}})^*(F(x))$ and Proposition 7.1.3 ensures $F(x) \in \partial H_x^{\mathrm{id}}(0; v)$. The subdifferentiability of H_x^{id} at 0 follows and this entails the claimed stability. □

These preparations allow us to construct a gap map to $(SVVI)$ by means of the set-valued objective map of the Fenchel set-valued dual $(DV^{SVVI}; x)$ as being $\gamma^{SVVI} : X \rightrightarrows V \cup \{+\infty_K\}$, defined by

$$\gamma^{SVVI}(x) = \bigcup_{\substack{\Gamma \in \mathcal{L}(X,V),\\ (F(x)-\Gamma)(X) \cap K = \{0\}}} \operatorname{Max} \{(\Gamma - F(x))(X) + F(x)x + G(x) - \bigcup_{y \in X}[\Gamma y + G(y)]\}.$$

Theorem 7.3.6. *The set-valued map γ^{SVVI} is a gap map of the set-valued variational inequality problem $(SVVI)$.*

Proof. We verify that γ^{SVVI} fulfills the properties (a) and (b) of Definition 7.3.1.

Let us arbitrarily chose $y \in X$ and $v^* \in \gamma^{SVVI}(y)$. Then there exists $\Gamma \in \mathcal{L}(X, V)$ fulfilling $(F(y) - \Gamma)(X) \cap K = \{0\}$ and

$$v^* \in \operatorname{Max}\left\{(\Gamma - F(y))(X) + F(y)y + G(y) - \bigcup_{z\in X}[\Gamma z + G(z)]\right\} = (\Phi_y^{\mathrm{id}})^*(0, \Gamma),$$

where $\Phi_y^{\mathrm{id}} : X \times X \rightrightarrows V \cup \{+\infty_K\}$ is defined as

$$\Phi_y^{\mathrm{id}}(u, z) = G(u) - G(y) + F(y)(u + z - y).$$

As $0 \in \Phi_y^{\mathrm{id}}(y, 0)$, by Corollary 7.1.12$(a)$ one has $v^* \not\leq_K 0$, which proves (a).

We come now to (b). Let $x \in X$ be a solution to $(SVVI)$. Due to our previous considerations $x \in \operatorname{dom} G$, x is an efficient solution and $(x, 0)$ is a minimal pair to $(PV^{SVVI}; x)$. As the problem $(PV^{SVVI}; x)$ is stable with respect to Φ_x^{id}, by Theorem 7.3.1 there exists an efficient solution $\Gamma \in \mathcal{L}(X, V)$ to $(DV^{SVVI}; x)$ with $(\Gamma, 0)$ a corresponding maximal pair, fulfilling $(F(x) - \Gamma)(X) \cap K = \{0\}$. In particular it holds

$$0 \in \operatorname{Min}\left\{(F(x) - \Gamma)(X) - F(x)x - G(x) + \bigcup_{y\in X}[\Gamma y + G(y)]\right\}$$

and, consequently, $0 \in \gamma^{SVVI}(x)$.

Conversely, let be $x \in X$ with $0 \in \gamma^{SVVI}(x)$. Hence there exists $\Gamma \in \mathcal{L}(X, V)$ such that $(F(x) - \Gamma)(X) \cap K = \{0\}$ and

$$0 \in \operatorname{Max}\left\{(\Gamma - F(x))(X) + F(x)x + G(x) - \bigcup_{y\in X}[\Gamma y + G(y)]\right\} = (\Phi_x^{\mathrm{id}})^*(0, \Gamma).$$

This implies that $x \in \operatorname{dom} G$. On the other hand, we have $0 \in F(x)(x - x) + G(x) - G(x) = \Phi_x^{\mathrm{id}}(x, 0)$ and applying Corollary 7.1.12(b) to this particular situation we obtain that x is an efficient solution and $(x, 0)$ a minimal pair to $(PV^{SVVI}; x)$, while Γ is an efficient solution and $(\Gamma, 0)$ a maximal pair to $(DV^{SVVI}; x)$. Consequently, x is a solution to $(SVVI)$. □

Let us pay attention to the vector variational inequality problem (VVI) as a particular case of $(SVVI)$, by considering $G : X \to V \cup \{+\infty_K\}, G = \delta_{\mathcal{A}}^V$. Then, for a fixed $x \in \mathcal{A}$, the family of set-valued minimum problems associated to (VVI) is

$$(PV^{VVI}; x) \quad \operatorname*{Min}_{y\in\mathcal{A}}\{F(x)(y - x)\}.$$

The corresponding Fenchel set-valued dual arises from $(DV^{SVVI}; x)$ as being

$$(DV^{VVI}; x) \quad \operatorname{Max} \bigcup_{\substack{\Gamma\in\mathcal{L}(X,V),\\ (F(x)-\Gamma)(X)\cap K=\{0\}}} \operatorname{Min}\left\{(F(x) - \Gamma)(X) - F(x)x + \Gamma(\mathcal{A})\right\}.$$

Thus $\gamma^{VVI} : \mathcal{A} \rightrightarrows V \cup \{+\infty_K\}$,

$$\gamma^{VVI}(x) = \bigcup_{\substack{\Gamma \in \mathcal{L}(X,V), \\ (F(x)-\Gamma)(X) \cap K = \{0\}}} \operatorname{Max}\{(\Gamma - F(x))(X) + F(x)x - \Gamma(\mathcal{A})\},$$

is the particularization of γ^{SVVI} to this special case. A particular instance of γ^{VVI} was delivered in [1] in finite dimensional spaces, the ordering cone being the nonnegative orthant of the image space. By Theorem 7.3.6 one obtains the following result.

Theorem 7.3.7. *The set-valued map γ^{VVI} is a gap map of the vector variational inequality problem (VVI).*

Remark 7.3.3. One can carry out similar considerations concerning gap maps for $(SVVI)$ and (VVI) based on the Fenchel set-valued dual problems to $(PV^{SVVI}; x)$ and $(PV^{VVI}; x)$, respectively, by using k-conjugation as developed in subsection 7.1.3, too (see [1] for a similar approach in finite dimensional spaces).

7.4 Conjugate duality based on weakly efficient solutions

Another conjugate duality approach for set-valued optimization problems can be developed if weakly efficient solutions are considered. Even more, the weak ordering defined in this sense allows also to develop a fruitful concept of infimum and supremum of a set in topological vector spaces. This makes it possible to establish a conjugate duality approach for set-valued optimization problems which is somewhat closer to the conjugate duality for scalar optimization problems than the ones considered in the sections 7.1-7.3. This concept has been introduced in topological vector spaces by Tanino in [178] and developed by Song in [168–170].

In the following we summarize this approach and give some extensions and applications of it.

7.4.1 Basic notions, conjugate maps and subdifferentiability

Within this section we generally assume that X and V are Hausdorff topological vector spaces with V partially ordered by the nontrivial pointed convex closed cone K with nonempty interior. In this section we work with the weak ordering "$<_K$" defined by means of $\operatorname{int}(K)$. On $\overline{V}$ we consider the addition and the multiplication with scalars as done in section 2.1, excepting the conventions in (2.1) which are not considered, as they will be in what follows avoided. On the other hand, we assume that for $x^* \in K^*$ it holds $\langle x^*, -\infty_K \rangle = -\infty$, as this situation can occur in the forthcoming investigations.

Let $M \subseteq \overline{V}$ be a given set. The sets $A(M) := \{v \in \overline{V} : \tilde{v} <_K v$ for some $\tilde{v} \in M\}$ and $B(M) := \{v \in \overline{V} : v <_K \tilde{v}$ for some $\tilde{v} \in M\}$ are called the *set of elements above* M and the *set of elements below* M, respectively. Let us observe that if $M \subseteq V$ is a nonempty set (i.e. $\pm\infty_K \notin M$), then $A(M) = \{M + \operatorname{int}(K)\} \cup \{+\infty_K\}$, $B(M) = \{M - \operatorname{int}(K)\} \cup \{-\infty_K\}$, while, obviously, $A(\emptyset) = B(\emptyset) = \emptyset$. In subsection 2.4.2 we have defined weakly minimal and maximal elements of M when this is a subset of V. Now we give a generalization of these notions whenever M is an arbitrary subset of $\overline{V}$.

Definition 7.4.1. *Let $M \subseteq \overline{V}$ be a given set.*

(a) An element $\bar{v} \in \overline{V}$ is said to be a weakly infimal element of M if there is no $v \in M$ fulfilling $v <_K \bar{v}$ and if for any $\tilde{v} \in \overline{V}$ such that $\bar{v} <_K \tilde{v}$ there exists some $v \in M$ satisfying $v <_K \tilde{v}$. The set of the weakly infimal elements of M is denoted by $\operatorname{WInf}(M, K)$ *and it is called the weak infimum of M.*
(b) An element $\bar{v} \in M$ is said to be a weakly minimal element of M if there is no $v \in M$ fulfilling $v <_K \bar{v}$. The set of the weakly minimal elements of M is denoted by $\operatorname{WMin}(M, K)$ *and it is called the weak minimum of M.*
(c) An element $\bar{v} \in \overline{V}$ is said to be a weakly supremal element of M if there is no $v \in M$ fulfilling $\bar{v} <_K v$ and if for any $\tilde{v}$ such that $\tilde{v} <_K \bar{v}$ there exists some $v \in M$ satisfying $\tilde{v} <_K v$. The set of the weakly supremal elements of M is denoted by $\operatorname{WSup}(M, K)$ *and it is called the weak supremum of M.*
(d) An element $\bar{v} \in M$ is said to be a weakly maximal element of M if there is no $v \in M$ fulfilling $\bar{v} <_K v$. The set of the weakly maximal elements of M is denoted by $\operatorname{WMax}(M, K)$ *and it is called the weak maximum of M.*

Remark 7.4.1. According to Definition 7.4.1 one has that $\operatorname{WInf}(\emptyset, K) = \{+\infty_K\}$, $\operatorname{WSup}(\emptyset, K) = \{-\infty_K\}$ and $\operatorname{WMin}(\emptyset, K) = \operatorname{WMax}(\emptyset, K) = \emptyset$.

Remark 7.4.2. Let $M \subseteq \overline{V}$ be a given set.

(*a*) One has $\bar{v} \in \operatorname{WInf}(M, K)$ if and only if $\bar{v} \notin A(M)$ and $A(\bar{v}) \subseteq A(M)$, while $\bar{v} \in \operatorname{WSup}(M, K)$ if and only if $\bar{v} \notin B(M)$ and $B(\bar{v}) \subseteq B(M)$.

(*b*) In case $M \subseteq V$ the notions weak minimum and weak maximum given here coincide with the classical ones given in section 2.4.

(*c*) It is straightforward to see that $\operatorname{WMin}(M, K) = M \cap \operatorname{WInf}(M, K)$ and $\operatorname{WMax}(M, K) = M \cap \operatorname{WSup}(M, K)$. Further, there is $B(M) = -A(-M)$, $\operatorname{WSup}(M, K) = -\operatorname{WInf}(-M, K)$ and $\operatorname{WMax}(M, K) = -\operatorname{WMin}(-M, K)$.

(*d*) It holds $\operatorname{WSup}(M, K) = \{-\infty_K\}$ if and only if $B(M) = \emptyset$. This is the case if and only if $M = \emptyset$ or $M = \{-\infty_K\}$. Moreover, $\operatorname{WInf}(M, K) = \{+\infty_K\}$ if and only if $A(M) = \emptyset$. This is the case if and only if $M = \emptyset$ or $M = \{+\infty_K\}$.

(*e*) It holds $\operatorname{WSup}(M, K) = \{+\infty_K\}$ if and only if $B(M) = V \cup \{-\infty_K\}$ and $\operatorname{WInf}(M, K) = \{-\infty_K\}$ if and only if $A(M) = V \cup \{+\infty_K\}$.

Remark 7.4.3. For two arbitrary sets $M_1, M_2 \subseteq \overline{V}$ it holds $M_1 \cap A(M_2) = \emptyset \Leftrightarrow B(M_1) \cap M_2 = \emptyset$.

As the ordering cone K is assumed fixed, from now on we write for convenience simply $\operatorname{WInf} M$, $\operatorname{WMin} M$, $\operatorname{WSup} M$ and $\operatorname{WMax} M$ instead of

$\operatorname{WInf}(M, K)$, $\operatorname{WMin}(M, K)$, $\operatorname{WSup}(M, K)$ and $\operatorname{WMax}(M, K)$, respectively. The weak infimum and weak supremum of a set satisfy some fundamental relations which are useful for developing a duality theory for set-valued optimization problems regarding the weak ordering.

Let us summarize a few of these relations. For the straightforward proofs we refer to [177] as far as the case V is finite dimensional is concerned. But the results and proofs may be transferred in a straightforward manner to partially ordered topological vector spaces (cf. [178]).

Proposition 7.4.1. *Let $M \subseteq \overline{V}$ be a given set. Then it holds*

(a) $\operatorname{WSup} M = \operatorname{WSup} B(M)$;
(b) $B(M) = B(\operatorname{WSup} M)$;
(c) $M \subseteq \operatorname{WSup} M \cup B(M) = \operatorname{WSup} M \cup B(\operatorname{WSup} M)$;
(d) $\overline{V} = \operatorname{WSup} M \cup A(\operatorname{WSup} M) \cup B(\operatorname{WSup} M)$ *and the three sets on the right-hand side are disjoint.*

One should notice that an analogous relations apply for the weak infimum, too.

Proposition 7.4.2. *(a) When $M_1, M_2 \subseteq \overline{V}$ are given sets it holds $B(M_1 + M_2) = B(M_1) + B(M_2)$, in case the sum $+\infty_K + (-\infty_K)$ does not occur.*
(b) When I is an arbitrary index set, for $M_i \subseteq \overline{V}$ given sets, $i \in I$, it holds $B(\cup_{i \in I} M_i) = \cup_{i \in I} B(M_i)$.

Proposition 7.4.3. *Let $F, G : X \rightrightarrows \overline{V}$ be set-valued maps. Then it holds*

$$\operatorname{WSup} \bigcup_{x \in X} [F(x) + G(x)] = \operatorname{WSup} \bigcup_{x \in X} [F(x) + \operatorname{WSup} G(x)],$$

where it is assumed that the sum $+\infty_K + (-\infty_K)$ does not occur.

Proof. By using Proposition 7.4.1(a), (b) and Proposition 7.4.2(a) we obtain

$$\operatorname{WSup} \bigcup_{x \in X} [F(x) + G(x)] = \operatorname{WSup} B\Big(\bigcup_{x \in X} [F(x) + G(x)] \Big)$$

$$= \operatorname{WSup} \bigcup_{x \in X} [B(F(x)) + B(G(x))] = \operatorname{WSup} \bigcup_{x \in X} [B(F(x)) + B(\operatorname{WSup} G(x))]$$

$$= \operatorname{WSup} B\Big(\bigcup_{x \in X} [F(x) + \operatorname{WSup} G(x)] \Big) = \operatorname{WSup} \bigcup_{x \in X} [F(x) + \operatorname{WSup} G(x)].$$

□

Remark 7.4.4. Concerning Proposition 7.4.3 let us refer to Proposition 7.1.7 where a similar result has been stated for maximal sets. There, in order to guarantee the coincidence of the sets $\operatorname{Max} \cup_{x \in X} [F(x) + G(x)]$ and $\operatorname{Max} \cup_{x \in X} [F(x) + \operatorname{Max} G(x)]$, external stability for $\operatorname{Max} G(x)$ for all $x \in X$ has been supposed. The advantage of using the weak supremum is given by the fact that such a restrictive condition can be omitted and this fact allows to derive a set-valued duality theory with less restrictions.

From Proposition 7.4.3 we obtain the following corollaries which play an important role in our considerations.

Corollary 7.4.4. *If $F : X \rightrightarrows \overline{V}$ is a set-valued map, then*

$$\operatorname{WSup} \bigcup_{x\in X} F(x) = \operatorname{WSup} \bigcup_{x\in X} \operatorname{WSup} F(x).$$

Corollary 7.4.5. *For $M_1, M_2 \subseteq \overline{V}$ given sets it holds*

$$\operatorname{WSup}(M_1 \cup M_2) = \operatorname{WSup}(\operatorname{WSup} M_1 \cup \operatorname{WSup} M_2).$$

Corollary 7.4.6. *For $M \subseteq \overline{V}$ a given set it holds* $\operatorname{WSup} M = \operatorname{WSup}(\operatorname{WSup} M)$.

Next, conjugate maps and subdifferentials for set-valued maps based on the weak supremum are introduced.

Definition 7.4.2. *Let $F : X \rightrightarrows \overline{V}$ be a set-valued map.*

(a) The set-valued map

$$F^* : \mathcal{L}(X, V) \rightrightarrows \overline{V},\ F^*(T) = \operatorname{WSup} \bigcup_{x\in X} [Tx - F(x)]$$

is called the conjugate map of F.

(b) The set-valued map

$$F^{**} : X \rightrightarrows \overline{V},\ F^{**}(x) = \operatorname{WSup} \bigcup_{T\in\mathcal{L}(X,V)} [Tx - F^*(T)]$$

is called the biconjugate map of F.

(c) The linear continuous operator $T \in \mathcal{L}(X, V)$ is said to be a subgradient of F at $(x, v) \in \operatorname{gph} F$, if

$$Tx - v \in \operatorname{WMax} \bigcup_{y\in X} [Ty - F(y)].$$

The set of all subgradients of F at (x, v) is called the subdifferential of F at (x, v) and it is denoted by $\partial F(x; v)$. Further, for all $x \in X$ denote $\partial F(x) := \bigcup_{v\in F(x)} \partial F(x; v)$. If for all $v \in F(x)$ we have $\partial F(x; v) \neq \emptyset$ then F is said to be subdifferentiable at x.

One can verify properties of conjugate maps and subgradients which are analogous to the ones given in subsection 7.1.1 for the corresponding notions based on Definition 7.1.2.

First, let us mention that the formulae (7.1) and (7.2) remain valid also for the conjugate maps as defined above. Moreover, one also has a *Young-Fenchel type inequality*.

Proposition 7.4.7. *Let $F : X \rightrightarrows \overline{V}$ be a set-valued map. For all $x \in X$ and all $T \in \mathcal{L}(X, V)$ there is $[F(x) - Tx] \cap B(-F^*(T)) = \emptyset$.*

Proof. Since $F^*(T) = \text{WSup} \cup_{x \in X} [Tx - F(x)]$, from Proposition 7.4.1(*c*) it follows

$$Tx - F(x) \subseteq \bigcup_{x \in X} [Tx - F(x)]$$

$$\subseteq \text{WSup} \bigcup_{x \in X} [Tx - F(x)] \cup B\Big(\text{WSup} \bigcup_{x \in X} [Tx - F(x)]\Big) = F^*(T) \cup B(F^*(T)).$$

Further, Proposition 7.4.1(*d*) implies $[Tx - F(x)] \cap A(F^*(T)) = \emptyset$, which is by Remark 7.4.2(*c*) equivalent to $[F(x) - Tx] \cap B(-F^*(T)) = \emptyset$. □

Corollary 7.4.8. *Let $F : X \rightrightarrows \overline{V}$ be a set-valued map. For all $x \in X$ and all $T \in \mathcal{L}(X, V)$ it holds $v + v^* \not<_K Tx$, whenever $v \in F(x)$ and $v^* \in F^*(T)$.*

Proof. Assume that the assertion fails. Then there exist $x \in X$, $T \in \mathcal{L}(X, V)$, $v \in F(x)$ and $v^* \in F^*(T)$ such that $v + v^* <_K Tx$, i.e. $v - Tx \in -v^* - \text{int}(K)$. But, by definition, $-v^* - \text{int}(K) \subseteq B(-F^*(T))$, therefore $v - Tx \in B(-F^*(T))$, contradicting the assertion of Proposition 7.4.7. □

Corollary 7.4.9. *Let $F : X \rightrightarrows \overline{V}$ be a set-valued map. If $v \in F(0)$ and $v^* \in -F^*(T)$, for $T \in \mathcal{L}(X, V)$, then $v \not<_K v^*$.*

Proposition 7.4.10. *Let $F : X \rightrightarrows \overline{V}$ be a set-valued map. Then $F(x) \cap B(F^{**}(x)) = \emptyset$ and $F(x) \subseteq F^{**}(x) \cup A(F^{**}(x))$ for all $x \in X$. In other words, for all $x \in X$ it holds $v \not<_K u$, whenever $v \in F(x)$ and $u \in F^{**}(x)$.*

Proof. Let be $x \in X$. From Proposition 7.4.7 it follows that $F(x) \cap B(Tx - F^*(T)) = \emptyset$ for all $T \in \mathcal{L}(X, V)$. Using Proposition 7.4.1(*b*) we get

$$B\Big(\bigcup_{T \in \mathcal{L}(X,V)} [Tx - F^*(T)]\Big) = B\Big(\text{WSup} \bigcup_{T \in \mathcal{L}(X,V)} [Tx - F^*(T)]\Big) = B(F^{**}(x))$$

and further, by Proposition 7.4.2(*b*) and Proposition 7.4.7, there is $F(x) \cap B(F^{**}(x)) = \emptyset$. The definition of the biconjugate map and Proposition 7.4.1(*d*) imply $F(x) \subseteq F^{**}(x) \cup A(F^{**}(x))$. This completes the proof. □

Proposition 7.4.11. *Let $F : X \rightrightarrows \overline{V}$ be a set-valued map and $x \in X$. For $v \in F(x)$ there is $T \in \partial F(x; v)$ if and only if $Tx - v \in F^*(T)$.*

Proof. Let $x \in X$ be fixed and $v \in F(x)$. By the definition of the subgradient there is $T \in \partial F(x; v)$ if and only if $Tx - v \in \text{WMax} \cup_{y \in X} [Ty - F(y)]$. By Remark 7.4.2(*c*) this is nothing else than

$$Tx - v \in \bigcup_{y \in X} [Ty - F(y)] \cap \text{WSup} \bigcup_{y \in X} [Ty - F(y)] = \bigcup_{y \in X} [Ty - F(y)] \cap F^*(T),$$

being further equivalent to $Tx - v \in F^*(T)$. □

The following result is an analog of Proposition 7.1.4.

Proposition 7.4.12. *Let $F : X \rightrightarrows \overline{V}$ be a set-valued map. If F is subdifferentiable at $x \in X$, then $F(x) \subseteq F^{**}(x)$. If, additionally, $F(x) = \text{WInf}\, F(x)$, then $F(x) = F^{**}(x)$.*

Proof. Let be $v \in F(x)$. The subdifferentiability of F at x entails by Proposition 7.4.11 the existence of some $\overline{T} \in \mathcal{L}(X,V)$ such that $\overline{T}x - v \in F^*(\overline{T})$. Thus there exists $\bar{v}^* \in F^*(\overline{T})$ fulfilling $v = \overline{T}x - \bar{v}^*$. Corollary 7.4.8 ensures that it does not exist any $T \in \mathcal{L}(X,V)$ and $v^* \in F^*(T)$ such that $v <_K Tx - v^*$. That means

$$v = \overline{T}x - \bar{v}^* \in \operatorname{WMax} \bigcup_{T \in \mathcal{L}(X,V)} [Tx - F^*(T)]$$

$$\subseteq \operatorname{WSup} \bigcup_{T \in \mathcal{L}(X,V)} [Tx - F^*(T)] = F^{**}(x)$$

and thus $F(x) \subseteq F^{**}(x)$. Now assume that $F(x) = \operatorname{WInf} F(x)$ and take an arbitrary $v \in F^{**}(x)$. By Proposition 7.4.1(*d*), $\overline{V} = F(x) \cup A(F(x)) \cup B(F(x))$ and, in view of Proposition 7.4.10, $v \notin A(F(x))$. Let us assume that $v \in B(F(x))$. Then there exists $\tilde{v} \in F(x)$ such that $v <_K \tilde{v}$. Because F is supposed to be subdifferentiable at x, there exists $\widetilde{T} \in \mathcal{L}(X,V)$ such that $\widetilde{T}x - \tilde{v} \in F^*(\widetilde{T})$. This means that $v <_K \tilde{v} \in \widetilde{T}x - F^*(\widetilde{T})$, i.e. $v \in B(\widetilde{T}x - F^*(\widetilde{T}))$. But this contradicts the assumption $v \in F^{**}(x) = \operatorname{WSup} \cup_{T \in \mathcal{L}(X,V)} [Tx - F^*(T)]$. Hence, $v \in F(x)$ and, consequently, $F^{**}(x) \subseteq F(x)$. □

For a set-valued map $F : X \rightrightarrows \overline{V}$ we define its K*-epigraph* in a modified manner than in Definition 7.1.5, namely being

$$\operatorname{epi}_K F = \{(x,v) \in X \times V : v \in (F(x) + K) \cup A(F(x))\}.$$

We say that F is K*-convex* if $\operatorname{epi}_K F$ is convex. In case $-\infty_K \in F(x)$ for some $x \in X$, then $(x,v) \in \operatorname{epi}_K F$ for all $v \in V$. Note that in case $-\infty_K \notin F(x)$ for all $x \in X$, $\operatorname{epi}_K F = \{(x,v) \in X \times V : v \in F(x) + K\}$ and the K-convexity is nothing else than the same notion as introduced in Definition 7.1.5(*b*). The following lemma provides conditions ensuring that $-\infty_K \notin F(x)$ for all $x \in X$.

Lemma 7.4.13. *Let* $F : X \rightrightarrows \overline{V}$ *be a* K*-convex set-valued map such that* $\bar{x} \in \operatorname{core}(\operatorname{dom} F)$ *with* $-\infty_K \notin F(\bar{x})$ *and* $\operatorname{WMin} F(\bar{x}) \neq \emptyset$. *Then* $-\infty_K \notin F(x)$ *for all* $x \in X$.

Proof. Without loss of generality suppose that $\bar{x} = 0$ and hence $0 \in \operatorname{core}(\operatorname{dom} F)$, $-\infty_K \notin F(0)$ and $\operatorname{WMin} F(0) \neq \emptyset$. Assume there exists $\tilde{x} \in \operatorname{dom} F$ such that $-\infty_K \in F(\tilde{x})$. Because of $0 \in \operatorname{core}(\operatorname{dom} F)$ there exists an $\varepsilon > 0$ such that $\varepsilon\tilde{x} \in \operatorname{dom} F$ and also $\hat{x} := -\varepsilon\tilde{x} \in \operatorname{dom} F$. Then 0 is representable as a convex combination of the elements $\tilde{x} \in \operatorname{dom} F$ and $\hat{x} \in \operatorname{dom} F$, namely $0 = (1/(1+\varepsilon))\hat{x} + (\varepsilon/(1+\varepsilon))\tilde{x}$. From $-\infty_K \in F(\tilde{x})$ follows $(\tilde{x}, v) \in \operatorname{epi}_K F$ for all $v \in V$. Consider $\hat{v} \in V$ such that $(\hat{x}, \hat{v}) \in \operatorname{epi}_K F$ and an arbitrary $v \in V$. We obtain by the convexity of $\operatorname{epi}_K F$

$$\frac{1}{1+\varepsilon}(\hat{x}, \hat{v}) + \frac{\varepsilon}{1+\varepsilon}(\tilde{x}, v) = \left(0, \frac{1}{1+\varepsilon}\hat{v} + \frac{\varepsilon}{1+\varepsilon}v\right) \in \operatorname{epi}_K F$$

and, consequently, $(0,v) \in \operatorname{epi}_K F$ for all $v \in V$. This implies $A(F(0)) = V \cup \{+\infty_K\}$ and therefore, by Remark 7.4.2(*e*), $\operatorname{WInf} F(0) = \{-\infty_K\}$. Since

$\operatorname{WMin} F(0) = F(0) \cap \operatorname{WInf} F(0) \neq \emptyset$, there must be $-\infty_K \in F(0)$ and in this way we obtain a contradiction. This leads to the desired conclusion. □

Next we introduce a result which ensures the subdifferentiability of F.

Proposition 7.4.14. *Let* $F : X \rightrightarrows V \cup \{+\infty_K\}$ *be a* K*-convex set-valued map with* $\operatorname{int}(\operatorname{epi}_K F) \neq \emptyset$. *If* $\bar{x} \in \operatorname{int}(\operatorname{dom} F)$ *and* $F(\bar{x}) \subseteq \operatorname{WInf} F(\bar{x})$, *then* F *is subdifferentiable at* $\bar{x}$.

This proposition has been formulated and proven in [178, Proposition 4.3] without assuming that $\operatorname{int}(\operatorname{epi}_K F) \neq \emptyset$. But this assumption is necessary because the separation theorem used within the proof requires such a condition as Song remarked in [168] (see also [170]). If, however, X and V are finite dimensional, this assumption can be omitted.

7.4.2 The perturbation approach

In subsection 7.1.2 we have presented the perturbation approach based on conjugate set-valued maps defined by means of minimality. Now we deal with analogous investigations based on weak minimality using the preliminaries from subsection 7.4.1.

Let X and V be Hausdorff topological vector spaces with X^* and V^* topological dual spaces, respectively. As in section 7.1 we consider for a set-valued map $F : X \rightrightarrows V \cup \{+\infty_K\}$ with $\operatorname{dom} F \neq \emptyset$ the general set-valued optimization problem

$$(PSVG_w) \quad \operatorname*{WInf}_{x \in X} F(x)$$

and this time we are interested in *weakly minimal elements* or even *weakly infimal elements* of the image set $F(X) = \cup_{x \in X} F(x)$ with respect to the nontrivial pointed convex closed cone $K \subseteq V$ with $\operatorname{int}(K) \neq \emptyset$. An element $\bar{x} \in X$ such that there exists $\bar{v} \in F(\bar{x})$ with $\bar{v} \in \operatorname{WMin} F(X)$ is called *weakly efficient solution* to $(PSVG_w)$, while the pair $(\bar{x}, \bar{v})$ is said to be a *weakly minimal pair* to the problem $(PSVG_w)$.

A particular instance of this problem arises when the set-valued map F is replaced with the vector-valued function $f : X \to V \cup \{+\infty_K\}$. In this case $(PSVG_w)$ becomes

$$(PVG_w) \quad \operatorname*{WInf}_{x \in X} f(x)$$

and we look for weakly efficient solutions $\bar{x} \in X$ characterized as fulfilling $f(\bar{x}) \in \operatorname{WMin} f(X)$.

Proposition 7.4.15. *The element* $\bar{x} \in X$ *is a weakly efficient solution to* $(PSVG_w)$ *if and only if there exists* $\bar{v} \in F(\bar{x})$ *such that* $0 \in \partial F(\bar{x}; \bar{v})$.

Similar to subsection 7.1.2 we associate a vector dual problem to $(PSVG_w)$ based on a corresponding perturbation approach. We introduce a set-valued perturbation map $\Phi : X \times Y \rightrightarrows V \cup \{+\infty_K\}$, where Y is another Hausdorff topological vector space with the topological dual space Y^*, such that $\Phi(x, 0) = F(x)$ for all $x \in X$. The space Y is called the *perturbation space*. Further, $(PSVG_w)$ is embedded into a family of perturbed problems

$$(PSVG_{wy}) \quad \operatorname*{WInf}_{x \in X} \Phi(x, y),$$

where $y \in Y$ is the *perturbation variable*. The problem $(PSVG_{w0})$ coincides with $(PSVG_w)$.

Like in subsection 7.1.2, the dual problem is defined by means of the conjugate of the perturbation map $\Phi^* : \mathcal{L}(X, V) \times \mathcal{L}(Y, V) \rightrightarrows \overline{V}$,

$$\Phi^*(T, \Lambda) = \operatorname{WSup} \bigcup_{x \in X, y \in Y} [Tx + \Lambda y - \Phi(x, y)].$$

To the primal problem $(PSVG_w)$ we attach the dual problem

$$(DSVG_w) \quad \operatorname*{WSup}_{\Lambda \in \mathcal{L}(Y,V)} \{-\Phi^*(0, \Lambda)\}.$$

We look for $\overline{\Lambda} \in \mathcal{L}(Y, V)$ such that there exists $\bar{v}^* \in -\Phi^*(0, \overline{\Lambda})$ fulfilling $\bar{v}^* \in \operatorname{WMax} \cup_{\Lambda \in \mathcal{L}(Y,V)} \{-\Phi^*(0, \Lambda)\}$. Such a mapping $\overline{\Lambda}$ is called a *weakly efficient solution* to $(DSVG_w)$ and $(\overline{\Lambda}, \bar{v}^*)$ is said to be a *weakly maximal pair* to $(DSVG_w)$. The next theorem expresses the weak duality for $(PSVG_w)$ and $(DSVG_w)$.

Proposition 7.4.16. *Let $x \in X$ and $\Lambda \in \mathcal{L}(Y, V)$ be given. Then $-\Phi^*(0, \Lambda) \cap A(\Phi(x, 0)) = \emptyset$ or, equivalently, $\Phi(x, 0) \cap B(-\Phi^*(0, \Lambda)) = \emptyset$.*

Proof. Assume that there exist $x \in X$ and $\Lambda \in \mathcal{L}(Y, V)$ such that $-\Phi^*(0, \Lambda) \cap A(\Phi(x, 0)) \neq \emptyset$. Let v be an element of this intersection. Then there exists $\tilde{v} \in \Phi(x, 0)$ fulfilling $\tilde{v} <_K v$. But this contradicts the statement in Corollary 7.4.8. The equivalent relation follows via Remark 7.4.3. □

The last result can be reformulated as the following corollary, where we denote

$$\operatorname{WInf}(PSVG_w) := \operatorname{WInf} \Big\{ \bigcup_{x \in X} \Phi(x, 0) \Big\}$$

and

$$\operatorname{WSup}(DSVG_w) := \operatorname{WSup} \Big\{ \bigcup_{\Lambda \in \mathcal{L}(Y,V)} -\Phi^*(0, \Lambda) \Big\}.$$

Corollary 7.4.17. *It holds $\operatorname{WSup}(DSVG_w) \cap A(\operatorname{WInf}(PSVG_w)) = \emptyset$ or, equivalently, $B(\operatorname{WSup}(DSVG_w)) \cap \operatorname{WInf}(PSVG_w) = \emptyset$.*

Proof. Assume the contrary and take an element $v \in \operatorname{WSup}(DSVG_w) \cap A(\operatorname{WInf}(PSVG_w))$. As $v \in A(\operatorname{WInf}(PSVG_w)) = A(\cup_{x \in X} \Phi(x, 0))$, there exists $v' \in \Phi(x', 0)$, for some $x' \in X$, satisfying $v' <_K v$. Since $v \in \operatorname{WSup}\{\cup_{\Lambda \in \mathcal{L}(Y,V)} - \Phi^*(0, \Lambda)\}$, there exist $\widetilde{\Lambda} \in \mathcal{L}(Y, V)$ and $\tilde{v} \in -\Phi^*(0, \widetilde{\Lambda})$, with $v' <_K \tilde{v}$. Thus $\tilde{v} \in -\Phi^*(0, \widetilde{\Lambda}) \cap A(\Phi(x', 0))$ and this contradicts the statement of Proposition 7.4.16. The equivalent relation follows again via Remark 7.4.3. □

Remark 7.4.5. If we assume that the objective function of the primal problem $(PSVG_w)$ is a vector-valued function $f : X \to V \cup \{+\infty_K\}$, in other words if we deal with (PVG_w), then the second condition in Proposition 7.4.16 may be reformulated as $f(x) = \Phi(x, 0) \notin B(-\Phi^*(0, \Lambda))$.

The following existence result for weakly efficient solutions is an easy consequence of the weak duality result.

Corollary 7.4.18. *Let be $\bar{v} \in F(\bar{x}) \cap \{-\Phi^*(0, \overline{\Lambda})\}$ for $\bar{x} \in X$ and $\overline{\Lambda} \in \mathcal{L}(Y, V)$. Then $\bar{x}$ is a weakly efficient solution and $(\bar{x}, \bar{v})$ is a weakly minimal pair to $(PSVG_w)$, while $\overline{\Lambda}$ is a weakly efficient solution and $(\overline{\Lambda}, \bar{v})$ is a weakly maximal pair to $(DSVG_w)$.*

Proof. Assume that $(\bar{x}, \bar{v})$ is not a weakly minimal pair to $(PSVG_w)$. Then there exist $x \in X$ and $v \in F(x) = \Phi(x, 0)$ such that $v <_K \bar{v}$. Thus $\bar{v} \in A(\Phi(x, 0))$ and $\bar{v} \in -\Phi^*(0, \overline{\Lambda})$ which contradicts Proposition 7.4.16. A similar argumentation shows the claimed assertion for $(\overline{\Lambda}, \bar{v})$. □

In analogy to section 7.1, where we have used the minimal value map, we consider here the so-called *infimal value map* $H : Y \rightrightarrows \overline{V}$ defined by

$$H(y) = \text{WInf}(PSVG_{wy}) = \text{WInf} \bigcup_{x \in X} \Phi(x, y) = \text{WInf}\, \Phi(X, y).$$

Then $H(0) = \text{WInf}(PSVG_w) = \text{WInf}\, \Phi(X, 0)$.

Lemma 7.4.19. *For all $\Lambda \in \mathcal{L}(Y, V)$ there is $H^*(\Lambda) = \Phi^*(0, \Lambda)$.*

Proof. By the definition of the conjugate map we have for all $\Lambda \in \mathcal{L}(Y, V)$ that

$$\begin{aligned} H^*(\Lambda) &= \text{WSup} \bigcup_{y \in Y} [\Lambda y - H(y)] = \text{WSup} \bigcup_{y \in Y} [\Lambda y - \text{WInf} \bigcup_{x \in X} \Phi(x, y)] \\ &= \text{WSup} \bigcup_{y \in Y} \text{WSup} \bigcup_{x \in X} [\Lambda y - \Phi(x, y)]. \end{aligned}$$

Further, by Corollary 7.4.4 we get $H^*(\Lambda) = \text{WSup} \cup_{x \in X, y \in Y} [\Lambda y - \Phi(x, y)] = \Phi^*(0, \Lambda)$. □

By using Lemma 7.4.19 $\text{WSup}(DSVG_w)$ may be rewritten as

$$\begin{aligned} \text{WSup}(DSVG_w) &= \text{WSup} \bigcup_{\Lambda \in \mathcal{L}(Y,V)} \{-\Phi^*(0, \Lambda)\} \\ &= \text{WSup} \bigcup_{\Lambda \in \mathcal{L}(Y,V)} \{-H^*(\Lambda)\} = H^{**}(0). \end{aligned}$$

In other words, $(DSVG_w)$ may be formally reformulated as

$$(DSVG_w) \quad \underset{\Lambda \in \mathcal{L}(Y,V)}{\text{WSup}} \{-H^*(\Lambda)\}.$$

Remark 7.4.6. As $\operatorname{WInf}(PSVG_w) = H(0)$, the duality for the pair $(PSVG_w)$–$(DSVG_w)$ can be expressed by means of $H(0)$ and $H^{**}(0)$. In particular, $\operatorname{WInf}(PSVG_w) = \operatorname{WSup}(DSVG_w)$ is nothing else than $H(0) = H^{**}(0)$. Throughout the last two sections of this chapter we understand under *strong duality* the situation when $H(0) = H^{**}(0)$.

The notion of stability of the primal problem with respect to the perturbation map plays a crucial role when delivering strong duality assertions.

Definition 7.4.3. *The problem $(PSVG_w)$ is called stable with respect to the perturbation map Φ if the infimal value map H is subdifferentiable at* 0.

We prove next that stability ensures strong duality for the primal-dual set-valued pair $(PSVG_w) - (DSVG_w)$.

Theorem 7.4.20. *If the problem $(PSVG_w)$ is stable, then*

$$\operatorname{WInf}(PSVG_w) = \operatorname{WSup}(DSVG_w) = \operatorname{WMax}(DSVG_w).$$

Proof. If the problem $(PSVG_w)$ is stable, then the infimal value map H is subdifferentiable at 0 and, by Proposition 7.4.12, there is $H(0) \subseteq H^{**}(0)$. On the other hand, from Corollary 7.4.6 and Remark 7.4.2(c) follows

$$\operatorname{WInf} H(0) = \operatorname{WInf} \operatorname{WInf} \bigcup_{x \in X} \Phi(x,0) = \operatorname{WInf} \bigcup_{x \in X} \Phi(x,0) = H(0).$$

Then Proposition 7.4.12 implies $H(0) = H^{**}(0)$ and, by Remark 7.4.6 this is equivalent to $\operatorname{WInf}(PSVG_w) = \operatorname{WSup}(DSVG_w)$. Since

$$\operatorname{WMax}(DSVG_w) \subseteq \operatorname{WSup}(DSVG_w) = \operatorname{WInf}(PSVG_w),$$

it remains to show that $\operatorname{WInf}(PSVG_w) \subseteq \operatorname{WMax}(DSVG_w)$. To this end take an arbitrary $\bar{v} \in \operatorname{WInf}(PSVG_w) = H(0)$. Since H is subdifferentiable at 0, there exists $\overline{\Lambda} \in \mathcal{L}(Y,V)$ such that $\overline{\Lambda} \in \partial H(0;\bar{v})$, in other words $\overline{\Lambda}0 - \bar{v} \in \operatorname{WMax} \cup_{y \in Y} [\overline{\Lambda} y - H(y)] \subseteq H^*(\overline{\Lambda}) = \Phi^*(0,\overline{\Lambda})$. Assuming that $\bar{v} \notin \operatorname{WMax}(DSVG_w)$ there would exist $\widetilde{\Lambda} \in \mathcal{L}(Y,V)$ and $\tilde{v} \in -\Phi^*(0,\widetilde{\Lambda})$ such that $\bar{v} <_K \tilde{v}$. From the definition of the weak infimum, using that $\bar{v} \in \operatorname{WInf} \cup_{x \in X} \Phi(x,0) = H(0)$, follows that there exist $\tilde{x} \in X$ and $v' \in \Phi(\tilde{x},0)$, with the property $v' <_K \tilde{v}$. Therefore $\tilde{v} \in -\Phi^*(0,\widetilde{\Lambda}) \cap A(\Phi(\tilde{x},0))$ which contradicts Proposition 7.4.16. Thus $\bar{v} \in \operatorname{WMax}(DSVG_w)$ and this completes the proof. □

Theorem 7.4.21. *Assume that the problem $(PSVG_w)$ is stable. Then for each weakly efficient solution $\bar{x} \in X$ to $(PSVG_w)$ and $\bar{v} \in F(\bar{x})$ such that $(\bar{x},\bar{v})$ is a weakly minimal pair to $(PSVG_w)$ there exists a weakly efficient solution $\overline{\Lambda} \in \mathcal{L}(Y,V)$ to $(DSVG_w)$ such that $\bar{v} \in -\Phi^*(0,\overline{\Lambda})$ and $(\overline{\Lambda},\bar{v})$ is a weakly maximal pair to $(DSVG_w)$.*

Proof. Because $(\bar{x}, \bar{v})$ is a weakly minimal pair to $(PSVG_w)$ there is $\bar{v} \in H(0)$. Since H is subdifferentiable at 0 there exists $\overline{\Lambda} \in \mathcal{L}(Y, V)$ such that $\overline{\Lambda} \in \partial H(0; \bar{v})$. Thus $\bar{v} \in -\Phi^*(0, \overline{\Lambda})$. On the other hand, by Theorem 7.4.20 we have $\bar{v} \in \mathrm{WMax}(DSVG_w)$ and so $\overline{\Lambda}$ is a weakly efficient solution to $(DSVG_w)$ with $(\overline{\Lambda}, \bar{v})$ corresponding weakly maximal pair. □

Next, we give necessary optimality conditions of subdifferential type.

Theorem 7.4.22. *The weakly minimal pair $(\bar{x}, \bar{v})$ to $(PSVG_w)$ and the corresponding weakly maximal pair $(\overline{\Lambda}, \bar{v})$ to $(DSVG_w)$ from Theorem 7.4.21 satisfy the optimality conditions $(0, \overline{\Lambda}) \in \partial \Phi(\bar{x}, 0; \bar{v})$, or, equivalently, $\overline{\Lambda} \in \partial H(0; \bar{v})$.*

Proof. The proof is completely analogous to the one of Theorem 7.1.16 if one applies Proposition 7.4.11 instead of Proposition 7.1.3. □

These necessary optimality conditions turn out to be sufficient for strong duality. The proof of the following statement can be done in the lines of the one of Theorem 7.1.17.

Theorem 7.4.23. *Let $\bar{x} \in X$, $\bar{v} \in F(\bar{x})$ and $\overline{\Lambda} \in \mathcal{L}(Y, V)$ fulfill $(0, \overline{\Lambda}) \in \partial \Phi(\bar{x}, 0; \bar{v})$. Then $\bar{x}$ is a weakly efficient solution and $(\bar{x}, \bar{v})$ a weakly minimal pair to $(PSVG_w)$, while $\overline{\Lambda}$ is a weakly efficient solution and $(\overline{\Lambda}, \bar{v})$ is a weakly maximal pair to $(DSVG_w)$.*

Remark 7.4.7. In Theorem 7.4.23, instead of $(0, \overline{\Lambda}) \in \partial \Phi(\bar{x}, 0; \bar{v})$ one can equivalently require that $\overline{\Lambda} \in \partial H(0; \bar{v})$.

Without assuming neither stability for $(PSVG_w)$, nor the existence of a weakly efficient solution to $(PSVG_w)$, Song gives in [170, Theorem 6.3] some equivalent conditions for strong duality. We state this result in the following.

Theorem 7.4.24. *Let $\Lambda \in \mathcal{L}(Y, V)$ be given. Then the following conditions are equivalent:*

(i) there exists $v \in -\Phi^*(0, \Lambda)$ with $v \in \mathrm{WInf}(PSVG_w) \cap \mathrm{WMax}(DSVG_w)$;
(ii) $\mathrm{WInf}(PSVG_w) \cap (-\Phi^*(0, \Lambda)) \neq \emptyset$;
(iii) $\Lambda \in \partial H(0)$.

Proof. It is clear that (i) implies (ii). To show that (ii) implies (iii), let be $v \in \mathrm{WInf}(PSVG_w) \cap (-\Phi^*(0, \Lambda))$. Then by the definition of the infimal value map H and by Lemma 7.4.19 it holds $v \in H(0) \cap (-H^*(\Lambda))$ and so $\Lambda 0 - v \in H^*(\Lambda)$. Via Proposition 7.4.11 the last relation is equivalent to $\Lambda \in \partial H(0; v)$, which is a subset of $\partial H(0)$. This proves (iii). Finally, (i) follows from (iii) as in the proof of Theorem 7.4.20, with the mention that here we do not need stability. We have that $\Lambda \in \partial H(0; v)$ for some $v \in H(0) = \mathrm{WInf}(PSVG_w)$, since $\partial H(0) = \cup_{u \in H(0)} \partial H(0; u)$, and as in the mentioned theorem it follows that $v \in -\Phi^*(0, \Lambda)$ and $v \in \mathrm{WMax}(DSVG_w)$. □

Since the stability of the primal problem implies strong duality, it is important to have criteria that entail this property. As in section 7.1, the so-called weakly K-upper boundedness introduced in Definition 7.1.6 in connection to a set-valued map plays a crucial role in this context.

Consider the set-valued map $\Psi : Y \rightrightarrows V \cup \{+\infty\}$, $\Psi(y) = \Phi(X, y)$. As stated in Lemma 7.1.23, Ψ is K-convex if and only if Φ is K-convexlike-convex. For the proof of the following lemma we refer to [168, Proposition 3.2].

Lemma 7.4.25. *If Ψ is K-convex or, equivalently, Φ is K-convexlike-convex, then H is K-convex.*

The next result and its proof can be found in [168, Proposition 3.3].

Lemma 7.4.26. *It holds* $\operatorname{epi}_K \Psi \subseteq \operatorname{epi}_K H \subseteq \operatorname{cl}(\operatorname{epi}_K \Psi)$.

The previous two lemmata allow to derive the following stability result (see [170, Theorem 6.4]).

Theorem 7.4.27. *Let the perturbation map $\Phi : X \times Y \rightrightarrows V \cup \{+\infty\}$ be K-convexlike-convex (or, equivalently, Ψ be K-convex). Assume that there exists $y' \in \operatorname{dom} \Psi$ such that Ψ is weakly K-upper bounded on some neighborhood of y'. If $0 \in \operatorname{int}(\operatorname{dom} \Psi)$, then the primal problem $(PSVG_w)$ is stable.*

Proof. First let us observe that by Lemma 7.4.25 the infimal value map H is K-convex. Whenever $H(0) = \operatorname{WInf} \Psi(0) = \{-\infty_K\}$, there is $H^* \equiv \{+\infty_K\}$, meaning that H is subdifferentiable at 0. Consider the case when $H(0) \neq \{-\infty_K\}$. The relation $0 \in \operatorname{int}(\operatorname{dom} \Psi)$ entails the existence of U, a neighborhood of 0 in Y, such that $U \subseteq \operatorname{dom} \Psi$. Thus $U \subseteq \operatorname{dom} H$ and so $0 \in \operatorname{int}(\operatorname{dom} H)$. By Corollary 7.4.6 one has $H(0) = \operatorname{WInf} \Psi(0) = \operatorname{WInf}(\operatorname{WInf} \Psi(0)) = \operatorname{WInf} H(0)$ and so $\operatorname{WMin} H(0) = H(0) \neq \emptyset$ and $-\infty_K \notin H(0)$.

Thus for H all the hypotheses of Lemma 7.4.13 are fulfilled and its application yields $-\infty_K \notin H(y)$ for all $y \in Y$. Consequently, H is a set-valued map which takes values in $V \cup \{+\infty_K\}$. Since Ψ is weakly K-upper bounded on some neighborhood of $y' \in \operatorname{dom} \Psi$, by Lemma 7.1.21 there is $\operatorname{int}(\operatorname{epi}_K \Psi) \neq \emptyset$ and this implies, by Lemma 7.4.26, that $\operatorname{int}(\operatorname{epi}_K H) \neq \emptyset$. With $H(0) = \operatorname{WInf} H(0)$ and $0 \in \operatorname{int}(\operatorname{dom} H)$, the assumptions of Proposition 7.4.14 are satisfied. Therefore H is subdifferentiable at 0, which is nothing else than that $(PSVG_w)$ is stable. □

Closely related to Theorem 7.4.27 is the following result, formulated by Song in [169].

Theorem 7.4.28. *Suppose that $\Phi : X \times Y \rightrightarrows V \cup \{+\infty\}$ is K-convexlike-convex (or, equivalently, Ψ is K-convex) and that the infimal value map H is weakly K-upper bounded on some neighborhood of 0. Then $(PSVG_w)$ is stable.*

Proof. Because H is weakly K-upper bounded on a neighborhood of 0 one has $0 \in \operatorname{int}(\operatorname{dom} H)$. Like in the proof of Theorem 7.4.27 we get $-\infty_K \notin H(y)$ for all $y \in Y$. Lemma 7.1.21 is now applicable and guarantees that $\operatorname{int}(\operatorname{epi}_K H) \neq \emptyset$. The conclusion follows again via Proposition 7.4.14. □

7.5 Some particular instances of $(PSVG_w)$

In this section we employ the approach described in section 7.4 to different particular instances of $(PSVG_w)$. We consider both the set-valued optimization problem with constraints and the one having the composition with a linear continuous mapping as objective map. An approach to gap functions for set-valued equilibrium problems is also given.

7.5.1 The set-valued optimization problem with constraints

In this subsection we investigate the duality for the set-valued optimization problem with constraints by using a similar procedure to the one considered in subsection 7.2, but by employing weak minimality and weak maximality. Again, we consider different perturbations of the primal problem which lead to several set-valued dual optimization problems.

The set-valued primal problem we study is the following

$$(PSV_w^C) \quad \operatorname*{WInf}_{x \in \mathcal{A}} F(x)$$
$$\mathcal{A} = \{x \in S : G(x) \cap (-C) \neq \emptyset\}$$

where X, Z and V are Hausdorff topological vector spaces, Z is partially ordered by the convex cone $C \subseteq Z$ and V is partially ordered by the nontrivial pointed convex closed cone $K \subseteq V$, $S \subseteq X$ is a nonempty set, while $F : X \rightrightarrows V \cup \{+\infty_K\}$ and $G : X \rightrightarrows Z$ are set-valued maps such that $\operatorname{dom} F \cap S \cap G^{-1}(-C) \neq \emptyset$. By X^*, Z^* and V^* we denote the corresponding topological dual spaces of X, Z and V, respectively.

We look for weakly efficient solutions to (PSV_w^C). These are elements $\bar{x} \in \mathcal{A}$ such that there exists an element $\bar{v} \in F(\bar{x})$ having the property $\bar{v} \in \operatorname{WMin} F(\mathcal{A})$. In this case the pair $(\bar{x}, \bar{v})$ is said to be a weakly minimal pair to (PSV_w^C).

We consider the set-valued perturbation maps Φ^{C_L}, Φ^{C_F} and $\Phi^{C_{FL}}$ as introduced in subsection 7.2.1 and formulate three set-valued duals to (PSV_w^C) via the conjugate maps of the perturbed maps. The calculations of the three conjugate maps is done in the lines of the ones in subsection 7.2.1. Thus for all $\Lambda \in \mathcal{L}(Z, V)$ and all $\Gamma \in \mathcal{L}(X, V)$ we obtain

$$(\Phi^{C_L})^*(0, \Lambda) = \operatorname{WSup} \bigcup_{x \in S} [\Lambda(G(x) + C) - F(x)],$$

$$(\Phi^{C_F})^*(0, \Gamma) = \operatorname{WSup}\{F^*(\Gamma) - \Gamma(\mathcal{A})\}$$

and

$$(\Phi^{C_{FL}})^*(0,\Gamma,\Lambda) = \mathrm{WSup}\left\{F^*(\Gamma) + \bigcup_{x\in S}[-\Gamma x + \Lambda(G(x))] + \Lambda(C)\right\}.$$

For determining $(\Phi^{C_F})^*$ and $(\Phi^{C_{FL}})^*$ Proposition 7.4.3 has been applied in a similar way as Proposition 7.1.7 for the corresponding calculations in subsection 7.2.1. Again with Proposition 7.4.3 and by considering the following conjugate map with respect to the set S

$$(-\Lambda G)_S^*(-\Gamma) = \mathrm{WSup} \bigcup_{x\in S}[-\Gamma x + \Lambda(G(x))],$$

we get

$$(\Phi^{C_{FL}})^*(0,\Gamma,\Lambda) = \mathrm{WSup}\{F^*(\Gamma) + (-\Lambda G)_S^*(-\Gamma) + \Lambda(C)\}.$$

Concerning the term $\Lambda(C)$ in the formulae of $(\Phi^{C_L})^*(0,\Lambda)$ and $(\Phi^{C_{FL}})^*(0,\Gamma,\Lambda)$ the following lemma is useful.

Lemma 7.5.1. *Let be $\Lambda \in \mathcal{L}(Z,V)$ and $\Gamma \in \mathcal{L}(X,V)$. If $\Lambda(C)\cap \mathrm{int}(K) \neq \emptyset$, then $(\Phi^{C_L})^*(0,\Lambda) = \{+\infty_K\}$ and $(\Phi^{C_{FL}})^*(0,\Gamma,\Lambda) = \{+\infty_K\}$.*

Proof. Let $\Lambda \in \mathcal{L}(Z,V)$ be fixed and assume that for $c' \in C$ it holds $\Lambda c' \in \mathrm{int}(K)$. We prove that $(\Phi^{C_L})^*(0,\Lambda) = \{+\infty_K\}$. By similar arguments it can be proven that for all $\Lambda \in \mathcal{L}(Z,V)$ and $\Gamma \in \mathcal{L}(X,V)$ one has $(\Phi^{C_{FL}})^*(0,\Gamma,\Lambda) = \{+\infty_K\}$.

Consider an arbitrary $\tilde{v} \in V$. We show that there exists $v \in \cup_{x\in S}[\Lambda(G(x)+C) - F(x)]$ with $v >_K \tilde{v}$. Then, by definition, $\mathrm{WSup}\cup_{x\in S}[\Lambda(G(x)+C) - F(x)] = \{+\infty_K\}$. Indeed, since $\Lambda c' \in \mathrm{int}(K)$ there exists an absorbing neighborhood W of 0 in V such that $\Lambda c' + W \subseteq \mathrm{int}(K)$. Thus $W \subseteq -\Lambda c' + \mathrm{int}(K)$ and hence $V = \cup_{\alpha>0}\alpha(-\Lambda c' + \mathrm{int}(K))$. For an $x' \in \mathrm{dom}\, F \cap S \cap G^{-1}(-C)$ take $v' \in F(x') \cap V$ and $z' \in G(x')$. By the above representation of V, there exist $\alpha' > 0$ and $k' \in \mathrm{int}(K)$ with the property $\Lambda z' - v' - \tilde{v} = -\Lambda(\alpha' c') + k'$. Thus for $v := \Lambda z' + \Lambda(\alpha' c') - v' \in \Lambda(G(x')) + \Lambda(C) - F(x') \subseteq \cup_{x\in S}[\Lambda(G(x)) + \Lambda(C) - F(x)]$, one has $v >_K \tilde{v}$. □

Taking into consideration the last lemma and the definition of the set-valued dual problem $(DSVG_w)$ given in subsection 7.4.2 the following *Lagrange set-valued dual problem* may be associated to (PSV_w^C)

$$(DSV_w^{C_L})\quad \mathrm{WSup} \bigcup_{\substack{\Lambda\in\mathcal{L}(Z,V)\\ \Lambda(C)\cap(-\mathrm{int}(K))=\emptyset}} \mathrm{WInf} \bigcup_{x\in S}[F(x) + \Lambda(G(x)+C)].$$

Analogously, the *Fenchel set-valued dual problem* reads as

$$(DSV_w^{C_F})\quad \mathrm{WSup} \bigcup_{\Gamma\in\mathcal{L}(X,V)} \mathrm{WInf}\{-F^*(\Gamma) + \Gamma(\mathcal{A})\},$$

while the *Fenchel-Lagrange set-valued dual problem* becomes

$$(DSV_w^{C_{FL}})\quad \mathrm{WSup}_{\substack{\Gamma\in\mathcal{L}(X,V),\Lambda\in\mathcal{L}(Z,V)\\ \Lambda(C)\cap(-\mathrm{int}(K))=\emptyset}} \mathrm{WInf}\{-F^*(\Gamma) - (\Lambda G)_S^*(-\Gamma) + \Lambda(C)\}.$$

Remark 7.5.1. The problem (PSV_w^C) has been considered also in [170] with respect to Lagrange duality. There the set-valued dual $(DSV_w^{C_L})$ can be found in some modified formulation.

The duality results obtained in subsection 7.4.2 for the primal-dual pair $(PSVG_w) - (DSVG_w)$ may be applied to the primal problem (PSV_w^C) and its set-valued dual problems introduced above. In particular, weak duality is fulfilled for (PSV_w^C) and $(DSV_w^{C_L})$, $(DSV_w^{C_F})$ and $(DSV_w^{C_{FL}})$, respectively. Concerning strong duality and optimality conditions, the general results from section 7.4 are also applicable. First, let us consider optimality conditions and strong duality for the Lagrange set-valued dual problem.

Theorem 7.5.2. *(a) Suppose that the problem (PSV_w^C) is stable with respect to the perturbation map Φ^{C_L}. Let $\bar{x} \in \mathcal{A}$ be a weakly efficient solution to (PSV_w^C) and $\bar{v} \in F(\bar{x})$ such that $(\bar{x}, \bar{v})$ is a weakly minimal pair to (PSV_w^C). Then there exists $\overline{\Lambda} \in \mathcal{L}(Z,V)$, a weakly efficient solution to $(DSV_w^{C_L})$, with $(\overline{\Lambda}, \bar{v})$ corresponding weakly maximal pair such that strong duality holds and the following conditions are fulfilled*

(i) $\bar{v} \in \operatorname{WInf} \cup_{x \in S}[F(x) + \overline{\Lambda}(G(x) + C)]$;

(ii) $\overline{\Lambda}(C) \cap (-\operatorname{int}(K)) = \emptyset$;

(iii) $\overline{\Lambda}(G(\bar{x}) + C) \cap (-\operatorname{int}(K)) = \emptyset$.

(b) Assume that for $\bar{x} \in \mathcal{A}$, $\bar{v} \in F(\bar{x})$ and $\overline{\Lambda} \in \mathcal{L}(Z,V)$ the conditions (i)−(iii) are fulfilled. Then $\bar{x}$ is a weakly efficient solution and $(\bar{x}, \bar{v})$ is a weakly minimal pair to (PSV_w^C), while $\overline{\Lambda}$ is a weakly efficient solution and $(\overline{\Lambda}, \bar{v})$ is a weakly maximal pair to $(DSV_w^{C_L})$.

Proof. The proof is similar to the proof of Theorem 7.2.1 with some obvious modifications. In particular, one has to use Theorem 7.4.21 instead of Theorem 7.1.15 as has been done in the proof of Theorem 7.2.1 □

Using the Fenchel set-valued dual problem $(DSV_w^{C_F})$ similar optimality conditions as in Theorem 7.2.2 are available.

Theorem 7.5.3. *(a) Suppose that the problem (PSV_w^C) is stable with respect to the perturbation map Φ^{C_F}. Let $\bar{x} \in \mathcal{A}$ be a weakly efficient solution to (PSV_w^C) and $\bar{v} \in F(\bar{x})$ such that $(\bar{x}, \bar{v})$ is a weakly minimal pair to (PSV_w^C). Then there exists $\overline{\Gamma} \in \mathcal{L}(X,V)$, a weakly efficient solution to $(DSV_w^{C_F})$, with $(\overline{\Gamma}, \bar{v})$ corresponding weakly maximal pair such that strong duality holds and*

$$\bar{v} \in \operatorname{WInf}[-F^*(\overline{\Gamma}) + \overline{\Gamma}(\mathcal{A})].$$

(b) Assume that for $\bar{x} \in \mathcal{A}$, $\bar{v} \in F(\bar{x})$ and $\overline{\Gamma} \in \mathcal{L}(X,V)$ one has $\bar{v} \in \operatorname{WInf}[-F^*(\overline{\Gamma}) + \operatorname{WInf} \overline{\Gamma}(\mathcal{A})]$. Then $\bar{x}$ is a weakly efficient solution and $(\bar{x}, \bar{v})$ a weakly minimal pair to (PSV_w^C), while $\overline{\Gamma}$ is a weakly efficient solution and $(\overline{\Gamma}, \bar{v})$ a weakly maximal pair to $(DSV_w^{C_F})$.

By means of the Fenchel-Lagrange set-valued dual problem $(DSV_w^{C_{FL}})$ analogous results can be derived.

Theorem 7.5.4. *(a) Suppose that the problem* (PSV_w^C) *is stable with respect to the perturbation map* $\Phi^{C_{FL}}$. *Let* $\bar{x} \in \mathcal{A}$ *be a weakly efficient solution to* (PSV_w^C) *and* $\bar{v} \in F(\bar{x})$ *such that* $(\bar{x}, \bar{v})$ *is a weakly minimal pair to* (PSV_w^C). *Then there exists* $(\overline{\Gamma}, \overline{\Lambda}) \in \mathcal{L}(X,V) \times \mathcal{L}(Z,V)$, *a weakly efficient solution to* $(DSV_w^{C_{FL}})$, *with* $(\overline{\Gamma}, \overline{\Lambda}, \bar{v})$ *corresponding weakly maximal pair such that strong duality holds and the following optimality conditions are fulfilled*

(i) $\bar{v} \in \mathrm{WInf}[-F^*(\overline{\Gamma}) - (\overline{\Lambda}G)_S^*(-\overline{\Gamma}) + \overline{\Lambda}(C)]$;
(ii) $\overline{\Lambda}(C) \cap (-\operatorname{int}(K)) = \emptyset$;
(iii) $\overline{\Lambda}(G(\bar{x}) + C) \cap (-\operatorname{int}(K)) = \emptyset$.

(b) Assume that for $\bar{x} \in \mathcal{A}$, $\bar{v} \in F(\bar{x})$ *and* $(\overline{\Gamma}, \overline{\Lambda}) \in \mathcal{L}(Z,V) \times \mathcal{L}(X,V)$ *the conditions* $(i) - (iii)$ *are fulfilled. Then* $\bar{x}$ *is a weakly efficient solution and* $(\bar{x}, \bar{v})$ *a weakly minimal pair to* (PSV_w^C), *while* $(\overline{\Gamma}, \overline{\Lambda})$ *is a weakly efficient solution and* $(\overline{\Gamma}, \overline{\Lambda}, \bar{v})$ *a weakly maximal pair to* $(DSV_w^{C_{FL}})$.

The final part of this subsection is devoted to the introduction of some conditions which guarantee the stability for the primal problem (PSV_w^C) in the interaction with its different duals we derived above and, consequently, the existence of strong duality for all these primal-dual pairs. First we deal with the perturbation map Φ^{C_L} having in mind the Lagrange duality.

Theorem 7.5.5. *Let* $F \times G$ *be* $K \times C$*-convexlike on* S *and* F *be weakly* K*-upper bounded on* $\operatorname{dom} F \cap S$. *If* $0 \in \operatorname{int}[G(\operatorname{dom} F \cap S) + C]$, *then* (PSV_w^C) *is stable with respect to the perturbation map* Φ^{C_L}; *hence* $\mathrm{WInf}(PSV_w^C) = \mathrm{WSup}(DSV_w^{C_L}) = \mathrm{WMax}(DSV_w^{C_L})$.

Proof. As follows from Remark 7.2.7, the map Φ^{C_L} is K-convexlike-convex. The consequence of the regularity condition is the same as in the proof of Theorem 7.2.7, namely it holds $0 \in \operatorname{int}(\operatorname{dom} \Psi^{C_L})$, where $\Psi^{C_L} : Z \rightrightarrows V \cup \{+\infty_K\}$, $\Psi^{C_L}(z) := \Phi^{C_L}(X, z)$. This map turns out to be weakly K-upper bounded on a neighborhood of 0 in Z. Then Theorem 7.4.27 is applicable and provides the desired stability. Further, Theorem 7.4.20 completes the proof. □

If the generalized Slater regularity condition is valid, namely $G(\operatorname{dom} F \cap S) \cap (-\operatorname{int}(C)) \neq \emptyset$, then the interior point regularity condition in Theorem 7.5.5 is verified. On the other hand, we observe that if S is convex, F is K-convex on S and G is C-convex on S, then $F \times G$ is $K \times C$-convex on S and therefore also $K \times C$-convexlike on S.

The previous result can be originally found in [170, Corollary 6.3]. As in subsection 7.2.3 (see Theorem 7.2.8) one can give an alternative result for stating strong duality for the Lagrange set-valued dual problem.

Theorem 7.5.6. *Let* $F \times G$ *be* $K \times C$*-convexlike on* S. *If* $G(\operatorname{dom} F \cap S) \cap (-\operatorname{int}(C)) \neq \emptyset$, *then* (PSV_w^C) *is stable with respect to the perturbation map* Φ^{C_L}; *hence* $\mathrm{WInf}(PSV_w^C) = \mathrm{WSup}(DSV_w^{C_L}) = \mathrm{WMax}(DSV_w^{C_L})$.

The next result concerns the stability with respect to the Fenchel perturbation map Φ^{C_F}.

Theorem 7.5.7. *Let $\mathcal{A} \subseteq X$ be a convex set and F a K-convex set-valued map. If there exists $x' \in \operatorname{dom} F \cap \mathcal{A}$ such that F is weakly K-upper bounded on some neighborhood of x', then (PSV_w^C) is stable with respect to the perturbation map Φ^{C_F}; hence* $\operatorname{WInf}(PSV_w^C) = \operatorname{WSup}(DSV_w^{C_F}) = \operatorname{WMax}(DSV_w^{C_F})$.

Proof. Under the hypotheses of the theorem, the perturbation map $\Phi^{C_F} : X \times X \rightrightarrows V \cup \{+\infty_K\}$, $\Phi^{C_F}(x,y) = F(x+y) + \delta_{\mathcal{A}}^V(x)$, is K-convex and, consequently, K-convexlike-convex. Like in the proof of Theorem 7.2.10, it follows that $\Psi^{C_F} : X \rightrightarrows V \cup \{+\infty_K\}$, $\Psi^{C_F}(y) := \Phi^{C_F}(X,y)$, is weakly K-upper bounded on a neighborhood of 0 in X and that $0 \in \operatorname{int}(\operatorname{dom} \Psi^{C_F})$. The conclusion follows via Theorem 7.4.27 and Theorem 7.4.20. □

Also here we give an alternative result, which can be proved in the lines of Theorem 7.2.11.

Theorem 7.5.8. *Let $\mathcal{A} \subseteq X$ be a convex set and F a K-convex set-valued map. If $\operatorname{dom} F \cap \operatorname{int}(\mathcal{A}) \neq \emptyset$, then (PSV_w^C) is stable with respect to the perturbation map Φ^{C_F}; hence* $\operatorname{WInf}(PSV_w^C) = \operatorname{WSup}(DSV_w^{C_F}) = \operatorname{WMax}(DSV_w^{C_F})$.

Finally, we treat the stability with respect to the Fenchel-Lagrange duality.

Theorem 7.5.9. *Let $S \subseteq X$ be a convex set, F a K-convex set-valued map and G a C-convex set-valued map. If F is weakly K-upper bounded on $\operatorname{dom} F$ and $U \times W$ is a neighborhood of $(0,0)$ in $X \times Z$ such that $W \subseteq \cap_{y \in U} G[(\operatorname{dom} F - y) \cap S] + C$, then (PSV_w^C) is stable with respect to the perturbation map $\Phi^{C_{FL}}$; hence* $\operatorname{WInf}(PSV_w^C) = \operatorname{WSup}(DSV_w^{C_{FL}}) = \operatorname{WMax}(DSV_w^{C_{FL}})$.

Proof. It is easy to verify that $\Phi^{C_{FL}}$ is K-convex and, consequently, K-convexlike-convex. Like in the proof of Theorem 7.2.13, it follows that $\Psi^{C_{FL}} : X \times Z \rightrightarrows V \cup \{+\infty_K\}$, $\Psi^{C_{FL}}(y,z) := \Phi^{C_{FL}}(X,y,z)$, is weakly K-upper bounded on $U \times W$ and $0 \in \operatorname{int}(\operatorname{dom} \Psi^{C_{FL}})$. The conclusion follows via Theorem 7.4.27 and Theorem 7.4.20. □

The following result can be proven in a similar way like Theorem 7.2.14.

Theorem 7.5.10. *Let $S \subseteq X$ be a convex set, F a K-convex set-valued map and G a C-convex set-valued map. If there exist $x' \in S$ such that $0 \in \operatorname{int}(G(x') + C)$ and a neighborhood U of x' in X such that F is weakly K-upper bounded on U, then (PSV_w^C) is stable with respect to the perturbation map $\Phi^{C_{FL}}$; hence* $\operatorname{WInf}(PSV_w^C) = \operatorname{WSup}(DSV_w^{C_{FL}}) = \operatorname{WMax}(DSV_w^{C_{FL}})$

Remark 7.5.2. In the papers [47, 123, 124] similar investigations concerning vector conjugate duality based on the weak infimum notion used here are provided and analogous perturbations as we have done here and in earlier works (cf. [1, 3, 4, 23]) are proposed. Inclusion relations between the image sets of the vector dual problems are given and saddle points and their relations to duality are also considered (cf. [124]). The stability assertions given in [124] concern the constrained vector optimization problem (PSV_w^C) with vector-valued objective and constrained functions and coincide in this particular case with Theorem 7.5.6 and Theorem 7.5.10. In particular, these assertions are special instances of the general assertions including more general stability criteria claimed in Theorem 7.5.5 and Theorem 7.5.9. We refer finally to the fact that already in [23] for problems even more general than (PSV_w^C) Fenchel-Lagrange type perturbations and corresponding set-valued dual problems have been considered including stability criteria based on regularity conditions of Slater type as in Theorem 7.5.10. In [47,123] some modified Lagrange and Fenchel-Lagrange dual problems are introduced in whose formulation the dual variables Λ are taken as being elements of the set of positive mappings $\mathcal{L}_+(Z,V)$. The weak and strong duality assertions follow analogously to the ones presented above.

7.5.2 The set-valued optimization problem having the composition with a linear continuous mapping in the objective map

In the following we treat the problem (PSV^A) as introduced in section 7.3, but with respect to weakly efficient solutions. Let X, Y and V be Hausdorff topological vector spaces, V being partially ordered by the nontrivial pointed convex cone $K \subseteq V$ with $\operatorname{int}(K) \neq \emptyset$, $A \in \mathcal{L}(X,Y)$ and $F : X \rightrightarrows V \cup \{+\infty_K\}$ and $G : Y \rightrightarrows V \cup \{+\infty_K\}$ be given set-valued maps such that $\operatorname{dom} F \cap A^{-1}(\operatorname{dom} G) \neq \emptyset$. The problem we treat here is

$$(PSV_w^A) \quad \underset{x \in X}{\operatorname{WInf}} \{F(x) + G(Ax)\}.$$

An element $\bar{x} \in X$ is said to be a weakly efficient solution to (PSV_w^A) if there exists a $\bar{v} \in F(\bar{x}) + G(A\bar{x})$ such that $\bar{v} \in \operatorname{WMin} \cup_{x \in X} \{F(x) + G(Ax)\}$. In this situation the pair $(\bar{x}, \bar{v})$ is said to be a weakly minimal pair to (PSV_w^A). Also here we take as set-valued perturbation map

$$\Phi^A : X \times Y \rightrightarrows V \cup \{+\infty_K\}, \Phi^A(x,y) = F(x) + G(Ax + y).$$

Further, let be $\Psi^A : Y \rightrightarrows V \cup \{+\infty_K\}$, $\Psi^A(y) = \Phi^A(X,y)$ and let the *infimal value map* $H^A : Y \rightrightarrows \overline{V}$ be defined by $H^A(y) = \operatorname{WInf} \Psi^A(y) = \operatorname{WInf} \Phi^A(X,y)$. Then $H^A(0) = \operatorname{WInf} \Phi^A(X,0) = \operatorname{WInf}(PSV_w^A)$. The calculation of the conjugate map $(\Phi^A)^*$ can be done similarly as in section 7.3, this time by using Proposition 7.4.3 instead of Proposition 7.1.7. For $\Gamma \in \mathcal{L}(Y,V)$ the conjugate looks like

$$(\Phi^A)^*(0,\Gamma) = \operatorname{WSup}\{F^*(-\Gamma \circ A) + G^*(\Gamma)\}.$$

By the general approach described in subsection 7.4.2 this leads to the following Fenchel set-valued dual problem to (PSV_w^A)

$$(DSV_w^A) \quad \operatorname{WSup} \bigcup_{\Gamma \in \mathcal{L}(Y,V)} \operatorname{WInf}\{-F^*(-\Gamma \circ A) - G^*(\Gamma)\}$$

For (DSV_w^A) we consider weakly efficient solutions and weakly maximal pairs in accordance to the general definition in subsection 7.4.2. The above approach verifies weak duality and this can be expressed via Corollary 7.4.17 in the form $A(\mathrm{WInf}(PSV_w^A)) \cap \mathrm{WSup}(DSV_w^A) = \emptyset$. The next result is concerning strong duality and follows by means of Theorem 7.4.21.

Theorem 7.5.11. *(a) Suppose that the problem (PSV_w^A) is stable with respect to the perturbation map Φ^A. Let $\bar{x} \in X$ be a weakly efficient solution to (PSV_w^A) and $\bar{v} \in F(\bar{x}) + G(A\bar{x})$ such that $(\bar{x}, \bar{v})$ is a weakly minimal pair to (PSV_w^A). Then there exists $\overline{\Gamma} \in \mathcal{L}(Y, V)$, a weakly efficient solution to (DSV_w^A), with $(\overline{\Gamma}, \bar{v})$ corresponding weakly maximal pair such that strong duality holds and*

$$\bar{v} \in \mathrm{WInf}\{-F^*(-\overline{\Gamma} \circ A) - G^*(\overline{\Gamma})\}.$$

(b) Assume that for $\bar{x} \in X$, $\bar{v} \in F(\bar{x}) + G(A\bar{x})$ and $\overline{\Gamma} \in \mathcal{L}(Y, V)$ one has $\bar{v} \in \mathrm{WInf}[-F^*(-\overline{\Gamma} \circ A) - G^*(\overline{\Gamma})]$. Then $\bar{x}$ is a weakly efficient solution and $(\bar{x}, \bar{v})$ a weakly minimal pair to (PSV_w^A), while $\overline{\Gamma}$ is a weakly efficient solution and $(\overline{\Gamma}, \bar{v})$ is a weakly maximal pair to (DSV_w^A).

The next result supplies a stability and strong duality statement, respectively, based on convexity and regularity assumptions for the functions occurring in the primal problem.

Theorem 7.5.12. *Let the set-valued maps F and G be K-convex. If there exists $x' \in \operatorname{dom} F \cap A^{-1}(\operatorname{dom} G)$ such that G is weakly K-upper bounded on some neighborhood of Ax', then (PSV_w^A) is stable with respect to the perturbation map Φ^A; hence* $\mathrm{WInf}(PSV_w^A) = \mathrm{WSup}(DSV_w^A) = \mathrm{WMax}(DSV_w^A)$.

Proof. The convexity assumptions guarantee that Φ^A is K-convex on $X \times Y$. Like in the proof of Theorem 7.3.3, one can show that $\Psi^A : Y \rightrightarrows V \cup \{+\infty_K\}, \Psi^A(y) = \Phi^A(X, y)$ is weakly K-upper bounded on a neighborhood of 0 in Y and that $0 \in \operatorname{int}(\operatorname{dom} \Psi^A)$. Thus Theorem 7.4.27 implies the stability, while Theorem 7.4.20 ensures the strong duality. □

Remark 7.5.3. In case $X = Y$ and $A = \mathrm{id}_X$ Theorem 7.5.12 becomes [170, Theorem 6.7].

Of particular interest is the study of the duality for the set-valued optimization problem

$$(PSV_w^S) \quad \underset{x \in S}{\mathrm{WInf}}\, F(x),$$

where $S \subseteq X$ is a nonempty set and $F : X \rightrightarrows V \cup \{+\infty\}$ a set-valued map such that $\operatorname{dom} F \cap S \neq \emptyset$. We reformulate this problem by making use of the vector-valued indicator function δ_S^V as being

$$(PSV_w^S) \quad \underset{x \in X}{\mathrm{WInf}}\{F(x) + \delta_S^V(x)\}.$$

Obviously, (PSV_w^S) is a particular case of (PSV_w^A), namely for $X = Y$, $A = \mathrm{id}_X$ and $G = \delta_S^V$. Via (DSV_w^A) we get the following Fenchel set-valued dual problem to (PSV_w^S)

$$(DSV_w^S) \quad \mathrm{WSup} \bigcup_{\Gamma \in \mathcal{L}(X,V)} \mathrm{WInf}\{-F^*(\Gamma) + \Gamma(S)\}.$$

A general assertion on strong duality can be given by particularizing Theorem 7.5.11, while via Theorem 7.5.12 one can give some verifiable conditions to this purpose. It is interesting to notice that one can give, in particular, two different conditions ensuring stability and therefore strong duality depending on whether we perturb F or δ_S^V. In other words one can consider the perturbation maps

$$\Phi_1^S : X \times X \rightrightarrows V \cup \{+\infty_K\},\ \Phi_1^S(x,y) = F(x+y) + \delta_S^V(x)$$

and

$$\Phi_2^S : X \times X \rightrightarrows V \cup \{+\infty_K\},\ \Phi_2(x,y) = F(x) + \delta_S^V(x+y),$$

both revealing the same dual problem (DSV_w^S). Employing Φ_1^S, we obtain the following strong duality result, which is a direct consequence of Theorem 7.5.12.

Theorem 7.5.13. *Let the set $S \subseteq X$ be convex and F a K-convex set-valued map. If there exists $x' \in \operatorname{dom} F \cap S$ such that F is weakly K-upper bounded on some neighborhood of x', then (PSV_w^S) is stable with respect to the perturbation map Φ_1^S; hence* $\mathrm{WInf}(PSV_w^S) = \mathrm{WSup}(DSV_w^S) = \mathrm{WMax}(DSV_w^S)$.

By using this time the perturbation map Φ_2^S, one can give another strong duality statement for (PSV_w^S) and (DSV_w^S).

Theorem 7.5.14. *Let the set $S \subseteq X$ be convex and F a K-convex set-valued map. If $\operatorname{dom} F \cap \operatorname{int}(S) \neq \emptyset$, then (PSV_w^S) is stable with respect to the perturbation map Φ_1^S; hence* $\mathrm{WInf}(PSV_w^S) = \mathrm{WSup}(DSV_w^S) = \mathrm{WMax}(DSV_w^S)$.

Proof. Since S is convex, δ_S^V is a K-convex map. For an $x' \in \operatorname{dom} F \cap \operatorname{int}(S)$, there exists U, a neighborhood of x' in X such that $U \subseteq S$. As $\delta_S^V(x) \cap (-K) \neq \emptyset$ for all $x \in S$, it follows that δ_S^V is weakly K-upper bounded on U. Thus the hypotheses of Theorem 7.5.12 are also in this case fulfilled and this provides the conclusion. □

7.5.3 Set-valued gap maps for set-valued equilibrium problems

In this subsection we deal with gap maps for set-valued equilibrium problems. As in subsection 7.3.2, where we have dealt with gap maps for set-valued variational inequalities, such maps give characterizations of the solutions of the equilibrium problem. Vector equilibrium problems are generalizations of scalar equilibrium problems (cf. [16]). Equilibrium problems include optimization, Nash equilibria, complementarity, fix point, saddle point and variational problems. They have practical applications in many fields like game theory, economics and mathematical physics.

Let us first recall the formulation of a scalar equilibrium problem in finite dimensional spaces. Given a a nonempty set $\mathcal{A} \subseteq \mathbb{R}^n$, assume that $f : \mathcal{A}\times\mathcal{A} \to \mathbb{R}\cup\{+\infty\}$ is a *bifunction* satisfying $f(x,x) = 0$ for all $x \in \mathcal{A}$. The equilibrium problem consists in finding an $x \in \mathcal{A}$ such that

$$(EP) \quad f(x,y) \geq 0 \quad \text{for all } y \in \mathcal{A}.$$

A function $\gamma : \mathcal{A} \to \mathbb{R}\cup\{+\infty\}$ is said to be a *gap function* for (EP) (cf. [129]) if it satisfies the following properties

(a) $\gamma(y) \geq 0$ for all $y \in \mathcal{A}$;
(b) $\gamma(x) = 0$ if and only if x is a solution to (EP).

Next we consider the extension of the scalar equilibrium problem (EP) to the weak set-valued equilibrium problem.

Let X and V be Hausdorff topological vector spaces, V being partially ordered by the nontrivial pointed convex cone $K \subseteq V$ with $\operatorname{int}(K) \neq \emptyset$, and $F : X \times X \rightrightarrows V \cup \{+\infty_K\}$ a set-valued map. Further, let $\mathcal{A} \subseteq X$ be a nonempty set with the property that $0 \in F(x,x)$ for all $x \in \mathcal{A}$. The weak set-valued equilibrium problem consists in finding an element $x \in \mathcal{A}$ such that

$$(SVEP_w) \quad 0 \notin F(x,y) + \operatorname{int}(K) \text{ for all } y \in \mathcal{A},$$

which means that for all $y \in \mathcal{A}$ there is no $v \in F(x,y)$ with $v <_K 0$.

Definition 7.5.1. *A set-valued map $\gamma : \mathcal{A} \rightrightarrows V \cup \{+\infty_K\}$ is said to be a gap map for $(SVEP_w)$ if it satisfies the following conditions*

(a) $0 \notin \gamma(y) + \operatorname{int}(K)$ for all $y \in \mathcal{A}$;
(b) $0 \in \gamma(x)$ if and only if x solves $(SVEP_w)$.

We observe that $x \in \mathcal{A}$ is a solution to $(SVEP_w)$ if and only if x is a weakly efficient solution and $(x,0)$ is a weakly minimal pair to the set-valued optimization problem

$$(PSV_w^{VEP}; x) \quad \operatorname*{WInf}_{y\in\mathcal{A}} F(x,y),$$

which can be rewritten as

$$(PSV_w^{VEP}; x) \quad \operatorname*{WInf}_{y\in X} \{F(x,y) + \delta_{\mathcal{A}}^V(y)\}.$$

Its Fenchel set-valued dual problem looks like

$$(DSV_w^{VEP}; x) \quad \operatorname{WSup} \bigcup_{\Gamma\in\mathcal{L}(X,V)} \operatorname{WInf}\{-F_x^*(\Gamma) + \Gamma(\mathcal{A})\},$$

with $F_x^*(\Gamma) = \text{WSup}_{y \in X}\{\Gamma y - F(x,y)\}$ being the conjugate map of the set-valued map $F_x : X \rightrightarrows V \cup \{+\infty_K\}$, defined by $F_x(y) = F(x,y)$. Let $x \in \mathcal{A}$ be fixed. We consider the following regularity conditions ensuring stability for $(PSV_w^{VEP}; x)$, the first one being based on Theorem 7.5.13

$$(RC_1^{VEP}; x) \left| \begin{array}{l} \exists y' \in \mathcal{A} \text{ such that } F_x \text{ is weakly } K\text{-upper bounded} \\ \text{on some neighborhood of } y' \text{ in } X. \end{array} \right.$$

while the other one on Theorem 7.5.14,

$$(RC_2^{VEP}; x) \mid \text{dom}\, F_x \cap \text{int}(\mathcal{A}) \neq \emptyset.$$

Proposition 7.5.15. *Let $\mathcal{A} \subseteq X$ be a convex set and F_x a K-convex set-valued map. If $(RC_1^{VEP}; x)$ or $(RC_2^{VEP}; x)$ is fulfilled, then* $\text{WInf}(PSV_w^{VEP}; x) = \text{WSup}(DSV_w^{VEP}; x) = \text{WMax}(DSV_w^{VEP}; x)$.

Now we are ready to define a gap map for $(SVEP_w)$ by means of the objective map of the Fenchel set-valued dual $(DV_w^{VEP}; x)$, as being $\gamma^{VEP} : \mathcal{A} \rightrightarrows V \cup \{+\infty_K\}$,

$$\gamma^{VEP}(x) = \bigcup_{\Gamma \in \mathcal{L}(X,V)} \text{WSup}[F_x^*(\Gamma) - \Gamma(\mathcal{A})].$$

Theorem 7.5.16. *Let $\mathcal{A} \subseteq X$ be a convex set and F_x a K-convex set-valued map for all $x \in \mathcal{A}$. If for all $x \in \mathcal{A}$ $(RC_1^{VEP}; x)$ or $(RC_2^{VEP}; x)$ is fulfilled, then γ^{VEP} is a gap map for the weak set-valued equilibrium problem $(SVEP_w)$.*

Proof. We have to show that γ^{VEP} satisfies the properties (a) and (b) of Definition 7.5.1.

Chose arbitrarily $y \in \mathcal{A}$ and $v^* \in \gamma^{VEP}(y)$. Then there exists $\Gamma \in \mathcal{L}(X,V)$ such that $v^* \in \text{WSup}[F_y^*(\Gamma) - \Gamma(\mathcal{A})]$. By the weak duality for $(PSV_w^{VEP}; y)$ and $(DSV_w^{VEP}; y)$ follows that $-v^* \notin A(F_y(y) + \delta_{\mathcal{A}}^V(y))$. Since $0 \in F_y(y) + \delta_{\mathcal{A}}^V(y)$, one has $0 \not<_K -v^*$ or, equivalently, $v^* \not<_K 0$.

We come now to (b) and consider an arbitrary $x \in \mathcal{A}$. If $x \in \mathcal{A}$ is a solution to $(SVEP_w)$, then x is a weakly efficient solution and $(x, 0)$ is a weakly minimal pair to $(PV_w^{VEP}; x)$. Thus $0 \in \text{WMax}(DV_w^{VEP}; x)$ and so there exists $\Gamma \in \mathcal{L}(X,V)$ with $0 \in \text{WInf}[-F_x^*(\Gamma) + \Gamma(\mathcal{A})]$. This implies that $0 \in \gamma^{VEP}(x)$. Conversely, let be $x \in \mathcal{A}$ such that $0 \in \gamma^{VEP}(x)$. Then there exists $\Gamma \in \mathcal{L}(X,V)$ fulfilling $0 \in \text{WSup}[F_x^*(\Gamma) - \Gamma(\mathcal{A})]$. Further, there is $0 \in F(x,x) + \delta_{\mathcal{A}}^V(x)$. Thus, by Theorem 7.5.11 follows that x is a weakly efficient solution and $(x, 0)$ is a weakly minimal pair to $(PV_w^{VEP}; x)$. Consequently, x turns out to be a solution to $(SVEP_w)$ and the condition (b) in Definition 7.5.1 is verified. □

We close the section by investigating a particular case of $(SVEP_w)$, namely the classical vector equilibrium problem. Let $\mathcal{A} \subseteq X$ be a nonempty set and $f : X \to V \cup \{+\infty_K\}$ a vector-valued function. For $F : X \times X \to V \cup \{+\infty_K\}$ defined by $F(x,y) = f(y) - f(x)$ the weak set-valued equilibrium problem $(SVEP_w)$ means to find an element $x \in \mathcal{A}$ such that

$$(VEP_w) \quad f(y) \not<_K f(x) \text{ for all } y \in \mathcal{A}.$$

One can see that $x \in \mathcal{A}$ is a solution to (VEP_w) if and only if x is a weakly efficient solution to

$$(PV_w^{\mathcal{A}}) \quad \underset{x\in\mathcal{A}}{\mathrm{WInf}}\, f(x).$$

Let $x \in \mathcal{A}$ be fixed. Then the corresponding optimization problem to (VEP_w) reads as

$$(PV_w^{VEP}; x) \quad \underset{y\in X}{\mathrm{WInf}}\{f(y) - f(x) + \delta_{\mathcal{A}}^V(y)\}.$$

Its Fenchel set-valued dual problem is

$$(DV_w^{VEP}; x) \quad \mathrm{WSup} \bigcup_{\Gamma\in\mathcal{L}(X,V)} \mathrm{WInf}\{-F_x^*(\Gamma) + \Gamma(\mathcal{A})\},$$

with

$$\begin{aligned} F_x^*(\Gamma) &= \underset{y\in X}{\mathrm{WSup}}\{\Gamma y - (f(y) - f(x))\} \\ &= f(x) + \underset{y\in X}{\mathrm{WSup}}\{\Gamma y - f(y)\} = f(x) + f^*(\Gamma). \end{aligned}$$

This leads to the following formulation of the dual

$$(DV_w^{VEP}; x) \quad \mathrm{WSup} \bigcup_{\Gamma\in\mathcal{L}(X,V)} \mathrm{WInf}\{-f^*(\Gamma) + \Gamma(\mathcal{A})\} - f(x).$$

Thus the gap map $\gamma^{VEP} : \mathcal{A} \rightrightarrows V \cup \{+\infty_K\}$ turns out to be

$$\gamma^{VEP}(x) = f(x) + \bigcup_{\Gamma\in\mathcal{L}(X,V)} \mathrm{WSup}[f^*(\Gamma) - \Gamma(\mathcal{A})].$$

It is easy to see that the property $0 \in \gamma^{VEP}(y) + \mathrm{int}(K)$ for all $y \in \mathcal{A}$ is equivalent to the weak duality for the vector optimization problem $(PV_w^{\mathcal{A}})$ and its Fenchel set-valued dual

$$(DV_w^{\mathcal{A}}) \quad \mathrm{WSup} \bigcup_{\Gamma\in\mathcal{L}(X,V)} \mathrm{WInf}\{-f^*(\Gamma) + \Gamma(\mathcal{A})\}.$$

On the other hand, the relation $0 \in \gamma^{VEP}(x)$, for $x \in \mathcal{A}$, expresses the strong duality for the primal-dual pair $(PV_w^{\mathcal{A}}) - (DV_w^{\mathcal{A}})$.

Bibliographical notes

Different to the vector duality concepts presented in the other chapters of this monograph that are mainly based on scalarization and scalar conjugacy, in the present final chapter we have developed a duality theory for set-valued optimization problems based on vector conjugation. We present this approach separately for both efficient and weakly efficient solutions because despite similar assertions and results, the basic notions and investigations differ in several aspects. The part dealing with efficiency and minimality has its roots in the paper [180] of Tanino and Sawaragi, where the notions of conjugate map and subdifferential for vector-valued functions taking values in a finite dimensional space $\mathbb{R}^k$ with $\mathbb{R}^k_+$ as ordering cone have been introduced. In their book [163] Sawaragi, Nakayama and Tanino extended the approach from [180] by working in a setting where the dual variables were represented via matrices. On that background some first investigations have been devoted to conjugate vector duality for finite dimensional multiobjective problems, in particular a vector Lagrange dual problem has been displayed.

Some aspects of conjugate maps for Hausdorff topological vector spaces and vector duality can be found in Luc's book [125] as well as some investigations on Lagrange duality and a kind of axiomatic duality. In the paper [1] Altangerel, Boţ and Wanka extended and applied Tanino's and Sawaragi's approach by introducing a Fenchel-Lagrange set-valued dual problem besides the Lagrange and Fenchel set-valued dual problems for constrained vector optimization problems in finite dimensional spaces. These results have been applied for the construction of gap functions for vector variational inequalities.

In this chapter a comprehensive and detailed theory for set-valued conjugate duality in infinite dimensional spaces is developed, covering, in particular, the results contained in the before mentioned works.

The second approach concerning vector conjugation is related to the notions of infimal and supremal sets of a given set based on the weak ordering introduced by a cone with nonempty interior. The basic notions associated with this infimum and supremum concept can be found in the paper [177] due to Tanino. This concept is similar to the one given earlier by Kawasaki (cf. [114]) in finite dimensional spaces. In [178] Tanino has used it for deducing a conjugate duality theory for vector optimization problems in topological vector spaces. We notice that several kinds of supremum and infimum definitions in multidimensional spaces have been provided in the preceding time mostly accompanied by considerations regarding conjugate vector as set-valued duality in different settings (see [40, 82, 94, 113, 146, 152–154, 210]).

We mention, in particular, Song's contributions [168–170], where the approach from Tanino [178] has been extended to Hausdorff topological vector spaces and set-valued maps, since our investigations are carried out in that framework (see also [1, 3, 4, 25]). Some recent results on this topic can be found in [47, 123, 124]. The reader is referred also to a some different conjugation concept for set-valued optimization problems regarding duality, as developed in [9, 10].

References

1. Altangerel L, Boţ RI, Wanka G (2007) Conjugate duality in vector optimization and some applications to the vector variational inequality. Journal of Mathematical Analysis and Applications 329(2):1010–1035
2. Altangerel L, Boţ RI, Wanka G (2007) On the construction of gap functions for variational inequalities via conjugate duality. Asia-Pacific Journal of Operational Research 24(3):353–371
3. Altangerel L, Boţ RI, Wanka G (2007) Variational principles for vector equilibrium problems related to conjugate duality. Journal of Nonlinear and Convex Analysis 8(2):179–196
4. Altangerel L, Wanka G (2007) Gap functions for vector equilibrium problems via conjugate duality. Preprint 26. Fakultät für Mathematik, Technische Universität Chemnitz
5. Antczak T (2006) A modified objective function method for solving nonlinear multiobjective fractional programming problems. Journal of Mathematical Analysis and Applications 322(2):971–989
6. Antczak T (2005) Saddle point criteria and duality in multiobjective programming via an η-approximation method. The ANZIAM Journal 47(2):155–172
7. Attouch H, Brézis H (1986) Duality for the sum of convex functions in general Banach spaces. In: Barroso JA (ed) Aspects of mathematics and its applications. North-Holland Publishing Company, Amsterdam, 125–133
8. Auslender A (1976) Optimisation: méthods numérique. Masson, Paris
9. Azimov AY (2008) Duality for set-valued multiobjective optimization problems I: Mathematical programming. Journal of Optimization Theory and Applications 137(1):61–74
10. Azimov AY (2008) Duality for set-valued multiobjective optimization problems II: Optimal control. Journal of Optimization Theory and Applications 137(1):75–88
11. Bector CR, Bector MK, Klassen JE (1977) Duality for a nonlinear programming problem. Utilitas Mathematica 11:87–99
12. Bector CR, Chandra S, Singh C (1990) Duality in multiobjective fractional programming. In: Cambini A, Castagnoli E, Martein L, Mazzoleni P, Schaible S (eds) Generalized convexity and fractional programming with economic applications. Proceedings of the International Workshop on Generalized Concavity, Fractional Programming and Economic Applications held at the University of

Pisa, Pisa, May 30–June 1, 1988. Lecture Notes in Economics and Mathematical Systems 345. Springer-Verlag, Berlin, 232–241
13. Bector CR (1973) Duality in nonlinear fractional programming. Zeitschrift für Operations Research 17(5):A183–A193
14. Ben-Israel A, Ben-Tal A, Zlobec S (1981) Optimality in nonlinear programming: a feasible directions approach. John Wiley & Sons, New York
15. Benson HP (1979) An improved definition of proper efficiency for vector maximization with respect to cones. Journal of Mathematical Analysis and Applications 71(1):232–241
16. Blum E, Oettli W (1994) From optimization and variational inequalities to equilibrium problems. The Mathematics Student 63(1–4):123–145
17. Borwein JM (1977) Proper efficient points for maximizations with respect to cones. SIAM Journal on Control and Optimization 15(1):57–63
18. Borwein JM (1980) The geometry of Pareto efficiency over cones. Mathematische Operationsforschung und Statistik Series Optimization 11(2):235–248
19. Borwein JM (1983) On the existence of Pareto efficient points. Mathematics of Operations Research 8(1):64–73
20. Borwein JM, Goebel R (2003) Notions of relative interior in Banach spaces. Journal of Mathematical Sciences (New York) 115(4):2542–2553
21. Borwein JM, Lewis AS (1992) Partially finite convex programming. I. Quasi relative interiors and duality theory. Mathematical Programming 57(1):15–48
22. Borwein JM, Lewis AS (2006) Convex analysis and nonlinear optimization: Theory and examples, Second edition. CMS Books in Mathematics/Ouvrages de Mathématiques de la SMC 3. Springer-Verlag, New York
23. Boţ RI (1999) The Fenchel duality in set-valued optimization. Master Thesis. Facultatea de Matematică şi Informatică, Universitatea "Babeş-Bolyai" Cluj-Napoca
24. Boţ RI (2003) Duality and optimality in multiobjective optimization. Ph.D. Thesis. Fakultät für Mathematik, Technische Universität Chemnitz
25. Boţ RI (2008) Conjugate duality in convex optimization. Habilitation Thesis. Fakultät für Mathematik, Technische Universität Chemnitz
26. Boţ RI, Csetnek ER, Moldovan A (2008) Revisiting some duality theorems via the quasirelative interior in convex optimization. Journal of Optimization Theory and Applications 139(1):67–84
27. Boţ RI, Csetnek ER, Wanka G (2008) Regularity conditions via quasi-relative interior in convex programming. SIAM Journal on Optimization 19 (1):217–233
28. Boţ RI, Dumitru A, Wanka G (2009) A new Fenchel dual problem in vector optimization. Proceedings of the Indian Academy of Sciences - Mathematical Sciences 119(2):251–265
29. Boţ RI, Grad S-M, Wanka G (2007) A general approach for studying duality in multiobjective optimization. Mathematical Methods of Operations Research (ZOR) 65(3):417–444
30. Boţ RI, Grad S-M, Wanka G (2007) A new constraint qualification and conjugate duality for composed convex optimization problems. Journal of Optimization Theory and Applications 135(2):241–255
31. Boţ RI, Grad S-M, Wanka G (2008) A new constraint qualification for the formula of the subdifferential of composed convex functions in infinite dimensional spaces. Mathematische Nachrichten 281(8):1088–1107

32. Boţ RI, Grad S-M, Wanka G (2009) New regularity conditions for Lagrange and Fenchel-Lagrange duality in infinite dimensional spaces. Mathematical Inequalities & Applications 12(1):171–189
33. Boţ RI, Grad S-M, Wanka G (2009) Generalized Moreau-Rockafellar results for composed convex functions. Optimization 58(6–7)
34. Boţ RI, Hodrea IB, Wanka G (2008) ε-optimality conditions for composed convex optimization problems. Journal of Approximation Theory 153(1):108–121
35. Boţ RI, Kassay G, Wanka G (2005) Strong duality for generalized convex optimization problems. Journal of Optimization Theory and Applications 127(1):45-70
36. Boţ RI, Wanka G (2004): An analysis of some dual problems in multiobjective optimization (I). Optimization 53(3):281–300
37. Boţ RI, Wanka G (2004): An analysis of some dual problems in multiobjective optimization (II). Optimization 53(3):301–324
38. Boţ RI, Wanka G (2006) A weaker regularity condition for subdifferential calculus and Fenchel duality in infinite dimensional spaces. Nonlinear Analysis Theory & Applications 64(12):2787–2804
39. Boţ RI, Wanka G (2006) An alternative formulation for a new closed cone constraint qualification. Nonlinear Analysis Theory & Applications 64(6):1367–1381
40. Brumelle S (1981) Duality for multiple objective convex programs. Mathematics of Operations Research 6(2):159–172
41. Burachik RS, Jeyakumar V (2005) A dual condition for the convex subdifferential sum formula with applications. Journal of Convex Analysis 12(2):279–290
42. Breckner W, Kolumbán I (1968) Dualität bei Optimierungsaugaben in Topologischen Vektorräumen. Mathematica 10(33):229–244
43. Breckner W, Kolumbán I (1969) Konvexe Optimierungsaufgaben in Topologischen Vektorräumen. Mathematica Scandinavica 25:227–247
44. Cambini R, Carosi L (2005) Duality in multiobjective optimization problems with set constraints. In: Eberhard A, Hadjisavvas N, Luc DT (eds) Generalized convexity, generalized monotonicity and applications. Proceedings of the 7th International Symposium on Generalized Convexity and Generalized Monotonicity held in Hanoi, August 27–31, 2002. Nonconvex Optimization and its Applications 77. Springer-Verlag, New York, 131–146
45. Carrizosa E, Fliege J (2002) Generalized goal programming: polynomial methods and applications. Mathematical Programming 93(2):281–303
46. Chandra S, Craven BD, Mond B (1986) Generalized fractional programming duality: a ratio game approach. Journal of the Australian Mathematical Society Series B 28(2):170–180
47. Chen CR, Li SJ (2009) Different conjugate dual problems in vector optimization and their relations. Journal of Optimization Theory and Applications 140(3):443–461
48. Chen GY, Goh CJ, Yang XQ (2000) On gap functions for vector variational inequalities In: Giannessi F (ed) Vector variational inequalities and vector equilibria. Mathematical theories. Kluwer Academic Publishers, Dordrecht, 55–72
49. Chen G, Huang X, Yang X (2005) Vector optimization. Set-valued and variational analysis. Lecture Notes in Economics and Mathematical Systems 541. Springer-Verlag, Berlin

50. Combari C, Laghdir M, Thibault L (1994) Sous-différentiels de fonctions convexes composées. Annales des Sciences Mathématiques du Québec 18(2):119–148
51. Combari C, Laghdir M, Thibault L (1999) On subdifferential calculus for convex functions defined on locally convex spaces. Annales des Sciences Mathématiques du Québec 23(1):23–36
52. Corley HW (1981) Duality theory for maximizations with respect to cones. Journal of Mathematical Analysis and Applications 84(2):560–568
53. Craven BD (1989) A modified Wolfe dual for weak vector minimization. Numerical Functional Analysis and Optimization 10(9–10):899–907
54. Craven BD (1977) Lagrangean conditions and quasiduality. Bulletin of the Australian Mathematical Society 16(3):325–339
55. Craven BD (1980) Strong vector minimization and duality. Zeitschrift für Angewandte Mathematik und Mechanik 60(1):1–5
56. Craven BD (1976) Sufficient Fritz-John optimality conditions. Bulletin of the Australian Mathematical Society 19(4):462–468
57. Craven BD, Mond B (1977) Sufficient Fritz John optimality conditions for nondifferentiable convex programming. Journal of the Australian Mathematical Society Series B 19(4):462–468
58. Deutsch F (2001) Best approximation in inner product spaces. CMS Books in Mathematics/Ouvrages de Mathématiques de la SMC 7. Springer-Verlag, New York
59. Dinkelbach W (1967) On nonlinear fractional programming. Management Science Series A 13:492–498
60. Edgeworth FY (1881) Mathematical psychics. Kegan Paul, London
61. Egudo RR (1989) Efficiency and generalized convex duality for multiobjective programs. Journal of Mathematical Analysis and Applications 138(1):84–94
62. Egudo RR (1987) Proper efficiency and multiobjective duality in nonlinear programming. Journal of Information & Optimization Sciences 8(2):155–166
63. Egudo RR, Weir T, Mond B (1992) Duality without constraint qualification for multiobjective programming. Journal of the Australian Mathematical Society Series B 33(4):531–544
64. Ehrgott M (2000) Multicriteria optimization. Lecture Notes in Economics and Mathematical Systems 491. Springer-Verlag, Berlin
65. Ehrgott M, Gandibleux X (2002) Multiple criteria optimization: state of the art annotated bibliographic surveys. International Series in Operations Research & Management Science 52. Kluwer Academic Publishers, Boston
66. Eichfelder G (2008) Adaptive scalarization methods in multiobjective optimization. Springer-Verlag, Berlin
67. Ekeland I, Temam R (1976) Convex analysis and variational problems. North-Holland Publishing Company, Amsterdam
68. Fenchel W (1949) On conjugate convex functions. Canadian Journal of Mathematics 1:73–77
69. Fuchssteiner B, Lusky W (1981) Convex cones. North-Holland, Amsterdam
70. Gale D, Kuhn HW, Tucker AW (1951) Linear programming and the theory of games. In: Koopmans TC (ed) Activity analysis of production and allocation. John Wiley & Sons, New York, 317–329
71. Geoffrion AM (1968) Proper efficiency and the theory of vector maximization. Journal of Mathematical Analysis and Applications 22(3):618–630

72. Gerstewitz C (1983) Nichtkonvexe Dualität in der Vektoroptimierung. Wissenschaftliche Zeitschrift. Technische Hochschule "Carl Schorlemmer" Leuna-Merseburg 25(3):357–364
73. Gerstewitz C, Göpfert A (1981) Zur Dualität in der Vektoroptimierung. Seminarbericht. Sektion Mathematik. Humboldt-Universität Berlin 39:67–84
74. Gerstewitz C, Iwanow E (1985) Dualität für nichtkonvexe Vektoroptimierungsprobleme. Wissenschaftliche Zeitschrift der Technischen Hochschule Ilmenau 31(2):61–81
75. Gerth C, Weidner P (1990) Nonconvex separation theorems and some applications in vector optimization. Journal of Optimization Theory and Applications 67(2):297–320
76. Giannessi F (1980) Theorems of alternative, quadratic programs and complementarity problems. In: Cottle RW, Giannessi F, Lions JL (eds) Variational inequalities and complementary problems. Theory and applications. Proceedings of an International School of Mathematics on Variational Inequalities and Complementarity Problems in Mathematical Physics and Economics held in Erice, June 19–30, 1978. John Wiley & Sons, Chichester, 151–168
77. Goh CJ, Yang XQ (2002) Duality in optimization and variational inequalities. Optimization Theory and Applications 2. Taylor & Francis, London
78. Göpfert A (1986) Multicriterial duality, examples and advances. In: Fandel G, Grauer M, Kurzhanski A, Wierzbicki AP (eds) Large-scale modelling and interactive decision analysis. Proceedings of the international workshop held in Eisenach, November 18–21, 1985. Lecture Notes in Economics and Mathematical Systems 273. Springer-Verlag, Berlin, 52–58
79. Göpfert A, Gerth C (1986) Über die Skalarisierung und Dualisierung von Vektoroptimierungsproblemen. Zeitschrift für Analysis und ihre Anwendungen 5(4):377–384
80. Göpfert A, Nehse R (1990) Vektoroptimierung: Theorie, Verfahren und Anwendungen. BG Teubner Verlagsgesellschaft, Leipzig
81. Göpfert A, Riahi H, Tammer C, Zălinescu C (2003) Variational methods in partially ordered spaces. Springer-Verlag, New York
82. Gros C (1978) Generalization of Fenchel's duality theory for convex vector optimization. European Journal of Operational Research 2(5):368–376
83. Guerraggio A, Molho E, Zaffaroni A (1994) On the notion of proper efficiency in vector optimization. Journal of Optimization Theory and Applications 82(1):1–21
84. Hamel AH, Heyde F, Löhne A, Tammer C, Winkler K (2004) Closing the duality gap in linear vector optimization. Journal of Convex Analysis 11(1):163–178
85. Hanson M (1981) On sufficiency of the Kuhn-Tucker conditions. Journal of Mathematical Analysis and Applications 80(2):545–550
86. Harker PT, Pang JS (1990) Finite-dimensional variational inequality and nonlinear complementarity problems: a survey of theory, algorithms and applications. Mathematical Programming 48(2):161-220
87. Hartley R (1978) On cone-efficiency, cone-convexity and cone-compactness. SIAM Journal on Applied Mathematics 34(2):211–222
88. Henig MI (1982) Proper efficiency with respect to cones. Journal of Optimization Theory and Applications 36(3):387–407
89. Hirriart-Urruty J-B (1982) ε-subdifferential calculus. In: Aubin JP, Vinter RB (eds) Convex analysis and optimization. Research Notes in Mathematics 57. Pitman, 43–92

90. Hiriart-Urruty J-B, Lemaréchal C (1993) Convex analysis and minimization algorithms I: Fundamentals. Springer-Verlag, Berlin
91. Hiriart-Urruty J-B, Lemaréchal C (1993) Convex analysis and minimization algorithms II: Advanced theory and bundle methods. Springer-Verlag, Berlin
92. Hiriart-Urruty J-B, Lemaréchal C (2001) Fundamentals of convex analysis. Springer-Verlag, Berlin
93. Hurwicz L (1958) Programming in linear spaces. In: Arrow KJ, Hurwicz L, Uzawa H (eds) Studies in linear and non-linear programming. Stanford University Press, Stanford, 38–102
94. Isac G, Postolică V (1993) The best approximation and optimization in locally convex spaces. Verlag Peter Lang, Frankfurt am Main
95. Isermann H (1974) Lineare Vektoroptimierung. Ph.D. Thesis. Rechts- und Wissenschaftliche Fakultät, Universität Regensburg
96. Isermann H (1978) Duality in multiple objective linear programming. In: Zionts S (ed) Multiple criteria problem solving. Lecture Notes in Economics and Mathematical Systems 155. Springer-Verlag, Berlin, 274–285
97. Isermann H (1978) On some relations between a dual pair of multiple objective linear programs. Zeitschrift für Operations Research 22(1):A33–A41
98. Isermann H (1974) Proper efficiency and the linear vector maximum problem. Operations Research 22(1):189–191
99. Islam MA (1994) Sufficiency and duality in nondifferentiable multiobjective programming. Pure and Applied Mathematika Sciences 39(1-2):31–39
100. Jagannathan (1973) Duality for nonlinear fractional programs. Zeitschrift für Operations Research 17(1):A1–A3
101. Jahn J (1983) Duality in vector optimization. Mathematical Programming 25(3):343–353
102. Jahn J (1984) Scalarization in vector optimization. Mathematical Programming 29(2):203–218
103. Jahn J (1986) Mathematical vector optimization in partially ordered linear spaces. Verlag Peter Lang, Frankfurt am Main
104. Jahn J (2004) Vector optimization - theory, applications, and extensions. Springer-Verlag, Berlin
105. Jameson G (1970) Ordered linear spaces. Lecture Notes in Mathematics 141. Springer-Verlag, Berlin
106. Jeyakumar V, Song W, Dinh N, Lee GM (2005) Stable strong duality in convex optimization. Applied Mathematics Report AMR 05/22. University of New South Wales
107. Kaliszewski I (1986) Norm scalarization and proper efficiency in vector optimization. Foundations of Control Engineering 11(3):117–131
108. Kaliszewski I (1994) Quantitative Pareto analysis by cone separation technique. Kluwer Academic Publishers, Boston
109. Kanniappan P (1984) Duality theorems for convex programming without constraint qualification. Journal of the Australian Mathematical Society Series A 36(2):252–266
110. Kantorovitch L (1937) Lineare halbgeordnete Räume. Recueil Mathématique 44:121-168
111. Kassem MAEH (2006) Symmetric and self duality in vector optimization problem. Applied Mathematics and Computation 183(2):1121–1126
112. Kaul RN, Lyall V (1989) A note on nonlinear fractional vector maximization. Opsearch 26(2):108–121

113. Kawasaki H (1982) A duality theorem in multiobjective nonlinear programming. Mathematics of Operations Research 7(1):95–110
114. Kawasaki H (1981) Conjugate relations and weak subdifferentials. Mathematics of Operations Research 6(4):593–607
115. Khurana S (2005) Symmetric duality in multiobjective programming involving generalized cone-invex functions. European Journal of Operational Research 165(3):592–597
116. Kim DS, Lee GM, Lee BS (1987) On multiobjective symmetric dual programs. Journal of Information & Optimization Sciences 18(1):113–120
117. Kinderlehrer D, Stampacchia G (1980) An introduction to variational inequalities and their applications. Academic Press, New York
118. Kornbluth JSH (1974) Duality, indifference and sensitivity analysis in multiple objective linear programming. Operational Research Quarterly 25:599–614
119. Kuhn HW, Tucker AW (1950) Nonlinear Programming. In: Neyman J (ed) Proceedings of the Second Berkeley Symposium on Mathematical Statistics and Probability. University of California Press, Berkeley and Los Angeles, 481–492
120. Kutateladze SS (1977) Changes of variables in the Young transformation. Soviet Mathematics Doklady 18(2):545–548
121. Lai HC, Ho CP (1986) Duality theorem of nondifferentiable convex multiobjective programming. Journal of Optimization Theory and Applications 50(3):407–420
122. Lampe U (1981) Dualität und eigentliche Effizienz in der Vektoroptimierung. Seminarberichte der Sektion Mathematik der Humboldt-Universität zu Berlin 37:45–54
123. Li SJ, Chen CR, Wu SY (2009) Conjugate dual problems in constrained set-valued optimization and applications. European Journal of Operational Research 196(1):21–32
124. Li SJ, Yao SF, Chen CR (2008) Saddle points and gap functions for vector equilibrium problems via conjugate duality in vector optimization. Journal of Mathematical Analysis and Applications 343(2):853–865
125. Luc DT (1989) Theory of vector optimization. Lecture Notes in Economics and Mathematical Systems 319. Springer-Verlag, Berlin
126. Mahajan DG, Vartak MN (1977) Generalization of some duality theorems in nonlinear programming. Mathematical Programming 12(3):293–317
127. Mangasarian OL (1969) Nonlinear programming. McGraw-Hill Book Company, New York
128. Martin DH (1985) The essence of invexity. Journal of Optimization Theory and Applications 47(1):65–76
129. Mastroeni G (2003) Gap functions for equilibrium problems. Journal of Global Optimization 27(4):411–426
130. Miettinen KM (1998) Nonlinear multiobjective optimization. Kluwer Academic Publishers, Boston Dordrecht London
131. Mishra SK, Wang S, Lai KK (2008) V-invex functions and vector optimization. Springer Optimization and its Applications 14. Springer-Verlag, New York
132. Mishra SK, Wang S, Lai KK (2009) Generalized convexity and vector optimization. Nonconvex Optimization and its Applications 90. Springer-Verlag, Berlin
133. Mitani K, Nakayama H (1997) A multiobjective diet planning support system using the satisficing trade-off method. Journal of Multi-Criteria Decision Analysis 6(3):131–139

134. Mititelu Ş (2002) The unitary character of Mond-Weir duality in nonlinear programming. Proceedings of the 5th Conference of the Romanian Society of Probability and Statistics (Bucharest, 2002). Analele Universităţii Bucureşti - Matematică-Informatică 51(1):91–98
135. Mititelu Ş (1992) Unitary and geometrical aspects of the Wolfe duality. Studii şi Cercetări Matematice 44(6):503–511
136. Mond B (1965) A symmetric dual theorem for non-linear programs. Quarterly of Applied Mathematics 23:265–269
137. Mond B, Hanson MA (1984) On duality with generalized convexity. Mathematische Operationsforschung und Statistik Series Optimization 15(2):313–317
138. Mond B, Weir T (1981) Generalized concavity and duality. In: Schaible S, Zemba WT (eds) Generalized concavity in optimization and economics. Proceedings of the NATO Advanced Study Institute, University of British Columbia, Vancouver, 1980. Academic Press, New York-London, 263–279
139. Mond B, Weir T (1991) Symmetric duality for nonlinear multiobjective programming. In: Kumar HI, Santosh K (eds) Recent developments in mathematical programming. CRC Press, 137–153
140. Mond B, Zlobec S (1979) Duality for nondifferentiable programing without a constraint qualification. Utilitas Mathematica 15:291–302
141. Nachbin L (1965) Topology and order. Van Nostrand, Princeton
142. Nakayama H (1984) Geometric consideration of duality in vector optimization. Journal of Optimization Theory and Applications 44(4):625–655
143. Nakayama H (1984) Duality theory in vector optimization: an overview. In: Haimes YY, Chankong V (eds) Decision making with multiple objectives. Proceedings of the sixth international conference on multiple-criteria decision making held at Case Western Reserve University, Cleveland, Ohio, June 4–8, 1984. Lecture Notes in Economics and Mathematical Systems 242. Springer-Verlag, Berlin, 109–125
144. Nakayama H (1996) Some remarks on dualization in vector optimization. Journal of Multi-Criteria Decision Analysis 5(3):218–255
145. Nehse R (1981) Duale Vektoroptimierungsprobleme vom Wolfe-Typ. Seminarberichte der Sektion Mathematik der Humboldt-Universität zu Berlin 37:55–60
146. Nieuwenhuis JW (1980) Supremal points and generalized duality. Mathematische Operationsforschung und Statistik Series Optimization 11(1):41–59
147. Nikodem K (1989) K-convex and K-concave set-valued functions. Zeszyty Naukowe Politechniki Łódźkiej 559. Łódź
148. Pareto V (1896) Cours d'economie politique. F. Rouge, Lausanne
149. Patriksson M (1999) Nonlinear programming and variational inequality problems. A unified approach. Applied Optimization 23. Kluwer Academic Publishers, Dordrecht
150. Penot JP, Théra M (1982) Semicontinuous mapping in general topology. Archiv der Mathematik 38(2):158–166
151. Peressini AL (1967) Ordered topological vector spaces. Harper & Row, New York
152. Ponstein J (1982) On the dualization of multiobjective optimization problems. Economics Institute Report 87. University of Groningen
153. Postolică V (1986) A generalization of Fenchel's duality theorem. Annales des Sciences Mathématiques du Québec 10(2):199–206
154. Postolică V (1986) Vectorial optimization programs with multifunctions and duality. Annales des Sciences Mathématiques du Québec 10(1):85–102

155. Precupanu T (1984) Closedness conditions for the optimality of a family of non-convex optimization problem. Mathematische Operationsforschung und Statistik Series Optimization 15(3):339–346
156. Ritter K (1970) Optimization theory in linear spaces. III. Mathematical programming in partially ordered Banach spaces. Mathematische Annalen 184(4):133–154
157. Rockafellar RT (1970) Convex analysis. Princeton University Press, Princeton
158. Rockafellar RT (1974) Conjugate duality and optimization. Regional Conference Series in Applied Mathematics 16. Society for Industrial and Applied Mathematics, Philadelphia
159. Rockafellar RT (1981) The theory of subgradients and its applications to problems of optimization. Convex and nonconvex functions. Heldermann-Verlag, Berlin
160. Rodrigues B (1990) The Fenchel duality theorem in Fréchet spaces. Optimization 21(1):13–22
161. Rödder W (1977) A generalized saddlepoint theory; its application to duality theory for linear vector optimum problems. European Journal of Operational Research 1(1):55–59
162. Rubinov AM, Gasimov RN (2004) Scalarization and nonlinear scalar duality for vector optimization with preferences that are not necessarily a pre-order relation. Journal of Global Optimization 29(4):455–477
163. Sawaragi Y, Nakayama H, Tanino T (1985) Theory of multiobjective optimization. Mathematics in Science and Engineering 176. Academic Press, Orlando
164. Schaible S (1976) Fractional programming. I. Duality. Management Science 22(8):858–867
165. Schandl B, Klamroth K, Wiecek MM (2002) Norm-based approximation in multicriteria programming. Global optimization, control, and games IV. Computers & Mathematics with Applications 44(7):925–942
166. Schechter M (1977) A subgradient duality theorem. Journal of Mathematical Analysis and Applications 61(3):850–855
167. Schönefeld P (1970) Some duality theorems for the non-linear vector maximum problem. Unternehmensforschung 14(1):51–63
168. Song W (1997) Conjugate duality in set-valued vector optimization. Journal of Mathematical Analysis and Application 216(1):265–283
169. Song W (1998) A generalization of Fenchel duality in set-valued vector optimization. Mathematical Methods of Operations Research (ZOR) 48(2):259–272
170. Song W (1998) Duality in set-valued optimization. Dissertations Mathematicae CCCLXXV. Polish Academy of Science, Institute of Mathematics, Warszawa
171. Suneja SK, Aggarwal S, Davar S (2002) Multiobjective symmetric duality involving cones. European Journal of Operational Research 141(3): 471–479
172. Tammer C (1994) A variational principle and applications for vectorial control approximation problems. Studii şi Cercetări Ştiinţifice. Seria Matematică. Universitatea din Bacău 4:85–98
173. Tammer C, Göpfert A (2002) Theory of vector optimization. In: Ehrgott M, Gandibleux X (eds) Multiple criteria optimization: state of the art annotated bibliographic surveys. International Series in Operations Research & Management Science 52. Kluwer Academic Publishers, Boston, 1–70
174. Tammer C, Tammer K (1991) Generalization and sharpening of some duality relations for a class of vector optimization problems. Zeitschrift für Operations Research 35(3):249–265

175. Tammer C, Winkler K (2003) A new scalarization approach and applications in multicriteria d.c. optimization. Journal of Nonlinear and Convex Analysis 4(3):365–380
176. Tanaka T, Kuroiwa D (1993) Some general conditions assuring $\operatorname{int} A + B = \operatorname{int}(A + B)$. Applied Mathematics Letters 6(3):51–53
177. Tanino T (1988) On supremum of a set in a multidimensional space. ournal of Mathematical Analysis and Application 130(2):386–397
178. Tanino T (1992) Conjugate duality in vector optimization. Journal of Mathematical Analysis and Application 167(1):84–97
179. Tanino T, Kuk H (2002) Nonlinear multiobjective programming. In: Ehrgott M, Gandibleux X (eds) Multiple criteria optimization: state of the art annotated bibliographic surveys. International Series in Operations Research & Management Science 52. Kluwer Academic Publishers, Boston, 71–128
180. Tanino T, Sawaragi Y (1980) Conjugate maps and duality in multiobjective optimization. Journal of Optimization Theory and Applications 31(4):473–499
181. Vogel W (1977) Vektoroptimierung in Produkträumen. Mathematical Systems in Economics 35. Anton Hain Verlag, Meisenheim am Glan
182. Wanka G, Boţ RI (2000) Multiobjective duality for convex-linear problems. In: Inderfurth K, Schwödiauer G, Domschke W, Juhnke F, Kleinschmidt P, Wäscher G (eds) Operations research proceedings 1999. Selected papers of the symposium (SOR '99) held at the Otto-von-Guericke University Magdeburg, Magdeburg, September 1–3, 1999. Springer-Verlag, Berlin, 36–40
183. Wanka G, Boţ RI (2001) Multiobjective duality for convex-linear problems II. Mathematical Methods of Operations Research (ZOR) 53(3):419–433
184. Wanka G, Boţ RI (2002) A new duality approach for multiobjective convex optimization problems. Journal of Nonlinear and Convex Analysis 3(1):41–57
185. Wanka G, Boţ RI (2002) Multiobjective duality for convex ratios. Journal of Mathematical Analysis and Applications 275(1):354–368
186. Wanka G, Boţ RI (2002) On the relations between different dual problems in convex mathematical programming. In: Chamoni P, Leisten R, Martin A, Minnemann J, Stadtler H (eds) Operations research proceedings 2001. Selected papers from the International Conference (OR 2001) held at the Gerhard-Mercator University, Duisburg, September 3–5, 2001. Springer-Verlag, Heidelberg, 255–262
187. Wanka G, Boţ RI, Vargyas ET (2007) Duality for location problems with unbounded unit balls. European Journal of Operational Research 179(3):1252–1265
188. Weidner P (1990) An approach to different scalarizations in vector optimization. Wissenschaftliche Zeitschrift der Technischen Hochschule Ilmenau 36(3):103–110
189. Weidner P (1994) The influence of proper efficiency on optimal solutions of scalarizing problems in multicriteria optimization. OR Spektrum 16(4):255–260
190. Weir T (1986) A duality theorem for a multiple objective fractional optimization problem. Bulletin of the Australian Mathematical Society 34(3):415–425
191. Weir T (1987) Proper efficiency and duality for vector valued optimization problems. Journal of the Australian Mathematical Society Series A 43(1):21–34
192. Weir T (1988) A note on invex functions and duality in multiple objective optimization. Opsearch 25(2):98–104

193. Weir T (1989) Duality for nondifferentiable multiple objective fractional programming problems. Utilitas Mathematica 36:53–64
194. Weir T (1989) On duality in multiobjective fractional programming. Opsearch 26(3):151–158
195. Weir T (1990) On efficiency, proper efficiency and duality in multiobjective programming. Asia-Pacific Journal of Operational Research 7(1):46–54
196. Weir T, Mond B (1987) Duality for generalized convex programming without a constraint qualification. Utilitas Mathematica 31:233–242
197. Weir T, Mond B (1989) Generalised convexity and duality in multiple objective programming. Bulletin of the Australian Mathematical Society 39(2):287–299
198. Weir T, Mond B (1991) Multiple objective programming duality without a constraint qualification. Utilitas Mathematica 39:41–55
199. Weir T, Mond B (1988) Symmetric and self duality in multiple objective programming. Asia-Pacific Journal of Operational Research 5(2):124–133
200. Weir T, Mond B, Craven BD (1987) Weak minimization and duality. Numerical Functional Analysis and Optimization 9(1–2):181–192
201. Wierzbicki AP (1977) Basic properties of scalarizing functionals for multiobjective optimization. Mathematische Operationsforschung und Statistik Series Optimization 8(1):55–60
202. Wolfe P (1961) A duality theorem for non-linear programming. Quarterly of Applied Mathematics 19:239–244
203. Yang XM, Teo KL, Yang XQ (2000) Duality for a class of nondifferentiable multiobjective programming problems. Journal of Mathematical Analysis and Applications 252(2):999–1005
204. Zălinescu C (1983) Duality for vectorial nonconvex optimization by convexification and applications. Analele Ştiinţifice ale Universităţii "Alexandru Ioan Cuza" Iaşi Secţia I-a Matematică 29(3):15–34
205. Zălinescu C (1987) Solvability results for sublinear functions and operators. Mathematical Methods of Operations Research (ZOR) 31(3):A79–A101
206. Zălinescu C (1999) A comparison of constraint qualifications in infinite-dimensional convex programming revisited. Journal of Asutralian Mathematical Society Series B 40(3):353–378
207. Zălinescu C (2002) Convex analysis in general vector spaces. World Scientific, River Edge
208. Zeidler E (1985) Nonlinear functional analysis and its applications III: Variational methods and optimization. Springer-Verlag, New York
209. Zeleny M (1974) Linear multiobjective programming. Lecture Notes in Economics and Mathematical Systems 95. Springer-Verlag, Berlin
210. Zowe J (1975) A duality theorem for a convex programming problem in order complete vector lattices. Journal of Mathematical Analysis and Applications 50(2):273–287

Index

MIX
Papier aus verantwortungsvollen Quellen
Paper from responsible sources
FSC® C105338

If you have any concerns about our products,
you can contact us on
ProductSafety@springernature.com

In case Publisher is established outside the EU,
the EU authorized representative is:
Springer Nature Customer Service Center GmbH
Europaplatz 3, 69115 Heidelberg, Germany

Printed by Libri Plureos GmbH
in Hamburg, Germany